Coastal structures and breakwaters

Proceedings of the conference organized by the Institution of Civil Engineers, and held in London on 6-8 November 1991

Thomas Telford, London

Conference organized by the Institution of Civil Engineers, and co-sponsored by the American Society of Civil Engineers, the Japan Society of Civil Engineers, the Ministry of Agriculture, Fisheries and Food, the National Rivers Authority, and the Permanent International Association of Navigation Congresses (UK)

Organizing committee: J. E. Clifford (Chairman), N. W. H. Allsop (HR Wallingford), A. Bell (Kirk McClure Morton), P. Hunter (Sir Alexander Gibb & Partners), P. Lacey (Ove Arup & Partners), F. O'Hara (Costain International), D. Richardson (MAFF), I. Whittle (National Rivers Authority), H. Burcharth (University of Aalborg), Y. Goda (Yokohama National University), H. W. Partenscky (University of Hannover), K. Pilarczyk (Rijkswaterstaat), W. A. Price (Rendel Palmer & Tritton), and R. M. Sorensen (Lehigh University)

A CIP catalogue record for this book is available from the British Library.

ISBN 0-7277-1672-7

First published 1992

Published on behalf of the Institution of Civil Engineers by Thomas Telford Services Ltd, Thomas Telford House, 1 Heron Quay. London E14 4JD.

Printed and bound in Great Britain by Redwood Press.

Contents

1. Impact of expected climate change effects on coastal structure design

R. M. SORENSEN, Lehigh University

SYNOPSIS. Increasing concentrations of carbon dioxide and other gases in the earth's atmosphere are expected to cause increased warming of the atmosphere. This, in turn, will have several effects of possible importance to the design of coastal structures, the foremost and most investigated effect being a significant rise in eustatic sea level. These effects are briefly discussed and the available projections of sea level rise for the next century are presented. Then, the important impacts of sea level rise on the design of coastal structures are presented as are the specific impacts on the various classes of structures. Finally, two case studies that demonstrate the potential impact of sea level rise are summarized.

INTRODUCTION

1. The concentration of carbon dioxide in the earth's atmosphere has increased by about 25 percent over the past century. This increase, along with related increases in trace gases, aerosols, and water vapor has caused an observed northern hemisphere temperature increase of approximately 0.5 degrees centigrade during this period (ref. 1). The temperature rise is caused by the atmosphere's greenhouse effect, i.e. the atmospheric gases limit the reradiation from the earth's surface of longwave (infrared) radiation which in turn heats the earth and ocean surfaces and the atmosphere. The concentrations of carbon dioxide and other gases in the atmosphere are expected to continue increasing in the future with a consequent continuing (accelerating?) increase in the earth's temperature.

2. It is expected that the earth's average surface atmospheric temperature will rise, but that this rise will not be uniform. Significant regional and seasonal average temperature variations are expected to occur. The general warming trend will lead to a rise in sea levels owing to thermal expansion of the ocean's water and to a melting of glacial snow and ice. The regional and seasonal variations in temperature rise will, in turn, alter precipitation and surface runoff patterns, the extent and distribution of sea ice,

ocean current patterns, continental weather systems, and tropical storm patterns and intensities. At present, it is not possible to make quantitative projections of any of these effects except for the expected rise of sea level, and sea level rise projections still need significant refinement.

WARMING PREDICTION AND EFFECTS

3. Becasue of the complex coupling and feedback effects of the physical and chemical processes in the climate system, it is not possible to make meaningful predictions of atmospheric and oceanic warming by extrapolation of historic empirical data. Recourse is required to mathematical models of the atmosphere-ocean system. For a discussion of some of these models and their results, see chapters 3 and 4 of reference 1.

4. A starting point for most of the mathematical models is estimates of carbon dioxide, trace gas and aerosol concentration increases in the atmosphere. These are fed into an atmospheric circulation numerical model which yields predictions of regional and seasonal temperature changes with time. Uncertainties develop in these models from a lack of full understanding of certain feedback mechanisms such as how much more water vapor the warmer atmosphere will have, what the change in the earth's albedo will be owing to changes in snow/ice cover, and the magnitude and consequences of changed levels of cloud cover. With appropriate transfer mechanisms the atmospheric models are coupled to oceanic models to predict ocean system temperature changes. With assumptions for ocean thermal expansion and glacial ice melt sea level changes predictions can be made.

5. The atmospheric circulation models virtually all show that projected atmospheric warming will be greater near the poles than around the equator, and that the temperature increases will be greater in the winter than in the summer (e.g. ref. 2). Consequently, future temperature differentials would be less from the poles to the equator. This will result in less vigorous ocean current and continental weather systems. Jet stream patterns and monsoonal weather patterns in the tropics will be altered. This will alter storm track patterns. Also, these atmsopheric/oceanic temperature pattern changes will result in less sea ice.

6. The average warming in the tropics will mean that hurricanes will be more plentiful as they will be able to form further north and during a longer season. Their rate of over-water dissipation will be less as they travel away from the equator owing to the warmer water. Consequently, increased storm surge levels (neglecting sea level rise effects) and design wave heights might be expected at a hurricane prone site for a given return period. Modified jet stream patterns will also modify average hurricane tracks.

7. Reduced sea ice will result in larger fetches for wave generation in polar latitudes and longer periods of shoreline

exposure to wave attack. However, less dynamic weather systems in northern latitudes will counter these effects owing to decreased wind speeds at a particular site for a given return period.

8. At present, only rough qualitative indications of the changes discussed in paragraphs 5 to 7 are available from the atmospheric/oceanic models. It will be a long time before quantitative information of value to the designer of a coastal structure will be available. The one area where quantitative predictions have been made is sea level rise. It appears that the potential impacts of increased sea levels will be more significant, on the average, to the design of coastal structures than the other factors mentioned above.

SEA LEVEL RISE

9. Historic sea level change trends to various coastal locations around the world typically show a wide variety of positive (mostly) and negative values. For example, Hicks and Hickman (ref. 3) report on yearly relative mean sea level trends for forty five U.S. coastal stations based on fifty to over a hundred years of tide gage records. The twenty five U.S. Atlantic coast relative sea level change values range from +1.79 to +4.33 mm/yr; whereas, four stations in Alaska had mean sea level change trends of -0.09 to -12.4 mm/yr. It appears that the wide range of values is owing to significant coastal uplift or subsidence superimposed on a smaller eustatic sea level rise. In a recently completed study, Douglas (ref. 4) looked at sea level change data from twenty one selected tide gage stations that were not in areas of converging tectonic plates (and thus not as active techtonically). Also he made corrections for postglacial subsidence or uplift so as to get the best possible estimates of eustatic sea level change. Based on the data from these stations, which had an average record length of 76.4 years, he found an average global eustatic sea level rise of 1.79 mm/yr (equivalent to 0.59 feet per century). Others have reported similar global values during the past century (e.g. see ref. 5, page 17 where seven different estimates typically between 1.1 and 1.5 mm/year are reported). Comparing the results from Hicks and Hickman with that from Douglas, it appears that during the past century there has been a gloval eustatic sea level rise but that at many locations local subsidence or uplift effects have been more dominant.

10. Since records indicate that carbon dioxide concentrations in the atmosphere are not only increasing, but at an accelerating rate (ref. 6), strong concern has developed during the past decade that eustatic sea level rise might also accelerate and become a significant future problem. Assuming a reasonable temperature rise and including the effects of oceanic thermal expansion and melting of Greenland and mountain glaciers (but not those in the Antarctic), Revelle (ref. 7) estimated a 700 mm (2.3 feet) eustatic sea level rise

by the year 2075.

11. A more thorough analysis of potential eustatic sea level rise was presented by Hoffman et al. (ref. 8) in 1983. Briefly, they modeled the entire system starting from estimates of world population/productivity growth with related energy use, to resulting carbon dioxide and other emissions, to global warming, to ocean volume expansion including thermal expansion and transfer of snow and ice from the land to the ocean. From various assumptions throughout their model, they developed the four eustatic sea level rise scenarios that are tabulated in Table 1. Of the four, they felt that the mid-range low scenario was most realistic because it chose the middle ground or most realistic assumptions throughout the analysis. This scenario produces a 912 mm (3.0 feet) projected rise within a century. In 1986 they somewhat revised their projections (ref. 9) to produce a high and low scenario based on revised input on glacial processes and future atmospheric emissions. These are also tabulated in Table 1.

Table 1. Eustatic sea level rise estimates (millimeters)

YEAR	2000	2025	2050	2075	2085	2100
Revelle (83)				700		
Hoffman et al. (83)						
Low	48	130	230	380		560
Mid-low	88	262	523	912		1440
Mid-high	132	393	786	1368		2166
High	171	549	1167	2127		3450
Hoffman et al. (86)						
Low	35	100	200	360	440	570
High	55	210	550	1910	2580	3680

12. Table 1 indicates the wide range of predicted sea level rise estimates produced by the best projections available to date. Values range from a half to over three and a half meters of eustatic sea level rise by the year 2100. Linear extrapolation of the historic rate determined by Douglas (paragraph 9) would yield a rise of two-tenths of a meter by 2100. Generally, the projected values well exceed the measured historic rates including the effects of subsidence (see paragraph 9).

13. For discussion purposes, it appears that eustatic sea level rise projections slightly lower than the Hoffman et al. (1983) mid-low values tabulated in Table 1 might be used (keeping in mind the wide range of possible values for any given date as suggested by Table 1). This would be just less than a half-meter rise during the next fifty years and about

a one meter rise during the next century (plus any uplift or subsidence).

14. For design purposes, until better sea level rise projections are made, a designer should consider the range of scenarios from the Hoffman et al. (83) low to mid-high values, with appropriate corrections for coastal uplift or subsidence. (Uplift or subsidence rates can be estimated by subtracting 1.5 mm/yr from relative sea level change rates measured at a particular location during the past several decades.)

IMPACTS ON COASTAL STRUCTURE DESIGN

15. Some less substantial coastal structures such as wood groins and bulkheads have typical design lives of twenty or thirty years. Typical design lives for most substantial coastal structures (e.g. stone mound groins and breakwaters, concrete seawalls and piers) are fifty years or longer. The actual service lives of these structures may often exceed one hundred years. A sea level rise of a meter or more over a century will have a significant impact on most of these structures.

16. It is of utmost importance for a coastal structure designer to decide whether the structure being designed can be easily retrofitted to respond to a rise in sea level if/when it occurs. With the strong likelyhood of a significant sea level rise during the next century, the designer should incorporate ease of retrofitting into the structure design in order to respond to a raised sea level as it occurs. If the type of structure being designed can not easily be designed for this retrofitting, then the structure should be designed for the ultimate sea levels that are likely to occur during its anticipated service life.

Sea Level Rise Impacts

17. A rise in relative sea level will impact on a coastal structure in a number of ways. Many coastal structures with significant ocean exposure have a design wave height limited by the wave breaking depth just seaward of the structure. Irrespective of the wave amplitude spectrum generated by the design storm, the maximum wave height at the structure will be about equal to the water depth at the structure. Consequently, as sea level rises the structure will be exposed to increasing wave heights. This in turn will lead to increased wave loadings on the structure. Wave runup, overtopping and energy transmission to the lee of the structure will also increase owing both to the raised sea level and to the increased incident wave height.

18. For those structures whose design wave height is not limited by breaking at the structure, the effects of atmospheric warming on the change in design wave height are difficult to predict as discussed in paragraphs 5 to 7. Increased water depths will slightly alter the effects of wave refraction, wave amplitude increases as the wave enters shallower water, and bed-induced decay (owing to bottom friction

and percolation). Changes in storm track and strength, as well as increased fetches in northern latitudes will also modify incident wave heights for design conditions.

19. A rise in sea level not only moves the MSL line landward due to the increased water level flooding further up the beach; but there is also an adjustment of the equilibrium shore profile resulting in beach erosion owing to the movement of sand from the beach face to offshore. Bruun (ref. 10) explains this phenomenon and provides a method for calculating its effect. Figure 1(a) shows these two components of the shoreward movement of MSL and the related increase in water depth at a particular point on the beach.

20. When a structure such as a seawall or bulkhead is in place at the waters edge, it restricts the offshore movement of sand from the portion of the beach face located landward of the structure. Beach profile adjustment in response to sea level rise then requires that an additional volume of sand be removed from the beach face just seaward of the structure (see Figure 1(b)). Dean and Maurmeyer (ref. 11) explain this phenemonon and present a modified application of the Brunn procedure for calculating the erosion that should occur at the toe of a structure owing to sea level rise. This profile adjustment not only causes erosion at the base of the structure, but the water depth in front of the structure is increased by more that the amount of the sea level rise, with the attendant consequences on the incident wave height as discussed in paragraphs 17 and 18.

21. A third consequence of raised sea level is that the probabilities of given storm plus tide water levels at a site are increased (return periods decreased), usually significantly (see ref. 12). This is demonstrated in Figure 2 which shows a combined storm surge plus tide stage versus return period curve (solid line) for Monmouth Beach, N.J. (ref. 13). A storm surge plus tide level of 10.5 feet above MSL now has a return period of fifty years. With a sea level rise of a half meter (approximately 1.5 feet) in less than fifty years, the return period for this water level (dashed line) will be less than twenty years. For sites that have flatter elevation versus return period curves the effect would be more dramatic (see ref. 12).

22. The above analysis of change in return period with a rise in sea level assumes that the half meter sea level rise has a negligible effect on the resulting tide and storm surge. As discussed above, atmospheric warming will change storm tracks and strengths slightly. And deeper water will cause a slight decrease in storm surge for a given wind condition. Tide wave amplification will also be slightly less with increased water levels. But these effects should be relatively small so the analysis shown in paragraph 21 should give a good indication of what will occur.

23. An example of the effect of sea level rise on flooding frequency may be found at the City of Venice. Since the beginning of this century, sea level has risen about 0.11 m

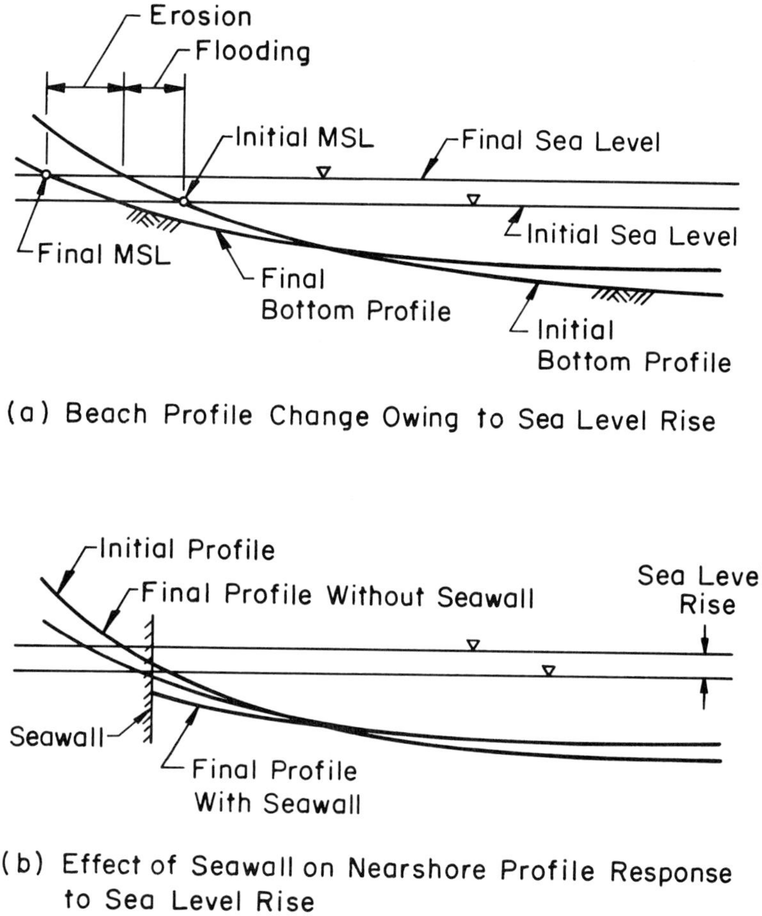

Figure 1. Beach profile response to sea level rise.

and, due to ground water removel, the City has sunk about 0.14 m. Storm surge flooding of the Piazza San Marco in Venice has increased from a few times each winter at the beginning of the century to a few dozen times per winter now.

24. Other sea level rise impacts that might be of some significance to certain coastal structures include: the equivalent rise in the ground water table along the coast; modified longshore sediment transport regimes owing to changes in the incident wave climate, nearshore refraction patterns, and possible modifications in the source of sediment for longshore transport; and lesser required dredging depths for navigation adjacent to port structures.

Specific Impacts on Common Types of Coastal Structures

25. It would be relatively easy to raise the crest elevation of stone mound structures such as seawalls and revetments that

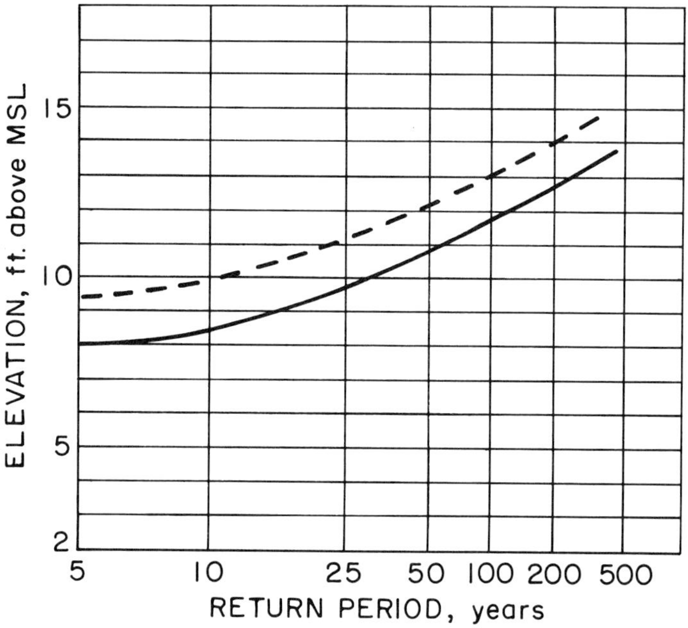

Figure 2. Storm surge plus tide stage versus return period, Monmouth Beach, N.J.

are built along the shoreline, as necessary to respond to a rise in sea level. But the higher incident design wave heights, owing to the raised sea level along with an increased erosion in front of the structure owing to shore profile adjustment, will require larger armor stone sizes to maintain structure stability.

26. For example, a half-meter rise in sea level could result in say a three-quarter meter increase in breaking wave height and, the required weight of armor stone for a stone mound structure is a function of approximately the cube of the wave height. So, if a three-quarter meter increase in wave height amounted to a significant percentage increase in the total incident wave height the structure would incur increased damage with time and might ultimately require placement of a larger stone size armor layer. Also, the toe erosion caused by shore profile adjustment would require continued maintenance and strenghtening of the structure's toe protection.

27. Groins, jetties and shore-connected breakwaters would suffer similar consequences at their seaward ends. As sea level rises, the shoreward retreat of the MSL line might make the structure susceptable to flanking during storms. Shore-parallel offshore breakwaters would decrease in effectiveness as the distance from the structure to the MSL line increases. Also, any significant changes in the alongshore sediment transport regime might affect the functional behavior of some of these structures and require structure modification.

28. Bulkheads and groins constructed of wood, metal or composite wood/metal usually have an effective service life of significantly less than fifty years and could be adjusted to sea level rise as they are replaced.

29. Harbor structures such as piers and docks generally are constructed in a way that makes it hard for them to be retrofitted for sea level rise. But, inside of a harbor where there is greatly reduced wave action, only the rise in sea level would be of concern. If there is no significant subsidence, and some future eustatic sea level rise is anticipated, it would be possible to design future pier and dock facilities for these changes.

30. Coastal zone land drainage control structures would be adversely affected by both the sea level rise and the ground water table rise, as well as by any increase in precipitation owing to atmospheric warming. Dikes and levees, which are not exposed to significant wave action can be relatively easily raised. The capacity of pumping facilities for low and below sea level installations can be increased without much difficulty.

Two Case Studies

31. Two published case studies that have evaluated the impact of potential sea level rise on coastal structures will be briefly reviewed. The first deals with a stone mound seawall built along a section of the north coast of New Jersey and the second considers coastal drainage systems in South Carolina and Florida.

32. Seawall. Kyper and Sorensen (ref. 14) evaluated the potential impact of two sea level rise scenarios - the Hoffman et al. (83) low and mid-high scenarios (Table 1) - on the stone mound seawall at Sea Bright. New Jersey. The historic subsidence in the Sea Bright area has been 3.3 mm/year. This constant value was added to the eustatic sea level rise predictions to yield the working total sea level rise scenarios.

33. This seawall has a crest elevation that varies between 3.4 and 5.2 meters above MSL. It is fronted by occasional small pocket beaches held by short groins. About 60 percent of the seawall length has no beach in front of it, but is fronted by an average water depth of about 1.1 meters below MSL. An analysis of future sea wall stability was done using this latter condition.

34. Using the Dean and Maurmeyer adaptation of the Brunn procedure (paragraph 20), the expected profile changes in front of the seawall were calculated for the two scenarios. For the low scenario, the depth at the toe of the structure would increase from the present average value of 1.1 meters to 1.7 meters and for the mid-high scenario the depth at the toe would increase to 1.95 meters by the year 2100.

35. Storm surge plus tide versus return period curves for the site were adjusted as described in paragraph 21 to obtain the fifty year return period water levels at any future time up to the year 2100, for the two scenarios.

36. From the analyses described in paragraphs 34 and 35, and assuming the breaking wave height at the structure was the design wave, plots of the design wave height for the two scenarios were developed from the present time to the year 2100. The design wave height for the present design water level at the toe of the structure is 4.6 meters. For the low scenario, this increases to a height of 5.5 meters by the year 2100. For the mid-high scenario, this increases to a height of 6.9 meters by the year 2100.

37. The average armor stone weight on the seawall is 6.8 metric tons which is unstable for the present design wave height of 4.6 meters. (The stable design wave height for this armor unit weight is only 3.4 meters.) This is consistent with the history of the seawall during the past two decades during which it has suffered relatvely frequent damage. Of course, with the expected increases in design wave height for the two sea level rise scenarios, the amount and frequency of damage would greatly increase.

38. An analysis was done to determine the required armor stone weight for the seawall to be stable in the year 2100 under the two scenarios. It was assumed that the present armor face slope of 1:1.5 would be maintained and that the new armor stone would be layed over this slope. The required armor stone sizes were so large (54 metric tons for the mid-high scenario) that concrete dolosse were used for the analysis. The result was a required 7.2 metric ton dolosse for the low scenario, and a required 14.1 metric ton dolosse for the mid-high scenario.

39. Kyper and Sorensen also investigated the required increase in seawall crest elevation by the year 2100 under the two sea level rise scenarios for different rates of wave overtopping. And, they investigated the volumes of sand required to maintain a protective beach in front of the seawall to the year 2100, for the two scenarios. With a protective beach having an adequate berm width and elevation, no modifications to the seawall would be required.

40. Coastal drainage. Titus et al. (ref. 15) review two case studies that they conducted on the impact of sea level rise and an increase in precipitation owing to atmospheric warming, on coastal watershed drainage systems in Charleston, South Carolina and Fort Walton Beach, Florida. Increased storm surge and tide elevations which reduce system head differentials and raised water tables which saturate the ground and decrease infiltration, both diminish the capacity of drainage structures. For gravity drainage systems, drainage channels need to be widened and have their sides raised and drainage pipes have to be increased in size. Systems that employ pumps have to have pump capacities increased.

41. For the two studies, the authors employed the Hoffman et al. (83) mid-low sea level rise scenario (Table 1) adjusted for local historic subsidence. Also, they assumed a future 10% increase in precipitation during the system design storms. Specifically, they estimated the increased flows, specified

the best option for improving the drainage systems to handle the increased design flows, and estimated the related cost increases.

42. The Grove Street drainage system in Charleston has a box culvert to the sea fed by lateral pipelines along each city street. The existing system is underdesigned for a five year storm. New systems were designed for a ten year storm. Larger drainage pipes are the best way to design for predicted year 2025 sea level/precipitation conditions, while pumping stations would have to be installed to satisfy year 2075 conditions. The estimated cost of designing for today's sea level/precipitation conditions and then retrofitting the system in 2025 would be ten times greater than the cost of designing now for anticipated 2025 conditions. Cost analyses indicate that if the anticipated sea level rise scenario has a 30% or greater probability of occurrence, it will be less costly to design now for the expected sea level rise.

43. The Gap Creek drainage system in Fort Walton Beach, which includes open drainage ditches, separate storm sewers, detention basins, and new sea outfall lines, has been improved during the past several years, but additional improvements are planned in the near future. The projected increase in sea level/precipitation would result in the flooding of several houses with water to a depth of a meter or so. The cost of upgrading the system now for projected 2025 conditions is the same as making the improvements later as time passes. This is because the system contains numerous components which can be sufficiently modified as time passes and actual contemporary conditions dictate. Thus, it is unwise to improve the system now for possible conditions in 2025.

CONCLUSION

44. It is hoped that this paper gives the reader a sense of the possible results of future climate change, the level to which these potential changes have been quantified, and their potential impact on the design of coastal structures. It is quite apparent that much more research is needed to improve our capability to predict future climate change impacts. It is also apparent that the designer of coastal structures that will have an effective life of a century or so should give serious consideration to the potential effects of climate change.

REFERENCES

1. MACCRACKEN M.C. and LUTHER F.M. Projecting the climatic effects of increasing carbon dioxide, DOE/ER-0237, U.S. Department of Energy, Washington, D.C., December, 1985.
2. MANABE S. and STROUFFER R.J. Sensitivity of a global climate model to an increase of CO2 concentration in the atmosphere. Journal of Geophysical Research, 1980, 5529-5554.
3. HICKS S.S. and HICKMAN L.E. United States sea level variations through 1986. Shore and Beach, July 1988, 3-7.

4. DOUGLAS B.C. Global sea level rise. Paper submitted to Journal of Geophysical Research, October, 1990.
5. NATIONAL RESEARCH COUNCIL Responding to changes in sea level: engineering implications, U.S. National Research Council, National Academy Press, 1987.
6. KEELING C.D., BACASTOW R.B. and WHORF T.P. Measurements of the concentration of carbon dioxide at Mauna Loa Observatory, Hawaii, in Carbon Dioxide Review, Oxford University Press, New York, 1982, 377-382.
7. REVELLE R. Probable future changes in sea level resulting from increased atmospheric carbon dioxide. In Changing Climate, National Academy of Science, Washington, D.C., 1983, 433-447.
8. HOFFMAN J.S., KEYES D. and TITUS J.G. Projecting future sea level rise; methodology, estimates to the year 2100, and research needs. U.S. Environmental Protection Agency, Washington, D.C., 1983.
9. HOFFMAN J.S., WELLS J.B. and TITUS J.G. Future global warming and sea level rise. In Iceland Coastal and River Symposium '85, Icelandic National Energy Authority, 1986.
10. BRUUN P. Sea-level rise as a cause of shore erosion. Journal, Waterways and Harbors Division, ASCE, Vol. 88, WW1, 1962, 117-133.
11. DEAN R.G. and MAURMEYER E.M. Models for beach profile response. Handbook of Coastal Processes and Erosion, CRC Press, Boca Raton, Florida, 1983, 151-165.
12. PUGH D.T. Is there a sea level problem? Proceedings, Institution of Civil Engineers, Part 1, Vol. 88, June 1990, 347-366.
13. U.S. ARMY CORPS OF ENGINEERS, NEW YORK DISTRICT General Design Memorandum: Atlantic coast of New Jersey, Sandy Hook to Barnegat Inlet beach erosion control project, section 1-Sea Bright to Ocean Township, New Jersey, May 1988.
14. KYPER T.N. and SORENSEN R.M. The impact of selected sea level rise scenarios on the beach and coastal structures at Sea Bright, N.J. Proceedings, Coastal Zone '85 Conference, Baltimore, MD, 1985.
15. TITUS J.G., KUO C.Y., GIBBS M.J., LAROCHE T.B., WEBB M.K., and WADDELL J.O. Greenhouse effect, sea leve rise, and coastal drainage systems. Journal, Water Resources Planning and Management Division, ASCE, Vol. 113, 1987, 216-227.

2. Evaluating the performance of seawalls

J. P. AHRENS, NOAA/Sea Grants, and T. BENDER, US Army Engineer District

SYNOPSIS. A study of the performance of seawalls is described. The study included the collection and evaluation of a considerable amount of laboratory data. Quantitative methods to evaluate the performance of seawalls are presented.

INTRODUCTION

1. There is ever increasing pressure in the United States to develop waterfront sites that are vulnerable to erosion and need shoreline protection. The pressure comes partly from increasing population coupled with a strong tendency for people to locate preferentially on or near the coast. In addition, many existing sites that were once safe from erosion are now experiencing problems due to reduction in sediment supply and rising sea level. Therefore, the need for properly designed and engineered shore protection, including seawalls and seawalls fronted by rubble, is increasing.

2. To provide insight on the performance of various seawall configurations this paper summarizes findings from an extensive study of seawall design conducted at the Coastal Engineering Research Center (CERC). The study was initiated by serious coastal flooding that occurred at Roughans Point and Revere, Massachusetts in February 1978, Hardy and Crawford (1986). Since coastal flooding behind seawalls has continued to be a wide spread problem in the United States the study has continued to the present. The study has included research to solve site specific problems and general research. During the investigation considerable effort has been directed at conducting laboratory tests with irregular waves to determine the effectiveness of various seawall/revetment configurations or cross sections, Ahrens and Heimbaugh (1988). Because physical model tests are expensive and time consuming, it was apparent early in the study that it was necessary to include findings from other laboratories. Research from other labs was used to confirm and extend the results of the CERC study.

BACKGROUND

3. The first comprehensive research on irregular wave overtopping of coastal structures was conducted by Owen (1980,1982a,1982b). During his study Owen investigated the influence of depth and length of fronting berms on the overtopping of sea dikes. Owen abandoned the awkward approaches to estimating overtopping rates that use design curves, overtopping coefficients which are functions of wave conditions, and potential runup, eg. the Shore Protection Manual, (SPM,1984). Potential runup is the runup that would occur

on the structure if it were high enough to prevent overtopping. The relation that Owen used to estimate overtopping rates was,

$$Q' = Q_0 \exp(C_1 F') \tag{1}$$

where Q' and F' are the dimensionless overtopping rate and dimensionless freeboard respectively. Q_o and C_1 are dimensionless overtopping coefficients determined by regression analysis of lab data.

4. Some very interesting insight on wave overtopping of breakwaters is given by Jensen and Juhl (1987). Using a simplified theoretical analysis, Jensen and Juhl show that an approximately exponential relationship between the overtopping rate, Q, and the dimensionless freeboard provides a reasonable conceptual model. This model can be written

$$Q = Q_0 \exp(C_1 F') \tag{2}$$

where the dimensionless freeboard used by Jansen and Juhl is given by F' = F/H_s, with F the freeboard of the structure and H_s the incident significant wave height. Other functional forms for the dimensionless freeboard work very satisfactorily in Eq. 2, eg. see Ahrens and Heimbaugh (1988). Jansen and Juhl show considerable laboratory data which supports their conceptual model. The laboratory results are interesting because they include a number of breakwaters with various crownwall configurations which are hydraulically similar to seawalls fronted by rubble. Jansen and Juhl indicate that they can find no uniquely superior way to normalize the overtopping rate and for that reason their data plots show Q versus F/H_s. This finding appears to supersede conclusions reached in earlier research where overtopping rates were presented in dimensionless form, Jensen (1984).

5. During the CERC study considerable effort was expended to find improved ways to normalize Q and to find functional forms to estimate overtopping rates that were superior to Eq.'s 1 or 2. However, no preferred way to normalize Q was discovered and no simple form to estimate overtopping rates was found superior to Eq. 2, Ahrens and Heimbaugh (1988). These findings strongly support the conclusions of Jansen and Juhl (1987) and are in turn supported by recent work by Hallermeier (1990). The advantages of using a simple model such as Eq. 2 to evaluate seawall performance are:

(a) The functional form of Eq. 2 is consistent with a considerable amount of laboratory data.

(b) The functional form is supported by simple conceptual models of Jansen and Juhl, and Hallermeier.

(c) There is no spurious correlation to confuse the relationship between Q and possible causative variables.

(d) Predicting Q is useful because it can be directly related to flooding potential and hazards or damage to the structure such as tabulated in Fukuda, et al. (1973) or Goda (1985).

6. There is another useful aspect of the simple functional form for overtopping given by either Eq. 1 or Eq. 2. Although it is difficult to relate individual overtopping coefficients to the physics of waves overtopping a structure, it is possible to develop a variable using the coefficients that helps rank the performance of a structure. A measure of the hydraulic performance of a seawall/revetment configuration is given by,

$$K_q = \int_{F'_{min}}^{\infty} Q_o \exp(C_1 F')dF'$$
$$= -\frac{Q_o}{C_1}\exp(C_1 F'_{min}) \qquad (3)$$

where K_q is the overtopping ranking coefficient and $F'_{(min)}$ is the lower limit of the relative freeboard which can be effectively tested in the lab. K_q is dimensionless if Eq 1 is used in the integrand and has units of overtopping rate if Eq.2 is used in the integrand. During lab tests a point is reached where the water levels are so high and the waves so large relative to the structure that it is inundated by the overtopping flow; at this point small features of the structure's cross section probably do not have much influence on the overtopping rate, Ahrens and Heimbaugh (1988). This inundation effect is used to determine the value of $F'_{(min)}$. Geometrically K_q is the area under the regression curve from $F'_{(min)}$ to infinity.

7. From the CERC laboratory test data it was found that the most effective dimensionless freeboard variable was given by,

$$F' = \frac{F}{(H_{mo}^2 L_p)^{1/3}} \qquad (4)$$

where the freeboard, F, is the vertical distance between the mean water level and the crest of the structure, H_{mo} is the incident zeroth moment wave height and L_p is the Airy wave length calculated using the water depth at the toe of the structure and T_p the period of peak energy density of the incident wave spectrum. In the following analysis of CERC data the method of evaluating the performance of specific seawall/revetment configurations is based on the functional form given by Eq. 2 using the dimensionless freeboard given by Eq. 4. Inundation effects for the Eq. 4 freeboard occur approximately at, $F'_{(min)} = 0.3$, Ahrens and Heimbaugh (1988).

STRATEGIES FOR IMPROVING THE PERFORMANCE OF SEAWALL/REVETMENT CONFIGURATIONS

Recurved Walls

8. Figure 1 shows profile views of three seawall/revetment configurations, Ahrens (1988). The revetment portions of the three structures are identical but one structure is topped by a vertical wall (designated VW), one by a mildly recurved wall (MR), and one by a severely recurved wall (SR). Also shown in Figure 1 are regression curves of the Eq. 2 type fit to data collected with the specific structure configuration and plotted versus the relative freeboard defined by Eq. 4. In Figure 1, it can be seen that overtopping rates are considerably higher for the structure topped by a vertical wall than they are for the structure topped with a severely recurved wall. The structure with the moderately recurved wall had overtopping rates between the other two extreme wall types. A figure such as Figure 1 provides a convenient way to evaluate the performance of a seawall/revetment configuration and to compare different configurations.

9. A limitation to the performance evaluation method given above can be seen in Figure 2. In Figure 2, the data used to calculate the regression curve for the seawall/revetment configuration topped by a vertical wall are shown with the

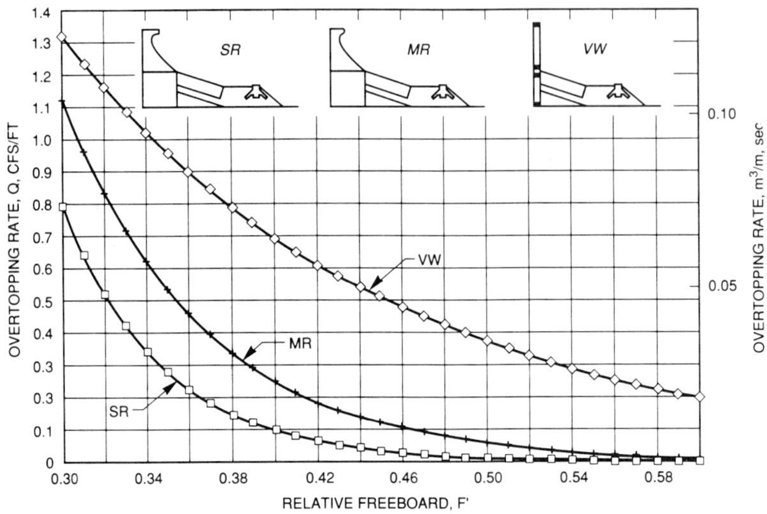

Figure 1, comparison of overtopping trends for seawall/revetment configurations having a vertical wall (designated VW), moderately recurved wall (MR), and a severely recurved wall (SR) respectively.

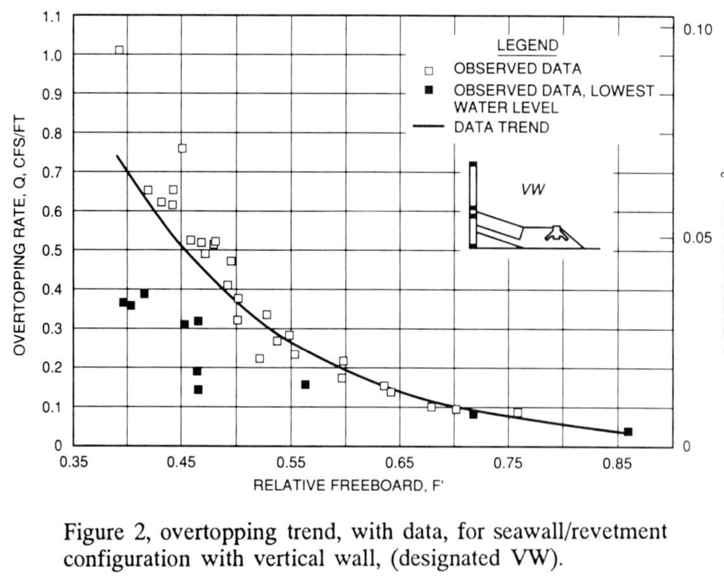

Figure 2, overtopping trend, with data, for seawall/revetment configuration with vertical wall, (designated VW).

regression curve, Ahrens (1988). A number of data points fall conspicuously below the overtopping trend established by the regression curve. These anomalous points are all associated with the lowest water level tested for this structure, i.e. ds = 18.5 cm; lowest water level points are shown shaded in Figure 2. The range of ds tested was from 18.5 to 22.7 cm. Since the dimensionless freeboard defined by Eq. 3 takes the water level into consideration, data collected at all water levels should all fall within the same scatter band in Figure 2 unless there is a strong influence from another source. Additional analysis indicates that when the water depth over the rubble fronting a vertical wall is small and the rubble is reasonably extensive the rubble can be quite effective in attenuating the waves and reducing the overtopping. However, there appears to be a point when a small increase in water depth will make the fronting rubble much less effective in reducing overtopping. This enhanced wave attenuation effect by the rubble berm is masked in the tests of structures with recurved walls by the considerably reduced overtopping due to the effectiveness of the recurve.

Rubble Fronting a Wall

10. Often a limited amount of rubble is placed near the base of a seawall to prevent toe scour. Typically the amount of rubble used solely for toe protection is inadequate and too close to the wall to reduce overtopping rates substantially.

11. In at least one location, the performance of a seawall appears to have deteriorated because the beach in front of the wall has disappeared. At Seabright, N.J., wave overtopping of the local seawall seems to be occurring more frequently and may be related to the loss of updrift sediment supply. Weggel (1988) suggests that seawalls function more effectively the more energy they dissipate, because dissipation reduces the intensity of the standing wave in front of the structure. Support for this concept comes from observations of some very successful rubble mound seawalls in Santa Cruz County, California, Magoon, et al. (1988). Small beaches have developed in front of some seawalls in location where there was no beach before the rubble wall was constructed. This beach building effect may be anomalous or it may be because the rubble seawall reflects less wave energy than the local sea cliffs that it is protecting.

12. For some profile configurations the reduction in overtopping due to rubble fronting a seawall can be disappointing. In Figure 3 two overtopping trends are shown, one for a vertical wall with a small recurve (designated RP–1) and another with the same wall fronted by a standard riprap revetment (RP–2). The revetment has cut the overtopping rate about in half but considering the size of the revetment the magnitude of the improvement was disappointing, Ahrens et al. (1986). Two deficiencies relating to the fronting revetment design can be seen in Figure 3, they are:

(a) The revetment is continued so high up the wall that the small recurve is rendered ineffective.
(b) The revetment is continued so high up the wall that it acts as a ramp for the waves to overtop the wall.

The latter deficiency is referred to as ramp effect and was also noted by Goda (1985) in lab tests of a seawall fronted by a concrete block revetment. Ramp effect also appears to explain the poor performance of an impermeable seawall with steps from toe to crest. The stepped seawall actually performed worse than an impermeable vertical seawall with the same freeboard, Ward (1991).

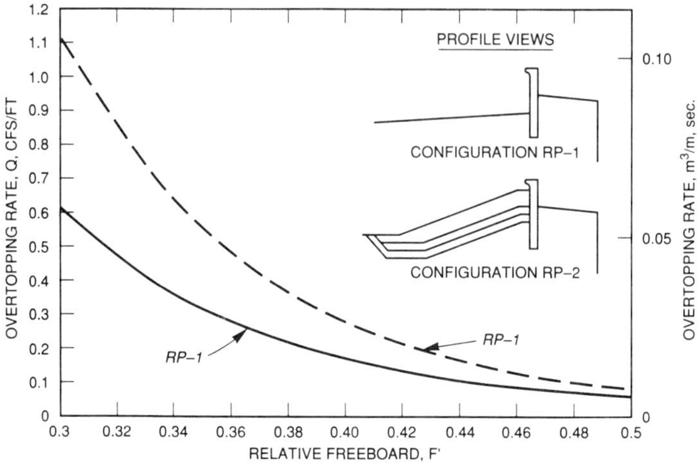

Figure 3, comparison of overtopping trends for a vertical wall with a small recurve (designated RP–1) and the same seawall fronted by a standard riprap revetment (RP–2).

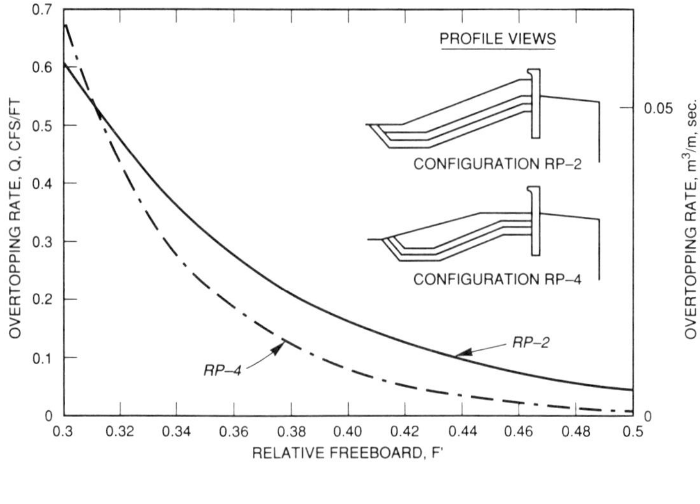

Figure 4, comparison of overtopping trends for a vertical seawall with small recurve, fronted by a standard riprap revetment (designated RP–2) and the same wall fronted by a bermed, absorber revetment (RP–4).

Influence of a Berm

13. An improved revetment configuration (designated RP–4) is shown with the standard riprap profile (RP–2) in Figure 4, Ahrens et al. (1986). The improved configuration uses a berm well below the wall crest to allow the recurve to be effective and the vertical wall itself to create a major discontinuity to the runup flow. Configuration RP–4 also uses some additional armor stone near the toe to initiate wave breaking earlier and to enhance energy dissipation. The overtopping data trend curves show that the bermed revetment performs considerably better than the standard revetment over almost the entire range of dimensionless freeboards. Based on these findings, a bermed revetment is planned to front the seawall at Roughans Point, Massachusetts, and a bermed revetment has been built to front the seawall at Temple Emanuel on Lake Michigan in Chicago, Illinois, Pheiffer and Ahrens (1989).

14. Ahrens and Ward (1991) conducted laboratory tests to determine the effectiveness of a rubble berm in reducing the maximum runup of irregular waves on a riprap revetment. The riprap armor and bedding/filter layers were on a plane, impermeable 1 on 2 slope and was fronted by a rubble berm composed of armor size stone. Test configurations included two sizes of berm as well as a no berm configuration for control. Water levels and wave conditions were also varied. One important finding from this study was that you could express the reduction in runup due to the berm as a reduction factor, r, given by

r = runup on berm fronted riprap/runup on plane riprap ;

r was found to be a function of berm width and height and the wave conditions. The runup reduction factor developed by Ahrens and Ward is similar to the reduction factor given by Battjes (1974) for monochromatic waves on a smooth, impermeable berm. Within the range of test conditions the reduction in runup was somewhat modest, up to 15 to 20 percent, corresponding to r = 0.85 to 0.80. However, modest reductions in runup appear to translate into very substantial reductions in overtopping. To estimate overtopping rates in this instance an overtopping rate formula based on potential runup is necessary. The potential runup overtopping formula given by Weggel (1976) and also in the SPM (1984) can be written,

$$Q = \sqrt{g Q'_o H'^3_o} \left(\frac{R - F}{R + F} \right)^{0.1085/\alpha} \tag{5}$$

where g is the acceleration of gravity, H_o' is the deep water wave height, R is the potential runup, and Q_o' and alpha are overtopping coefficients. If Eq. 4 is used in a ratio to determine overtopping rates for structures with berms to structures without berms, the resulting equation can be written,

$$\frac{Q(\text{berm})}{Q(\text{no berm})} = \left[\frac{(R_b - F)\,(R_n + F)}{(R_b + F)\,(R_n - F)} \right]^{0.1085/\alpha} \tag{6}$$

where R_n and R_b are the potential runup on a riprap revetment with no berm and with a berm respectively.

15. To appreciate the implications of Eq.6 consider some typical values of the relevant variables, i.e. 0.045 <= alpha <= 0.090, R_n/F = 1.2, and r = 0.85, 0.95. The range of the alpha was based on the range shown in the SPM (1984) for

riprap revetments, a potential runup 20 percent higher than the revetment crest represents a severe overtopping condition which could destroy the revetment, and reduction factors ranging from 0.85 to 0.95 are within the range observed by Ahrens and Ward (1991). Based on the typical values given above Eq.6 indicates that a 5 percent reduction in runup could reduce overtopping rates by about one–half to one–third and a reduction in runup of 15 percent could reduce overtopping rates by more than an order of magnitude.

16. Owen (1980) investigated the use of berms to reduce irregular wave overtopping of seadikes. Owen varied the wave conditions and the height and length of the berm fronting the dike, as well as the slope of the dike. Figure 5 shows the logarithm of the overtopping rank coefficient, K_q, versus the cross sectional area of the fronting berm. Data points are denoted by numbers which indicate the slope of the structure, eg. 4 denotes a 1 on 4 slope, etc. The figure indicates that as the cross sectional area of the berm increases overtopping rates or K_q decrease; the correlation between the two variables seem surprisingly high. The high correlation suggests that generally overtopping rates are influenced more by the size of the berm than by it's geometry. Figure 5 supports the implications of Eq.6 by indicating that berms can easily reduce overtopping rates by an order of magnitude or more. Dikes with a 1 on 4 slope have the lowest rates for a given berm cross sectional area, probably because of the relatively low runup on such a flat slope. Dikes with slopes of 1 on 1 and 1 on 2 generally show about equally poor performance for a given cross section. The trend of the data in Figure 5 tends to be steeper for small berms than for large berms. This trend is interpreted as indicating that as impermeable berms get wider wave ponding on the berm tends to reduce its effectiveness. K_q was computed for Owen's data using $F'_{(min)}$ = 0.05, Owen (1990).

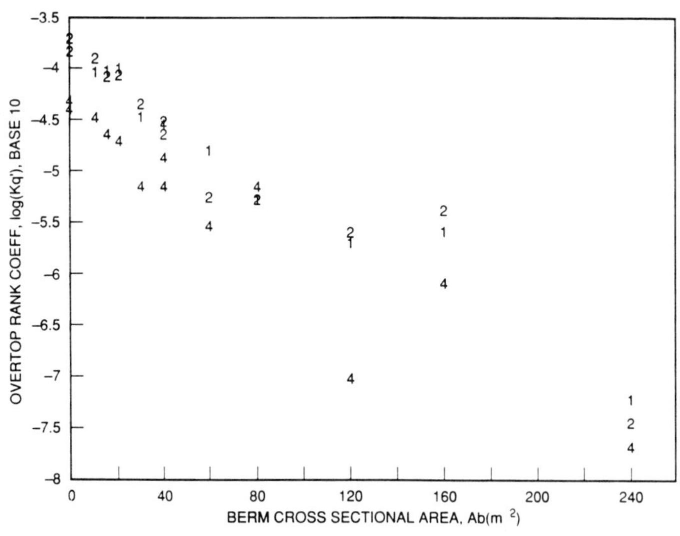

Figure 5, logarithm of the overtopping ranking coefficient, K_q, (Eq. 3) versus the cross sectional area of the berm. Numbers indicate slope of dike face, eg. 4 indicates 1:4 (vertical : horizontal) slope.

Raise the Freeboard or Build a Parapet

17. Another method to reduce wave overtopping is to increase the crest elevation of a seawall. Because overtopping rates increase approximately exponentially with freeboard, the value of increasing the height of the wall can be readily appreciated. Tests of the Roughans Point seawall included configurations with a 30 and 61 cm cap on the wall. Figure 6 shows overtopping trends for four seawall/revetment configurations, including the wall with a 30 cm cap (designated RP–7) and the wall with a 61 cm cap (RP–8); dimensions given are prototype scale based on a 1 : 16 (model : prototype) undistorted Froude scale. In Figure 6 configurations RP–4, RP–7, and RP–8 all have the same wide berm, absorber revetment fronting the wall so the effectiveness of the caps can be readily compared. In many locations raising the height of a seawall would be unacceptable but Figure 6 shows that relatively small caps are equivalent, in terms of reducing overtopping rates, to a great deal of stone placed in front of a seawall. To help the comparison of the regression trends shown in Figure 6, the relative freeboards for configurations RP–7 and RP–8 were computed as if they had the same freeboards as configurations RP–1 and RP–4.

18. Laboratory tests by Bradbury, Allsop, and Stephens (1988) were conducted to evaluate the performance of various breakwater crownwall configurations. Because of the geometry, some of Bradbury's et al. configurations are ideal for determining the effectiveness of using a parapet wall to reduce wave overtopping of a riprap protected embankment. Questions relating to the expected performance of parapet walls are quite common in the United States but the tests by Bradbury et al. appear to be the only relevant laboratory work. Based on the data of Bradbury, et al. Ward and Ahrens (1990) developed an equation which explicitly used the parapet height to estimate overtopping rates. Bradbury's et al. study shows that relatively low parapets can make substantial reductions in overtopping but a point is reached where further increases in height are not beneficial. It was also shown that a recurved parapet is more effective than a vertical parapet with the same height.

19. Increases in sea level will do two things to decrease the effectiveness of seawalls. Most important sea level rise reduces the freeboard, but it can also allow slightly larger waves to reach the structure. The combination of these effects can cause a surprisingly great deterioration in the performance of a seawall. Consider the original Roughans Point seawall, near Boston, MA., as an example. Current estimates of the net sea level rise in the Boston area are somewhat over 1 mm per year, Douglas (1990). Over the last fifty years, the approximate age of the structure, the total rise would be roughly 5 cm. For a water level which might occur every year or so, say 2.45 m NGVD, a 5 cm increase in water level will increase the overtopping rate by about 30 percent. A water level of 2.45 m gives a water depth at the toe of the wall of about 2.6 m and a freeboard of 2.92 m. The above analysis is based on Equations 2 and 4 and could be applied equally well to evaluating the performance of seawalls subject to the fluctuating water levels of the Great Lakes.

CONCLUSIONS AND IMPORTANT FINDINGS

(a) Results of an extended study of seawall design show that regression curves fit to laboratory data provide a logical way to evaluate the effectiveness of a seawall configuration and to compare the performance of different configurations. Data trends always show an approximately exponential relationship between the overtopping rate and the dimensionless freeboard.

(b) Findings from this study indicate that there is no completely satisfactory way to normalize the overtopping rate. This finding is consistent with research conducted by Jansen and Juhl (1987).

(c) Using a riprap revetment with a berm fronting a seawall was found to be a practical way to improve the performance of an existing vertical seawall.

(d) Recurved walls and parapets have been found to be an effective way to reduce overtopping rates. A recurved wall will reduce overtopping over a similar vertical wall with the same freeboard. However, the cost effectiveness of recurved walls is difficult to demonstrate.

(e) Research on seawalls could benefit from more cooperation among researches and the associated institutions. Publishing basic information and data on test setup, test procedures, wave conditions, structure geometry, and overtopping rates would be very useful. Wave data should include both statistical and energy based parameters.

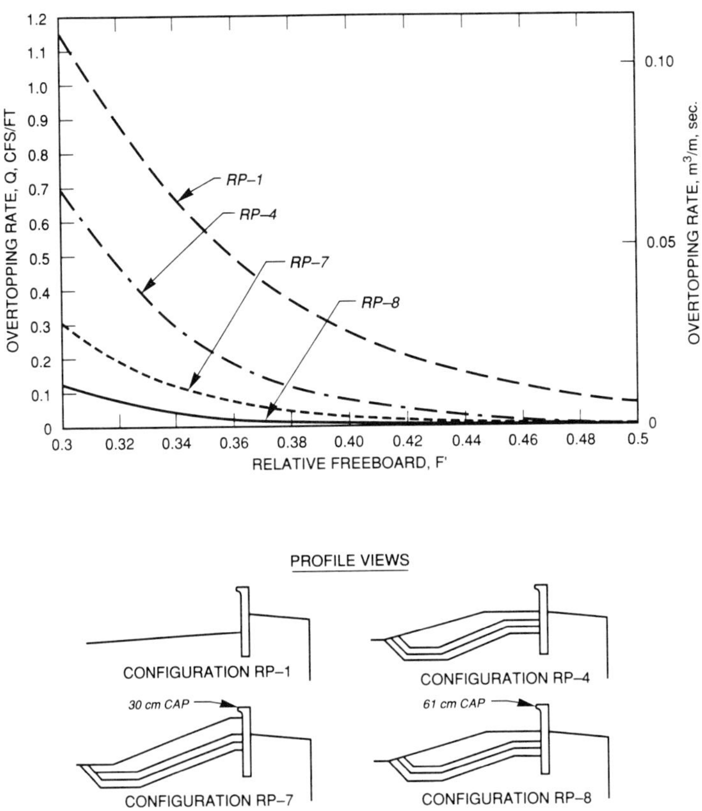

Figure 6, overtopping trend curves for configurations, RP-1, RP-4, RP-7, and RP-8, to illustrate the value of increasing the height of the seawall.

ACKNOWLEDGMENTS
20. The authors would like to thank Don Ward, civil engineer, and John Heggins, civil engineering technician, who provided essential information on the seawall study at CERC. Opinions expressed in this paper are those of the authors and do not necessarily reflect the opinions or policies of NOAA or the U.S. Army Corps of Engineers.

REFERENCES

1. Ahrens, J.P., and Heimbaugh, M.S., "Seawall Overtopping Model", Proceedings 21st Coastal Engineering Conference, Malaga, Spain, June 1988.
2. Ahrens, J.P., Heimbaugh, M.S., and Davidson, D.D., "Irregular Wave Overtopping of Seawall/Revetment Configurations, Roughans Point, Massachusetts", Technical Report CERC-86-7, U.S. Army Engineer Waterways Experiment Station, Vicksburg, MS, 1986.
3. Ahrens, J.P. and Ward, D.L., "Performance of Bermed Revetments", ASCE Journal of Waterway, Port, Coastal, and Ocean Engineering, scheduled for Sept. 1991.
4. Ahrens, J.P., "Methods to Reduce Wave Runup and Overtopping of Existing Structures", Technical Report REMR-CO-7, Coastal Engineering Research Center, U.S. Army Waterways Experiment Station, Vicksburg, MS, Oct. 1988.
5. Battjes, J.A., "Wave Runup and Overtopping", Report to the Technical Advisory Committee on Protection Against Inundation, Rijkswaterstaat, the Hague, the Netherlands, 1974.
6. Bradbury, A.P., Allsop, N.W.H., and Stephens, R.V., "Hydraulic Performance of Breakwater Crownwalls", Report SR 146, Hydraulic Research Limited, Wallingford, England, March 1988.
7. Douglas, B.C., "Global Sea Level Rise", submitted to Journal Geophysical Research - Oceans, Oct. 1990.
8. Fukuda, N., Uno, T., and Irie, I., "Field Observations of Wave Overtopping of Wave Absorbing Revetment", Coastal Engineering in Japan, Vol. 17, 1974.
9. Goda, Y., Random Seas and the Design of Maritime Structures , University of Tokyo Press, Tokyo, Japan, 1985.
10. Hallermeier, R.J., personal communication, 1990.
11. Hardy, T.A. and Crawford, P.L., "Frequency of Coastal Flooding at Roughans Point, Broad Sound, Lynn Harbor, and the Saugus-Pines River System", Technical Report 86-8, Coastal Engineering Research Center, U.S. Army Waterways Experiment Station, Vicksburg, Ms., Sept. 1986.
12. Jansen, O.J., "A Monograph on Rubble Mound Breakwaters", a book published by the Danish Hydraulic Institute, Nov. 1984.
13. Jensen, O.J. and Juhl, J., "Wave Overtopping on Breakwaters and Sea Dikes", International Conference on Coastal and Port Engineering in Developing Countries, Beijing, China, Sept. 1987.
14. Magoon, O.T., Pope, J.L., Sloan, R.L. and Treadwell, D.D., "Long Term Experience with Seawalls on an Exposed Coast", Proceedings of 21st Coastal Engineering Conference, Malaga, Spain, June 1988.
15. Owen, M.W., "Design of Seawalls Allowing for Wave Overtopping", Report No. EX 924, Hydraulics Research Station, Wallingford, England, June 1980.
16. Owen, M.W., "The Hydraulic Design of Sea-Wall Profiles", Proceedings International Coastal Engineering Conference, University of Southhampton, England, Sept. 1982a.

17. Owen, M.W., "Overtopping of Sea Defenses", Proceedings of International Conference on Hydraulic Modeling of Engineering Structures, Coventry, England, Sept. 1982b.
18. Owen, M.W., personal communication, 1990.
19. Pfeiffer, H., and Ahrens, J.P., "A Practical Application of a Low Berm Revetment", The REMR Bulletin, Vol 7, No. 1, U.S. Army Waterways Experiment Station, Vicksburg, MS, Jan 1990.
20. Shore Protection Manual, 4th Ed., U.S. Army Engineer Waterways Experiment Station, Coastal Engineering Research Center, U.S. Government Printing Office, Washington, D.C. 1984.
21. Ward, D.L., and Ahrens, J.P. "Reduction of Wave Overtopping by Parapets", REMR Draft Technical Note, CO-RR-1.5, Coastal Engineering Research Center, 1990.
22. Ward, D.L., and Ahrens, J.P., "Use of a Rubble Berm for Reducing Runup and Overtopping on a 1V to 2H Riprap Slope; Experimental Model Investigation", Coastal Engineering Research Center, REMR Technical Report, in review 1991.
23. Ward, D.L., "Laboratory Tests of Irregular Wave Overtopping of Vertical and Stepped Seawalls", Technical Report, Coastal Engineering Research Center, U.S. Army Waterways Experiment Station, in preparation 1991.
24. Weggel, J.R., "Wave Overtopping Equation", Proceedings 15th Coastal Engineering Conference, Honolulu, Hawaii, July 1976.
25. Weggel, J.R., "Seawalls: The Need for Research, Dimensional Considerations, and a Suggested Classification", Journal of Coastal Research, Special Issue No. 4, Autumn 1988

3. Wave transmission at low-crested structures

J. W. VAN DER MEER, Delft Hydraulics, and
K. ANGREMOND, Delft University of Technology

SYNOPSIS. Existing data on wave transmission have been supplemented by recent model experiments. The results have been analysed, and formulae were developed to predict wave transmission over low-crested rubble mound breakwaters.

INTRODUCTION

1. Basic research on rubble mound structures is often limited to stability, runup, rundown and reflection of and on structures with no or only minor overtopping. In many cases, however, a certain degree of overtopping is acceptable, and will lead to considerable savings on the quantity of material used. The stability of these structures is often higher than for non-overtopped structures, due to the fact that wave energy can pass over the crest, giving lower wave forces on the armour layer of the seaward slope.

2. Low-crested structures have two main design aspects: stability and the amount of wave transmission. Stability of low-crested and reef breakwaters was described by Van der Meer and Pilarczyk (ref. 1). They re-analysed extensive data sets of Ahrens (ref. 2), Powell and Allsop (ref. 3), Givler and Sϕrensen (ref. 4) and Van der Meer (ref. 5) and came up with three categories of structures (reef-type, low-crested above SWL (stil water level) and submerged structures) and with practical design formulae for each category.

3. The other design aspect, wave transmission, will be treated in this paper. Re-analysis of above given references was supplemented with a new series of experiments. This has resulted in a practical design formula for wave transmission, including the effect of wave period or wave steepness, and crest width.

EXISTING DATA SETS

4. A number of studies on wave transmission has been published with enough details to make a comparison with other studies useful and possible. All data selected were obtained with random-wave testing. A short description of the data sets will be given in the following.

<u>Van der Meer (ref. 5)</u>

5. A very extensive investigation on stability of rock slopes and gravel beaches was performed at Delft Hydraulics between 1983 and 1987. The (basic) background and all test

data were described in Van der Meer (ref. 5). A part of the studie was focussed on stability of low-crested breakwaters.

6. Besides the stability, the wave transmission was measured in a number of cases. Three crest heights were tested, one with the crest well above the water level, one with the crest at the water level and one with the crest well below the water level. This research was not very extensive (only 31 tests with 2 wave periods), but it covered almost the complete range of crest heights. This means that it can serve as a connection between studies which only concerned crests above or below the water level.

Seelig (ref. 6)

7. Seelig (ref. 6) has measured wave transmission for a large number of structure cross-sections, mostly with monochromatic waves, but also with random waves. Only random wave data have been considered.

Ahrens (ref. 2)

8. The tests on reef-type breakwaters of Ahrens (ref. 2) concerned both stability and transmission. Hearn (ref. 7) gives a more extensive analysis of Ahrens' data, and she developed a design formula for wave transmission. A reef breakwater is a homogeneous structure without filter layers or core. The design parameter for this type of structure is the lowering of the crest under wave attack. During a test on wave transmission it may well possible that the crest height had changed, which makes it difficult to choose the correct crest height for that test. The crest height used in this paper is the height measured after the test.

Hydraulics Research (ref. 8) and Allsop (ref. 9)

9. Powell and Allsop (ref. 3) describe the hydraulic performance of low-crest breakwaters, including transmission. They refer to Hydraulics Research (ref. 8) and Allsop (ref. 9) and give combined results in figures. The report of Hydraulics Research is confidential, but Allsop was willing to hand over the relevant data on wave transmission together with the tested cross-sections.

Daemrich and Kahle (ref. 10)

10. Three types of structures were tested by Daemrich and Kahle (ref. 10), all with the crest at or below the water level.

GOVERNING VARIABLES

11. The most important variables with respect to wave transmission are summarized here and explained in Fig. 1.

12. The crest height above the water level, i.e. the crest freeboard is given by R_c. In particular, when the size of the armour units is large and the crest level is close to SWL (i.e. R_c close to 0), the definition of the crest level is crucial. From the existing data sets, it could not be verified in what way the crest level was defined. For the new test series, described more in detail in this paper, the crest has been defined as the plane through the upper edges of the

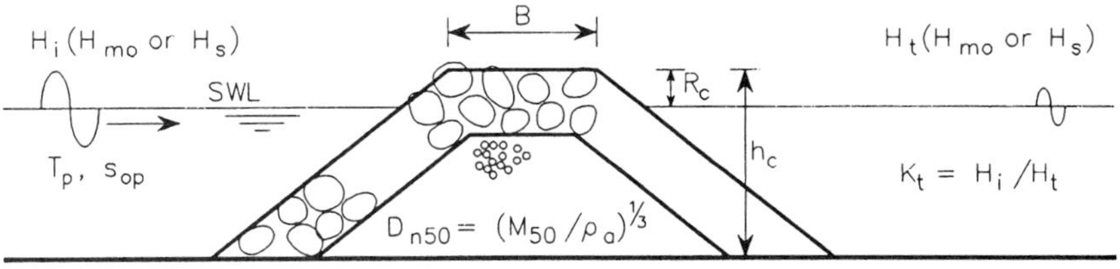

Fig. 1 Governing variables related to wave transmission

armour units. The height of the structure is defined by h_c. The crest width is defined by B. The water depth in front of the structure is given by h. The relationship between the parameters is: $R_c = h_c - h$.

13. The size of the armour rock is introduced as the nominal diameter D_{n50} ($D_{n50} = (M_{50}/\rho_a)^{1/3}$, where M_{50} is the average rock mass and ρ_a is the mass density).

14. The wave heights are given by H_i and H_t, representing the incoming and transmitted wave height respectively. Both values are expressed as H_s (mean of highest one third of the waves), or H_{m0} (based on spectrum, $4\sqrt{m_0}$). It must be mentioned here, that H_t is not always Rayleigh-distributed. The wave period, T_p, is used throughout this paper being the peak period of the spectrum; the wave steepness s_{op} is defined as $2\pi H_s/gT_p^2$, with H_s defined at the toe of the structure. The transmission coefficient K_t is given by $K_t = H_t/H_i$.

15. When discussing transmission of wave energy, it might be necessary to take into account the permeability of the structure. For this purpose the P-values have been used as introduced by Van der Meer (ref. 5).

FIRST ANALYSIS

16. Van der Meer (ref. 11) has attempted to analyse the existing data starting from the assumption that the transmission coefficient K_t would largely depend on the dimensionless crest height above SWL. The dimensionless crest height is defined in two ways, as:

$$R_c/H_i;$$

namely only related to the wave height (see Fig. 2) and as the parameter:

$$R^*_p = R_c/H_i * (s_p/2\pi)^{0.5} \qquad (1)$$

which includes both wave height and steepness, used by Powell and Allsop (ref. 3). They used T_m instead of T_p, however, in

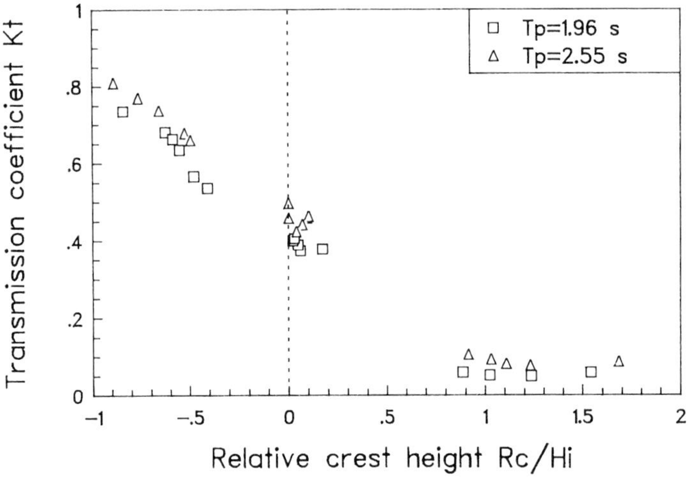

Fig. 2 Wave transmission as a function of R_c/H_i. Data of Van der Meer (ref. 11)

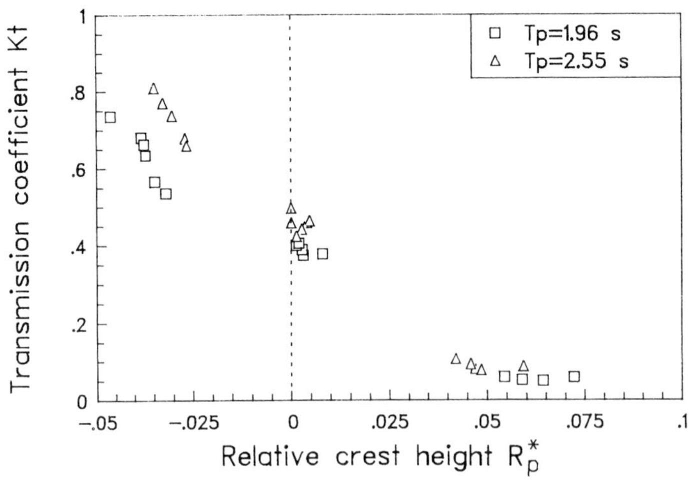

Fig. 3 Wave transmission as a function of R_p^*. Data of Van der Meer (ref. 11)

calculating the wave steepness. It is stressed again that in this report only the peak period is used.

17. Figures 2 and 3 give the transmission data of Van der Meer (ref. 11) as a function of these two dimensionless expressions for the crest height. From these figures it can be concluded that:

- for the same R_c/H_i a longer period yields more transmission.
- with R_p^* as parameter the transmission coefficients form one line for high values of R_p^* ($R_p^* \geq 0.025$).

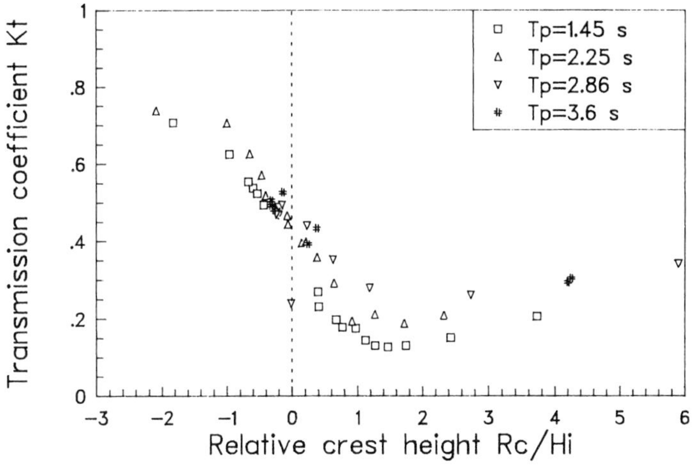

Fig. 4 Influence of the wave period on transmission. Data of Ahrens (ref. 2), subset 5/6

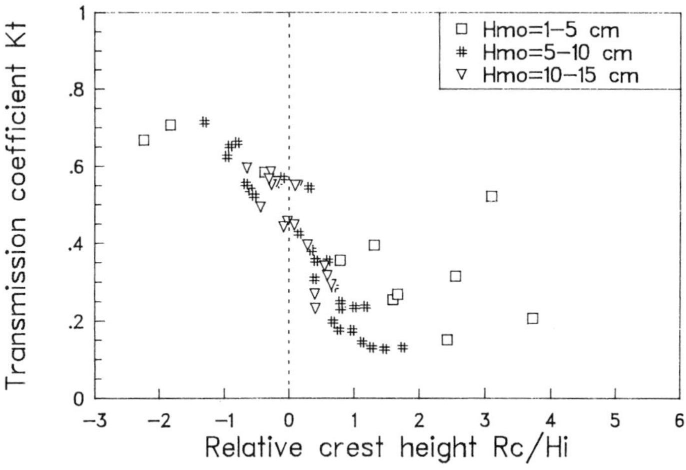

Fig. 5 Influence of wave height on transmission. Data of Ahrens (ref. 2), $T_p = 1.45$ s

- for $R_c/H_i = 0$ and $R_p^* = 0$ the same difference exists between the wave periods.
- for negative values of R_c, the use of R_p^* as parameter leads to less satisfactory results than the use of R_c/H_i.

18. The general conclusion is that R_p^* is not a better parameter than R_c/H_i as long as the whole range of relative crest levels is considered.

19. The above mentioned influence of the wave period is also clear from the data of Ahrens (ref. 2), as shown in Fig. 4. This is remarkable as his results are obtained with gradually deforming (reef type) breakwaters.

20. Plotting all data of Ahrens obtained for one particular wave period against the relative crest height, a wide scatter of K_t is observed for high values of R_c/H_i. A closer analysis shows that this scatter is mainly due to the occurrence of low wave heights, having roughly the same dimensions as the rock, see Fig. 5. Apparently the low-wave travels easily through the top of the structure.

21. Combining all data, and plotting the transmission coefficient against the relative crest height, Fig. 6 is obtained. As expected, the result shows considerable scatter, but a clear trend can be observed. Part of the scatter can be attributed to the influence of the wave period and partly to the influence of extremely small waves, while crest width and permeability may also have some influence.

22. The average value of K_t for $-2 < R_c/H_i < -1$ is about 0.8. Except for the triangles (Ahrens' data, small wave heights), the average value of K_t for $1 < R_c/H_i < 2$ is about 0.1. Between these ranges the value of K_t decreases almost linearly with R_c/H_i. Based on this simple analysis the following formula for wave transmission can be proposed:

$$\begin{aligned} &\text{For: } -2.0 < R_c/H_i < -1.13 && K_t = 0.80 \\ &\text{For: } -1.13 < R_c/H_i < 1.2 && K_t = 0.46 - 0.3R_c/H_i \\ &\text{For: } 1.2 < R_c/H_i < 2.0 && K_t = 0.10 \end{aligned} \qquad (2)$$

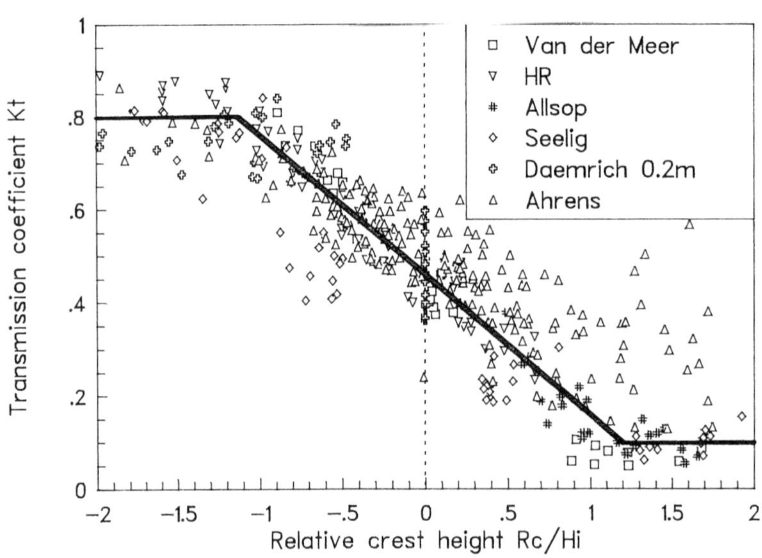

Fig. 6 Wave transmission versus relative crest height. All data

23. This curve is shown in Fig. 6 as well. The spreading is large which means that the formula may be simple. It means also that for application one should take into account this large spreading. The standard deviation of K_t was assessed from the graph, assuming a normal distribution and that amounted to $\sigma(K_t) = 0.09$. This means that the 90 per cent confidence levels are given by $K_t \pm 0.15$. It is evident that for large negative values of R_c/H_i, K_t should approach 1, and for large positive values of R_c/H_i a value close to 0.

24. For design purposes, however, the scatter is a serious drawback, as an accurate forecast of wave transmission may lead to considerable savings by reducing the total height of the structure. Therefore, it was decided to perform additional tests, and to continue the efforts towards a better expression for wave transmission.

RECENT TESTS

25. In order to study the wave transmission process further, additional tests were carried out by Daemen (ref. 12), as part of his masters thesis at the Delft University of Technology. The tests and the thesis were prepared at Delft Hydraulics under the supervision of authors.

26. The tests for this study were performed in a flume at Delft Hydraulics with a length of 50 m, a width of 1.0 m and a depth of 1.2 m. Irregular waves were generated conform to a preset energy density spectrum. Reflected waves were electronically compensated at the wave generator. With this system standing waves and basin resonance were avoided.

27. The incident wave height was measured together with the structure in the flume by means of two wave gauges placed apart about a quarter of a wave length. In this way the incident and reflected spectra were assessed. Wave heights were measured by the spectral method and by time analysis, which means that the wave height (H_{m0}) is defined as: four times the root mean square of the first moment of the wave energy density spectrum ($4\ (m_0)^{1/2}$). In deep water H_{m0} does not differ from the statistically defined wave height, H_s (average of the highest 1/3 of the waves).

28. At the end of the tests, the data from the two gauges behind the structure were filtered to remove long wave components which could be present behind the structure. These long wave components might influence the process because their existence is not a wave transmission phenomenon, but merely a result of the circumstances during testing. They occurred mainly when the crest of the structure was above the water level and could be traced because it was observed that the wave heights measured by the two wave gauges relative to each other differed considerably.

29. The structure consisted of a core of angular stone with a nominal diameter (D_{n50}) of 0.028 m and an armour layer. The armour layer consisted of angular stone with a nominal dia-

meter of 0.040 m, except for a few tests, where the nominal diameter amounted to 0.061 m.

30. In total 53 tests were performed, 34 tests with a wave steepness of 0.02, a nominal armour diameter of 0.040 m and with 6 values of the crest freeboard. Another 6 were also performed with the same wave steepness, but with a nominal armour diameter of 0.061 m and 3 values of the crest freeboard. Ten tests were performed with a wave steepness of 0.04, a nominal armour diameter of 0.040 m and 5 values of the crest freeboard.

31. The results of the tests are presented in Fig. 7 where a distinction has been made between the two wave steepness values of 0.02 (long waves) and 0.04 (short waves).

SECOND ANALYSIS

32. Until now wave transmission has been described in the conventional way as a function of R_c/H_i. It is not clear, however, that the use of this combination of crest freeboard and wave height produces similar results with on the one hand constant R_c and variable H_i and on the other variable R_c and constant H_i. Moreover, when R_c becomes zero, all influence of the wave height is lost which leads to a large spreading in the figure at $R_c = 0$. Therefore, it was decided to separate R_c and H_i in the second analysis.

33. The mass or nominal diameter of the armour layer of a rubble mound structure is determined by the extreme wave attack that can be expected during the life time of the structure. There is a direct relationship between the design wave height and the size of armour stone, which is often given as the stability factor $H_s/\Delta D_{n50}$, where Δ is the relative buoyant density. It can be concluded that the nominal diameter of the armour layer characterises the rubble mound structure. It is,

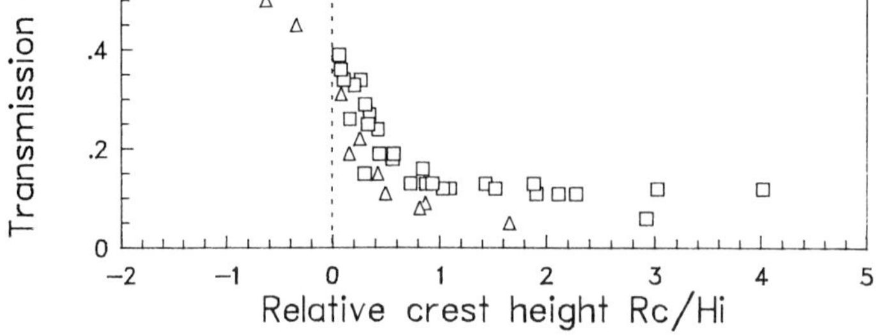

Fig. 7 Wave transmission data of present tests

therefore, also a good parameter to characterise the wave height and crest height in a dimensionless way.

34. The relative wave height can then be given as H_i/D_{n50}, in accordance with the stability factor, and the relative crest height as R_c/D_{n50}, the number of rocks that the crest level is above or below SWL. Finally, D_{n50} can be used to describe other breakwater properties as the crest width B. This yields the parameter B/D_{n50}.

35. The primary parameters for wave transmission can now be given as:

Relative crest height	: R_c/D_{n50}
Relative wave height	: H_i/D_{n50}
Fictitious wave steepness	: s_{op}

36. Secondary parameters are relative crest width, B/D_{n50}, permeability factor, P, and slope angle cotα. Furthermore, it should be noted that reef type breakwaters differ considerably from the conventional rubble mound structure.

Derivation of formula

37. Fig. 8 shows the wave transmission versus R_c/D_{n50} for the data of Van der Meer (ref. 11) and the present tests of Daemen (ref. 12). The data are grouped by constant relative wave steepness, s_{op}. Straight lines are drawn through the points with the same wave steepness. Fig. 8 makes clearly visible that a lower wave steepness (or a longer period) results in a higher transmission coefficient. This is true for

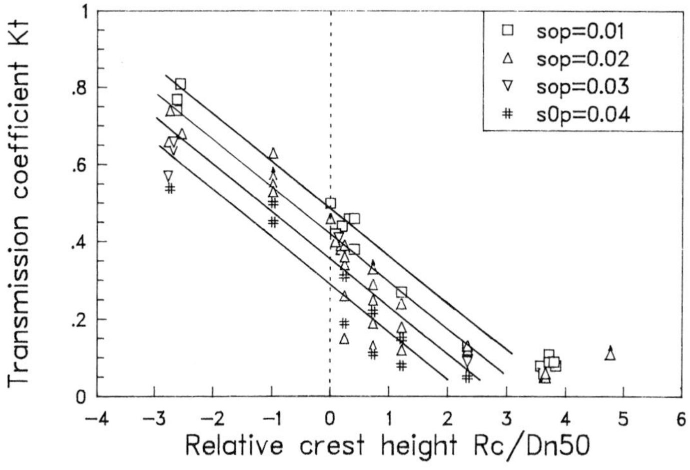

Fig. 8 Wave transmission versus R_c/D_{n50}. Data of Van der Meer (ref. 11) and Daemen (ref. 12)

the whole area of R_c/D_{n50}, except for large positive and negative values. Furthermore, the lines in Fig. 8 are parallel to each other.

38. The general trend of the wave transmission coefficient, going from high positive values of R_c/D_{n50} to high negative values, is that the transmission coefficient first stays low, then increases in the area of R_c/D_{n50} between +2 and -2 and finally remains high. Theoretically, the increase in wave transmission coefficient will be expressed by a smooth curve from 0 for very high crest heights to 1 for very low crest heights. The most important area, however, is the area where the transmission increases rapidly. For the sake of simplicity it is assumed that this area can be described by straight lines as shown in Fig. 8.

39. This means that a linear relationship is assumed in the area of R_c/D_{n50} roughly between +2 and -2. The wave transmission can now be described as:

$$K_t = aR_c/D_{n50} + b \tag{3}$$

40. In this equation "a" determines the slope of the line and "b" gives the value of K_t at $R_c/D_{n50} = 0$. From Fig. 8 it can already be concluded that the wave steepness s_{op} is only present in the coefficient "b" and not in the coefficient "a" (the lines in Fig. 8 are parallel).

41. Fig. 9 shows the data of Daemen (ref. 12) with a constant wave steepness of $s_{op} = 0.02$ and various classes of relative wave heights. In fact, Fig. 9 shows the influence of the relative wave height H_i/D_{n50} on wave transmission. From this Fig. 9 it can be concluded that for $R_c/D_{n50} < -1$ a larger H_i/D_{n50} produces larger wave transmission. For $R_c/D_{n50} > -1$ the opposite occurs: a larger H_i/D_{n50} gives smaller wave transmission.

42. This phenomenon can be explained in a physical way. On a low-crested breakwater, where R_c/D_{n50} is positive, the transmission is primarily determined by overtopping and thus by wave runup. In this area of R_c/D_{n50} a larger relative wave height gives a higher runup, thus more overtopping and hence a larger transmission coefficient. On a submerged breakwater, where R_c/D_{n50} is negative, higher waves will be more affected by the structure whereas small waves pass unhindered. In this case a larger relative wave height results in a smaller transmission coefficient.

43. The lines in Fig. 9 all have different angles and different intersections with the vertical axis $R_c/D_{n50} = 0$. This means that both coefficients "a" and "b" in Eq. 3 are influenced by the relative wave height H_i/D_{n50}. Only "b" is also influenced by the wave steepness s_{op}. To study the relationship

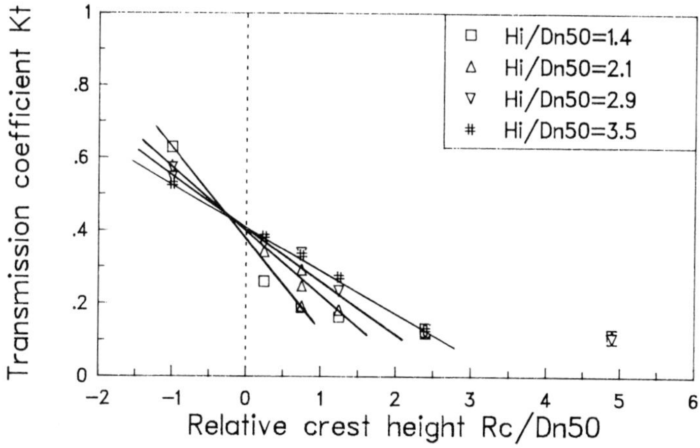

Fig. 9 Influence of relative wave height on wave transmission for constant wave steepness of $s_{op} = 0.02$

between "a" and H_i/D_{n50}, lines were fit through each data set with constant s_{op} and H_i/D_{n50}. In total 58 lines were drawn with constant wave steepness and relative wave height. The range of s_{op} was 0.01 to 0.04, that of H_i/D_{n50} 1.3 to 5.

44. All these coefficients "a" are shown in Fig. 10 as a function of the relative wave height H_i/D_{n50}. From this plot the following relationship for "a" was derived, which is also shown in Fig. 10:

$$a = 0.031\ H_i/D_{n50} - 0.24 \qquad (4)$$

Note that also the data of the reef type breakwaters of Ahrens fit in the figure. By substitution of Eq. 4 into Eq. 3, the formula for wave transmission so far looks like:

$$K_t = (0.031\ H_i/D_{n50} - 0.24)\ R_c/D_{n50} + b \qquad (5)$$

45. To assess the coefficient "b", again lines were drawn trough the data sets for constant wave steepness and relative wave heights, but now with the imposed values of "a" from Eq. 4. These coefficients are shown in Fig. 11, now versus the wave steepness s_{op}. The data of Daemrich and Kahle (ref. 10) and Powell and Allsop (ref. 3) did not fit in the figure. Daemrich and Kahle used structures with tetrapods (more porous than rock) and with a very narrow and a very wide crest. These data will be used later in this paper. Powell and Allsop used homogeneous structures with large rock in depth- limited situations. Some uncertainty exists about the given wave heights at the structure, which seem to be too high. Allsop could not clarify this point (personal communication). Therefore, these data were left out.

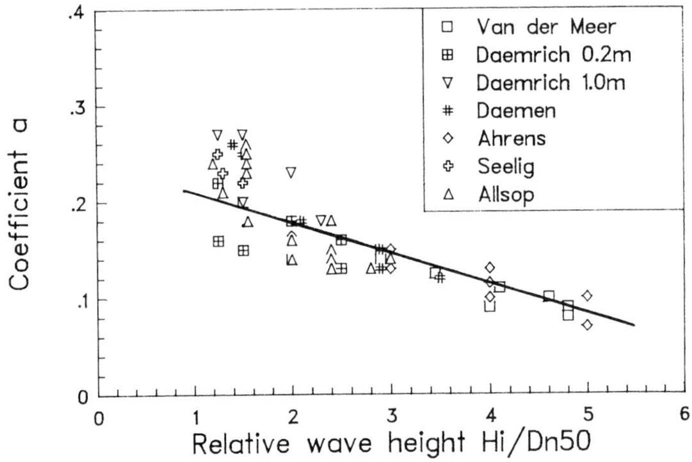

Fig. 10 The coefficient "a" in Eq. 3 versus the relative wave height H_i/D_{n50}

46. Fig. 11 shows very clearly that also the data of Ahrens (ref. 2) do not fit with the other data. In the figure the points are higher and show a smaller effect of the relative wave height. This might be caused by the fact that reef breakwaters are subject to deformation which results in a more or less rounded crest. Here the highest point of the crest is used for R_c. A more average crest height (say one or more diameters lower) would result in lines in graphs such as Figs. 8 and 9 which would be shifted to the left and would be more in agreement with the conventional breakwaters. However, from now on the reef breakwaters will be treated seperately.

47. Fig. 11 shows a clear influence of the wave steepness on the coefficient "b". The relative wave height has also influence on this coefficient, as stated earlier. A more in depth analysis, partly based on Fig. 11, gave the following relationship for the coefficient "b":

$$b = -5.42\ s_{op} + 0.0323\ H_i/D_{n50} + 0.44 \qquad (6)$$

48. A more or less similar result was obtained for reef type breakwaters:

$$b = -2.6\ s_{op} - 0.05\ H_i/D_{n50} + 0.85 \qquad (7)$$

49. A closer look at Eq. 6 showed that the value of 0.44 was not really a constant, but that it was also a function of the crest width, specially when the data of Daemrich and Kahle (ref. 10) with the narrow and wide crest were taken into account. Fig. 12 gives the "constant 0.44" as a function of the relative crest width B/D_{n50}. For the data of Van der Meer (ref. 11) and Daemen (12) indeed the value of 0.44 is found.

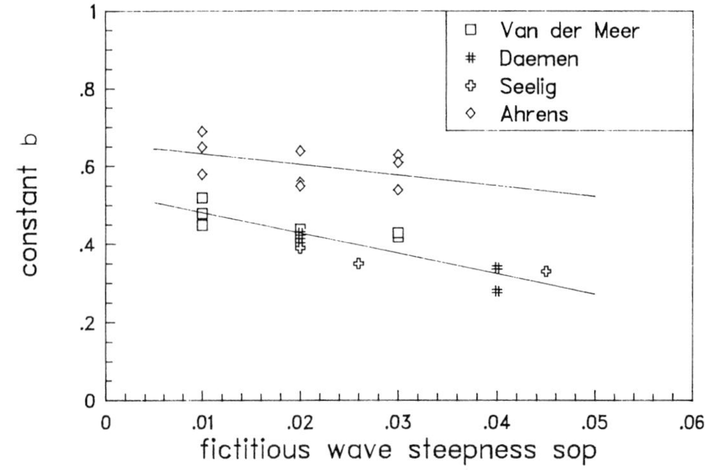

Fig. 11 The coefficient "b" in Eq. 5 versus the wave steepness s_{op}

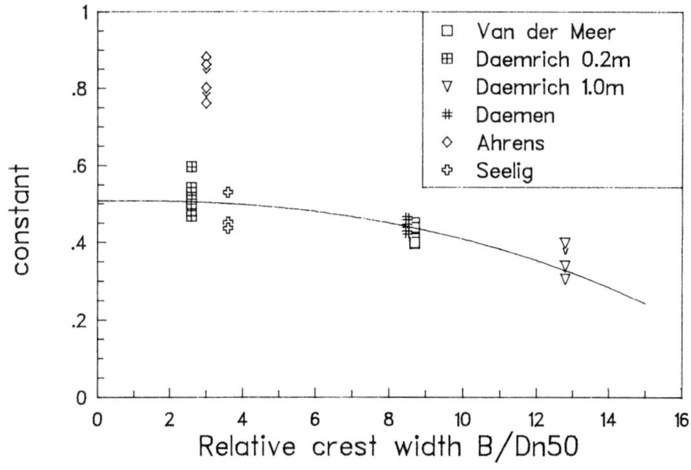

Fig. 12 Influence of crest width on coefficient in Eqs. 6 and 7

For a smaller crest width the coefficient is higher and for a wider crest lower. This constant can be described by:

$$\text{constant} = 0.51 - 0.0017\ (B/D_{n50})^{1.84} \qquad (8)$$

which is only valid for conventional breakwaters (Ahrens´ data give a value of 0.85, see Eq. 7 and Fig. 12). Substitution of Eq. 8 in Eq. 6 gives the final relationship for b:

$$b = -5.42\ s_{op} + 0.0323\ H_i/D_{n50} - 0.0017\ (B/D_{n50})^{1.84} + 0.51 \qquad (9)$$

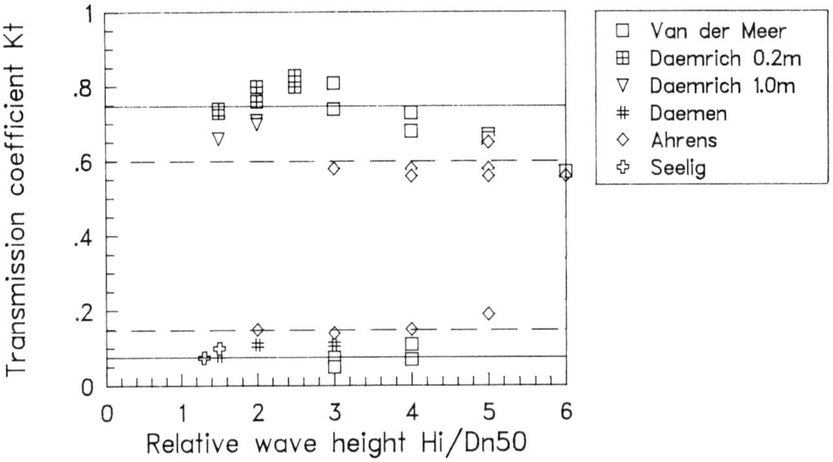

Fig. 13 Minimum and maximum wave transmission

50. Eq. 3 with related Eqs. 7 and 9 produce the transmission coefficient in the area where the transmission changes rapidly with changing crest height. The upper and lower boundaries will be treated here. Fig. 13 gives the values of the transmission coefficient which holds for high and low relative crest heights, outside the range given by the curve of Eq. 3. Fig. 13 shows that both the maximum and minimum transmission are independent of the relative wave height H_i/D_{n50}. Based on Fig. 13 the following minimum and maximum values were derived:

Conventional breakwaters:
Minimum: $K_t = 0.075$; maximum: $K_t = 0.75$ (10)

Reef type breakwaters:
Minimum: $K_t = 0.15$; maximum: $K_t = 0.60$ (11)

FINAL RESULTS

51. The outcome of the second analysis on wave transmission, including the data of Daemen (ref. 12), was a linear relationship between the wave transmission coefficient K_t and the relative crest height R_c/D_{n50}, which is valid between minimum and maximum values of K_t. In fig. 14 the basic graph is shown. The linearly increasing curves are presented by:

$$K_t = a\, R_c/D_{n50} + b \tag{3}$$

$$\text{with: } a = 0.031\, H_i/D_{n50} - 0.24 \tag{4}$$

Eq. 4 is applicable for conventional and reef type breakwaters. The coefficient "b" for conventional breakwaters is described by:

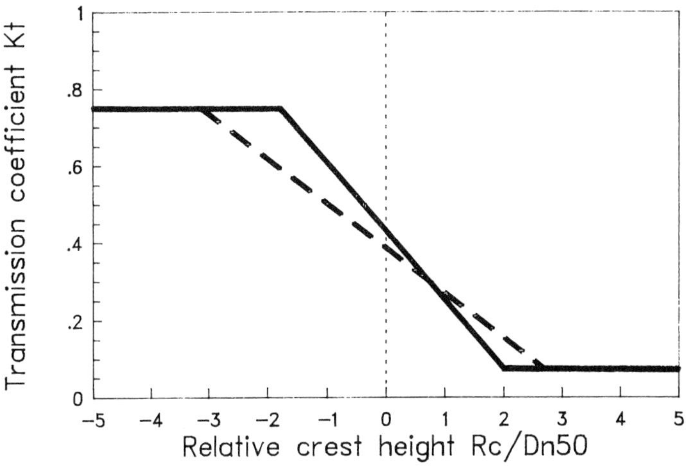

Fig. 14 Basic graph for wave transmission

$$b = -5.42\ s_{op} + 0.0323\ H_i/D_{n50} - 0.0017\ (B/D_{n50})^{1.84} + 0.51 \quad (9)$$

and for reef type breakwaters by:

$$b = -2.6\ s_{op} - 0.05\ H_i/D_{n50} + 0.85 \quad (7)$$

The following minimum and maximum values were derived:

Conventional breakwaters:
Minimum: K_t = 0.075; maximum: K_t = 0.75 (10)

Reef-type breakwaters:
Minimum: K_t = 0.15; maximum: K_t = 0.60 (11)

VALIDITY AND RELIABILITY OF FORMULA

52. The analysis was based on various groups with constant wave steepness and a constant relative wave height. The validity of the wave transmission formula (Eq. 3) corresponds, of course, with the ranges of these groups that were used. The formula is valid for:

$$1 < H_i/D_{n50} < 6 \text{ and } 0.01 < s_{op} < 0.05$$

Both upper boundaries can be regarded as physically bound. Values of $H_i/D_{n50} > 6$ will cause instability of the structure and values of $s_{op} > 0.05$ will cause waves breaking on steepness. In fact, boundaries are only given for too low wave heights relative to the rock diameter and for very low wave steepnesses.

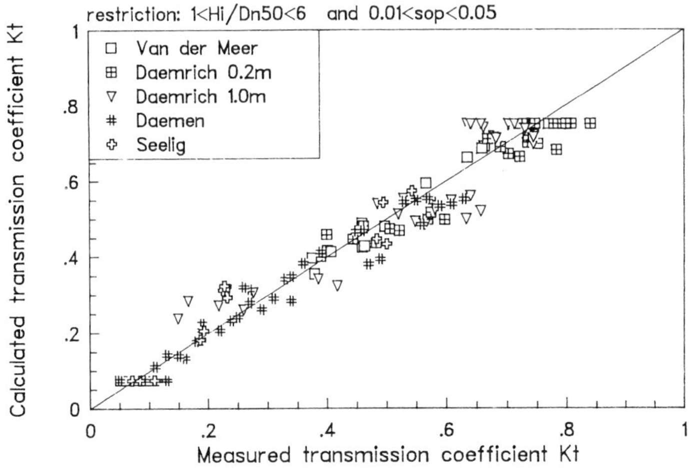

Fig. 15 Calculated versus measured wave transmission for conventional breakwaters

53. The formula is applicable outside the range given above, but the reliability is low. Fig. 15 shows the measured wave transmission coefficient versus the calculated one from Eq. 3, for various data sets of conventional breakwaters. The reliability of the formula can be described by assuming a normal distribution around the line in Fig. 15. With the restriction of the range of application given above, the standard deviation amounted to $\sigma(K_t) = 0.05$, which means that the 90 per cent confidence levels can be given by $K_t \pm 0.08$. This is a remarkable increase in reliability compared to the simple formula given by Eq. 2 and Fig. 6, where a standard deviation of $\sigma(K_t) = 0.09$ was given.

54. The reliability of the formula for reef-type breakwaters is more difficult to describe. If only tests are taken where the crest height had been lowered less than 10 per cent of the initial height h_c, and the test conditions lie within the range of application, the standard deviation amounts to $\sigma(K_t) = 0.031$. If the restriction on the crest height is not taken into account the standard deviation amounts to $\sigma(K_t) =$ 0.054.

REFERENCES

1. VAN DER MEER, J.W. and PILARCZYK, K.W. Stability of low-crested and reef breakwaters. ASCE, Proc. 22th ICCE, Delft, 1990.
2. AHRENS, J.P. Characteristics of reef breakwaters. WES, CERC, Vicksburg, 1987, Technical Report CERC-87-17.
3. POWELL, K.A. and ALLSOP, N.W.H. Low-crest breakwaters, hydraulic performance and stability. Hydraulics Research, Wallingford, 1985, Report SR 57.

4. GIVLER, L.D. and SϕRENSEN, R.M. An investigation of the stability of submerged homogeneous rubble-mound structures under wave attack. Lehigh University, H.R. IMBT Hydraulics, 1987, Report #IHL-110-86.
5. VAN DER MEER, J.W. Rock slopes and gravel beaches under wave attack. Delft University of Technology, Delft, 1988, PhD-thesis. Also Delft Hydraulics Publication No. 396.
6. SEELIG, W.N. Two-dimensional tests of wave transmission and reflection characteristics of laboratory breakwaters. WES, CERC, Vicksburg, 1980, Technical Report NO. 80-1.
7. HEARN, J.K. An analysis of stability of a wave field modification due to low-crested, sacrificial breakwaters. University of Florida, Miami, 1987, MSc-thesis.
8. ANONYMOUS. Seaford frontage - model tests of low crest breakwaters, performance and stability. Hydraulics Research, Wallingford, 1985, Report EX 1346.
9. ALLSOP, N.W.H. Low-crest breakwaters, studies in random waves. Proc. Coastal Structures ´83, Arlington, 1983.
10. DAEMRICH, K.F. and KAHLE, W. Schutzwirkung von Unterwasserwellenbrechern unter dem einfluss unregelmässiger Seegangswellen. Eigenverlag des Franzius-Instituts für Wasserbau und Küsteningenieurwesen, Heft 61, 1985.
11. VAN DER MEER, J.W. Data on wave transmission due to overtopping. Delft Hydraulics, Delft, 1990, Report H 986.
12. DAEMEN, I.F.R. Wave transmission at low-crested breakwaters. Delft University of Technology, Faculty of Civil Engineering, Delft, 1991, Masters Thesis.

Discussion

H. F. BURCHARTH, University of Aalborg
In which way did Mr Ahrens compensate for reflected waves in the flume tests?

C. T. BROWN, Seabee Developments
I have five points jointly to both Mr Ahrens and Dr van der Meer.

Mr Ahrens discusses the relative merits of reflector walls above porous revetments. In 1974, a vertical seawall at Balmoral Beach was severely damaged by waves of 16-20 s period which penetrated deep into the harbour. It was originally proposed to replace the wall with a stepped wall, which had better reflection and run-up characteristics. However, under extreme conditions the overtopping was markedly worse, due to the 'ramp effect'. With a vertical wall, the stresses induced on the wall are quite small; the wall is much cheaper than a recurved wall, and is very amenable to architectural treatment.

This leads on to the parameterization of overtopping criteria (flow, transmission). In reanalysing their data I found that the dimensionless crest elevation ((FBD+ds)/ds) (height of crest above bed divided by water depth) achieved the best condensation of the data. Unfortunately the spread of wave periods was not extensive.

In conjunction with Peter Fraenkel BMT I recently designed and tested a low-crest breakwater at Fairisle. This structure was required to improve the berthing conditions at the existing jetty, until waves through the entrance made the berth untenable. I am pleased to say that the model tests carried out at Hydraulics Research Wallingford appear to have confirmed the interpretation made by us of Allsop's data.

However, I am concerned at possible scale effects and three-dimensional effects. During testing for the Skegness seawall (at another establishment), a slight convexity in the nearshore approach focused the waves from the 2.4 m wide flume onto an area of about 1.2 m of the seawall under test, markedly increasing local overtopping. This is in reality quite typical of what may happen in reality, and this sort of condition should be borne in mind when evaluating and using the results of two-dimensional testing.

In conclusion I would like to know if anyone has data on the behaviour of berms (solid, not porous) above water level, to add to Owen's data, or data regarding the effects of roughness and porosity.

I would also appreciate the Authors' comments on the scale and three-dimensional effects on transmission, overtopping and stability of low-crested structures.

N. W. H. ALLSOP, HR Wallingford

One of the great advantages of the Owen formula is the ease with which the relative roughness parameter r can be adjusted to take account of slopes of different construction, and sea states. This allows the extension of the simple method evolved in 1980 to rough and rubble slopes, provided that the response of r to structure and wave parameters can be identified. All of the parameters that are input in predicting overtopping are conventionally used as single characteristic values.

A software package under evaluation at HR Wallingford allows each main parameter to be expressed as a stochastic variable with given mean value and probability distribution. The computer then calculates the response, such as the overtopping, as a mean value with its associated probability distribution. Happily, initial trials gave an example variability of the calculated discharge within a range of no more than a factor of 2, quite a narrow range in relation to the exponential nature of the overtopping response. The use of such simple packages will allow much more widespread evaluation of the probabilities of damage, provided that resources are available to generate appropriate response functions with attendant probability distributions.

J. W. VAN DER MEER, Delft Hydraulics

At Delft Hydraulics we have worked on overtopping recently, especially on dikes.

You use the local wave length L_p which is difficult to use. We often use L_{op}, the deep-water wave length. This gives in a steepness s_{op} the effect of the wave period rather than local wave length. Was the difference in results between L_p and L_{op} so large that L_p was really better?

Why do you use a combined parameter $F/H_{mo}{}^{2/3}L_p{}^{1/3}$? Why not $F/H_{mo}s_p{}^b$ and look for the 'best fit' b? A reduction coefficient for run-up (berms, roughness, etc.) gave on dikes a good prediction of overtopping.

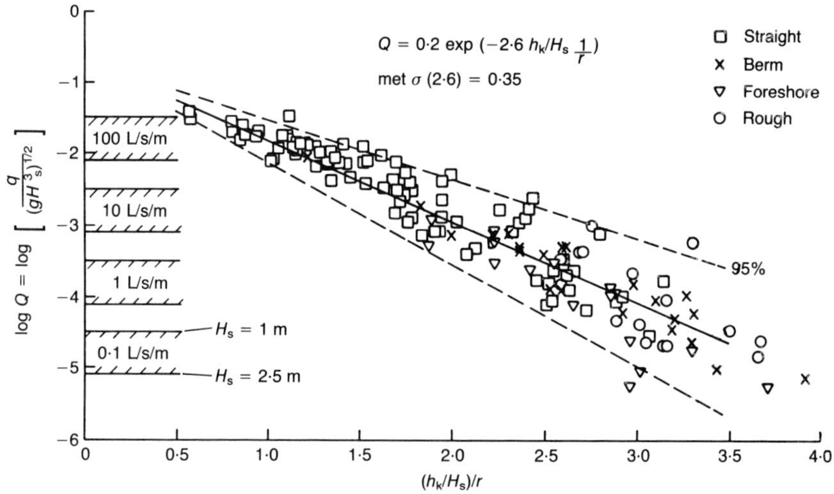

Fig. 1. Surging waves

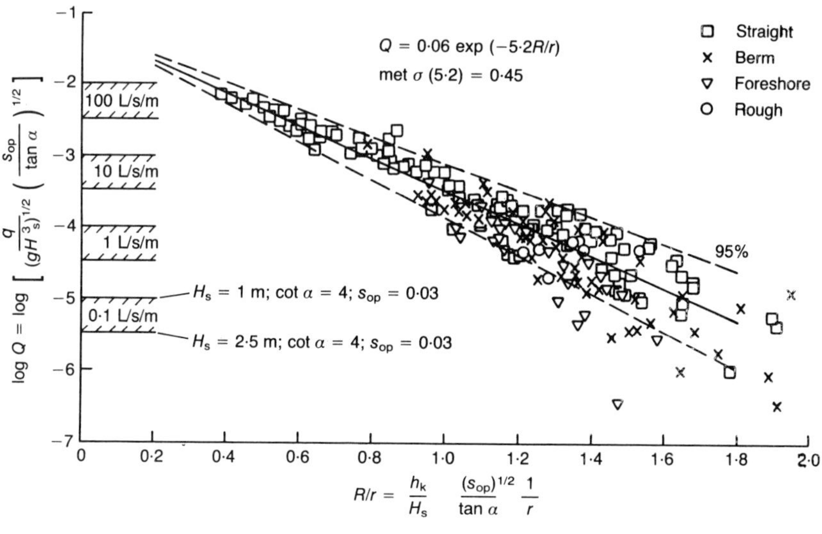

Fig. 2. Plunging waves

The data on overtopping for dikes had to be divided in plunging and surging waves. This affected the dimensionless freeboard and discharge. For plunging waves the wave steepness and slope angle are required, whereas for surging waves it was only the wave height. Figs 1 and 2 summarize all the test data. Note that berms and rough slopes could be included with a reduction coefficient in the formula.

H. W. PARTENSCKY, University of Hannover
My first question is for Mr Ahrens. The predicted rise of the mean sea level requires a raising of the existing sea dikes along the German coast line. A rise of these dikes of about 1.0 m will cost about 3 billion DM. As an alternative solution, the placing of crown walls on the existing dikes is considered.

Therefore I should like to ask if you have made tests with seaward slopes of 1:6 or 1:8 in front of a seawall The results would be helpful to us, since we are planning large-scale tests in our large wave channel in Hannover in 1993/94.

My other question is for Dr van der Meer. Is the D_{n50}-value referred to the porosity of the cover layer only? How did you take into account the energy transfers through the entire breakwater, i.e. through the underlayers and the core of the structure?

M. A. LOSADA, University of Cantabria
In the analysis of breakwater performance it is useful to split the involved phenomena into the categories reflection, dissipation and transmission. The presentation of laboratory data or formulae also keeps the splitting of the phenomena. When talking about the reliability of formulae, this can be shown only by comparing the results of the three involved phenomena. Therefore, it would be more efficient to show simultaneously reflection, transmission and dissipation coefficients.

Is scattering in a surging wave a phenomenon related to long wave seiching in the flume?

W. LEEUWESTEIN, CUR
Mr Ahrens and Dr van der Meer have presented data on wave overtopping and transmission, respectively. The design parameters are physically closely related; therefore it is surprising that the authors use different freeboard parameters (F divided by $3^{1/3}$ of a 'water volume' and by D_{n50}, respectively). It seems that the first is more justified because with H and L only the external hydraulic conditions are involved, whereas by using D_{n50} a structural parameter is introduced. For example, what happens when D_{n50} becomes small?

In my opinion D_{n50} should be introduced as a secondary parameter related to a permeability parameter, rather than as the basic parameter for transmission (although transmission through the wall would be minor for high freeboards and short periods).

F. VASCO COSTA, Consulmar
My question is for Dr van der Meer. Among the forces affecting water motion, the viscosity forces are the

only ones that give occasion to the dissipation of energy. Can we ignore the contribution of such forces retarding and dissipating the kinetic energy of waves **passing over low-crested structures?**

It must be noted that in tests run in accordance to the Froude criterion the viscosity forces are represented several times larger than they should be. In tests run on a model at scale 1/50 they are represented 356 times larger than the gravity forces, which are supposed to be the only forces affecting the motion (Vasco Costa forces associated to different fluid properties as affected by scaling, Coastal Engineering pp. 371-377, 1981).

H. OUMERACI, University of Hannover

I would like to comment on the fundamental question of flow to define the transmission coefficient. Until now it has been generally defined as $K_T = H_i/H_t$ where H_i is the incident wave height and H_t is the transmitted wave height. However, everybody has already seen a couple of times aerial photographs showing that waves of say 15 s period are transmitted over a reef as shorter waves of 3-5 times smaller periods. (This can be explained by soliton theory.) But we know that the effect of a 15 s period wave is expected to be significantly different from that of a wave of 3 s period; this means that we have to rethink the concept and definition of the transmission coefficient.

J. P. Ahrens, Paper 2

In reply to Professor Burcharth, the wave flume was divided in half at the end with the seawall test section. One half had a gravel wave absorber beach with a slope of about 1:10. Rubberized horse hair was also used around the waterline on the absorber beach. Seiching did not appear to be a problem for the data collected at CERC and discussed in this Paper.

I agree with Dr Brown that a vertical wall may perform better than some other wall configurations if the 'ramp effect' is a severe problem.

I am not sure about wave run-up or overtopping data on a coastal structure with solid berms above the water level, that could be used to add to Owen's data. There may be some data collected in the 1960s by Professor Battjes of Delft Technical University which would satisfy your criteria; this data would almost certainly be from monochromatic waves. A CERC technical report contains data on irregular wave run-up on riprap revetments (Ahrens and Heimbaugh 1988).

Unfortunately most scale effects research on rubble structures appears to be for structures which are not overtopped. I am not aware of any scale effects studies specifically directed towards low-crested rubble mounds.

I agree with Mr Allsop that Michael Owen's research on seawall overtopping is an extremely valuable contribution to the state-of-the-art. There may, however, be problems in generalizing parameter r. You have heard Dr van der Meer's reservations about the roughness parameter. In the United States I know that Dr Robert Hallermeier also has reservations about using the roughness parameter to generalize Owen's smooth slope results.

In reply to Dr van der Meer, I started using the local wavelength as a variable in the spectral stability number as part of my study of reef breakwaters. Because the local wavelength was so effective in the spectral stability number, this encouraged me to use it in related research where overtopping was involved. The local wavelength seems to fit the physics better than the deep-water wavelength, since overtopping would be expected to increase both with increasing wave period and increasing water depth at the structure; this is also the way the local wavelength behaves. There are a number of non-iterative ways to calculate the local wavelength which can easily be incorporated into a spread sheet, so there really is not much of a problem in using this variable anymore. Unfortunately I can not provide a quantitative answer to your question since I have not specifically compared the two wavelengths. Since your point is quite logical I will try and have some of my former colleagues at CERC conduct a comparison.

I find the linking of the relative freeboard and the wave steepness in one parameter confusing. My impression is that after I have normalized the freeboard with $(H_{mo} \times H_{mo} \times L_p)^{1/3}$ that wave steepness does not explain much additional variance in the overtopping data, and therefore there is little reason to include it in the final prediction equation.

In response to Professor Partenscky's question, we have not conducted tests of sea dikes with slopes as flat as 1:6 or 1:8; structures of this type are rarely used to protect the coast in the United States.

I agree with Professor Losada that, where possible, it is very advantageous to present wave reflection, wave energy dissipation, and wave transmission data together in one figure to facilitate intercomparison. In the data presented in Paper 2 there appeared to be little influence of seiching in the wave tank. Also, note my reply to Professor Burcharth. We have, however, encountered seiching in other wave tanks, and have had difficulty either suppressing the seiche or in accounting for its influence in the data analysis.

I agree with Dr Leeuwestein that normalizing the

freeboard with just wave variables seems more logical than normalizing with a mixture of wave and stone variables.

J. W. VAN DER MEER, Paper 3

Wave transmission and wave overtopping are closely related. The amount of overtopping water may be important for dikes and seawalls where the crest is still well above sea water level. Wave transmission becomes important when the crest is so low that overtopping waves generate waves at the rear of the structure. In that case the structure may even be submerged. Therefore, it was good to have Papers 2 and 3 on wave overtopping and on wave transmission in one session.

Questions during the discussion concerned the influence of berms, roughness and three-dimensional effects on overtopping (Brown) and the various dimensionless parameters used for the relative crest height (Leeuwestein). Two extensive model investigations on wave run-up and wave overtopping at sea dikes were finished recently at Delft Hydraulics. These investigations included the effects of berms, roughness, depth limitation, long crested oblique wave attack and short crested perpendicular and oblique wave attack. The research has not yet been finalized, and it will be published elsewhere. The Rijkswaterstaat and the Technical Committee on Water Defences (TAW) in the Netherlands gave permission to present here a short summary of the (preliminary) results on wave overtopping.

Wave overtopping on dikes can be described by an exponential equation similar to the one used by Ahrens (Paper 2, equation (1)) and many others.

$$Q = a \exp (bR/\gamma) \tag{1}$$

where Q is a dimensionless discharge, a and b are coefficients, R is a dimensionless freeboard, and γ is a total reduction factor for the influence of berms, roughness, depth limitation and oblique wave attack.

Wave overtopping on dikes has to be divided into overtopping by plunging (breaking) and surging (non-breaking) waves. The transition between plunging and surging has been defined at $\xi_{op} = 2$, where $\xi_{op} = \tan\alpha/(s_{op})^{1/2}$ and $s_{op} = 2\pi H_s/gT_p^2$. Equation (1) becomes, for plunging waves with $\xi_{op} < 2$

$$Q = 0.06 \exp (-5.2R/\gamma) \tag{2}$$

with

$$Q = \frac{q}{gH\backslash_s} \quad \frac{s_{op}}{\tan\alpha} \qquad R = \frac{h_k}{H_s} \quad \frac{s_{op}}{\tan\alpha} \tag{3}$$

where q is the overtopping rate (m^3/s per m) and h_k is the crest freeboard.

For surging waves with $\xi_{op} > 2$, Equation (1) becomes

$$Q = 0.2 \exp(-2.6R/\gamma) \tag{4}$$

with

$$Q = \frac{q}{(gH_s^3)^{1/2}} \qquad R = \frac{h_k}{H_s} \tag{5}$$

The total reduction factor γ is defined by

$$\gamma = \gamma_b \gamma_f \gamma_\beta \gamma_h \tag{6}$$

where γ_b is the reduction factor for berms, γ_f is the reduction factor for roughness, γ_β is the reduction factor for oblique wave attack, and γ_h is the reduction factor for depth-limited wave attack.

The reduction factor for berms γ_b is defined by

$$\gamma_b = 1 - g_B + 0.5 g_B (d_h/H_s)^2 \quad \text{(min. 0.6 and max. 1.0)} \tag{7}$$

$$g_B = 0.06\xi_{op} B/H_s \qquad \text{for } \xi_{op} < 4.4(\tan\alpha)^{2/3} \tag{8}$$

$$g_B = 5B/L_{op} \qquad \text{for } \xi_{op} > 4.4(\tan\alpha)^{2/3} \tag{9}$$

where d_h is the depth of the berm below sea water level, B is the width of the berm and L_{op} is the wave length $gT^2_p/2\pi$.

The reduction factor for a rough slope γ_f for the range $\xi_{op} < 3 - 4$ is given by the following table.

Slope	Reduction factor γ_f
Smooth concrete, asphalt	1.0
Smooth block revetment	1.0
Grass (3 cm)	0.9
Blocks on smooth slope; height 10% lower than width f_b $f_b/H_s = 0.12 - 0.19$. Surface covered by blocks: 1/25	0.7
1/9	0.6
Ribs on smooth slope; $f_b/H_s = 0.12 - 0.19$ and relative distance of ribs $f_L/f_b = 7$ (optimum)	0.55
1 layer of rock armour layer ($H_s/D = 1.5 - 3$)	0.5

The reduction factor for oblique short-crested wave attack γ_β (as we have in nature) is given by

$$\gamma_\beta = 1 - 0.0033 \tag{10}$$

with β the angle of wave attack from normal (in degrees). The reduction factor for striking waves (β= 90°) becomes 0.7.

The reduction factor for depth-limited situations γ_h (shallow water on a sloping foreshore) is given by

$$\gamma_h = \frac{H_{2\%}}{1.4 H_s} \tag{11}$$

where $H_{2\%}$ is the wave height at the toe of the structure exceeded by 2% of the waves and H_s is the wave height at the toe of the structure. For deep water (Rayleigh distribution)

$$H_{2\%}/H_s = (0.5 \ln 50)^{0.5}/(1+H_s/h)^{1/3} \tag{12}$$

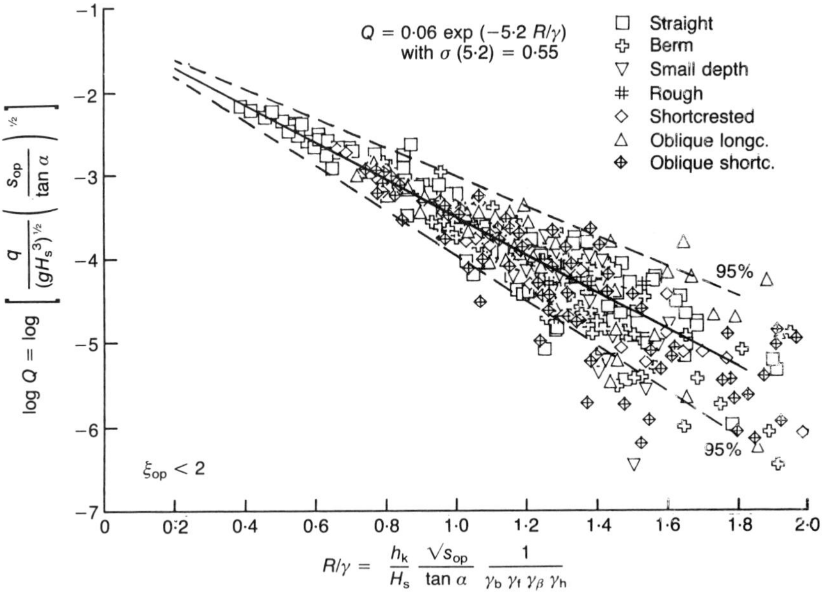
Q = 0·06 exp (−5·2 R/γ)
with σ (5·2) = 0·55
Straight
Berm
Small depth
Rough
Shortcrested
Oblique longc.
Oblique shortc.
95%
95%
ξop < 2
log Q = log [q / (gHs³)^½ (sop / tan α)^½]
R/γ = hk/Hs √sop/tan α 1/γb γf γβ γh

Fig. 3

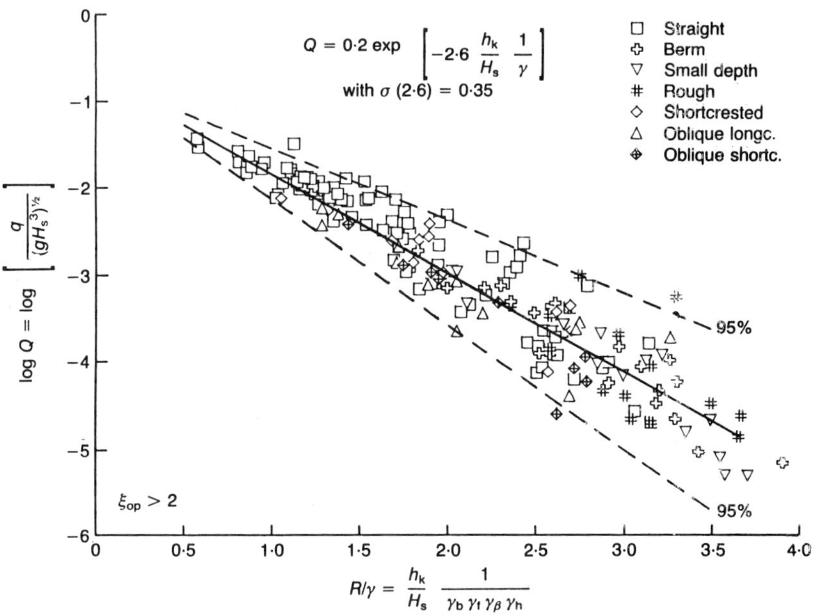
Q = 0·2 exp [−2·6 hk/Hs 1/γ]
with σ (2·6) = 0·35
Straight
Berm
Small depth
Rough
Shortcrested
Oblique longc.
Oblique shortc.
95%
95%
ξop > 2
log Q = log [q / (gHs³)^½]
R/γ = hk/Hs 1/γb γf γβ γh

Fig. 4

The formulae for wave overtopping (equations (2) - (5)) with the reduction factors given in equations (6) - (12) are shown in Figs 3 and 4 together with all available data (including USA, UK and German data). Fig. 3 shows about 400 data points and Fig. 4 about 190. The 95% confidence bands are given by treating the coefficients 5.2 and 2.6 in equations (2) and (4) as stochastic variables with normal distributions and standard deviations of, respectively, $\sigma(5.2) = 0.55$ and $\sigma(2.6) = 0.35$.

Most of Mr Brown's questions and remarks have been covered by the introductory remarks to this discussion. The only question that remains is the one on possible scale effects. Delft Hydraulics tested a berm breakwater both on a scale of 1 : 35 and on a scale of 1 : 7 in the large Delta flume. The objective was to determine possible scale effects. The structure had significant overtopping and the wave transmission was measured behind the structure. The 1 : 7 test consistently showed a slightly larger wave transmission. Most of the transmission went through the armour layer at the crest, causing different flow in the two models.

The conclusion on wave transmission was that for this structure (with relatively low wave transmission K_t = 0.1 - 0.2) there were indeed small-scale effects. One may expect scale effects on wave transmission if the transmission coefficient is small. For large wave transmission, say $K_t > 0.3$, the transmission goes over the crest instead of (partly) through the crest, and in that case one may expect no or smaller scale effects.

In reply to Mr Allsop, the formulae on wave overtopping given in the introductory remarks are similar to Owen's one, including the relative roughness parameter. Note that the formula has to be divided into one part for plunging and one part for surging waves. Note also that not only a relative roughness parameter (γ_f in the introductory remarks) was included, but also γ_b, γ_β and γ_h.

The test data of Owen were never published. Owen, however, was willing to transfer the data on smooth straight slopes to Delft Hydraulics, and these data have been included in Figs 3 and 4. The data of Owen on berms were never released.

It is useful to have user-friendly software packages for design purposes. Using the formulae in the introductory remarks, the software package at HR Wallingford may be improved. A second release of Delft Hydraulics programme BREAKWAT will, of course, also include wave overtopping.

In response to Professor Partenscky's question, how we came to the choice of D_{n50} to define the relative wave height H_s/D_{n50} and relative crest height R_c/D_{n50} is given in the paper in sections 32-35. With this choice it is fairly easy to distinguish various cases. For example, consider two cases with $R_c/H_s = 1$ and a stone size $D_{n50} = 1$ m.

Case 1: $H_s = 5$ m	Case 2: $H_s = 0.5$ m
$R_c = 5$ m	$R_c = 0.5$ m
$H_s/D_{n50} = 5$	$H_s/D_{n50} = 0.5$
$R_c/D_{n50} = 5$	$R_c/D_{n50} = 0.5$

With only R_c/H_s as parameter, one finds (see Fig. 6 in the paper) that K_t varies between 0.05 and 0.5: a fairly rough estimate. With the relative parameters with D_{n50} it is clear that Case 1 is a structure under design conditions (high waves, high structure) and one would expect low K_t-values of 0.05 - 0.10. Case 2 is a structure with a very low crest (only half a stone diameter above sea water level) subjected to low waves that travel easily through the crest layer. These low waves (probably long swell) may give K_t-values of the order of 0.3 - 0.5.

The only drawback of using D_{n50} is that it can only be applied to rubble mound structures and that a solid non-porous structure cannot be included.

Ahrens, in his papers on reef type structures, gave a nice graph of reflection and transmission versus relative crest height. Professor Losada is right that the phenomena are divided at the structure; waves reflect from the structure and overtop the structure, giving transmission, and energy dissipates on the structure. The problem is now to measure wave dissipation; it is the term left when reflection and transmission are known and can only be computed. The designer will probably not be interested in the dissipation and even not in the reflection, as reflection at low-crested structures is very low. Therefore, stability and wave transmission are the main design aspects at low-crested structures.

To Mr Leeuwestein, the question on using D_{n50} for low-crested rubble mound structures was answered above. The effect of permeability of the structure is only noticeable for relatively high crest heights. Then the energy will be transmitted through the structure. As soon as there is much wave overtopping and wave transmission, the effect of porosity will diminish.

Professor Vasco Costa is right when he states that viscous forces may cause scale effects in small-scale models when wave transmission is concerned. Again, scale effects are probably only significantly present in cases where most of the transmission goes through the armour layer on the crest or through the structure.

In reply to Dr Oumeraci, the case of very long and probably low swell at a low-crested structure is an exceptional case; it may be compared with Case 2 above. Probably it is out of the (safe) range of 1 < H_s/D_{n50}. Only for low wave steepness may the transmitted waves be divided into shorter ones. The question raised by Dr Oumeraci means that in cases where we have wave transmission by long swell, we have also to take into account the wave period.

4. Design innovations, including recent research contributions

H. F. BURCHARTH, University of Aalborg

INTRODUCTION

1. The dramatic failures of large rubble mound breakwaters in the late seventies and the early eighties caused many problems and did indeed intensify the research activities. The organizers of the 1983 "Conference on Breakwaters" at The Institution of Civil Engineers", London, asked me to present a theme speech titled "The Way Ahead", in which I identified a large number of areas where important information was lacking and pointed out ways of solving some of the problems. In order to evaluate some of the progress made since then, the 1983 paper will be used as a background for the present paper on recent developments.
In the design process considerations must be given to:

- the performance of the structure given by the objective of the structure
- determination of design loads from information on the environmental boundary conditions
- the performance of the structural elements by analysis of the relevant failure modes and the material aspects
- the construction procedure
- the maintenance
- reliability and cost assessments

Significant progress has taken place within some of these areas, but white spots are still present as will be discussed in the following sections.

ENVIRONMENTAL CONDITIONS

2. Standard notation

The increasing number of parameters involved in wave analyses made a standardization of the notation necessary. In 1986 IAHR/PIANC published the "List of Sea State Parameters", the notation of which should be followed when possible.

Shallow water waves

3. Coastal structures and breakwaters are generally in shallow water. For design wave conditions where wave breaking takes place due to limited water depths and the wave height distribution will differ from the Rayleigh distribution in that there exist relatively fewer large waves heights. With respect to the distribution of extreme wave and crest heights there is a lack of reliable model formulations. Thornton and Guza, 1983, show that at least op to $H_{10\%}$, i.e. the wave height exceeded by 2% of the waves, the Rayleigh distribution is valid. For higher waves in water depth d, Stive, 1986, proposed the following empirical correction to the Rayleigh distribution based on model tests but roughly checked against some prototype data

$$\hat{H}_{1\%} = H_{mo}\left(\frac{ln100}{2}\right)^{\frac{1}{2}}\left(1+\frac{H_{mo}}{d}\right)^{-\frac{1}{3}}$$

$$\hat{H}_{0.1\%} = H_{mo}\left(\frac{ln1000}{2}\right)^{\frac{1}{2}}\left(1+\frac{H_{mo}}{d}\right)^{-\frac{1}{2}}$$

where $\hat{H}_{1\%}$ means the 1% exceedence value of the wave height determined by zero crossing analysis, whereas the significant wave height H_{mo} is determined from the spectral moments. Stive also discusses the reliability of the corrections. The correction formulae are very useful for checking the wave height distribution in small scale physical model tests, cf. Fig. 1.

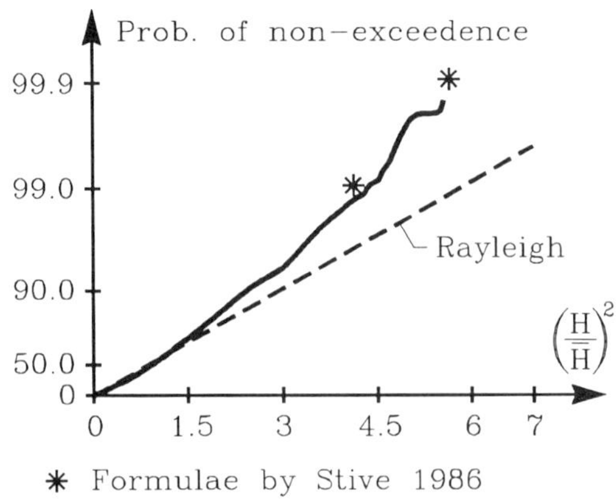

Fig. 1. Comparison of the expression by Stive, 1986, for shallow water wave height distribution with model test results. Aalborg Hydraulics Laboratory 1990.

4. Klopman et al., 1989, based on suggestions by Glukhovskiy and Battjes, proposed a semi-empirical expression for the wave height distribution which essentially comes down to a Weibull distribution with $\bar{H}/d$, i.e. mean wave height over water depth, as a parameter. Examples of some characteristic wave height values are given in Table 1, in which $H_{1/3}$ is the significant wave height based on zero crossing analysis and $H_{rms} = \left[\Gamma\left(2 - \frac{\bar{H}}{d}\right)\right]^{\frac{1}{2}}$ where Γ is the gamma function.

Table 1. Characteristic extreme wave height values in shallow water (Klopman et al. 1989).

$\frac{\bar{H}}{d}$	$\frac{H_{1/3}}{d}$	$\frac{H_{1/3}}{\sqrt{2}\,H_{rms}}$	$\frac{H_{1\%}}{\sqrt{2}\,H_{rms}}$	$\frac{H_{0.1\%}}{\sqrt{2}\,H_{rms}}$
0.0	0.0	1.42	1.52	1.86
0.2	0.30	1.36	1.35	1.59
0.4	0.55	1.30	1.18	1.34
0.6	0.75	1.21	1.02	1.10

There is still a strong need for further research into reliable models for shallow water waves.

5. For shallow water wave conditions Hughes, 1984, proposed an empirical spectrum, the so-called TMA-spectrum. It is a one dimensional spectrum (2-dimensional wave) for fetch limited sea where shallow water effects are included. The analytical expression is given by the JONSWAP spectrum expression multiplied by a shallow water wave effect function $\Phi(f, h)$

$$\Phi = \begin{cases} 2\pi^2 f^2 h/g & \text{for } f \leq 1/\left(2\pi(h/g)^{0.5}\right) \\ 1 - 1/2\left(2 - 2\pi f(h/g)^{0.5}\right)^2 & \text{for } f > 1\left(2\pi(h/g)^{0.5}\right) \end{cases}$$

where f is the frequency and h the water depth. At the same time the α-factor in the JONSWAP expression should be replaced by $\alpha = 0.076\left(gF/U^2\right)^{-0.22}$, where F is the fetch and U the wind velocity.

6. Numerical models for estimation of shallow water waves were based on 2-dimensional waves, i.e. waves with energy transport only in the direction of wave propagation. This resulted in too strong variations in wave heights in areas where, according to conventional refraction analysis, the orthogonals converged or diverged. In some cases unrealistic waveheight variations along a breakwater were predicted. Recently a third generation of models has been developed based on 3-dimensional waves, i.e. waves with horizontal spread of energy relative to the mean direction of wave propagation. The most advanced of these models takes

into account dissipation due to wave breaking, bottom induced dissipation, wave-current and wave-wave interactions and energy input from the wind. An example of such a model is the HISWA model by Holthuijsen et al., 1988, which is based on an Eulerian approach. In order to obtain high computational efficiency, two simplifications were introduced in this model. The first one is the removal of time as an independent variable to obtain a stationary model. This can be justified by the relatively short travel time of waves in most shallow water coastal regions. The second simplification is the parameterization of the basic balance equation of the spectrum using the zero-th and the first-order moments but still retaining the directional characteristics of the waves. The applied modelling of shallow water wave breaking due to large wave steepness or direct bottom effects is the one proposed by Battjes et al. 1979. Although the model contains several simplifications it seems to be very efficient and it represents a great step forward in reliable and easy prediction of shallow water sea states. Improvements can and will be introduced in the coming years. For waves in deeper water the WAM model, by the WAMDI Group, 1988, modified by Tolman, 1990, represents some of the latest developments.

Characterization of incident waves

7. The determination of the response of a coastal structure or a breakwater to waves is almost always determined by model tests or by formulae fitted to model test results. The large scatter in the results are mainly due to the stochastic nature of the loads and in case of rubble mound structures also due to the random placement of the armour blocks. However, significant scatter is also caused by incomplete characterization of the incident waves. Because the wave load on a structure is determined by the kinematics of the incident wave, a breaking wave causes a load different from a non-breaking wave of equal wave height and steepness. Consequently, the degree of instability of the incident waves is of importance. However, the conventional spectral description of incident waves does not include such information. This is acceptable for deep water wave situations, but in shallow water the differences in sea bed topography can cause rather different wave kinematics, and information on this is only weakly inherent in the normal one-dimensional amplitude variance power spectrum. In order to reduce the scatter and thereby the uncertainty related to breakwater respons it is proposed to seek an additional overall parameter which characterizes the degree of wave instability for the incident waves in front of the structure. A suitable parameter must contain information on both the horizontal and the vertical asymmetry of the waves, e.g. a combination of a surf similarity parameter $tan\,\alpha(H/L)^{-0.5}$ and an Ursell number $H\,L^2/d^3$ where H and

L are some characteristic values of incident wave height (e.g. $H_{1/10}$) and wave length at the position of the structure, and d is the local water depth at the structure. In case of a complex sea bed topography the slope α must be an average slope characterizing the sea bed in front of the structure.

Part of the variability in model test results are due to differences in wave grouping, which in flume test experiments are not described by the spectrum. Medina et al., 1990, found from large scale experiments that the hydraulic stability of rock armour is dependent on an envelope exceedence coefficient and a groupiness factor. The authors propose that existing stability formulae are modified accordingly.

Uncertainty on wave climate estimates

8. The dominating loads on breakwater and coastal structures are caused by waves. Moreover, the response of the structure is very sensitive to the size of the waves. Therefore, it is essential to evaluate the uncertainty inherent in the estimates on the design wave characteristics. The following elements are the main contributors to the total uncertainty:

1. Errors related to wave data sets, e.g. errors related to significant wave heights, H_s, measured by Waverider buoys or estimated from numerical hindcast models.

2. Statistical uncertainties related to extrapolation from short data samples to design events of low probability of occurrence, i.e. a 200-year return period H_s.

3. Lack of knowledge of the true long-term distributions, e.g. for H_s. Many theoretical distributions, e.g. a Weibull and a Gumbel, often fit a data sample equally well but might provide different extreme value estimates.

Procedures for estimating these and other uncertainties are now described in the PIANC PTC II Working Group 12, Subgroup-B report (Burcharth, 1989). For example related to item 1, Table 2 shows typical variational coefficients given for the most important sea state parameters, which can be used when more precise information is lacking. Many researchers have discussed the problem related to item 2, e.g. Kimbal 1949, Petrauskas et al. 1971, Lawless 1974, Challenor 1979, Wang et al. 1983, Rosbjerg 1985, Carter et al. 1986, Burcharth 1986, Goda 1988. The report provides for various distributions the formulae for the variational coefficient for extreme values of H_s as function of the return period and the number of wave data.

Table 2. Typical variational coefficients $\sigma' = \sigma/\mu$ (standard deviation over mean value) for measured and calculated sea state parameters (Burcharth, 1989).

Parameter	Methods of determination	estimated typical values σ'	Bias	Comments
Significant wave height, offshore	Accelerometer buoy, pressure cell, vertical radar	0.05-0.1	~ 0	
	Horizontal radar	0.15	~ 0	
	Hindcast, num. models	0.1-0.2	0-0.1	Very dependent on quality of weather maps.
	Hindcast, SMB method	0.15-0.2	?	Valid only for storm conditions in restricted sea basins.
	Visual observations from ships	0.2	0.05	
Mean wave period offshore on condition of fixed significant wave height	Accelerometer buoy records	0.02-0.05	~ 0	
	Estimates from amplitude spectra	0.15	~ 0	
	Hindcast, num. models	0.1-0.2	~ 0	
Duration of sea state with significant wave height exceeding a specific level	Direct measurements	0.02	~ 0	
	Hindcast, num. models	0.05-0.1	~ 0	
Spectral peak frequency offshore	Measurements	0.05-0.15	~ 0	
	Hindcast, num. models	0.1-0.2	~ 0	
Spectral peakedness offshore	Measurements and hindcast, num. models	0.4	~ 0	
		Degrees		
Mean direction of wave propagation offshore	Pitch - roll buoy	5°		
	Measurements μ, u, v and p, u, v	10°		
	Hindcast, num. models	15 – 30°		
		σ'		
Significant wave height nearshore determined from offshore significant waveheight taking into account typical shallow water effects (refraction, diffraction, shoaling, ...)	Numerical models	0.1-0.20	0.1	σ' can be much larger in some cases
	Manual calculations	0.15-0.35		
Astro tides	Prediction from constants	0.001-0.07	~ 0	
Storm surge	Numerical models	0.1-0.25	±0.1	

9. Even if we knew the true long-term distribution for the sea state there would still be some uncertainty as to the largest waves which could occur during any period of time (structural life) – simply because of the statistical vagaries of the elements. The probability p that the R-year return period event will be exceeded once or more during a structural life of T years is given by

$$p = 1 - \left(1 - \frac{1}{R}\right)^T \qquad (1)$$

which is depicted in Fig. 2 (p is often denoted the encounter probability).

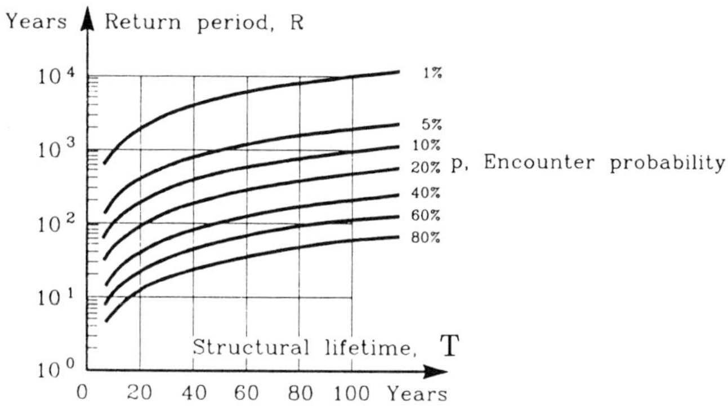

Fig. 2. Encounter probability.

10. Combining the encounter probability with the uncertainties given by items 1 - 3 makes it possible to estimate the probability of exceedence of a chosen design sea state within structural life, or to put it in another way, a design sea state corresponding to a given exceedence probability can be found. This is not a new invention, but information on uncertainty estimates is now more easily available, and consequently more general use by design offices is to be expected.

WAVE LOADS AND FAILURE MODES

11. The highly dynamic and stochastic nature of the problems makes applicable deterministic calculations almost impossible. Nevertheless, development of the deterministic numerical models should proceed, as discussed later. But for many years we have to rely on results produced by two different procedures illustrated in Figs. 3 and 4. The wave loads on a rubble mound are so complex that they are generally bypassed in the design process. Thus the normal procedure is to relate the characteristics of the incident waves, e.g. H_s and T_z, directly to the response of the structural elements, e.g. displacement of armour units, cf. Fig. 3.

Wave characteristics given by spectra or H_s , T_z	⇒	**Determination of structural respons (e.g. armour layer stability, run-up, etc.) by semi-empirical formulae or model tests**

Fig. 3. Conventional determination of structural response without load determination.

Only in cases where the response cannot be studied very well in small scale models because of significant scale and model effects it is the normal procedure to determine the wave load. The most important example concerns parapet walls where the wave loads must be known in order to calculate the stresses in the wall and its overall stability. This is because the internal pressure field, and consequently the base plate uplift pressure, cannot be studied without significant scale effects in normal size models. In this case the procedure is as shown in Fig. 4, i.e. similar to the conventional procedure related to other civil engineering structures.

Wave characteristics given by spectra or H_s , T_z	⇒	**Determination of wave loads by semiempirical formulae or model test results**	⇒	**Calculation of stresses and deformation**

Fig. 4. Conventional determination of structural response based on determination of the loads.

12. Fig. 5 shows the failure modes of a rubble mound breakwater. Analyses of these failure modes have been ongoing for several decades.

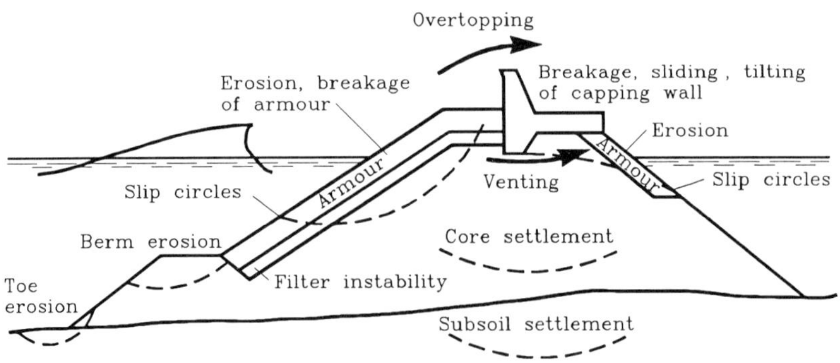

Fig. 5. Failure modes for a rubble mound breakwater.

In the following sections will be summarized some of the recent progress and some of the difficulties, which we are still facing.

ARMOUR UNITS AND ARMOUR LAYER STABILITY

Parameters for hydraulic stability under irregular wave conditions

13. Brorsen et al. 1974 demonstrated that the Hudson formula for armour stability is not valid for complex interlocking types of armour blocks with respect to the slope angle influence. Instead of the Hudson stability coefficient K_D they used the stability number $\frac{H}{(\rho_s/\rho_w - 1)(w/\rho_s)^{1/3}}$ $= (K_D cot\alpha)^{1/3}$, which for constant mass densities is proportional to $\frac{H}{h}$, where H is the wave height, h is the block height, w is the mass of the block, ρ_s and ρ_w the mass density of block and water, respectively, and α is the slope angle. In the CERC Shore Protection Manual, 1984, the same stability number is denoted N_s. Bruun et al., 1976 used for regular waves a functional relationship between the stability number $(K_D cot\alpha)^{1/3}$, denoted N_{ZD} for zero damage, and the surf similarity number $Ir = tan\alpha(H/L)^{-\frac{1}{2}}$. Burcharth, 1979, proposed this relationship to be used in tests with irregular waves in which case some characteristic values of H (e.g. H_S or $H_{1/10}$) and L (e.g. mean wave length) in front of the structures should be used. The functional relationship should be obtained from experiments in which duration of wave attack and core and filter layer permeability should be specified. Such experiments were actually later performed by Delft Hydraulics, and in 1988 Van der Meer presented the results in terms of stability formulae using the notation $\frac{H_s}{\Delta D_{n50}}$ for the above mentioned stability number, cf. paragraph 15. In cases where a relevant slope angle cannot be defined, the wave steepness cannot be implemented in a surf similarity number, for which reason a modified stability number including the wave steepness is used. As a forerunner Gravesen et al., 1977, replaced H_s^3 in the Hudson formula by $H_s^2 L$, where L is a characteristic wave length. Ahrens, 1987, in his work on reef breakwaters introduced $N_s^* = N_s\, s_P^{-\frac{1}{3}} = H_s/\Delta D_{n50} \cdot s_P^{-\frac{1}{3}} = (H_s^2 L_P)^{\frac{1}{3}} / (\Delta D_{n50})$, where L_P is the wave length related to the spectral peak period at the toe of the structure. Related to S-shape profiles or broken slopes Van der Meer uses $H_s/\Delta D_{n50} \cdot T_m\,(g/D_{n50})^{0.5}$, which is given the notation $H_o T_o$. With reference to the discussion in paragraph 7 it is to be expected that more precise stability parameters, which include information on the stage on instability of the incident waves, can be developed.

Classification and properties of rock armour

14. The stability of rock armour depends on the size and the shape of the blocks. Consequently, shape classification and strength and dura-

bility properties of quarry rock are important points to be considered in the design. The just published CIRIA/CUR "Manual on the use of rock in coastal and shoreline engineering" contains a comprehensive chapter on these matters written by Latham and Poole.

Rock armour stability

15. Based on extensive parametric large and small scale model tests by Delft Hydraulics, Van der Meer, 1988, presented the following empirical formulae for the stability of normal graded rock armour in head-on waves.

$$\frac{H_s}{\Delta D_{n50}} = \begin{cases} 6.2\, P^{0.18} Ir^{-0.5} \left(S/\sqrt{N}\right)^{0.2} & for Ir \leq Ir_c \\ 1.0\, P^{-0.13} Ir^{P} \left(S/\sqrt{N}\right)^{0.2} cot\alpha^{0.5} & for Ir > Ir_c \end{cases}$$

where

P	=	factor characterizing the permeability
S	=	factor signifying the damage level
N	=	number of waves
Ir	=	$tan\alpha\,(s_m)^{-0.5}$, where the wave steepness s_m is related to the mean period
Ir_c	=	$[6.2\, P^{0.31}(tan\alpha)^{0.5}]^{(P+0.5)^{-1}}$

The formulae are very well documented including reliability estimates but are valid only for straight slopes with crests exceeding the maximum run-up level.

16. Recently Bradbury et al., 1990, proposed modifications of the coefficients 6.2 and 1.0 in the formulae in order to take into account also the influence of the shape of the blocks. For a very limited range of conditions, i.e. slope 1 : 2 and impermeable core, they proposed the coefficients given in Table 3. At the same time $\left(S/\sqrt{N}\right)^{0.2}$ in the formulae should be changed to $\left(S/\sqrt{N}\right)^{0.25}$

Further studies are needed in order to obtain values covering the whole range of conditions.

Table 3. Modified coefficients in Van der Meers's rock armour stability formulae for certain conditions (Bradbury et al. 1990).

Shape of stones	Factor in formula for $Ir \leq Ir_c$	$Ir > Ir_c$
Elongate/tabular	6.7	1.3
Irregular, fresh	6.3	0.8
Equant	6.2	1.1
Semi-round	6.0	1.0
Very round	5.9	0.8

17. The hydraulic stability of rock armour reef breakwaters (reshaping) and statically stable low crested breakwaters (overtopped) and statically stable submerged breakwaters were studied by Givler et al. 1986, Ahrens 1987 and Van der Meer et al. 1990. The last reference contains proposals for stability formulae for these types of structures.

Toe stability

18. The stability of a toe or a submerged berm constructed of rock armour cannot yet be described by a reliable formula covering the range of realistic geometries. This is because sufficient parametric studies have not been performed. Van der Meer proposed the following preliminary formula corresponding to marginal damage

$$\frac{H_s}{\Delta D_{n50}} = 8.6 \left(\frac{h_t}{h}\right)^{1.43}$$

where h is the water depth in front of the structure and h_t is the water depth at the crest of the toe. The formula is based on test results mainly with depth limited wave conditions and the geometrical range $0.4 \leq \frac{h_t}{h} \leq 0.7$.

Reshaping (berm) breakwaters

19. The fact that natural reshaping as a self defence procedure leads to the most stable profile under the given conditions is the reason for the efficiency of breakwaters where the armour rocks are allowed to move. The stability factor range for these breakwaters is typically $3 < \frac{H_s}{\Delta D_{n50}} < 5$. Fig. 6 shows a typical cross section before and after reshaping in head on waves.

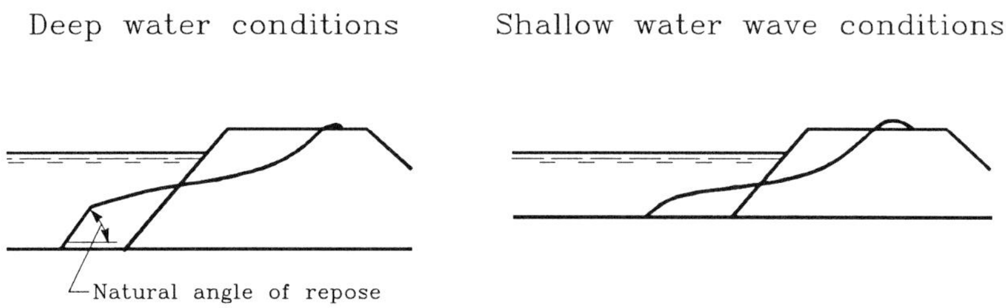

Fig. 6. Typical berm breakwater profiles before and after reshaping in deep and shallow water conditions.

20. The effenciency of the S-profile has been known for many years and used for dikes and breakwaters with conventional armour layers. Within recent years quite a lot of attention has been given to the concept of berm breakwaters starting with papers by Baird et al., e.g. 1984, discussing mainly the general advantage of the concept. Van der Meer, 1988, presented a comprehensive parametric study of the deformation of the seaward profile in head-on waves and produced on this basis an empirical computer program BREAKWAT which predicts the profile under given conditions. The program has been verified against data by others, cf. Fig. 7.

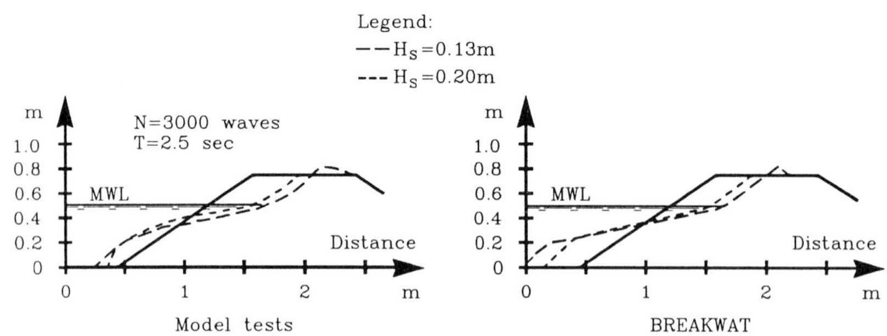

Fig. 7. Example of verification of the BREAKWAT computerprogram against data by Burcharth and Frigaard, 1988.

Burcharth et al., 1987, 1988, studied also the front profile development but concentrated on the 3-dimensional effects, i.e. erosion and deformation of roundheads plus the transport of eroded stones along the structure when exposed to oblique waves, (15° and 30° angles of incidence). During the same period institutes such as NRC, Canada, DHI, Denmark, and

DH, Holland, performed model tests with a number of site specific berm breakwaters. The experiences from these tests are seeked collected in the ongoing EEC MArine and Science Technology programme (MAST 1).

21. The following stability factor values were proposed by Burcharth et al., 1987-88, as typical threshold conditions for the start of significant transport of stones along the structure (angle of wave incidence for trunk $\leq 30°$)

	$\frac{H_s}{\Delta D_{n50}}$
Trunks exposed to steep oblique waves	< 4.5
Trunks exposed to long oblique waves	< 3.5
Roundheads	< 3

22. Vrijling et al., 1991, combined the results of Burcharth et al. and results from case studies at DH to obtain a first formula for the transport of stones along a trunk exposed to oblique waves after the reshaped profile has formed

$$S = 4.8 \cdot 10^{-5}\,(H_o T_o - 100)^{2.0} \quad \text{for} \quad H_o T_o > 100$$

$$S = 0 \quad \text{for} \quad H_o T_o < 100$$

where

S is the number of stones per wave,

$H_o = H_s/(\Delta D_{n50}), T_o = T_p(g/D_{n50})^{0.5}, T_p =$ spectral peak period.

However, more work has to be done in order to get closer design formulae for the stability of the trunk in oblique waves and for the roundhead stability. Also tests on crest and rear slope erosion in head-on waves are needed because only such erosion can cause failure in head-on waves. Tests to explore the effect of larger cores must also be performed in order to be able to optimize the cross section designs.

Stability of concrete crown walls

23. Crown wall stability is one of the more complicated failure modes to deal with because the variation in time of the wave force on the wave wall and the uplift pressure on the base plate must be known in order to calculate the stresses and the stability or movements. The wave load on the wall is in many cases a mix of quasi static pressure variation and extremely short duration slamming (shock) forces. The two types of loads must be treated separately because the superstructure is resting on rock material with elastic-plastic deformation properties which might contribute to dynamic response. Another reason is the different scale

effect of quasi static wave loads and shock loads. Moreover, the shock loads have a much stronger 3-dimensionality than the quasi-static loads. For this reason tests also in short crested seas (3-dimensional waves) are needed. Although we are dealing with a problem where dynamic amplifications can be significant it has been the tradition to treat it as a static problem.

24. No reliable and generally applicable formulae for calculation of the wave forces exist. The reason is first of all the very large number of geometrical parameters, cf. Fig. 8.

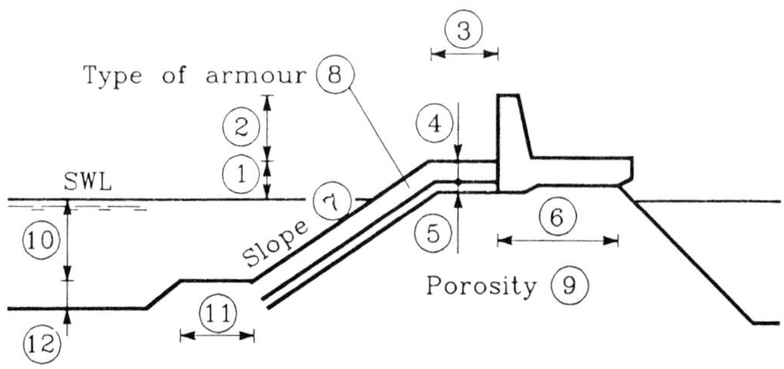

Fig. 8. Geometrical parameters in crown wall tests.

Moreover, the existing model test results very seldom contain usable information on shock forces, either because the sampling frequency was too low or because measurements were performed with strain-gauge mounted load cells, the natural frequency of which is almost always too close to the frequency of the shock loads to prevent biasing dynamic amplification, cf. Fig. 9 which illustrates the basic dynamic response of a simple linear undamped single degree of freedom system.

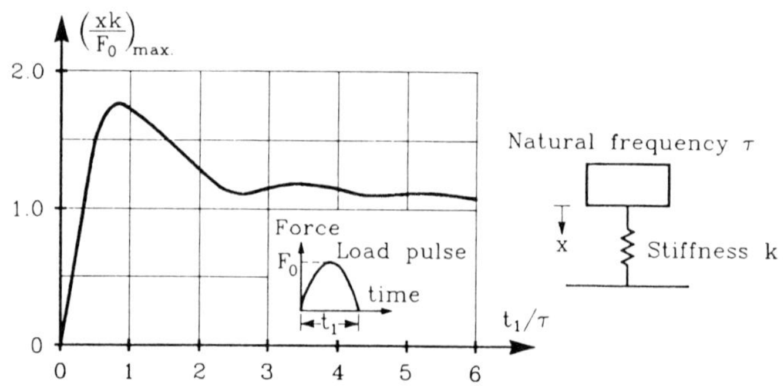

Fig. 9. Dynamic response of an undamped single degree of freedom spring-mass system exposed to shock load.

25. To avoid dynamic biasing the load should be recorded by means of pressure cells densely spaced over the full height and over widths larger than those of the crown wall sections. Because of strong scale effects related to pore pressures in fine granular material under highly non-steady conditions it is not possible to measure the true prototype distribution of the uplift pressures on the base plate in a normal size small scale model. As a safe solution a linear distribution between the measureable pressures at the front and the rear edges of the plate can be assumed. The correlation between wave wall forces and uplift forces must be contained in the statistical analysis. The stochastic nature of the loads calls for long test series and high frequency recordings. The whole phenomenon must be considered a dynamic problem where much attention must be paid to the dynamic characteristics of the model.

26. A true dynamic scaling of the model is not possible first of all because the properties of prototype size rock fill under dynamic loading are not known. Only information on the static and quasi static characteristics is available, for example from Kobayashi et al., 1987, who performed tests in large size triaxial apparatus and full scale experiments with the bearing capacity of a caisson on a rubble foundation in Onahama Port. However, investigations of the dynamic properties of full size rock fill are needed.

27. Scale effects related to entrained air bubbles and air pockets in waves hitting the solid structure must also be studied. As a minimum some tests should be performed to verify how sensitive pressures are to the amount and distribution of saturated air. This might be done by adding air bubbles in front of the structure, cf. Fig. 10.

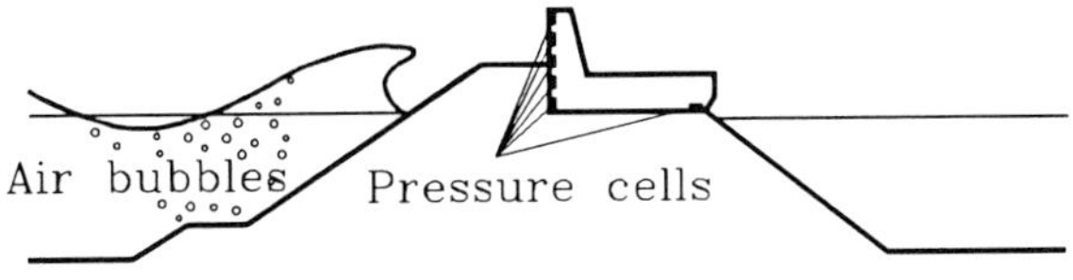

Fig. 10. Test of sensitivity of wave loads to air content (Burcharth, 1983).

The scale effects related to porous flow and air content were discussed in Burcharth, 1983, and a related comprehensive research project were proposed. Although some research has been ongoing since then no significant brake-through has appeared so far.

28. Armour unit placement patterns

It is to be expected that a regular pattern, where no isolated armour unit is much higher than the general surface, reduces high velocity hydrodynamic forces on single units and thereby increases the stability. Also a better initial interlocking or neighbour support by regular pattern placement contributes to the stability. This has been confirmed by recent experiments with Antifer cubes (model tests for Zeebrugge harbour design by HAECON), Tetrapods (Simm et al., 1988) and parallelopipedic blocks (comparative model tests of CEPYC and Aalborg Hydraulics Laboratory, 1990). However, it is important to notice that regular patterns normally increase the run-up and overtopping and cause a more sudden breakdown of the armour layer once movement does commence. From a construction point of view regular placement is, for some armour units, easier than a random placement where careful placement following prescribed coordinates most often is necessary anyway. More systematic knowledge about the influence of placement patterns on the hydraulic stability is not available.

CONCRETE ARMOUR UNITS

29. The reason for many of the recent breakwater failures was breakage of slender armour units like Dolosse and Tetrapods. This caused a renaissance in the use of bulky units like cubes and Antifer blocks and it also caused invention of new units such as the HARO(R) by Haecon and the ACCROPODE(R) by Sogreah, Fig. 11. Both units are robust and economical because of a relatively small effective volume of concrete, which for the HARO is due to a high porosity and for the ACCROPODE due to the one-layer concept.

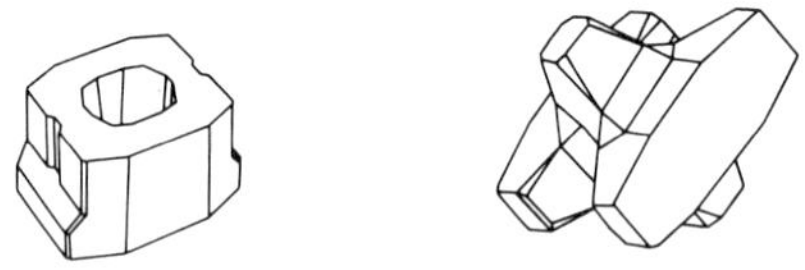

Fig. 11. HARO(R) and ACCROPODE(R).

Both units have a good hydraulic stability but different failure characteristics, the failure of an ACCROPODE layer being very brittle, i.e. at a certain level a small increase in wave height will cause a collapse of an otherwise visually stable armour. Fig. 12, indicates the characteristic differences in damage development for various types of rubble structures. The figure illustrates the importance of evaluation of confidence limits related to the estimated design sea state, especially in case of structures with brittle failure characteristics.

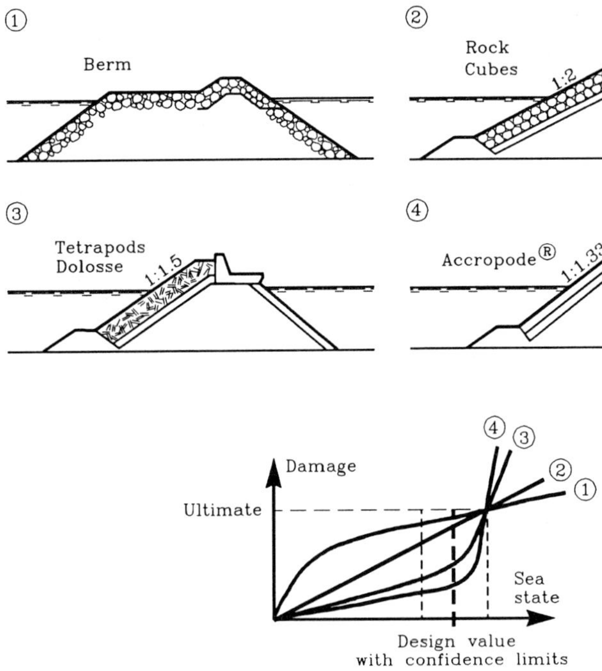

Fig. 12. Characteristic damage development for various types of rubble mound breakwaters.

30. Hollow single layer armour units like the COB and the SHED (hollow cubes) and the SEABEE have successfully been used since the early 1970's, Dunster et al. 1988. The units form an extremely stable armour layer when placed in close contact with neighbour units.

Fig. 13. Examples of hollow single layer armour units.

The high stability of units like COB and SHED is due to the large permeability and the "plane" surface, which reduces the hydraulic liftforces.

Moreover, the rectangular shape prevents flow forces parallel to the surface from generating displacement of a unit out of the layer as long as close contact with the neighbour units is intact. Allsop, 1988, reports for 2t SHED units a stability number value $H_s/\Delta D_n$ as high as 5.7 for conditions corresponding to minimal movements. (For comparison a typical design value for Dolos would be 2.8 or less.) No general hydraulic stability criteria for the hollow cubes have been presented so far. It is the opinion of the author of the present paper that hydraulic stability performance of the armour layer itself is not the limiting design criterion. The limitations are the stability of the boundaries at the toe, top and ends of the armour layer plus the long-term structural durability which sets lower limits for the dimensions of the concrete elements. Obtainable accuracy in construction of the layer under various site specific conditions is another important limiting factor. More research is needed to explore these design limitations.

31. The number of different concrete armour units is tremendous, but generally their perfomance are not well documented. In order to assure rational use of existing units it is much more important to fully document their hydraulic stability and integrity characteristics than the development of "new" units.
The breakage of slender armour units naturally directed research into the area of stress generation and strength of concrete. Key references are Burcharth 1980, 1981a and b, Burcharth et al. 1988 and 1991, Timco 1983, Endo 1985, Scott et al. 1986, Lin et al. 1986, Hall et al. 1987, Howell 1985, 1988, Howell et al. 1990, Markle 1990, Melby et al. 1990, Van der Meer et al. 1990, Zwamborn et al. 1990, Nishigori et al. 1986, McDougal et al. 1988, Ligteringen et al. 1990, Anglin et al. 1990, Bürger et al. 1990, Van Mier et al. 1990. Some of the most recent results will be summarized.

32. A Dutch research programme initiated 1983 by a group of institutes, consultants and contractors focused on the stresses in colliding (rocking) units and possible improvements of the impact resistance, Ligteringen et al., 1990. The most valuable practical results of the study are development of a simulation model for rocking and related breakage of Tetrapods, and innovative use of modified (e.g. corrugated) armour unit surfaces to reduce impact forces, cf. Fig. 14.

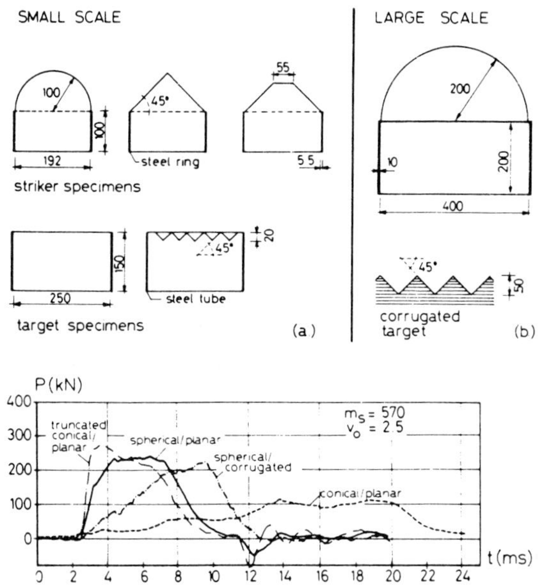

Fig. 14. Example of results from load-time response tests with impacting concrete specimens, Van Mier et al. 1991.

33. A coordinated research programme with the participation of CERC and University of Aalborg focuses on the development of design diagrams for Dolos, where the idea is that the strength of the units can be changed according to load conditions by changing the waist ratio of the unit, i.e. the Dolos cross section is a variable like in conventional structural elements. Moreover, the design diagrams should cover not only stresses from rocking (impacting) units, but also the quasi static wave load generated stresses as well as the often dominant static stresses caused by gravity and wedging. The study contains parametric studies of stresses and movements of small scale 200 g instrumented model units and calibration of these units against the instrumented 38 t Crescent City prototype Dolos as well as 200 kg instrumented units and prototype impact tests with 1.5 - 30 t Dolosse (Burcharth et al. 1991). An important result is the clarification of various scale effects related to small tests, which for practical and economical reasons must be used in parametric studies. The main results are the determination of the tensile stress distributions as function of wave load and structural geometry which form the basis for the design diagrams, cf. the example shown in Fig. 15.

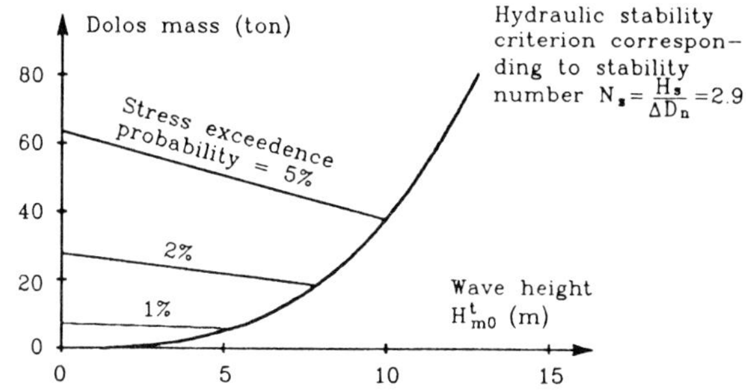

Fig. 15. Example of Dolosse design diagram covering both hydraulic stability and structural integrity for non-moving units.

Another important result relates to the influence of the surface roughness of the armour units. Smooth units, as often used in models, experience significantly larger stresses than units with normal concrete roughness and have less hydraulic stability, cf. also the comments on macro roughness in Burcharth, 1983.

34. Holtzhausen et al. 1990 studied the hydraulic stability of Dolosse with different waist ratios. Their results will be a valuable contribution to the final design diagrams.

35. A tremendous amount of work is still needed before complete design diagrams including the important fatigue effect (Burcharth 1984) have been produced. It is clearly an object for a coordinated effort within the coming second part of the EEC programme for Marine and Science Technology (MAST 2).

Concrete Technology

36. The permeability of the concrete should be kept as low as possible in order to obtain a good durability and high tensile strength of unreinforced concrete armour units. The permeability is dependent both on the porosity of the concrete matrix and the degree of cracking.

37. Crack is often due to the temperature differences created by the hydration process during curing. The related tensile stresses can easily exceed the very limited strength of the fresh young concrete thus causing formation of microcracks. The process is very complicated and can only be dealt with in an approximative manner, mainly because the descriptions of creep and relaxation processes of the hardening concrete are not precise enough to avoid large uncertainties in the calculation. The

calculations are performed by the use of special finite element computer programs for 3-dimensional bodies. However, a rule of thumb is that the temperature differences during curing should not exceed 20°C in concrete elements (which are not cast against a neighbour element).
There are several ways of dealing with the thermal stress problem, see for example Burcharth 1983. Besides the use of low heat cement also a simple insulation of the units during the first period of curing can help to avoid the problem which increases with the dimensions of the units. Insulation by means of a plastic cover was used by the author in 1981 related to trial casting of 90 t Antifer cubes applied for the emergency repair of the west breakwater at Sines. Since then the problem of thermal stress has been considered in many breakwater projects, among others the ongoing rehabilitation of the Sines west breakwater where insulation of the Antifer cubes is applied. Another way of dealing with the problem is to keep the effective dimensions of the units small. For cubes it can be done by making a hole as was done in the hot-climate Bosaso Harbour project in Somalia. Fig. 16 taken from Burcharth et al. 1991 illustrates the positive effect of the hole.

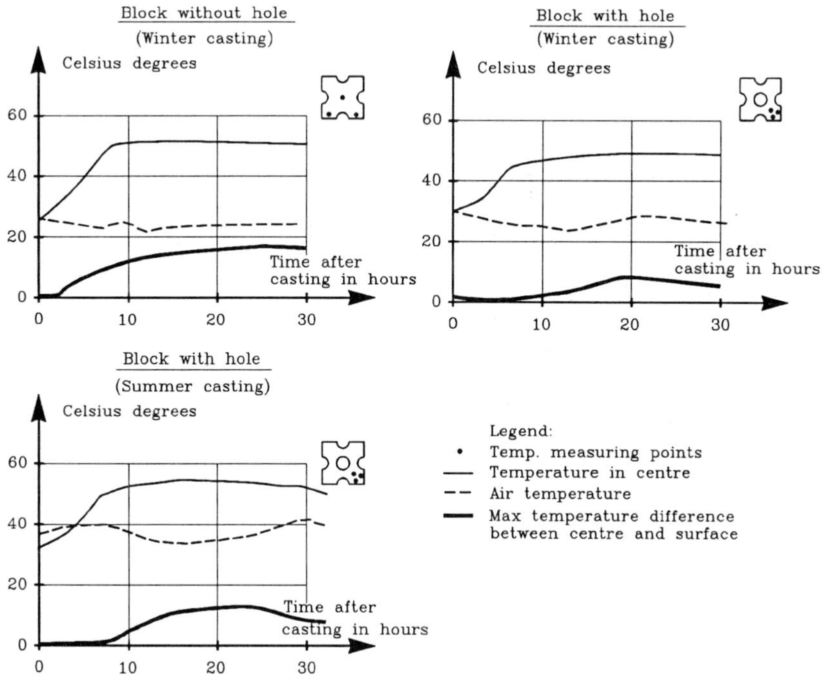

Fig. 16. Examples of temperature development in 30 t modified cubes with and without a hole.

The effect of a hole in a cube was also studied in the Dutch research programme, cf. paragraph 32. It confirmed the effectiveness of the hole in reducing temperature differences and thereby increasing the concrete strength. However, it was also conluded that the reduced cross section at the same time caused an even larger reduction of the overall strength of the units. This might be true for some cases, but in many cases the most serious problem is the long term deterioration of the concrete rather than the short term overall strength, which is more than sufficient where bulky units like cubes are applied in accordance with normal moderate hydraulic stability criteria.

36. Dense concretes attract much attention due to the increased strength and high resistance against cloride penetration and better freeze-thaw resistance. The concept of dense concretes is based on the use of very fine particles, e.g. Microsilica, and superplasticisers to compensate for the applied very low water-cement ratios of $w/c = 0.25 - 0.4$. However, already for $w/c =$ app. 0.4 the pore water is completely used, and consequently for lower w/c values a chemical selfdesiccation shrinkage takes place. This can cause very high stresses in cases where the hardening body is restricted against contractions.

Obviously the use of dense concretes is most advantageous for reinforced structures. For unreinforced armour units the benefits seem limited, mainly because the tensile strength is not increased significantly, i.e. the material is more brittle and will generally have an impact resistance comparable with conventional concretes. Moreover, the fatigue properties are not improved. In order to improve the tensile strength and increase the ductility various types of fibre reinforcement have been studied already for many years. Still, the application is very limited. The reason is that compared to the extra costs the advantages obtained are small because we are generally dealing with bulky large-volume structural elements where the gravity effect is the main stabilizing force. For such cases strength improvements do not result in significant reduction in volume. The use of fibre or conventional reinforcement might pay for certain slender complex types of armour units, but it is the belief of the author that a cheaper solution would be to change the shape of slender complex types of armour units by increasing the most stressed cross sections areas in order to comply with the loads, cf. the discussion in section 33.

OVERTOPPING

37. Restrictions on overtopping are often a vital part of the design criteria for coastal structures and breakwaters. Although formulae covering all common parameter ranges are not yet available some progress has been made, for example through parametric studies at HR by Owen,

Allsop, Bradbury and others. One of the results is the following formula (Bradbury et al., 1988) for the overtopping of crown walls in head-on waves. The time averaged discharge per unit width of the structure $\bar{Q}$ is given by

$$Q = A\,F^B$$

where $Q = \frac{\bar{Q}}{T_m\,g\,H_s}$ is a dimensionless overtopping, T_m is the mean wave period and $F = \left(\frac{Rc}{H_s}\right)^2 \left(\frac{s_p}{2\pi}\right)^{1/2}$ is a dimensionless freeboard, where Rc is the vertical distance from mean water level to the crest of the crown wall and s_p the wave steepness corresponding to the spectral peak period. A and B are empirical coefficients, the value of which depends on the geometry. For the tested sections the correlation coefficient ranges from 0.61 to 0.96 with an average of 0.80. Fig. 17 shows an example of a comparison with a case study. Considering the usual large scatter in overtopping test results the agreement is quite good.

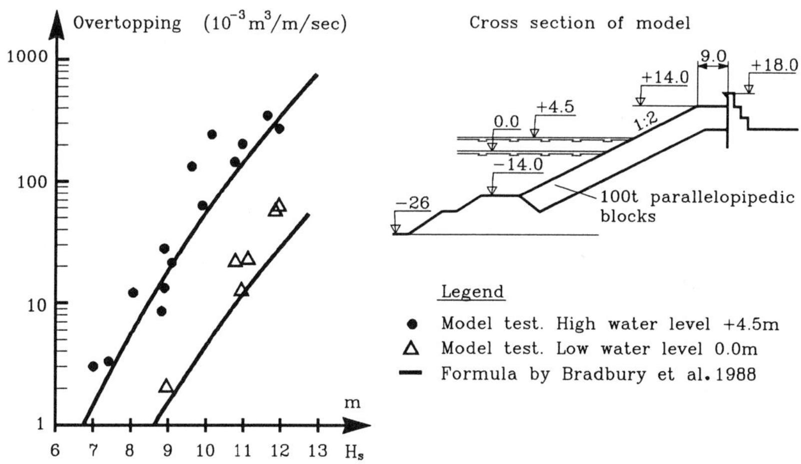

Fig. 17. Comparison between the Bradbury overtopping formula and a case study at Aalborg Hydraulics Laboratory.

The importance of overtopping of individual waves in determining the level of inconvenience or danger is discussed in Juul Jensen et al. 1987. The paper also contains valuable information on overtopping for a number of site specific cases.

38. Run-up, overtopping, wave transmission

Although a lot of results on these phenomena have been produced, no well documented generally applicable formulae for these phenomena are yet available. This is due to the lack of comprehensive parametric model test studies. However, work is ongoing and some important results related to transmission are presented at this conference in a paper by Van der Meer and d'Angremond.

FILTER STABILITY

39. Usually conventional, rather conservative, geometrical filter criteria like $D_{f15}/D_{b50} \leq 4$ are used in breakwater engineering. However, in some cases less restrictive criteria might save a filter layer or might allow the use of cheaper local materials for the filters. Such criteria, which allow some penetration of the base material into the filter, take into account the hydraulic load in terms of the hydraulic pressure gradient, Graauw 1984. The filter stability can be characterized by a critical hydraulic gradient, i_{cr}, beyond which significant migration starts. Fig. 18 shows an example of a filter design diagram for placed block revetments presented by Bezuijen et al. 1987. The diagram is based on permeameter tests with stationary and cyclic flow and covers only small grain sizes.

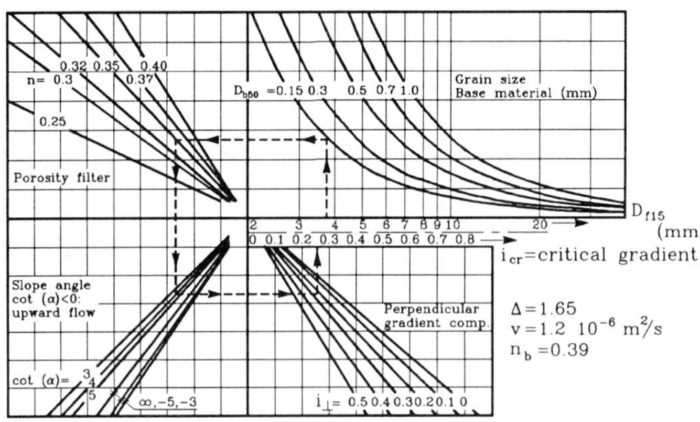

Fig. 18. Design diagram for granular filters in placed block revetments, Bezuijen et al. 1987.

40. The magnitude of the hydraulic gradients must be known in order to make use of the hydraulic filter criteria. The development of numerical models for internal flow in breakwaters will provide the tool for calculation of the hydraulic gradients. Among other studies the statistically

based transport model for filtration by Den Adel et al., 1987, gives insight in the parameter influence on the function of hydraulicly designed filters.

NUMERICAL MODELLING OF PHYSICAL PROCESSES IN WAVE-STRUCTURE INTERACTION

41. Numerical models are generally applicable only if they model the determining physical processes correctly with respect to parameter relationships. If this is not the case then a model cannot be used outside the parameter range where it has been calibrated against model test or prototype data. As opposed to many physical models the advantages of a correct numerical model are economical efficiency and the absence of scale effects (e.g. viscous scale effects in porous flow).

42. The problems related to numerical modelling rubble mound structures are

1. Modelling of the non-stationary porous flow and pressure field inside the structure, taking into account internal boundaries, anisotropy and water-air mixture.
2. Modelling of the outer flow field along the seaward slope. This serves as boundary conditions for item 1. It involves complications due to breaking wave action, air entrainment and air pockets, and interactions between incoming and reflected waves.
3. Modelling of structural response, e.g. the stability of individual armour units and armour unit layers. This includes modelling of the hydrodynamic forces as well as reaction forces.

43. Several researchers have performed small scale permeameter tests in order to study stationary porous flow. Most of the results are covering the laminar flow regime and cannot be extrapolated to the turbulent regime which is the most relevant in case of rubble mound breakwaters. Based on physical reasoning Burcharth et al. 1991 proposed the following relationship between the hydraulic gradient I and the discharge velocity V for the turbulent flow regime

$$I = I_c + \beta \frac{1-n}{n^3} \frac{1}{g\, d_{50}} (V - V_c)^2$$

where $V_c \quad = \frac{\boldsymbol{R}_c \nu}{d_{50}}$ is the transition velocity corresponding to the typical transition Reynolds' number, $\boldsymbol{R}_c = 300$.

$I_c \quad = \frac{\nu}{g\, d_{50}} \frac{1-n}{n^3} \left[\alpha_F (1-n) \boldsymbol{R}_c + \beta_F \boldsymbol{R}_c^2\right]$

ν = is the viscosity, n the porosity, α_F and β_F are shape-gradation coefficients in the Forchheimer equation for laminar flow. β is a coefficient signifying the shape and the gradation of the stones in the rough turbulent flow regime.

For quarry rock with medium gradation α_F and β_F will be approximately 400 and 4, respectively. Based on parameter test results mainly from University of Aalborg (Burcharth et al. 1991) and HR (Shih, 1990) β for angular stones was found to be 2.7 for narrow gradations and 2.9 for medium gradations.

44. The non-steady porous flow was studied by Hanoura et al. 1978 and Burcharth et al. 1991 using a free fall U-tube technique in order to see if any significant inertia effect in terms of added mass is present. However, the last authors concluded that the technique was unsuitable and the results too uncertain to make any conclusions. A theoretical model indicated that the added mass effect in practice is of minor importance. However, Smith et al. 1990 presented preliminary results from tests in a pulsating water tunnel indicating some significant effect at higher accelerations. Consequently, it must be concluded that we do not have a model, even for the simple one-dimensional non-steady (oscillatory) flow case. Experiments are urgently needed and should be performed in the ongoing MAST I program. Despite this situation, quite a number of engineering models have been developed. An overview is given in the following paragraphs.

45. McCorquodale, 1970, presented a two-dimensional model for unsteady non-darcy flow inside porous media. The model is based on the continuity equation and Navier-Stokes equations, which are solved by a finite element technique. Free surface flows can be modelled. The effect of a weak inertial term is included. The model has been compared to experimental results.

46. Nasser and McCorquodale 1975 presented a one-dimensional model for the propagation of long shallow water waves through a porous medium. The model is based on the depth integrated equations of continuity and motion, which are solved by the method of characteristics. The breakwater is schematized by a homogeneous and rectangular section. The model covers phenomena such as reflection, transmission, dissipation, run-up and rush-down. The dissipation is described by the Forchheimer expression. The numerical model has been compared to wave flume tests.

47. Di Natale 1988 presented a similar model. Run-up and rush-down are not included in this model. The model presented by Hannoura and McCorquodale 1985 is a mixed numerical model. A one-dimensional

finite difference model solves the continuity and momentum equations to obtain the internal water levels, and a two-dimensional finite element model is applied in order to predict the flow properties below the water level. The porous flow description is based on the Forchheimer expression together with an expression for the added mass. Air entrainment due to internal wave breaking is included.

48. The HADEER model has been developed at Delft Geotechniques (DG) through a cooperation between Barends, DG, Hannoura and McCorquodale, University of Windsor, Canada. HADEER is based on the work done by Hannoura and McCorquodale during the 1970's. The following summary is based on Hölscher et al., 1988. The model is a two-dimensional model applicable to the interior flow in a breakwater. The equations of continuity and momentum are solved by a combination of a one-dimensional finite difference method, giving the shape of the phreatic surface, and a two-dimensional finite element model, giving the velocity field below the phreatic surface. The results from the finite element model are used to calculate the motion of the phreatic surface. HADEER can be applied for breakwaters with sloping seaward face and with layers of varying porosity. The model describes laminar and turbulent friction, and inertia of the water is included. The wave run-up and the pressure distribution on the exterior slope are used as boundary conditions. The numerical model has been compared to flume tests.

49. The model by Austin and Schlueter 1982 solves the continuity equation and Navier-Stokes equation in two dimensions by a volume of fluid (VOF) method capable of modelling free surface flow. The original computer code, SOLA-VOF was developed at Los Alamos Scientific Laboratories. The present version predicts the flow field in a porous armour layer schematized as a rectangular block system. The drag and inertial forces acting on the individual blocks are derived subsequently. Finally, the response of the blocks is determined by a discrete element code based on the equations of motion.

50. The SAVOF model presently being developed at Delft Hydraulics 1991 is based on the SOLA-VOF model by Hirt and Nichols, 1981. The model, which solves the continuity equation and Navier-Stokes equations in two-dimensions, is based on the volume of fluid method.
This model is able to compute free surface flow, when the fluid domain becomes multiply-connected. Air entrainment is not included, the model simulates either fluid or vacuum. The seaward face of the structure must be vertical. At present, the model is not able to simulate flow inside a porous structure.

51. Kobayashi et al., 1986, 1987, 1988, presented a one-dimensional hydrodynamic model for determination of armour stability on rough im-

permeable slopes exposed to regular waves. The model solves the equations of continuity and momentum by a finite difference method. Free surface flow is included. The model covers the flow outside the structure, including reflection, run-up and down-rush. The subsequent evaluation of the armour stability is based on Morison type forces with constant empirical coefficients for drag, lift, and inertia forces. Recently the model was expanded to cover also permeable slopes, Kobayashi et al., 1989 and irregular wave reflection and run-up on rough impermeable slopes, Kobayashi et al. 1990.

52. Recent work at Imperial College London is directed towards the development of a model which is able to predict global slope changes by simulating individual armour displacement. A numerical model for slopes of spheres has been developed so far. Details will be presented in a paper by Norton and Holmes at this conference.

53. Thompson 1988 presented a hybrid numerical model for flow on and in breakwaters. Outside the structure the vertically integrated equations of continuity and momentum are solved by a finite difference technique. Phenomena such as reflection, run-up and down-rush are included. Inside the structure the flow field can be determined by either a two-dimensional finite difference model or from a simplified model, assuming that flow takes place only in a thin layer just below the surface of the slope. The model has been compared to experimental results.

54. Danish Hydraulic Institute 1989, 1990, has developed a general three-dimensional hydrodynamic model, SYSTEM 3. The model solves the continuity equation and Navier-Stokes equations by a finite difference method. Free surface flow can be modelled outside as well as inside a porous structure. The dissipation is modelled by the Forchheimer expression. At present, a two-dimensional example with a sloping, multilayered structure can be modelled.

55. Despite the many activities in numerical modelling it will take many years before a numerical breakwater model covering the various responses to irregular waves is applicable. To ensure a high quality of the models more research has to be devoted to correct mathematical descriptions of the physical processes rather than to numerical problems.

MONITORING OF PROTOTYPE STRUCTURES

56. Documentation of the as-built structure plus systematic monitoring during structural life are essential parts of any maintenance scheme. It is not an easy task in case of rubble mound breakwaters, where large parts of the slopes are under water and almost always facing some wave action. This together with poor underwater visibility makes inspection by divers impossible in many cases. Even conventional profile surveying

by means of a crane operated rod fitted with a basket at its bottom or by means of a narrow-beam echo-sounder can be impossible as discussed by Read, 1988.

57. Many breakwater designs involve such construction difficulties that a not-as-designed structure is the inevitable result. Fracture of fragile armour units during placement and not-as-designed distribution of materials in underwater sections are the most serious problems. Although we can improve the strength of armour units and put more attention to constructability in the design, the two mentioned problems will still remain to some extent. Fracture of armour units by underwater placement is due to high impact velocities. The velocity must therefore be recorded, for example by placing accelerometers on the sling (grab), and transmitted to the crane operator.

58. To secure a uniform distribution and placing density of underwater armour for some types of units, it is necessary to use a technique of visual observation. In large water depths it is not sufficient to place units to a co-ordinated grid, because even a moderate swell might displace the unit a couple of metres, which cannot be registered by the operator. Moreover, a satisfactory placement also requires a certain orientation of the unit, when placed to fit in the pack. The orientation and the rotation of the unit should therefore be visualized and controlled by the operator. Underwater low light colour video cameras mounted on a direction stabilized frame above the sling (grab) might solve the problem, at least where the visibility is good.

59. The inspection of the breakwater profile and the control of coverage density (gabs in the pack) are very important but difficult, especially in the case of big complex types of blocks. Side-scan sonar has been tried with moderate success some years ago, Patterson et al.,1983. More recent experiment by CERC revealed that in many cases it is very difficult to make a meaningful interpretation of the picture due to shading effects. Devices for the monitoring of underwater works for rubble mound breakwaters can be developed further. Since the costs involved will be minor compared to the construction costs, it is obvious that the designer should stimulate such a development, simply by specifying strict procedures for the control of underwater works in the tender documents. Without development in this field it is also impossible to follow the needed programmes for underwater maintenance throughout the lifetime of the structure.

60. Inspection of the dry parts of the breakwater is much easier. Land-based photographic recordings are often used, but is labour intensive and inaccurate in areas where the armour units are inaccessible, or even out of sight. Areal photogrammetry using photographs taken from air crafts is successfully used, e.g. in Zeebrugge Harbour, Van Damme

et al. 1990. Photogrammetric digitization makes it possible to establish a very accurate data base for the positions of the armour units and subsequent registration of any displacements.

PROBABILISTIC DESIGN, PARTIAL COEFFICIENTS

61. For quite some years papers on probabilistic design of breakwaters have been published, e.g. Nielsen et al. 1983, Le Mehaute et al. 1985. In design calculations it is now more common to take into account the uncertainties related to the involved parameters, at least in larger projects. The probability of failure (reliability) can be estimated once the uncertainty on each parameter has been defined, e.g. in terms of a standard deviation and normal distribution if the real distribution is unknown. The calculations are almost without exception based on so-called level II reliability analysis, see Madsen et al., 1986.
Special computer programmes for breakwaters and sea walls, BWREL (Break Water RELiability) and BWCODE (Break Water CODE) have been developed by Dalsgaard Sørensen, University of Aalborg. The programmes estimate the reliability of single failure modes, cf. Fig. 5, as well as series systems of failure modes and optimize and calibrate partial coefficients. The programmes were used in the work of the PIANC PTC II Working Group (No. 12) on Reliability of Rubble Mound Breakwaters, Burcharth 1990. It was found, as expected, that existing structures generally have a low reliability, i.e. a high probability of failure within structural life. This is mainly due to the tradition of adopting a 50-year or a 100-year return period design sea state, cf. the encounter-probability diagramme, Fig. 2. The analysis also revealed that existing structures and their environmental conditions are generally poorly documented, Ligteringen 1989. Moreover, it was found that the description of failure modes in terms of formulae covering geometries of real breakwaters is still very incomplete, Allsop, 1989.

62. The main objective of the PIANC working group was to propose ways of dealing with safety problems in a practical way. A conventional civil engineering partial coefficient system was rejected because there is no acceptable breakwater design tradition which could be used for the necessary calibration of the coefficients. Moreover, the quality of information about the wave climate varies so much that meaningful standard loads cannot be defined. Finally, it seems unacceptable to use a system where the real probability of failure within structural life is unknown.

63. A new safety coefficient was developed, where the partial coefficients are determined corresponding to a given failure probability, P_f, within structural life, T_{year}, and the coefficients take into account the uncertainty related to the site specific estimated long term wave height

contribution, Burcharth 1990.

64. Only two partial coefficients are used for each failure mode formula, γ_{H_s} on H_s and γ_z on the rest of the formula. The formulae are formulated as a design equations, e.g for the Hudson equation

$$\text{Block volume,}\quad V = \frac{(\gamma_{H_s} \cdot H_s)^3}{\frac{1}{\gamma_z} K_D \Delta^3 cot\alpha}$$

γ_{H_s} and γ_z are determind from the formulae

$$\gamma_{H_s} = \frac{H_s^{T_{P_f}}}{H_s^T} + \sigma'_{FH_s}{}^{\left(1+\left(\frac{H_s^{3T}}{H_s^T}-1\right)k_\beta P_f\right)} + \frac{0.05}{\sqrt{P_f N}}$$

$$\gamma_z = 1 - k_\alpha ln\, P_f$$

where H_s^T is the H_s corresponding to the T-year return period, T being the structural life.

H_s^{3T} is the H_s corresponding to the $3T$-year return periode.

$H_s^{T_{P_f}}$ is the H_s corresponding to an equivalent return period, T_{P_f}, defined as the return period corresponding to exceedence of $H_s^{T_{P_f}}$ with probability P_f within structural life T. T_{P_f} is found from the encounterprobability formula as $T_{P_f} = (1-(1-P_f)^{\frac{1}{T}})^{-1}$.

H_s^T, H_s^{3T} and $H_s^{T_{P_f}}$ are found from the site specific long term distribution of H_s, e.g. a Weibull distribution.

N is the number of H_s data values on which the long term distribution is based.

σ'_{FH_s} is the variational coefficient signifying the uncertainty related to the N data. Typical values are listed in Table 2, e.g. 0.05 - 0.1 for accelerometer buoy measurements.

k_α and k_β are coefficients with specific values for each failure mode formula, e.g for the Van der Meer formula for rock armour in plunging waves, $k_\alpha = 0.027$ and $k_\beta = 38$.

The concept is simple because only one set of k_α, k_β values is needed for each failure mode, and the partial coefficients are easily calculated when the long term wave height distribution has been found.

So far the k_α, k_β coefficients have been calculated only for the relative few failure modes because of the lack of general formulae for many failure modes. However, it is easy to expand the system as more formulae appear. It is important to notice that the reliability of the formulae must be documented (e.g. in terms of a standard deviation) in order to implement them in the partial coefficient system. More details about the system is given by Burcharth in another paper of this conference.

ACKNOWLEDGEMENT

The many inspiring discussions with researchers, designers, and contractors working in the coastal enginering field are gratefully acknowledged.

REFERENCES

AHRENS, J.P. 1987. Characteristics of reef breakwaters. CERC, Vicksburg, Technical Report CERC-87-17.

ALLSOP, N.W.H. 1988. Concrete armour units for rubble mound breakwaters and sea walls: Recent progress. Report SR 100, HR, Wallingford, U.K., 1988.

ALLSOP, N.W.H. 1989. Draft report of Subgroup A, PIANC PTC II, Working group 12 on Rubble mound breakwaters.

ANGLIN, C.D., SCOTT, R.D., TURCKE, D.I., TURCKE, M.A. 1990. The development of structural design criteria for breakwater armour units. Proc. Seminar Stresses in Concrete Armor Units, ASCE, Vicksburg, U.S.A.

BAIRD, W.F., HALL, K.R. 1984. The design of breakwaters using quarried stones. Proc. 19th Int. Conf. on Coastal Engineering, Houston, USA.

BARENDS, F.B.J., HOLSCHER, P. 1988. Modelling interior processes in a breakwater. Proc. Breakwaters '88, Eastbourne, U.K., 1988.

BATTJES, J.A., JANSEN, I.P.F.M. 1979. Energy loss and set-up due to breaking of random waves. Proc. 16th Coastal Eng. Conf., 1978, Hamburg.

BEZUIJEN, A., KLEIN BRETTELER, M., BAKKER, K.I. 1987. Design criteria for placed block revetments and granular filters. Proc. Int. Conf. on Coastal and Port Engineering in Developing Countries, Beijing, China, 1987.

BRADBURY, A.P., ALLSOP, N.W.H. 1988. Hydraulic effects of breakwater crown walls. Proc. Conf. Breakwaters '88, Eastborne, U.K., 1988.

BRORSEN, M., BURCHARTH, H.F., LARSEN, T. 1974. Stability of Dolos Slopes. Proc. 14th Coastal Eng. Conf., 1974, Copenhagen.

BRUUN, P., GÜNBAK, A.R. 1976. New design principles for rubble mound structures. Proc. 15th Int. Coastal Eng. Conf., Honolulu, Hawai, 1976.

BURCHARTH, H.F. 1979. The effect of wave grouping on on-shore structures. Coastal Engineering, 2 (1979), 189-199.

BURCHARTH, H.F. 1980. Full scale trials of dolosse to destruction. Proceeding of the 17th International Conference on Coastal Engineering.

BURCHARTH, H.F. 1981a. A design method for impact loaded slender armour units. Paper presented at the ASCE International Convention, New York. (Laboratoriet for Hydraulik og Havnebygning, Bulletin No.18, University of Aalborg.)

BURCHARTH, H.F. 1981b. Full-scale dynamic testing of dolosse to destruction. Coastal Engineering, Vol.4.

BURCHARTH, H.F. 1983. Materials, Structural Design of Armour Units. Proc. Seminar on rubble mound breakwaters. Royal Institute of Technology, Stockholm, Sweden, 19. April 1983. Bulletin No. TRITA-VBI-120 (Ed. Gösta Werner).

BURCHARTH, H.F. 1984. Fatigue in breakwater armour units. Proceeding of the 19th International Conference on Coastal Engineering, Houston, Texas.

BURCHARTH, H.F. 1986. On the uncertainties related to the estimation of extreme environmental condition. Proc. Seminar on Uncertainties related to the Design and Construction of Offshore Jacket Structures, Copenhagen, May 1986. Danish Society of Hydraulic Eng.

BURCHARTH, H.F. 1988. On methods of establishing design diagrams for structural integrity of slender complex types of breakwater armour units. Proceeding of 2nd International symposium on Wave Research and Coastal Engineering. Hannover, Leibnizhaus, Germany.

BURCHARTH, H.F., FRIGAARD, P. 1988. On the stability of berm breakwater roundheads and trunk erosion in oblique waves. Proc. ASCE Seminar on unconventional rubble mound breakwaters, Ottawa, Canada.

BURCHARTH, H.F., FRIGAARD, P. 1988. On 3-dimensional stability of reshaping breakwaters. Proc. 21st Int. Conf. on Coastal Engineering, Malaga, Spain.

BURCHARTH, H.F. 1989. Uncertainties related to environmental data and estimated extreme events. Report of subgroup B under the PIANC PTC II Working Group 12 on rubble mound structures. 3. draft.

BURCHARTH, H.F. 1990. Introduction of a partial coefficients system for the design of rubble mound breakwaters. Report of subgroup F under the PIANC PTC II Working Group 12 on rubble mound structures.

BURCHARTH, H.F., CHRISTENSEN, CLAUS 1991. On stationary and non-stationary porous flow in coarse granular materials. Report of the MAST G6-S Project I, Wave Action on and in Coastal Structures. University of Aalborg, Department of Civil Engineering, May 1991.

BURCHARTH, H.F., HOWELL, G.L., LIU, Z. 1991. On the determination of concrete armour unit stresses including specific results realted to Dolosse. Coastal Engineering, 15 (1991) 107-165.

BURCHARTH, H.F., TOSCHI, P.B., TURRIO, E., BALESTRA, T., NOLI, A., FRANCO, L., BETTI, A., MEZZEDIMI, S. 1991. Bosaso Harbour, Somalia. A new hot-climate port development. Third Int. Conf. on Coastal & Port Engineering in Developing Countries (COPEDEC III), Mombasa, Kenya, Sept. 1991.

BÜRGER, OUMERACI, H., PARTENSCKY, H.W. 1990. Impact Strain Investigations on Tetrapods Results of Dry and Hydraulic Tests. Proc. Seminar Stresses in Concrete Armor Units, ASCE, Vicksburg, U.S.A.

CARTER, D.J.T. & M.J. TUCKER 1986. Uncertainties in environmental design criteria. Underwater Technology, 12, pp 2-8 + 33.

CHALLENOR, P.G. 1979. Confidence limits for extreme value statistics. Institute of Oceanographic Science, Report no. 82.

DANISH HYDRAULIC INSTITUTE 1989. Numerical Geohydrodynamic Methods. Status Report.

DANISH HYDRAULIC INSTITUTE 1990. Numerical Geohydrodynamic Methods. Status Report 2.

DEN ADEL, H., BAKKER, K.I., SCHAMINCE, P.E.L. 1987. A transport model for filtration. Proc. 9th European Conference on Soil Mechanics and Foundation Engineering, Dublin, Ireland, 1987. 1987.

DI NATALE, M. 1988. Energy Dissipation in Rubble Mound Breakwaters. Modelling Soil-Water-Structure Interactions, Kolkman et al. (eds), pp. 417-426.

ENDO, T. 1985. Outline of study on structural strength of armour blocks. Hydraulic Laboratory, Nippon Tetrapods Co., Ltd.

GIVLER, L.D., SØRENSEN, R.M. 1986. An investigation of the stability of submerged homogeneous rubble mound structures under wave attack. Lehigh University, H.R. IMBT Hydraulics, Report IHL-110-86.

GRAAUW, A.F.F. DE, MEULEN , T. VAN DER, DOES DE BYE, M.R. VAN DER, 1984. Granular filters: design criteria. J. Waterw. Port, Coast. Ocean Eng. 110 (1).

GRAVESEN, H., SØRENSEN, T. 1977. Stability of rubble mound breakwaters. Proc. 24th Int. Navigation Congress.

GODA, Y. 1988. On the methodology of selecting design wave height. Proc. 21st Int. Conf. on Coastal Engr., Spain.

HALL, R.L., BEVINS, T.L., HOWELL, G. 1987. Drop Tests of Dolos Armor Units. Proc. of 1987 ASCE Structures Congress, Dynamics of Structures.

HANOURA, A.A., McCORQUADALE, J.A. 1978. Virtual mass of coarse granular media. Proc. ASCE, Journal Waterway, Port, Coastal and Ocean div., Vol. 104, WW2, 1978.

HANOURA, A.A., McCORQUADALE, J.A. 1985. Rubble mounds: Numerical modeling of wave motion. ASCE Journal Waterway, Port, Coastal and Ocean Engineering, Vol. 111, No. 5, pp. 800-816.

HIRT, C.W., NICHOLS, B.D. 1981. Volume of Fluid Method for the Dynamics of Free Boundaries. Journal of Computational Physics 39, pp. 201-225.

HOLTHUIJSEN, L.N., BOOIG, N., HERBERS, T.H.C. 1989. A prediction model for stationary, short crested waves in shallow water with ambient currents. Coastal Engineering, 13 (1989) pp 23-54.

HOLTZHAUSEN, A.H., ZWAMBORN, J.A. 1990. Stability of Dolosse with different waist thicknesses for irregular waves. Proc. 22nd Int. Conf. Coastal Engineering, Delft, Holland.

HOWELL, G.L., (ed) 1985. Proceeding of Workshop on Measurement and Analysis of Structural Response in Concrete Armor Units, January 23-24 Waterways Experiment Station, coastal Engineering Research Center, Vicksburg, Mississippi, USA,

HOWELL, G.L. 1988. Measurement of forces on dolos armour units at prototype scale. Proceeding of the 21st International Conference on Coastal Engineering, Spain.

HOWELL, G.L., RHEE, J.P. and ROSATI, J. 1990. Stresses in Dolos Armor Units due to Waves. Proc. Seminar Stresses in Concrete Armor Units, ASCE, Vicksburg, U.S.A.

HUGHES, S.A. 1984. The T.M.A. shallow water spectrum description and applications. U.S. Army Corps of Engineers, Tech. Report 84-7.

HÖLSCHER, P., DE GROOT, M.B., VAN DER MEER, J.W. 1988. Simulation of Internal Water Movement in Breakwaters. Modelling Soil-Water-Structure Interactions, Kolkman et al. (eds), pp. 427-433.

IAHR/PIANC 1986. List of sea state parameters. Supplement to PIANC Bulletin No. 52, January 1986.

JUUL JENSEN, O., JUHL, J. 1987. Wave overtopping on breakwaters and sea dikes. Proc. COPEDEC 1987, Beijing, China.

KIMBALL, B.F. 1949. An approximation to the sampling variances of an estimated maximum value of given frequency based on the fit of doubly exponential distribution of maximum values. Ann. Math. Statist., 20 : pp 110-113.

KLOPMANN, G., STIVE, M.J.F. 1989. Extreme waves and wave loading in shallow water. E & P Forum, Report No. 3.12/156.

KOBAYASHI, N., COX, D.T., WURJANTO, A. 1990. Irregular wave reflection and run-up on rough impermeable slopes. Journal of Waterway, Port, Coastal and Ocean Engineering, Vol. 116, No. 6.

KOBAYASHI, N., ROY, I., OTTA, A.K. 1986. Numerical simulation of Wave Run-up and Armor Stability. Offshore Technology Conference, 5088, pp. 51-60.

KOBAYASHI, M., MASAAK, T., TAKAHASHI, K. 1987. Bearing Capacity of a rubble mound supporting a gravity structure. Report of the Port and Harbour Research Institute Vol. 26, No. 5, Dec. 1987.

KOBAYASHI, N., OTTA, A.K., ROY, I. 1987. Wave reflection and run-up on rough slopes. Journal of Waterway, Port, Coastal and Ocean Engineering, Vol. 113, No. 3, pp. 282-298.

LATHAM, I.P., MANNION, M.B., POOLE, A.B., BRADBURY, A.P., ALLSOP, N.W.H. 1988. The influence of armour stone shape and rounding on the stability of breakwater armour layers. Report No. 1. Coastal Engineering Research Group, University of London and Hydraulics Research, U.K.

LAWLESS, J.F. 1974. Approximations to confidence intervals for parameters in the extreme value and Weibull distributions. Biometrika, 61, pp 123-129.

LE MEHAUTE, B., WANG, S. 1985. Wave statistical uncertainties and design of breakwaters. Journal of Waterway, Port, Coastal and Ocean Engineering, ASCE, Vol. 111, No. 5.

LIGTERINGEN, H. 1989. Description of selected breakwaters. Report of subgroup E under the PIANC PTC II Working Group 12 on rubble mound structures.

LIGTERINGEN, H., ALTINK, H., van ORSCHOT, J.H. 1990. Strength of Concrete Armour Units A Joint Industry Research. ASCE Proc. Seminar Stresses in Concrete Armor Units, CERC, Vicksburg, 1990, U.S.A.

LIN, W.M., RAU, C., SU, R.-L. 1986. The structural responses of dolos armour units under the dynamic loading. Proceeding of the 20th International Conference on Coastal Engineering, Taipei, Taiwan.

MADSEN, H.O., KRENK, S., LIND, N.C. 1986. Methods of Structural Safety. Prentice-Hall, 1986.

MARKLE, D.G. 1990. Crescent City Instrumented Model Dolos Study. Proc. Seminar Stresses in Concrete Armor Units, ASCE, Vicksburg, U.S.A.

McCORQUODALE, J.A. 1970. Finite Element Analysis of Non-Darcy Flow. Ph.D. thesis presented to the University of Windsor, Canada.

McDOUGAL, W.G., MALBY, J.A., TEDESCO, J.W. 1988. Wave forces on concrete armour units. Journal of Waterway, Port, Coastal and Ocean Engineering, Vol. 114, No. 5.

MEDINA, J.R., FASSARDI, C., HUDSPETH, R.T. 1990. Effects of wave groups on the stability of rubble mound breakwaters. Proc. 22nd International Conference on Coastal Engineering, ICCE, July 2-6, 1990, Delft, The Netherlands.

MELBY, J.A., ROSSEN, B.T., TEDESCO, I.W. 1990. An Analytical Investigation of Static Stresses in Dolosse. Proc. Seminar Stresses in Concrete Armor Units, ASCE, Vicksburg, U.S.A.

NASSER, M.S., McCORQUODALE, J.A. 1975. Wave Motion in Rockfill. Journal of Waterways, Harbors and Coastal Engineering Division, Vol. 101, No. WW2, pp. 145-159.

NIELSEN, S.R.K., BURCHARTH, H.F. 1983. Stochastic design of rubble mound breakwaters. Proc. 11th IFIP Conference on System Modelling and Optimization, Copenhagen, Denmark, 1983.

NISHOGORI, W., ENDO, T., SHIMEDA, A. 1986. On stress in tetrapods under wave action. Proceeding of the 20th International Conference on Coastal Engineering, Taipei, Taiwan.

PATTERSON, D.R., POPE, J. 1983. Coastal applications of side scan sonar. Proc. Coastal Structures '83. Washington 1983.

PETRAUSKAS, C. AAGAARD, P.M. 1971. Extrapolation of historical storm data for estimating design wave heights. Journ. of Society of Petroleum Engineers, Vol. 2, 1971, pp 25-35.

READ, J. 1988. The control of rubble-mound construction, with particular reference to Helguvik Breakwater in Iceland. Proc. Breakwater '88, Eastbourne, U.K., 1988.

ROSBJERG, D. 1985. Estimation in partial duration series with independent and dependent peak values. Journ. of Hydrology, 76, pp 183-195.

SCOTT, R.D., TURCKE, D.J., BAIRD, W.F. 1986. A unique instrumentation scheme for measuring loads in model dolos units. Proceeding of the 20th International Conference on Coastal Engineering, Taipei, Taiwan.

SIMM, J.D., HEDGES, T.S. 1988. Pore pressure response and stability of rubble-mound breakwaters. Proc. Breakwaters '88, Eastbourne, U.K., 1988.

STIVE, M.J.F. 1986. Extreme shallow water wave conditions. Delft Hydraulics. Internal report H533.

THOMPSON, A.C. 1988. Numerical Model of Breakwater Wave Flows. 21st International Conference on Coastal Engineering, pp. 2013-2027.

THORNTON, E.B., GUZA, R.T. 1983. Transformation of wave height distribution. J. Geophys. Research 88(C10), pp. 5925-5938.

TIMCO, G.W. 1983. On the structural integrity of Dolos units under dynamic loading conditions. Coastal Engineering. Vol. 7, No. 1.

TOLMAN, H.L. 1990. Wind wave propagation in tidal seas. Communications on Hydraulic and Geotechnical Engineering, Delft University of Technology, ISSN 0169-6548, Rep. No. 90-1.

VAN DAMME, L., DE WOLF, P., FRANSAGER, D., BRAET, K., VAN RENSBERGEN, J. 1990. Breakwater armour unit monitored using areal remote sensing. PIANC Bulletin No. 70, 1990.

VAN DER MEER, J.W. 1988. Rock slopes and gravel beaches under wave attack. Delft Hydraulics Communication No. 396.

VAN DER MEER, J.W. and HEYDRA, G. 1990. Impact Velocities of Rocking Armour Units. Proc. Seminar Stresses in Concrete Armor Units, ASCE, Vicksburg, U.S.A.

VAN DER MEER, J.W. and PILARCZYK, K.W. 1990. Stability of low-crested and reef breakwaters. Proc. 22nd Int. Conf. on Coastal Engineering, Delft, 1990.

VAN MIER, J.G.M., LENOS, S. 1991. Experimental analysis of the load-time histories of concrete to concrete impact. Coastal Engineering, 15 (1991).

VRIJLING, J.K., SMIT, E.S.P., DE SWART, P.F. 1991. Berm breakwater design. The longshore transport case. A probabilistic approach. Proc. Coatal STructures and Breakwaters '91, Nov. 1991, London.

WAMDI Group 1988. The WAM model. A third generation ocean wave prediction model. Journal of Physical Oceanography, 18, 12, pp 1775-1810.

WANG, S. & LE MEHAUTE, B. 1983. Duration of measurements and long-term wave statistics. Journal of Waterway, Port, Coastal and Ocean Engineering, ASCE, Vol. 109, No. 2, pp 236-247.

ZWAMBORN, J.A. and PHELP, D. 1990. Structural Tests on Dolosse. Proc. Seminar Stresses in Concrete Armor Units, ASCE, Vicksburg, U.S.A.

5. New types of structures for attenuating waves and breakwaters

R. BOUCHET, Government of Monaco

With regard to the lack of space and the important depths which lie immediatly off its coast, the PRINCIPALITY OF MONACO has often had to make innovations in the field of maritime works.

This paper presents the results of studies which have led to the designing of new types of structures for attenuating waves, and breakwaters.

It also presents the project of the new breakwater of LA CONDAMINE and a new conception of offshore urbanisation.

1) INTRODUCTION

In order to accommmodate its growth, the PRINCIPALITY OF MONACO, the smallest country in the world after the STATE OF VATICAN : only 2 square kilometres, has recovered land from the sea in several stages. The area of reclamations represents about 30% of the whole territory.

Owing to the importance of depths which lie immediately off the coast, MONACO has often had to make innovations in maritime works.

The first artificial beaches were built, thirty years age, at Larvotto, in depths of 5 to 7 metres.

At the time, the breakwater of Fontvieille (Fig. 1), constructed at depths varying from 30 to 40 metres, was the deepest in the world.

Figure 1

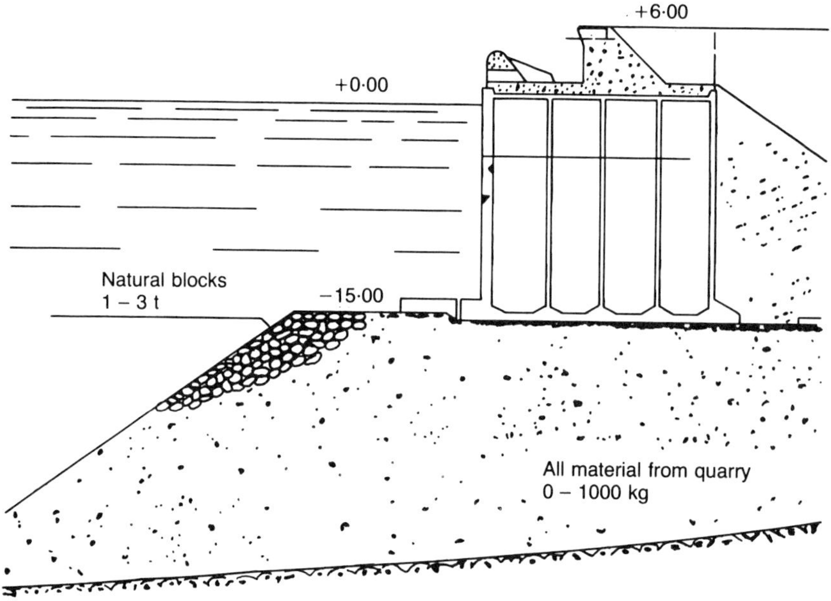

This record is currently held by the breakwater of the Port of KAMAISHI which our Japanese colleagues are building in depths reaching 63 metres. You will shortly see that we are preparing to take the record back again.

In the immediate future, our problem is to protect and enlarge the harbour of the Condamine, which is fully open to the East, before which are located depths reaching 50 metres and more.

Figure 2

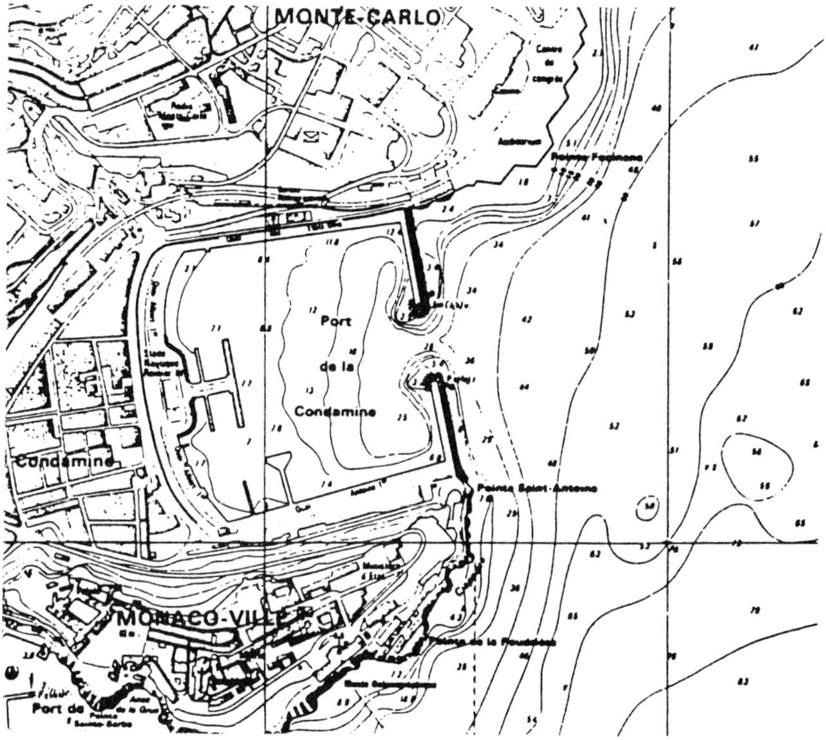

In view of the importance which the PRINCIPALITY OF MONACO has always attached to the protection of the marine environment, we have been impelled to study new types of ecological and economic structures, the particular feature of which is to oppose the waves with obstacles which are essentially horizontal, sunk in shallow water, instead of the massive vertical or sloping works used up to the present, whose volume and impact on the marine environment increase rapidly with depth.

These investigations were centered on the behaviour of the mass of water located between a horizontal slab and the bottom of the sea.

They have led to the designing of two types of structure, one a wave attenuator, the other a breakwater.

2) WAVE ATTENUATOR ("OSCILLATING SEAWALL")

Figure 3

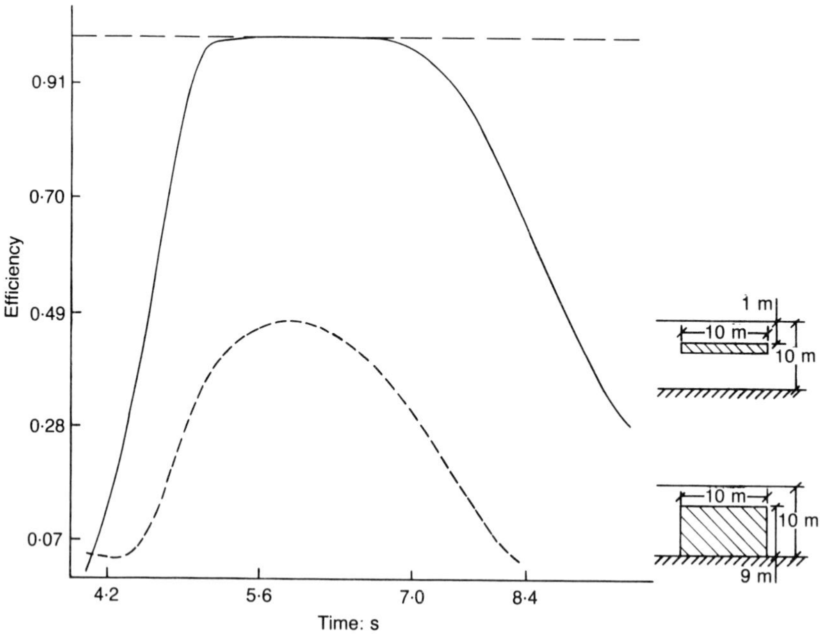

The basic experiment (Fig. 3) consists of comparing the energy performance of a block of submerged concrete, here a concrete block 12 metres wide with its upper surface 1 metre below the surface of the water, with a horizontal slab obtained by removing all the concrete from the lower part and retaining only a slab attached to the upper surface of this block.

It appears that the performance of the slab is superior to that of the block. So water may be more efficient than concrete in attenuating waves.

How can we explain this phenomenon?

Excited by the wave off shore, (Fig. 4), the mass of water located between the slab and the bottom of the sea immediately transmits onwards the variations of pressure to which it is subjected while the fraction of energy of the wave which passes over the slab takes a certain time to cross it.

Figure 4

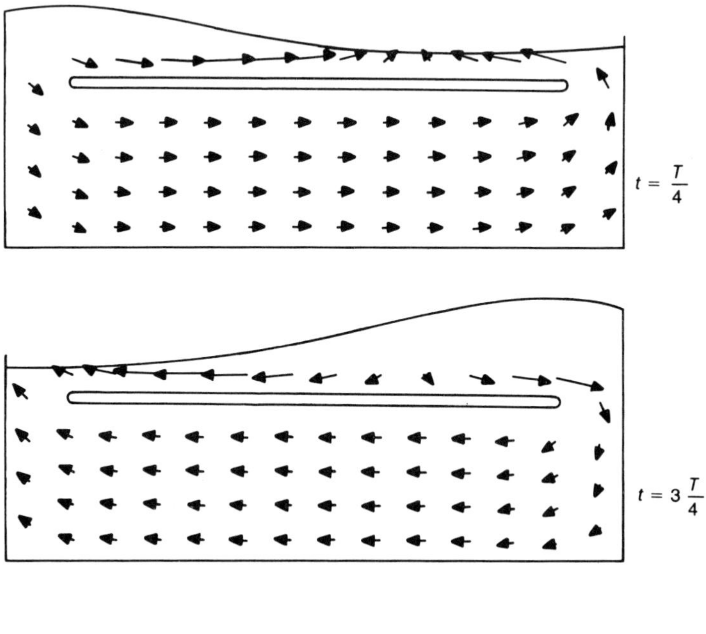

It can be seen that because of this difference in phase and over a given range of periods of time which is a function of the width of the slab and the depth at which it is submerged, it is possible to arrange that the wave transmitted by the oscillation of the mass of water located below the slab is equal and opposite to the wave which passes over it.

Calculations on a mathematical model show that it is possible to have an energy output equal to I over a relatively extensive period of time.

Figure 5 shows the theoretical performance of a 10-metres slab sunk in I metre of water.

Figure 5

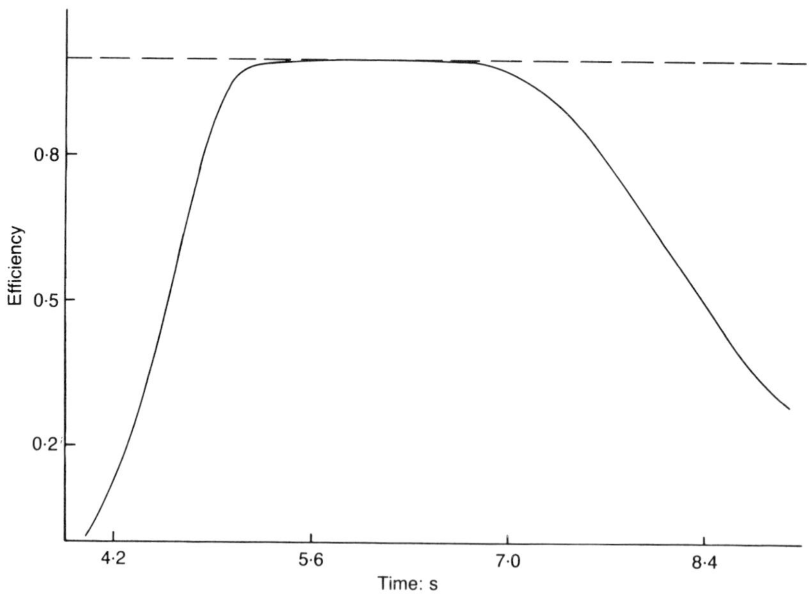

In natural surroudings, the phenomena observed are of course more complex. At the moment of opposition of the two waves mentioned above, harmonics appear and a slight agitation is set up behind the slab.

With two slabs of different widths and depths of immersion placed in series, one can obtain better performances over a longer range of time.

Thus, as an example, with a 16-metre slab sunk in 2 metres of water followed by an 8-metre slab sunk in I metre of water, the output measured in a swell channel exceeds 90% between 4 seconds and 7.5 seconds (Fig. 6).

Figure 6

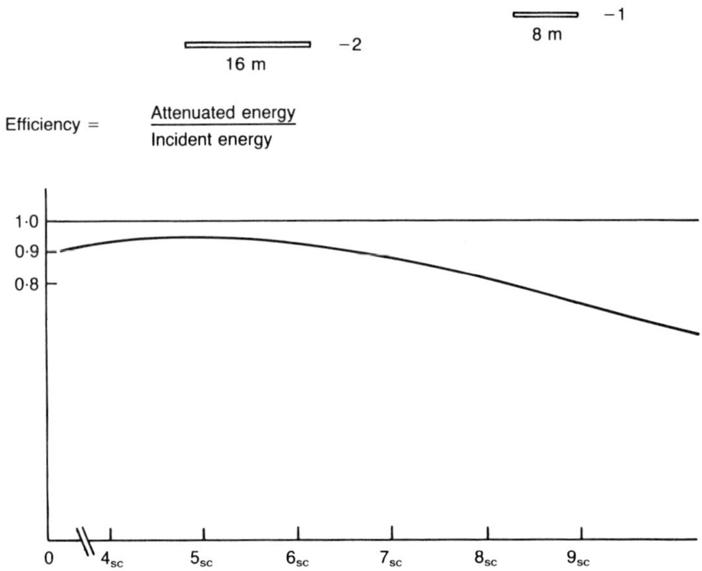

A first practical application of this type of wave attenuator which is the subject of a patent taken out jointly by the PRINCIPIA Company and the PRINCIPALITY OF MONACO has been put into operation in BARCELONA to calm the area of water at right angles to the outlet of a drain serving the land taken over for part of the Olympic Village.

The study and the development of this new system have recently been placed on the EuropeaOn research programme EUREKA.

3) BREAKWATER ("FIXED SEAWALL")

This consists of large pontoons with a flat base including an immersed part and mobilising the inertia of the mass of water located between the bottom of the pontoon and the sea bed.

It is known that when waves strike the vertical surface of a pontoon and are reflected from it, the amplitude of the lapping which is formed is twice as great as that of the waves off shore. It is these corresponding pressure variations which excite the water mass trapped under the pontoons (Fig. 7).

Figure 7

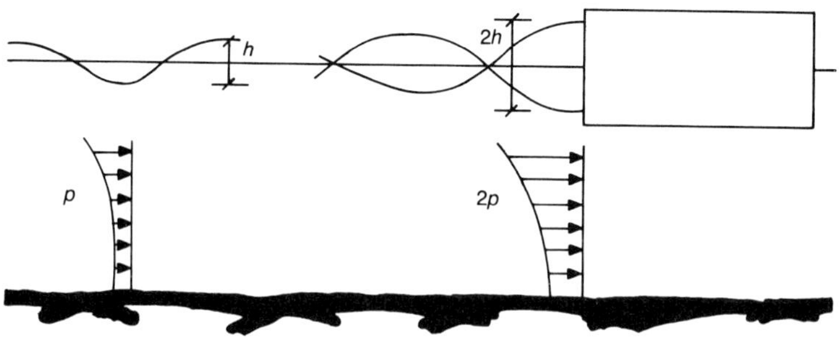

How can they be reduced ?

Quite simply, by extending the base of the pontoon in order to place the leading side of this mass of water no longer at right angles to the ventral segment but against of knot of laps.

Figure 8

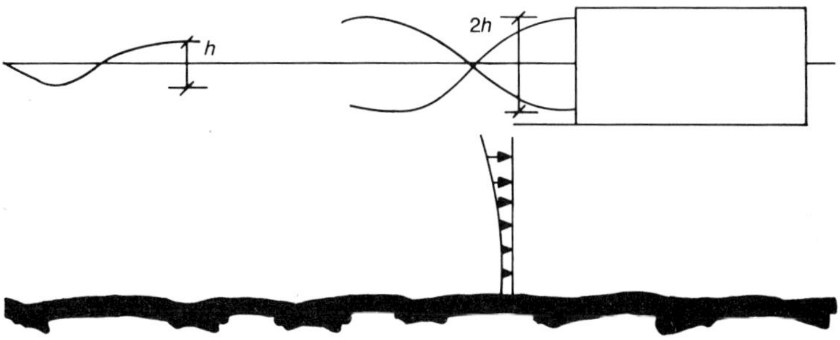

Over a long period we have studied with the CHALMERS TECHNICAL INSTITUTE of GOTEBORG the effect of the extension seawards, by means of a slab, of the base of the pontoons, using mathematical and hydraulic models.

The graph given here (Fig. 9) shows the effect of the width of the slab on the coefficient of transmission as a function of period of the incident waves. It confirms the wave-like nature of the phenomenon.

Figure 9

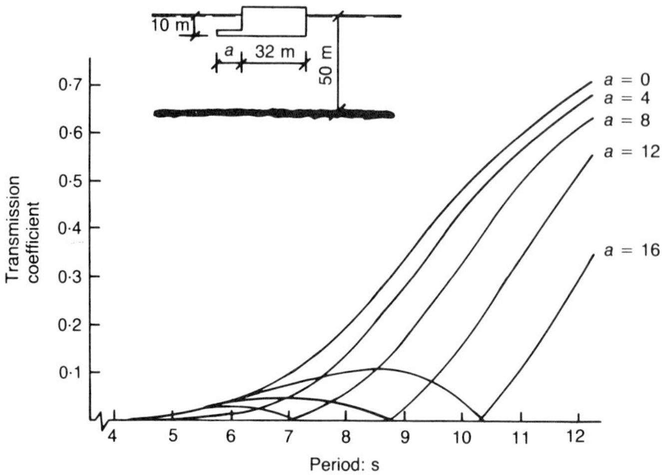

To take an example, with waves of 8 seconds and over, a pontoon 30 metres wide fitted with a 10-metres slab has a performance twice as great as that of a pontoon 40 metres wide (Fig. 10).

Figure 10

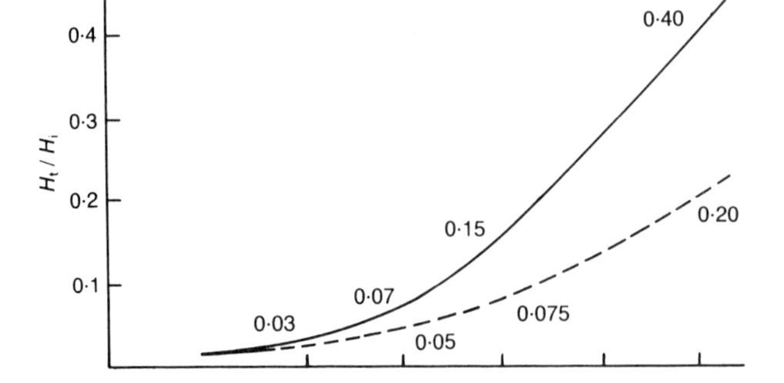

3.1 - Project of the new breakwater of LA CONDAMINE

Application of these results, which are the subject of a patent taken out by the PRINCIPALITY OF MONACO, has determined the dimensions of the future breakwater of the harbour of the Condamine (fig. 11).

Figure 11

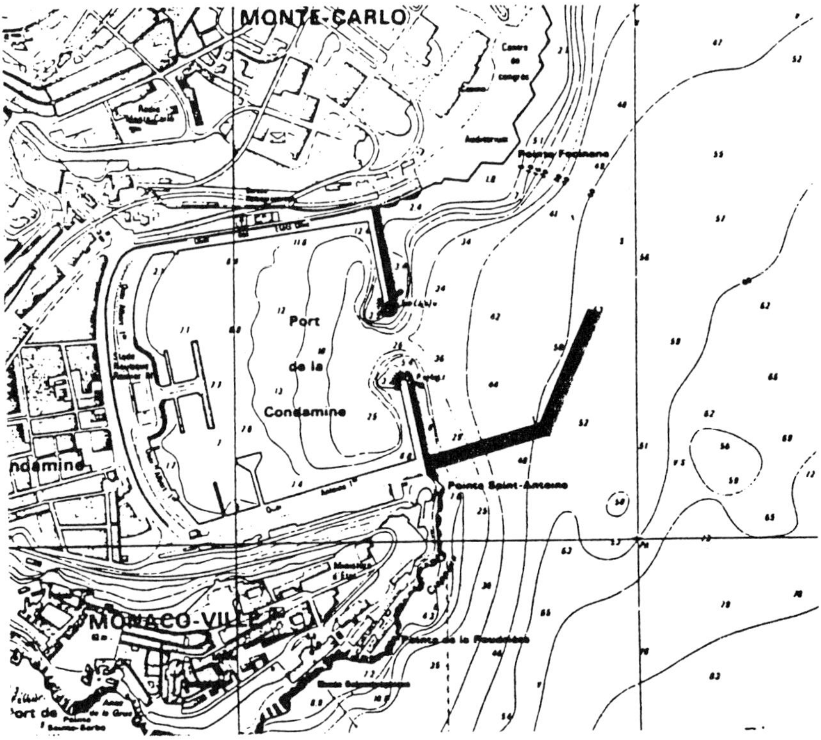

Facing the East, where the waves have a maximum significant period of 7 seconds and a maximum significant height of 3,5 metres the pontoons, the interiors of which will be equipped for use as car parks, will be fitted with slabs 8 metres wide (Fig. 12).

Figure 12

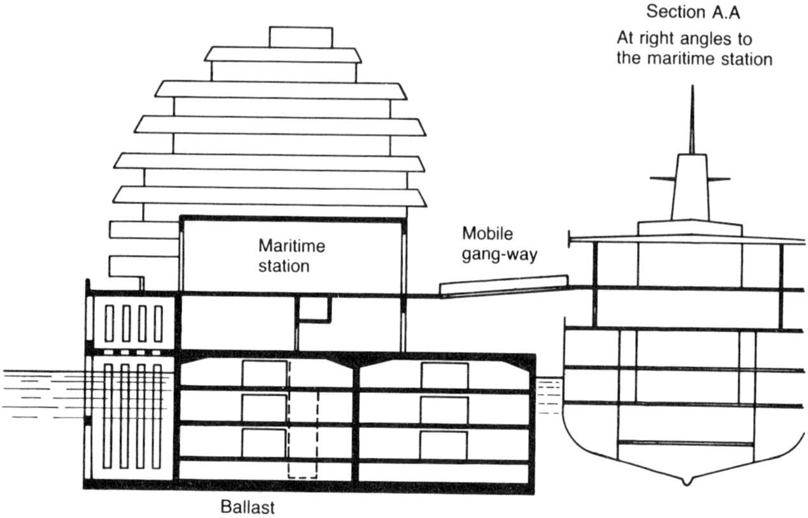

Facing South, where the waves have a maximum significant period of 10 seconds and a maximum significant height of 5 metres, the pontoons will be fitted with slabs 12 metres wide (Fig. 13).

Figure 13

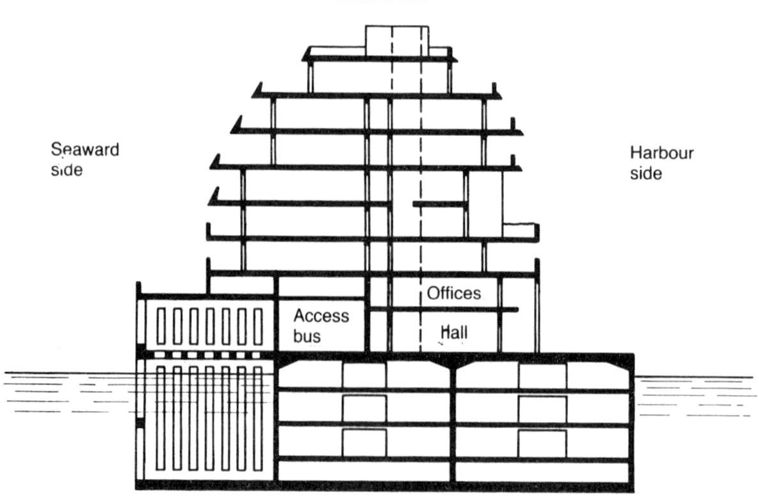

In addition to a maritime station, it is planned to build on the pontoons about 30,000 square metres of floor space to be used for shops, offices and housing.

3.2 - Project of urbanisation offshore Fontvieille 2

Due to the lack of space - we have no more available land left - an important project of urbanisation, offshore FONTVIEILLE quarter, has been recently presented to the Gouvernement.

It is concerned with the building of a breakwater 1600 meters long (fig. 14) in water 40 to 80 meters deep.

Figure 14

Sheltered by this breakwater, and on the artificial lagoon thus created, we plann to develop a program of about 200.000 square metres of floor space.

With the aim of protecting the marine environment which has always been the goal of the PRINCIPALITY OF MONACO. This development would be carried out without having to discharge filling material, but by mixing the techniques used by offshore petrol engineering, ship building and maritime works, all well known elsewhere.

The breakwater (figure 15) whose profile is specially designed to reduce both the forces acting on it and the forces acting on the mass of water located under it, would be constituted of pontoons, 60 metres long, 45m wide, 11 meters high, built in a dry-dock towed to the site and fixed on a jacket.

Figure 15

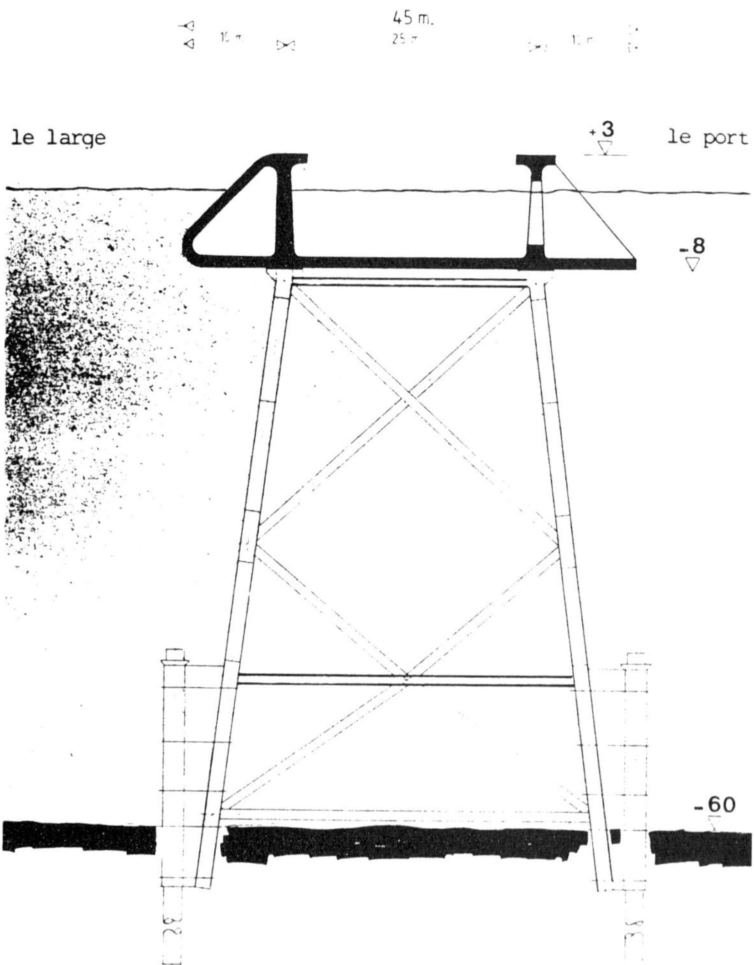

The buildings would too be constructed, first in a dry dock for the infrastructure (Figure 16) then along a quay for the superstructure (Figure 17).

Figure 16

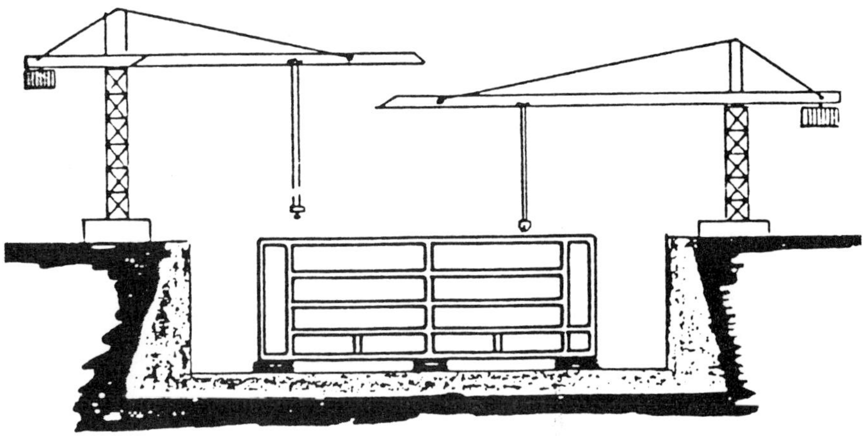

Figure 17

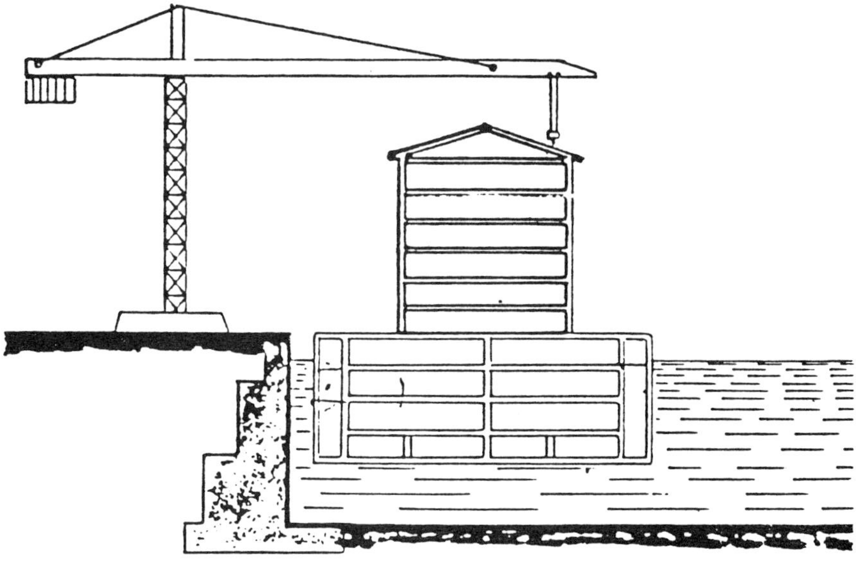

When completed, the whole unit containing car parks, fittings for the ground and buildings, i.e a quarter of a town, would be towed to MONACO, ballasted, and set on preinstalled piles (Figure 18).

Figure 18

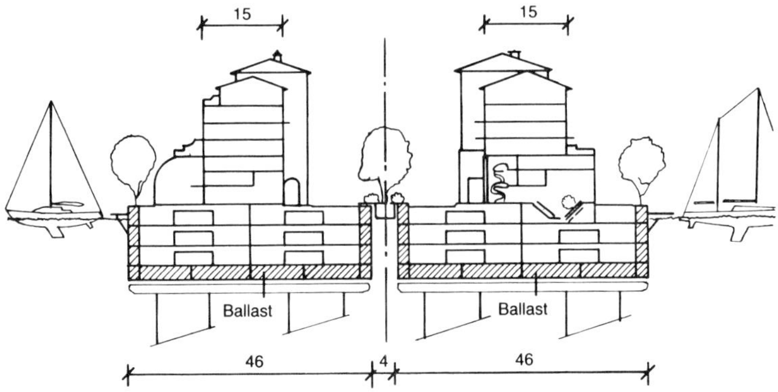

It may be noted in passing, that by adjusting the ballast, one may, to a large extent, make oneself master of the loads supported by the foundation and that the resistence to earthquakes of this type of building is very easily provided.

We think that, in this way, we shall tomorrow be able to reconcile the requirements of the PRINCIPALITY OF MONACO in the area of development, with those of the ecology and respect for the marine environment.

6. Design development of a tandem breakwater system for Hammond Indiana

J. C. COX, CH2M HILL, and G. R. CLARK, EWI Engineering Associates

SYNOPSIS. An innovative breakwater concept consisting of a conventional breakwater and a submerged-reef breakwater operating in tandem has been designed for the southern end of Lake Michigan. The tandem breakwater system was found to cost $1 million less to build than a single structure designed to meet the same operating criteria. Because of the depth-limiting behavior of the reef, the tandem design possesses a lower design risk for extreme events. This paper provides a summary of the design development and physical model testing results for the concept.

INTRODUCTION

1. There is high demand for protected moorage for recreational craft in the Chicago metropolitan area. However, there are few, if any, natural harborages. Consequently, to meet the demand, man-made small-craft harbors must be created along the exposed shoreline. Currently under construction in Hammond, Indiana, a neighboring community to Chicago, is the third-largest marina to be built in the United States. It is also the largest purpose-built small-craft harbor to be constructed at an exposed site in the country.

2. Hammond is situated at the southern tip of Lake Michigan. It has an exposed fetch of several hundred miles over deep water, so waves that develop on the lake and reach the site are essentially fully developed. In addition, the lake level can fluctuate 1 to 2 ft[1] annually, with the annual mean varying up to 4 ft on roughly 7-year cycles. Storm surges at the southern end of the lake can add 2 to 3 ft to the local water level.

3. The marina basin is situated in roughly 15 ft of water relative to a low water datum. Deepwater hindcast waves near the site range from 19.4 ft, T_p=10.3 sec, for the 20-year event to 23.6 ft, T_p=11.7 sec, for the 50-year event. Normal practice on the Great Lakes is to design for no damage in the 20-year event.

DESIGN DEVELOPMENT

4. To develop an appropriate breakwater design, a basin tranquility standard was first developed for the dockage and moored boats. The standard adopted followed the provisional

[1]To convert feet to meters, multiply feet by 0.305.

criteria developed by the Canadian Department of Fisheries and Oceans, which is presented in Table 1. In essence, the standard calls for waves less than 1 ft under normal conditions, but can allow up to 2-ft waves in extreme storm events.

Table 1. Provisionally Recommended Criteria for a "Good"[a] Wave Climate in Small-Craft Harbors

Direction and Peak Period of Design Harbor Wave	Wave Event Exceeded Once in 50 Years	Wave Event Exceeded Once a Year	Wave Event Exceeded Once Each Week
Head seas less than 2 seconds	These conditions not likely to occur during this event	Less than 1-foot wave height	Less than 1-foot wave height
Head seas between 2 and 6 seconds	Less than 2-foot wave height	Less than 1-foot wave height	Less than 0.5-foot wave height
Head seas greater than 6 seconds	Less than 2-foot wave height or 4-foot horizontal wave motion	Less than 1-foot wave height or 2-foot horizontal wave motion	Less than 0.5-foot wave height or 1.5-foot horizontal motion
Beam seas less than 2 seconds	The conditions not likely to occur during this event	Less than 1-foot wave height	Less than 1-foot wave height
Beam seas between 2 and 6 seconds	Less than 0.75-foot wave height or 2-foot horizontal motion	Less than 0.5-foot wave height or 1-foot horizontal motion	Less than 0.25-foot wave height or 0.75-foot horizontal motion

[a]For criteria for an "excellent" wave climate, multiply by 0.75, and for a moderate wave climate, multiply by 1.25.

5. With these standards established and the waves hindcast, basic parameters for a breakwater section could be defined. By examining the runup, overtopping, and transmission of a traditionally configured bi-layer breakwater, a section consisting of a 15-ft freeboard with 1:1.5 side slopes was defined. Required armor stone sizes ranged to 8 tons[2]. Such a section was found to be uneconomical due to cost and lack of availability of required material.

6. Alternative breakwater sections that would deliver similar performance were then explored. These alternatives fall into three generic types: graded stone (berm), parallel sheet pile wall, and gravity caisson structures.

---

[2]To convert tons (force) to neutons (N), multiply tons by 8,900.

All three of these options, when sized appropriately for the design conditions, proved more expensive than the conventional design.

7. To arrive at workable design, a value engineering technique was applied. The various breakwater components were examined in terms of the function they had to perform. Following the work of Danel (Ref. 1), it was recognized that failure of non-overtopped structures essentially occurs near the waterline, with the material slumping downslope forming an extended toe. It is interesting to note that this is in essence how a berm structure stabilizes, forming an "S"-shaped profile with a bench. Waves break on the toe and propagate across the bench before finally dying on the structure.

8. Danel recognized that the bench was really excess filler material and could be removed, with little impact on the structure performance, forming a wave stilling basin behind the toe. The result was a tandem breakwater system with an outer submerged reef and an inner main breakwater. Figure 1 explains the design evolution. The tandem breakwater concept was explored for the final design.

FINAL DESIGN

9. The tandem breakwater characteristics were developed to give a least-total-cost for the breakwater system. Working backwards from the main breakwater design gave parameters for the required reef performance.

10. A reasonable-cost main breakwater could be built if the freeboard could be reduced to 10 ft and the armor sized down to 3 tons with the same harbor tranquility requirements. This dictated that the transmitted wave from the reef could not exceed 9 ft in the design storm. The reef therefore had to be designed to break (depth limit) all waves in excess of this height.

11. Analytical methods do not exit to predict attenuation of waves by breaking over an abrupt submerged structure. Further, most theories on breaking consider only cases of incident wave heights that break at the traditional limit of 78 percent of water depth. Little information is available on breaking of waves whose incident height is significantly greater than this limit, as happens on a reef. Data offered by Aherns (Ref. 2), Tanaka (Ref. 3) and Horikawa and Kuo (Ref. 4) served to place initial bounds and performance expectations on the geometric description, spacing, and stability of the reef. However, physical model tests were ultimately used to develop design-quality data.

12. Since the performance of the tandem breakwater system is contingent on the performance of the reef, the primary modeling focus centered on reefs of various depths of submergence, crest widths, and reef porosities. The data were acquired experimentally in a laboratory wave flume, using stable, conoidal, incident wave forms. Time histories of the transforming breaking wave form were collected at various locations behind the reef.

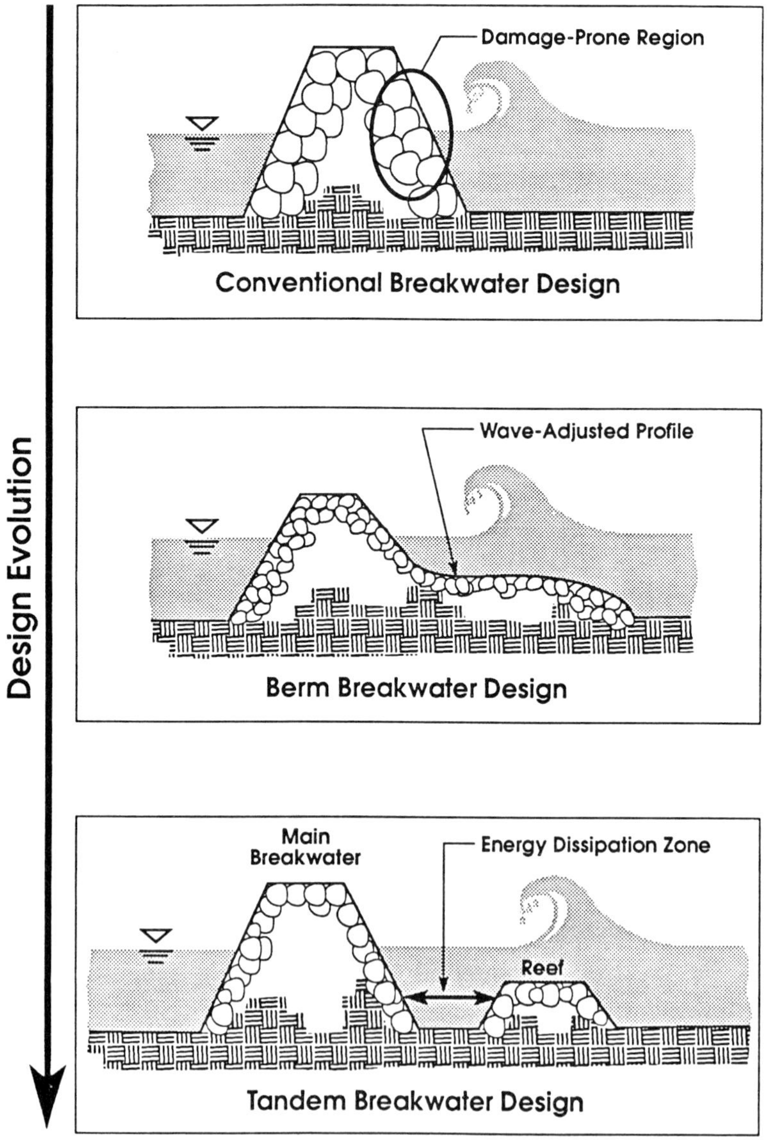

Figure 1
Tandem Breakwater Design Evolution

EFFECT OF RELATIVE REEF SUBMERGENCE

13. Various reef submergence depths were tested for wave heights beginning with incipient breaking across the reef and extending to wave heights up to six times the depth of reef submergence. The test revealed several fundamental principles about the wave transmission of a reef-type structure. Shown in Figure 2 is a plot of transmitted versus incident wave heights for three different reef submergence depths with a common cross section.

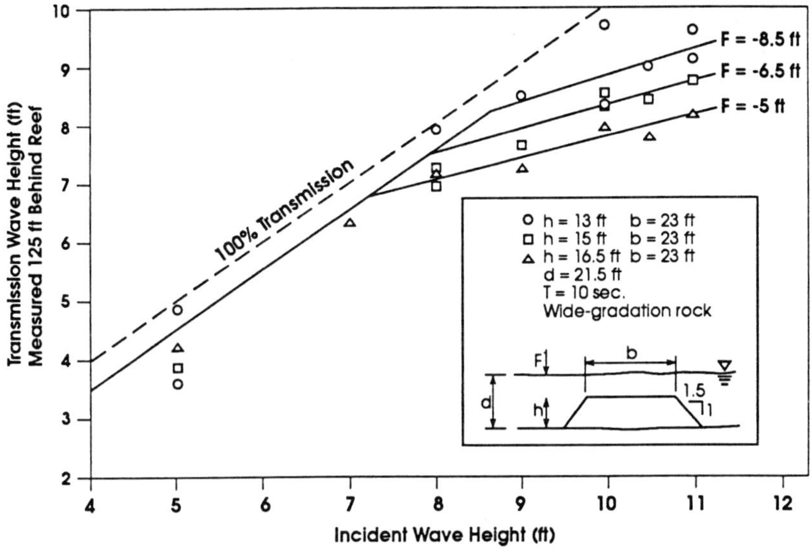

Figure 2
Effect of Reef Submergence on Wave Transmission

14. What is interesting to note is that, although breaking was first observed on the reef at wave-height-to-reef-submergence ratios consistent with standard breaking criteria (i.e., H_b-0.8 d_s), no significant attenuation of wave energy is observed until incident wave heights have gone far above the initial threshold for breaking. In fact it can be seen that, until the incident wave heights exceed the reef submergence depth, their transmission is adequately predicted by linear transmission theory in spite of breaking. In this range, because of the relative water depth over the reef, the wave is reacting in a more traditional shoaling and breaking process. For very shallow reef conditions, the wave response appears much different. The transmission does finally become depth limited and can be described by the empirical expression:

$$K_T = 0.8\left(-\frac{F}{H} + 0.5\right)^{1/2} \qquad (1)$$

where H = incident wave height
F = freeboard (negative for submerged)

EFFECT OF CREST WIDTH

15. The effect of the crest width of the reef was also explored. Tanaka (Ref. 5) suggested that increases in crest width at small ratios of crest width to wave length would cause a substantial reduction in wave transmission. As the crests become relatively wide, the rate of increased efficiency declined. However, the results of these tests, presented in Figure 3, reveal no such improvement, even though crest widths were trebled between the widest and the narrowest. Comparing Figures 2 and 3, it would clearly appear that the greatest efficiency for the same section volume is achieved by increasing height versus breadth.

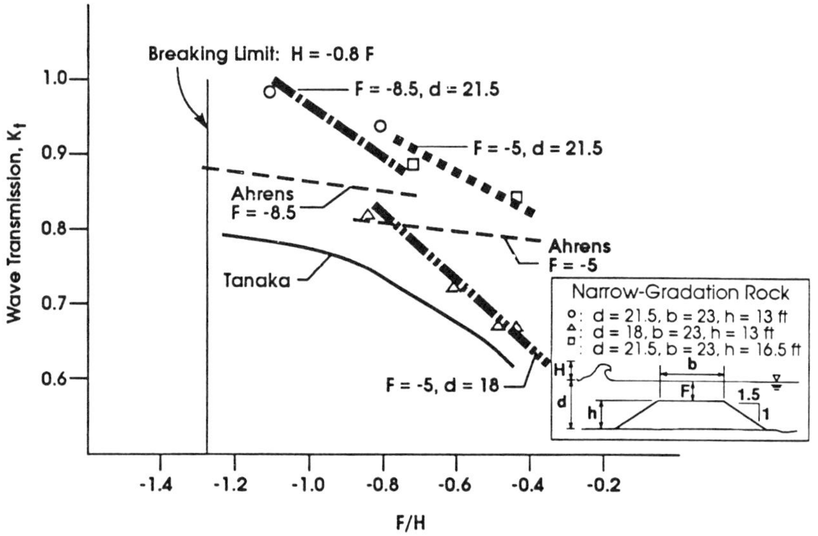

Figure 3
Comparison of Measured and Theoretical Reef Submergence Effects

EFFECT OF REEF POROSITY

16. Plotted in Figure 4 is a comparison of wave transmission past two model reefs constructed from stones of different size gradations, but with essentially the same mean stone diameter, d_{50}, as shown in Figure 5. The data suggest that porosity of the stone matrix, even for the deepest submergence, plays a significant role in the transmission efficiency. Decreasing the porosity by widening the stone gradation appears to lower transmission by 10 to 15 percent. By example, sand bars and coral reefs make the best attenuators because of their more solid nature.

COMPARISON OF DATA TO THEORIES

17. Plotted in Figure 6 is a comparison of the various existing transmission theories versus the data assembled

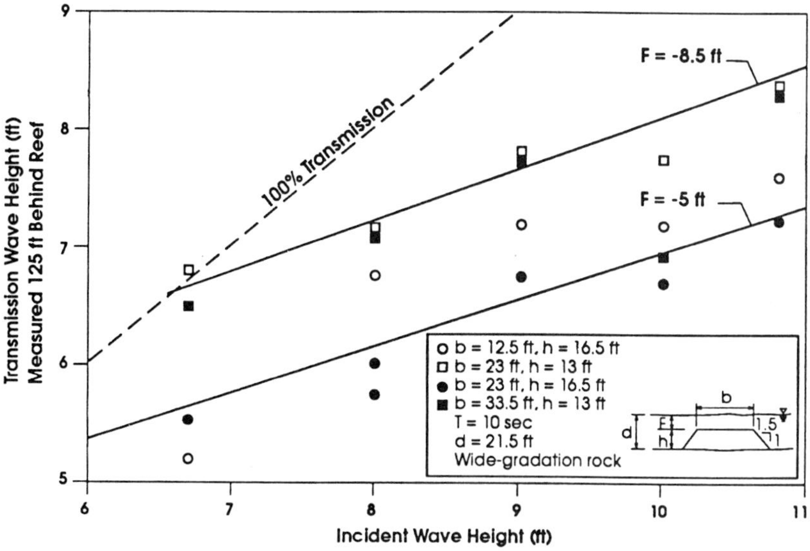

Figure 4
Effect of Crest Width on Wave Transmission

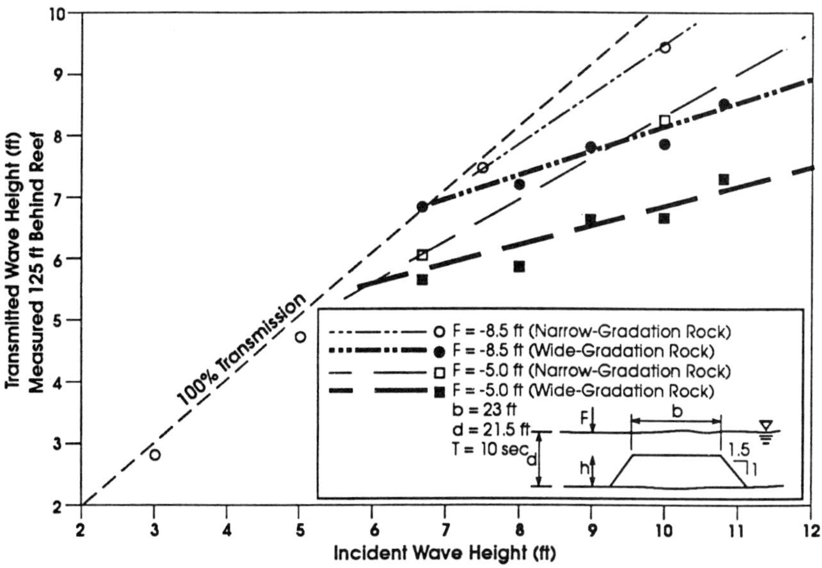

Figure 5
Effect of Reef Porosity on Wave Transmission

from these tests. Note the generally different rends in the data versus the theory. Tanaka's theory (Ref. 6) appears to underestimate transmission until the relative freeboard becomes small. However, the trend is correct. Ahrens' theory (Ref. 7) seems to bisect the data, making it more conservative at small freeboards, but, like Tanaka, still less conservative at deeper relative freeboards. Since the objective of this study generally was to create the minimum structure necessary to achieve the desired transmission, the theories need to be refined further to better describe behavior at deeper freeboards.

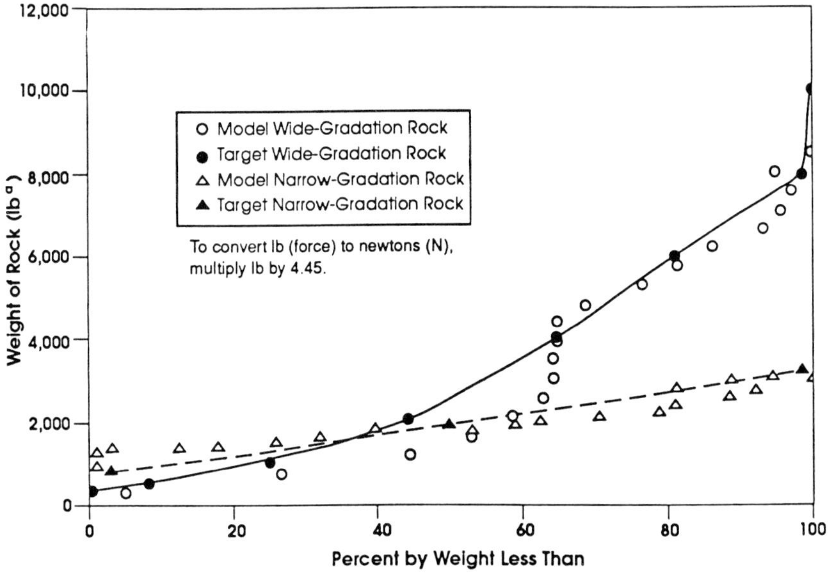

Figure 6
Definition of Model Reef Rock Gradations

BREAKWATER SPACING

18. Reef structures tend to have relatively narrow crests, typically being only a small percentage of the wavelength. The breaking process is therefore interrupted as the overtopping wave re-enters deeper water on the lee side. Battjes and Janssen (Ref. 8) in observing this phenomenon on natural bars, suggest that the attenuation process of breaking waves over bars may extend as much as 40 wave heights behind the initial breakpoint, reaching a stable transmitted height of approximately 50 percent on the incident wave height. This is generally consistent with the findings of Horikawa and Kuo (Ref. 9), who present an exponential rate of decay in wave height from the initial onset of breaking, which occurs at roughly 80 percent of the water depth over a shelf to an asymptotic stable height limit of roughly 35 percent of the water depth (50 percent reduction in wave height).

19. The data set presented by Horikawa and Kuo included only wave heights large enough to just break at the shelf edge. However, assuming the shape of the attenuation curve is the same for waves initially much higher than the break-point threshold, a non-dimensional attenuation curve versus break distance was developed. This curve is shown in Figure 7, and was used to guide the initial model test and design. Comparison of this curve with Figure 2 shows general agreement, with Figure 6 projecting slightly more attenuation. This would be expected since the wet stilling basin would be expected to be somewhat less efficient in attenuating the wave energy than a rock bench.

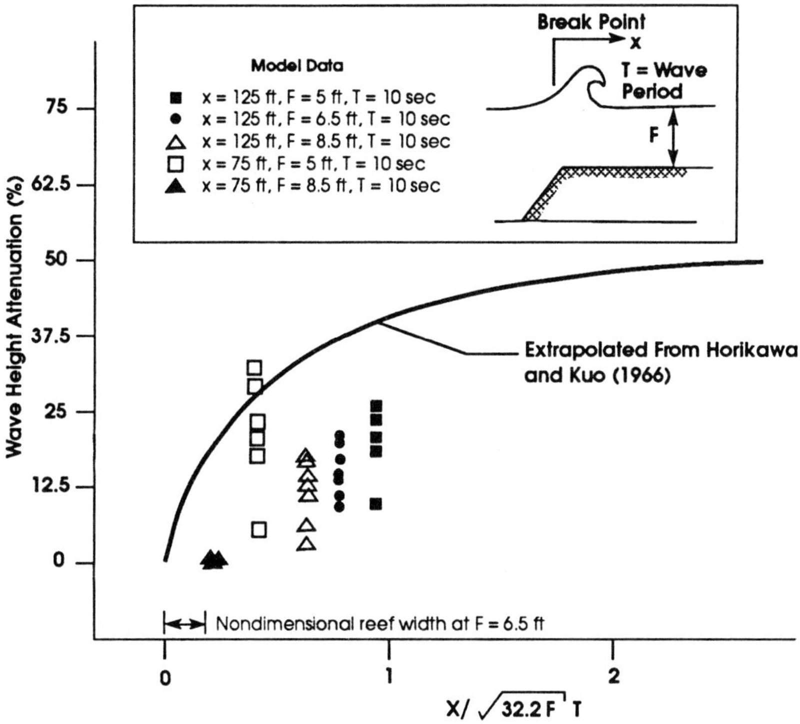

Figure 7
Nondimensional Wave Attenuation Distance

REEF STABILITY

20. Reef stability was predicted first using Ahrens' theory for graded stone structures (Ref. 10). The reef section was then physically tested for a broad range of submergences ranging from slightly emergent to deeply submerged. The tests indicated that the structure remained stable under all design waves and water levels. Further, Ahrens' stability theory was verified.

SUMMARY AND CONCLUSIONS

21. The final tandem breakwater design used for the Hammond breakwater is shown in Figure 8. For scale, the equivalently performing "conventional" breakwater is overlaid. The engineer's estimated cost difference between these two designs was nearly $1 million. The reef concept was readily accepted by environmental interests because it would create a new fish habitat area. The U.S. Coast Guard did not consider the reef as a navigation hazard if it has adequate signage. Fishing piers extending outward from the main breakwater and straddling the reef further divert boaters away from the reef and provide fishermen direct access to the reef.

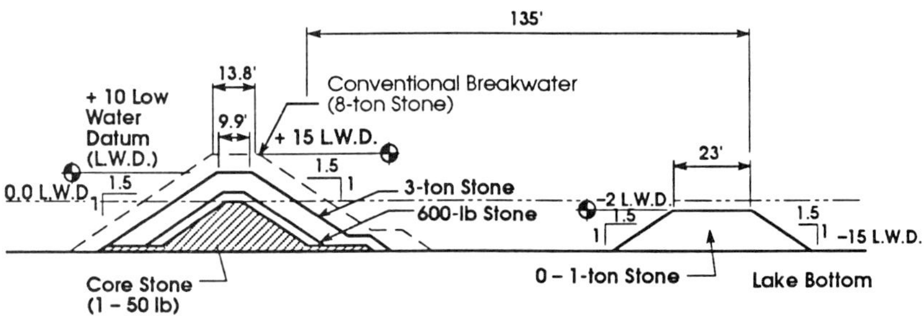

Figure 8
Tandem Breakwater Concept Schematic Cross Section

22. The tandem breakwater may not always provide the least-cost alternative, as section costs are strongly related to material costs and water depths. However, the tandem breakwater does offer significantly reduced design risk since high-return-period events are also reduced to a more manageable level, reducing the chance of structure failure in storm events outside the design envelope.

REFERENCES

1. DANEL P. Wave Theory Applied to Shore and Harbor Structures. Centennial Convention Paper #54. ASCE/ASME. 1952.

2. AHRENS J. Characteristics of Reef Breakwaters. USAE CERC TR 87-17. 1987.

3. TANAKA N. Effects of Submerged Rubble-Mound Breakwater on Wave Attenuation and Shoreline Stabilization. Proceedings of Japanese Coastal Engineering Conference. 1976.

4. HORIKAWA K and O. KUO. A Study of Wave Transformation Inside Surf Zone. Proceedings of Coastal Engineering Conference. 1966.

5. TANAKA N. Effects of Submerged Rubble-Mound Breakwater on Wave Attenuation and Shoreline Stabilization. Proceedings of Japanese Coastal Engineering Conference. 1976.

6. TANAKA N. Effects of Submerged Rubble-Mound Breakwater on Wave Attenuation and Shoreline Stabilization. Proceedings of Japanese Coastal Engineering Conference. 1976.

7. AHRENS J. Characteristics of Reef Breakwaters. USAE CERC TR 87-17. 1987.

8. BATTJES J. and J. JANSSEN. Energy Loss and Setup Due to Breaking of Random Waves. Proceedings of Coastal Engineering Conference. 1978.

9. HORIKAWA K and O. KUO. A Study of Wave Transformation Inside Surf Zone. Proceedings of Coastal Engineering Conference. 1966.

10. AHRENS J. Characteristics of Reef Breakwaters. USAE CERC TR 87-17. 1987.

7. Mathematical modelling of wave interaction with rubble mound breakwaters

K. R. HALL, Queen's University, Ontario, and
S. HETTIARACHCHI, Ove Arup & Partners

SYNOPSIS. The interaction of a wave with a rubblemound breakwater results in a complex flow field which is both non-linear and turbulent, particularly within a region close to the surface of the structure. Numerical models describing internal flow in a rubblemound breakwater are becoming increasingly important, particularly as the influence of scale effects on internal flow in physical hydraulic models are becoming understood as important. A number of numerical models to predict the internal breakwater flow kinematics have been produced in the past two decades. This paper provides a review of the state-of-the-art of numerical modelling of wave interaction with rubblemound breakwaters. Details of the theoretical development and the resulting numerical solution techniques are presented. Methods for incorporating secondary effects such as two phase (air-water) flow, inertia, and unbalanced boundary conditions are discussed. Limitations of the models resulting from the validity of the assumptions made in order to effect a numerical solution are discussed.

INTRODUCTION

1. When a water wave interacts with a porous rubblemound breakwater, transmission and reflection of the wave occur at the wave structure interface. Rubblemound breakwaters are generally composed of materials which are randomly shaped and randomly placed, making it impractical if not impossible to specify the local instantaneous velocity and pressure fields within the structure. However, an approximate description of the flow may be achieved by investigating representative pore averaged velocities obtained using procedures developed to analyse porous media flow.

2. The internal flow in a rockfill structure is unsteady, one or two phase non-Darcy flow and is described by non-linear partial differential equations with non-linear boundary conditions.

3. Two phase flow may occur as a result of external and internal wave breaking. The influence of this and other factors on the numerical solutions of fluid flow in a porous medium will be discussed later in this paper. A better understanding of the internal flow within the various layers of a porous rubblemound breakwater will assist in the estimation of breakwater armour layer stability, the effects of energy dissipation on wave run-up and run-down levels and wave transmission through breakwaters.

4. Existing models which describe the internal breakwater flow can be separated into several categories, based on the assumptions made, the method of describing the internal flow and the analytical solution techniques used:

(i) Internal flow is described by a potential flow which resists fluid motion according to Darcy's law (Biesel (1950), Le Mehaute (1957)).
(ii) Energy loss across the porous medium is approximated by using a model which assumes linear waves propagating past a representative field of cylinders (Goda and Ippen (1963), Straub and Herbich (1956), Keulegan (1968)).
(iii) Internal flow is described by a linearised Forchheimer resistance law based on Lorentz' principle which requires equivalent energy loss across the structure for linear and non-linear flow (Sollit and Cross (1972), Madsen and White (1975)).
(iv) Internal flow is described by a set of hyperbolic partial differential equations with the resistance described by the non-linear Forchheimer law (McCorquodale (1970), Nasser (1974), Hannoura (1978)).
(v) Internal flow is represented by flow through a vertical and horizontal pipe network (Mizurma (1985)).
(vi) Flow is described by the Navier Stokes equation and the resistance is obtained by considering the drag and inertial resistance of a rectangular block system (Austin and Schlueter (1982)).

Of the above methods, only types (iv) and (vi) solutions provide a description of the velocity and pressure fields within the porous structure.

5. Most of the models available are limited by the assumptions required to obtain an analytical solution. Typically, these assumptions include:

(i) Periodic long waves which are normally incident to the structure.
(ii) Waves do not break on the structure.
(iii) Two phase flow effects are negligible.
(iv) Complex structure geometry must be replaced with an 'equivalent' homogeneous rectangular structure.

(v) External energy dissipation (wave breaking) can be treated separately from internal energy dissipation (internal fluid resistance).

(vi) Internal effects can be ignored.

Specific details of the individual studies are provided in the following section.

SUMMARY OF PREVIOUS INVESTIGATIONS

6. The initial development in understanding wave interaction with a permeable rubblemound structure was directed towards understanding the behaviour of wave filters and wave absorbers (Biesel (1950), Le Mehaute (1957), Straub and Herbich (1956), Goda and Ippen (1963), Lean (1967) and Kondo (1970)).

7. Biesel (1950) coupled an external driving force represented by linear wave theory with an unsteady potential flow equation of motion for an infinitely long wave filter. The fluid resistance was governed by Darcy's law. Biesel was able to show that the wave decay in the direction of wave propagation followed an exponential law.

8. Le Mehaute (1957) developed an analytical solution similar to that of Biesel (1950) with the exception that the resistance term was assumed to be linearly proportional to the local velocity. The constant of proportionality in the theoretical resistance law was determined from the results of a series of model tests. Surprisingly, Le Mehaute observed that the reflection coefficients in his experimental mode were constant at a value of approximately 60%. This cannot be compared with the results of other investigators because of the lack of information available on the hydraulic properties of the granular media used in his studies.

9. Straub and Herbich (1956), Goda and Ippen (1963), Lean (1965) and Keulegan (1968) analysed wave damping (energy dissipation) on composite wave filters by analysing the flow resistance (described by linear wave theory) past a structure comprised of a series of vertical and horizontal cylinders.

10. Straub and Herbich (1956) proposed an artificial viscosity theory to account for wave damping in a wave filter. They coupled the Navier Stokes laminar regime dissipation function with a velocity field described by linear progressive wave theory to determine the energy dissipated within the porous filter. It must be noted that linear wave theory is not valid within the porous structure. The kinematic viscosity in the solution was replaced by an artificial viscosity. The value of the artificial viscosity was adjusted in order to provide agreement between the theoretical and experimental results. Wave reflection off the components of the wave filter was ignored. The method has very limited application primarily due to the difficulty in predicting the value of the artificial viscosity.

11. Goda and Ippen (1963) followed a similar semi-empirical approach analysing the change in energy flux across a wire mesh filter. The filter was represented by a series of horizontal and vertical cylinders. Spatial and temporal variation in velocity was based on linear progressive wave theory. Energy losses at each segment were assumed to occur as a result of drag and therefore the success of the scheme was contingent upon providing a suitable set of drag coefficients. The change in energy flux was found by the summation of the values of energy losses across each component screen. They found excellent agreement with the experimental results; however this should be expected since the drag coefficients which govern the energy loss at each screen were determined in the same experiments. The linear wave theory provided an adequate description of the velocity field in this type of structure having a high overall porosity, but will not adequately describe the flow in less porous structures.

12. Keulegan (1968) analysed wave damping in composite wave screens, based on the energy dissipation across the screen. He developed a differential equation expressing the rate of change of wave amplitude as a function of the energy lost to the resistive forces. Both linear and cnoidal wave theories were used to describe the amplitude and velocity functions in space and time. No account was taken of reflection off the front face of each segment of the screen. Experimental drag coefficients must be determined for each case.

13. One of the first investigations aimed at analysing wave interaction with a vertical faced rubblemound structure was undertaken by Kondo (1970). He developed an analytical approach to long wave interaction with homogeneous vertical faced structures by solving the one dimensional equation of motion for periodic, linearly damped free surface flow in a porous medium. Damping was based on the Lorentz principle which states that the realistic time-averaged turbulent flow resistance can be replaced by an equivalent time-averaged linear resistance. Kondo (1970) reported results similar to those of Biesel (1950) in which the decay of the wave amplitude in the direction of wave propagation was represented by an exponential function. In Kondo's model linear wave theory provided the external driving force which was coupled to the internal flow assuming continuity of pressure and velocity at the water-structure interfaces.

14. Sollitt and Cross (1972) followed a similar approach to that of Kondo. They used an analytical approach for determining the wave transmission through a breakwater. The internal flow was described by the unsteady equations of motion of flow in the pores of a coarse granular medium. The flow resistance was described by a linearised Forchheimer equation. Linearisation was achieved by employing the Lorentz principle. This yielded a linear potential flow problem. Linear wave theory was applied outside the breakwater. An eigen value solution was developed by ensuring the continuity of horizontal mass flux and pressure at the water-breakwater interfaces. Four unknown wave components were determined; external reflected (off the front face), externally transmitted (out of the back face), internally reflected (off the inside of the back face) and internally transmitted (from the inside of the front face). The amplitudes and phases of these wave components were solved from the boundary conditions.

15. The scheme can be applied only to vertical faced, homogeneous, rectangular breakwaters but has been adapted for use with sloping structures by developing an equivalent rectangular section based on submerged volume. The method does not include wave breaking and energy dissipation on the seaward slope of the breakwater. Because of the assumptions made in the solution, the method is not applicable if the wave amplitude and resulting wave orbital displacements are of the same order of magnitude as the individual rock diameters, since the waves would then be interacting with individual stones as opposed to a continuous porous slope.

16. Madsen (1974) provided a solution for linear periodic waves normally incident on a porous structure of rectangular cross section. In order to obtain an analytical solution he made the following assumptions:

(i) Homogeneous, rectangular structure.
(ii) Incident waves normal to the structure were defined by linear wave theory.
(iii) Flow resistance in the porous structure was linearly proportional to the flow velocity, i.e. is a Darcy type flow.

However, realising it was more appropriate to describe the flow using the Forchheimer equation, Madsen provided a linearised form as follows:

$$F = a + b\,|\bar{q}| \qquad (1)$$

where
F = resistance term
a, b = coefficients
$\bar{q}$ = effective bulk velocity.

17. Madsen and White (1976) advanced the solution of Madsen (1974) for predicting reflection and transmission coefficients through a homogeneous, rectangular structure, incorporating energy dissipation on the front slope of the structure. Flow within the structure was described by a linearised Forchheimer equation which was obtained utilising Lorentz' principle of equivalent work. Energy dissipation on the outer slope was calculated using methods for predicting attenuation due to bottom friction. The remaining incident wave energy was partitioned between reflected, transmitted and internally dissipated components.

18. Using this method, reflection and transmission characteristics of a structure can be determined from a knowledge of the incident wave characteristics, breakwater geometry, and porous media characteristics. Some discrepancy between predicted and observed reflection characteristics was noted by Madsen and White (1976) but the authors stated that these differences may be due to experimental errors in determining reflection coefficients. In fact, non-linear waves propagating within the media may have caused the poor comparison. Accurate prediction of the transmission coefficients was obtained.

19. The methods presented Sollitt and Cross (1972) and Madsen (1974), are valid only for homogeneous rectangular breakwaters. Madsen and White (1976) provided a method of representing a multi-layered trapezoidal breakwater by a rectangular breakwater which was hydraulically equivalent. Equal internal energy dissipation was assumed for steady flow. These methods cannot be used to establish a description of the spatial and temporal variation in velocity and pressure in the original trapezoidal structure.

20. The solutions presented above were all completed using the assumption of monochromatic waves. However, these theories are only approximate for the shallow coastal zone where the energy of the incident wave is concentrated in the long wave area of the spectrum. For a more realistic representation of the interaction between the real sea state and a rubblemound breakwater, consideration of the random nature of the wave motion is required.

21. Massel and Mei (1977) presented an analytical theory for random waves passing a porous breakwater. They followed the same procedures as Madsen and White (1976) but replaced the quadratic damping term with a stochastically equivalent linearisation technique. Transmission and reflection coefficients were reported statistically in terms of the standard deviations of wave spectral properties.

22. Massel and Butowski (1980) investigated wave interaction with a rectangular porous breakwater under the assumption that the incident wave spectrum was arbitrary. They also assumed normally incident waves, constant depth, a homogeneous and rectangular breakwater and incompressible and inviscid flow. The quadratic damping term due to the internal frictional resistance was replaced with an equivalent linear term. The standard deviations of the horizontal and vertical wave velocity components were calculated assuming they are non-correlated random functions with zero mean values. The method uses a description of the friction factor developed by Arbhabhirama and Dinoy (1973).

23. Considerable effort was made to solve the hyperbolic partial differential equations of motion for fluid in a rockfill structure at the University of Windsor during the 1970's (McCorquodale (1970), Nasser (1974), Hannoura (1978) and Hannoura and McCorquodale (1985 a,b)). These developments occurred in various stages and involved solution techniques utilising finite element methods, finite difference methods and hybrid methods (combined finite element and finite difference).

24. McCorquodale and Nasser (1974) described two techniques used for solving wave interaction in rockfill structures. The first method used a finite element approach to solve the hyperbolic differential equations governing the non-Darcy type flow in a porous medium (McCorquodale (1970)). The solution was developed for the case where the inertial term is small compared to the influence of turbulence. McCorquodale (1970) showed that this assumption is valid for unsteady flow in rockfill up to sizes of approximately 40mm. Numerous other investigators, however, have shown that the inertial term can be quite significant, particular for larger material such as that encountered in the armour layer.

25. Nasser (1974) solved the same set of governing equations which describe wave motion in a rockfill embankment with an impervious core using the method of characteristics and a finite difference technique to advance the solution. Discretisation of the solution domain was controlled by the characteristic directions to satisfy stability and convergence requirements.

26. In both models, the motion of external non-linear, long shallow water waves was coupled with the internal unsteady, non-Darcy flow. The models predicted the internal velocity and pressure fields and provided a spatial and temporal description of the free surface profile within the structure. It was found that the finite element model provided a better approximation of the two dimensional velocity field. In addition, the finite element solution was more flexible in treating irregular boundaries. The finite difference solution incorporated larger macroscopic internal effects. In both models, incompressible fluid and media were assumed and air entrainment was neglected.

27. Hannoura (1978) developed a hybrid model to solve the continuity and momentum equations for internal flow. Time integration was treated by a finite difference scheme and the spatial integration was achieved using a finite element solution. By combining the methodologies of McCorquodale and Nasser, Hannoura minimized the limitations of their individual solutions. For example, the major inertial effects were included in the finite difference method and the effects of non-homogeneity, complex boundary conditions and spatial variation in velocity and pressure were treated by the finite element solution, which was used to update the finite difference solution.

28. A series of experimental investigations was undertaken by Hannoura and McCorquodale (1979) to verify the results of the numerical model and to investigate the inertial effects and the effect of air entrainment on the hydraulic conductivity of the porous medium. The results indicated the presence of large inertial effects in coarse porous media. The virtual mass of the material was found to increase for larger grain sizes. A range of grain sizes of 16 to 144 mm was used in the tests. Porosity ranged from 0.34 to 0.51 and the shape factor, defined as the surface area of a stone divided by the surface area of an equivalent sphere, varied from 1.31 to 1.92.

29. It was found that air entrained in the flow in the porous structure can drastically reduce the effective conductivity of the structure, thus reducing wave transmission and increasing reflection. By using values of hydraulic conductivity which allowed for air entrainment, better agreement between numerical and experimental flume studies was achieved.

30. An extensive model system developed by Abbot, Skovgaard and Petersen (1978) was used by Madsen and Warren (1984) to investigate reflection and transmission from vertical porous structures. This system was based on the time-dependent, vertically integrated Boussinesq equations of conservation of mass and momentum and is able to simulate unsteady two-dimensional flows in vertically homogeneous fluids. The use of these equations is of particular importance in the simulation of short waves. They account for, amongst other phenomena, the deviation from a hydrostatic pressure distribution due to vertical accelerations and make it possible to consider a wide range of water waves which are not restricted by linear assumptions. The equations also include porosity which makes it possible to simulate the reflection - transmission phenomena in porous breakwaters.

31. Koutitas (1982) presented a numerical model of the stability of rubble mound breakwaters. The model which used finite difference methods describes the flow on the sloping face and inside the structure. It consists of two parts, the first considers a quantitative description of wave run-up on a permeable slope and the second describes the unsteady two dimensional unconfined flow in the voids of a rubblemound section. The first phase was based on a model presented by Gopalakrishnan and Tung (1980) and the second phase was based on a model previously developed by Koutitas. The coupling of the two models for simultaneous application was achieved through a staggered numerical solution of the open channel and groundwater models at successive locations in the time domain. It should be noted that this study adopted Darcy's assumption with special attention being paid to the estimation of the permeability coefficient. This model was not supported by experimental investigations.

32. An alternative approach to solving the internal flow conditions in a breakwater was taken by Mizurma (1985). He assumed that the breakwater could be represented by a number of horizontal and vertical flow channels between a system of rectangular blocks. The unsteady flow in the structure was described using the equation of motion for flow in a pipe. The resulting hyperbolic partial differential equations were solved using the method of characteristics advanced using a finite difference scheme. Mizurma found that the horizontal component of wave motion in the incident wave train (described by linear theory) at the seaward edge of the structure was transformed to vertical motion within the first few layers of blocks. Consequently, the characteristics of the rubblemound in a vertical direction greatly influence the flow.

33. The major drawbacks of this method which severely limit its application are the requirements to model rectangular structures, estimate channel separations (or equivalent pipe diameters) and estimate friction factors. In addition, inertial effects are not considered and flow in a porous medium is quite different from flow in an equivalent network of pipes due to the variability and spatial distribution of pore geometry.

34. Austin and Schlueter (1982) presented a numerical model of wave-structure interaction which contained two distinct components; (1) Determination of the forces resulting from a specified wave field, and (ii) Determination of the resulting armour unit motion.

35. The wave forces were calculated using a finite difference code, called SOLAVOF (Nicholas and Hirt (1976)), which uses the Navier Stokes equation and a form of the continuity equation, which includes limited fluid compressibility, as the constitutive equations.

36. The method uses a stationary network of rectangular cells to divide the calculation region into a finite number of elements with which fluid variables are associated. The primary field variables are the velocity components and the pressure. The final velocity and pressure distribution in each cell is obtained iteratively. The method cannot be used to describe breaking wave motion since the free surface slope must be less than the slope of the mesh cell diagonal (Nicholas and Hirt (1976)).

37. The forces acting on an obstacle are found by integrating the pressure around it; therefore, the inertial effects are directly included in the calculation. The method is limited to simple shaped armour units such as cubes and rectangles. Furthermore, channels between each unit must be provided (no direct unit to unit contact) because of the two dimensional nature of the solution.

38. The second component of Austin and Schlueter's (1982) solution involved determining the individual armour unit response to the applied pressure forces. This response and the constraints on surrounding units were calculated using procedures developed to simulate jointed rock behaviour. A dynamically explicit scheme was used to solve the equations of motion allowing time response of units and interactions to be followed. Verification or calibration for this unit movement model is not available and therefore the accuracy of this solution cannot be assessed.

39. Packwood (1983) developed a numerical model to describe the influence of a porous bed on the run-up of a single bore on a gently sloping sandy beach. The effects of the bottom shear stress and a porous bottom boundary condition were included. The flows into the porous bed was considered to be Darcy type flow. Packwood found that the porous bed influences the water in motion in two ways:

(i) Increases in bed permeability increased the attenuation of the incoming waves.

(ii) The roughness of the beach slope face influence the flow of the fluid.

40. The methodology used and the assumptions made by Packwood limit the applicability of this model to relatively small particles. Considerable modification to Packwood's model would be required to enable it to adequately describe wave interaction on a breakwater slope.

41. Kobayashi and Jacobs (1985) presented an analytical method for evaluating the stability of rip-rap under wave action. The characteristics of the flow on the surface layer of the breakwater slope during wave downrush were determined using a mathematical model based on a standing wave solution of the finite amplitude shallow water equations with an implicit account of the affects of wave breaking. Subsequently, the forces on a unit were determined to predict the critical condition for the initiation of erosion of stones on the surface layer. The stability of the stones was determined from a simplistic treatment of the force equilibrium equation and included the effects of drag, inertia, uplift and gravity acting on a unit on the slope.

42. The mathematical treatment of the solution of the wave motion on the slope is applicable to non-breaking waves only; however, Kobayashi and Jacobs (1985) applied the analysis to the case of breaking waves by limiting the analysis to the downrush period. They assumed that this is the most critical period for determining the forces on a unit, which is not necessarily the case.

43. The results were compared with the experimental work of Ahrens (1975) and a fairly good agreement was achieved. However, the method has a number of limitations including:

(i) The effects of frictional energy dissipation on the breakwater slope are not included.

(ii) The structure is considered impermeable so that flow into the structure is not considered.

(iii) The effects of depth in front of the structure are not included. Many investigators have found that run-up and run-down are a function of depth.

(iv) Experimental results to directly validate the solution are not available.

44. Further details of previous investigations relating to flow through porous structures are discussed in Hall (1987), Hettiarachchi (1987) and Hannaura and Barends (1982).

SUMMARY OF THEORETICAL ANALYSES

45. In this section the constitutive equations used to describe the fluid motion in several of the analytical solutions described earlier are reviewed. Complete solution techniques are not given and the reader is referred to either the original publication or Hall (1987) for further details.

46. The most versatile model described is the type utilised by McCorquodale (1970), Nasser (1974), Hannoura (1978) and Hannoura and McCorquodale (1985). Although they all utilised different solution techniques, the same constitutive equations governing fluid motion were used.

47. The resistance term, F, adopted by Nasser for the non-Darcy flow is governed by the Forchheimer equation (equation (1)), and is reduced to a non-dimensional form by invoking the Dupuit-Forchheimer assumption of parallel flow.

48. The bulk velocity term, $\overline{q}$, comprised of both horizontal and vertical components, where,

$$\overline{q} = \left(u_n^2 + \left(\frac{v_{ns}}{2}\right)^2\right)^{1/2} \tag{2}$$

u_n = horizontal velocity component
v_{ns} = vertical component of the surface macroscopic velocity

The equation for $\overline{q}$ is weighted to account for the fact that the vertical velocity is maximum at the free surface and zero at the bottom.

49. Assuming that the fluid is incompressible, air entrainment is negligible and the granular skeleton is incompressible, then the non-Darcy flow is governed by the following hyperbolic system of differential equations (Nasser and McCorquodale (1974)).

$$\frac{\partial u_n}{\partial x}\left(\frac{h_o+\eta_n}{m}\right) + \frac{\partial \eta_n}{\partial x}\frac{u_n}{m} + \frac{\partial \eta_n}{\partial t} = 0 \qquad \text{(Continuity)} \tag{3}$$

$$\frac{\partial u_n}{\partial x}\frac{u_n}{m} + \frac{\partial u_n}{\partial t} + \frac{\partial \eta_n}{\partial t}gm = -gm\,Fu_n \qquad \text{(Momentum)} \tag{4}$$

$$du_n = \frac{\partial u_n}{\partial x}dx + \frac{\partial u_n}{\partial t}dt \tag{5}$$

$$d\eta_n = \frac{\partial \eta_n}{\partial x}dx + \frac{\partial \eta_n}{\partial t}dt \tag{6}$$

(Total differentials of dependent variables) (5), (6)

where
η_n = free surface perturbation
u_n = depth averaged horizontal velocity
h_o = mean water level
m = porosity
F = non-Darcy friction term

50. Sollitt and Cross (1972) used the incompressible equation of motion where:

$$\frac{\partial q}{\partial t} = \frac{-1}{\rho} \nabla (p + \gamma Z) + \textbf{resistance forces} \tag{7}$$

where
q = instantaneous velocity vector at any point
P = corresponding pressure
γ = fluid weight density
Z = vertical coordinate
t = time

and

$$\nabla . q = 0 \tag{8}$$

where
∇ = gradient operator

51. The term for convective acceleration was ignored because finite amplitude waves are dissipated quickly within coarse granular media. The resistive forces were evaluated by combining steady and unsteady stress relationships so that,

$$\frac{-1}{\rho} \nabla (p + \gamma_Z) = \frac{\nu}{K_p} mq \frac{C_f}{(K_p)^{1/2}} m^2 |q| q + \frac{1-m}{m} C_m \frac{\partial q}{\partial t} \tag{9}$$

where
ν = kinematic viscosity
m = porosity
K_p = intrinsic permeability
C_f = dimensionless turbulent resistance coefficient
C_m = virtual mass coefficient

52. The linear resistance term governs at low Reynolds number flow, while the quadratic term governs at high Reynolds number flow. Therefore equation (7) becomes:

$$\frac{\partial q}{\partial t} = \frac{-1}{\rho} \nabla (p + \gamma Z) - \frac{\nu m\, q}{K_p} - \frac{C_f m^2 q|q|}{(K_p)^{1/2}} - \frac{1-m}{m} Cm \frac{\partial q}{\partial t} \tag{10}$$

53. Sollitt and Cross linearised the resistance by allowing:

$$\frac{nmq}{K_p} + \frac{C_f m^2 \, q|q|}{(K_p)^{1/2}} \quad \text{to approach } f\sigma q \tag{11}$$

where
σ = angular frequency of periodic motion
f = dimensional friction or damping coefficient

54. The value of f was assessed using Lorentz' condition which stipulates that both the non-linear and linear friction terms must account for the same amount of energy dissipation in one wave cycle.
55. Substituting the linearised damping into the equation (10) gives:

$$S \frac{\partial q}{\partial t} = \frac{-1}{\rho} \nabla (p + \gamma Z) - f\sigma q \tag{12}$$

where
S = 1 + Cm (1 - m)

56. Madsen and White presented the following equations governing the emotion of fluid within the porous structure:

$$m \frac{\partial \eta}{\partial t} + h_o \frac{\partial U}{\partial x} = 0 \tag{13}$$

$$S \frac{\partial U}{\partial t} + g \frac{\partial \eta}{\partial x} + \frac{f\omega}{m} U = 0 \tag{14}$$

U = horizontal discharge velocity
ω = wave frequency in radians

and

$$\frac{f\omega}{m} = a + b\,|U| \tag{15}$$

57. Austin and Schlueter (1982) used the Navier Stokes (eq. 16) and continuity (eq. 17) equations as the constitutive equations.

$$\frac{\partial u}{\partial t} + u\frac{\partial u}{\partial x} + v\frac{\partial u}{\partial y} = \frac{-1}{\rho}\frac{\partial p}{dx} + gx + \nu\left[\frac{\partial^2 u}{\partial x^2} + \frac{\partial^2 u}{\partial y^2}\right] \tag{16}$$

$$\frac{1}{\rho c^2}\frac{\partial p}{\partial t} + \frac{\partial u}{\partial x} + \frac{\partial v}{\partial y} = 0 \tag{17}$$

58. A concept known as the fraction volume of fluids (VOF) was introduced to describe the existence and position of the free surface within a finite difference cell, where:

$$\frac{\partial F_o}{\partial t} + u\frac{\partial F_o}{\partial x} + v\frac{\partial F_o}{\partial y} = 0 \qquad (18)$$

and

$F_0 = 0$ in an empty cell
$F_0 = 1$ in a full cell
$0 < F_0 < 1$ if a free surface exists in that cell.

59. Kobayashi and Jacobs (1985) used the finite amplitude shallow water wave equations to describe the flow on a breakwater slope. These equations are good approximations to inviscid flow when the horizontal length scale of the free surface and velocity variations are much greater than the water depth. The approximation is valid if both vorticity and vertical accelerations are negligible. The equations are:

$$\frac{\partial}{\partial t}(\eta + h) + \frac{\partial}{\partial x} u(\eta + h) = 0 \qquad (19)$$

$$\frac{\partial u}{\partial t} + u\frac{\partial u}{\partial x} + g\frac{\partial \eta}{\partial x} = 0 \qquad (20)$$

where
h = depth above slope (function of x)

60. Kondo (1976) used the following head loss equation which consisted of a laminar and a turbulent flow term:

$$\frac{\Delta h}{L} = \frac{v^2}{2gm^5 D}\left(\frac{C_2}{Dv\nu} + C_3\right) \qquad (21)$$

where
$\Delta h/L$ = head loss
C_2 = laminar flow coefficient
C_3 = turbulent flow coefficient
v = steady discharge velocity
D = representative diameter

For a breakwater, Kondo found that the ratio of $\left(\frac{\Delta h}{L}\right)_{turbulent}$ for the armour and core layers was approximately 1 to 15.

SOLUTION ANOMALIES

61. In order to provide a solution to the system of equations proposed by various investigators presented in the previous section, a number of physical assumptions were required. For example, some researchers were obliged to assume homogeneous rectangular cross-sections, that inertial effects could be neglected, that the effects of air entrainment were negligible, and the porous media skeleton was incompressible. Some of these factors have been found to make a significant contribution to controlling the motion of a fluid through a porous medium. A description of the methodologies used by various investigators to include these effects theoretically or empirically are given in the following sections.

CONSIDERATION OF INERTIAL EFFECTS

62. The results of an investigation by Hannoura and McCorquodale (1979) indicated the presence of inertial effects in coarse porous media. They undertook a series of experimental studies to evaluate the value of the inertial coefficient, C_m, of common construction materials for rubblemound breakwaters. Expanding the Forchheimer equation to include inertial effects results in:

$$i = (a + b\,|q|)q + \frac{1 + C}{g}\frac{\partial q}{\partial t} \qquad (22)$$

where

i = slope of the hydraulic grade line
$(a + b\,|q|)$ = steady flow resistance at velocity q
q = Darcy velocity
a, b = empirical Forchheimer coefficients of the porous medium
C = acceleration coefficient = $[(1 - m)/m]\, C_m$

Hannoura and McCorquodale represented the influence of inertia by the term:

$$\frac{1 + C}{g}\frac{\partial q}{\partial t} \qquad (23)$$

63. Estimation of the drag and virtual mass coefficients was made in a series of steady and unsteady flow experiments (Hannoura and McCorquodale (1985)).

EFFECT OF ENTRAINED AIR

64. Entrainment of air in the flow, which results as waves break onto the structure, may reduce the effective conductivity of the structure, and may affect wave transmission and reflection. Hannoura and McCorquodale (1970) conducted an experimental study to investigate these effects using granular materials ranging in size from 16 mm to 144 mm. Porosity ranged from 0.34 to 0.51.

65. From theoretical considerations, Hannoura and McCorquodale (1979) defined the relative conductivity in the unsaturated flow regime as:

$$\frac{K}{K_o} = \frac{q_w \, \Delta Z \, (a + b\,|q_w|)_o}{\Delta H_f} \qquad (24)$$

where

K = two phase conductivity (unsaturated)
K_o = saturated conductivity
q_w = Darcy velocity
$(a + b\,|q_w|)_o$ = saturated flow resistance
ΔH_f = two phase friction head loss through a distance ΔZ

66. Flow tests were made in a steady flow closed loop flume utilising a coarse-bubble air diffuser (Hannoura and McCorquodale (1978)). Both co-current and counter current flows were investigated. The influence of air entrainment on hydraulic conductivity is shown in Figure 1. The reduction in conductivity was found to be greater for counter current flow due to blockage of the flow by entrained air. They found that empirically:

$$K \sim K_o \left(1 - \frac{\%\ \text{air}}{100}\right)^3 \qquad (25)$$

67. The area of the unsaturated flow, A_{aw}, was found to be approximated by the expression:

$$A_{aw} = 7.46\, A_o \left(\frac{V_1}{V_{max}}\right)^{1/2} (1 - \sin\theta)^{1.15} \qquad (26)$$

where

A_o^2 = $A_1 / (2 \tan\theta)$
A_1 = run-up height
V_1 = run-up velocity
V_{max} = maximum seepage velocity through the rubblemound
θ = front slope of the breakwater

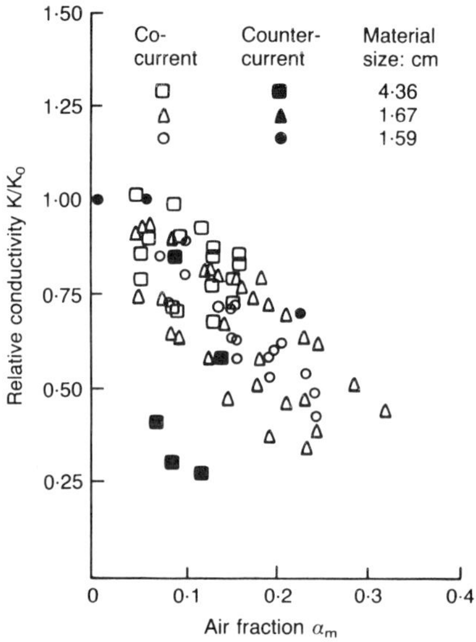

Figure 1 Influence of Air Content on Hydraulic Conductivity

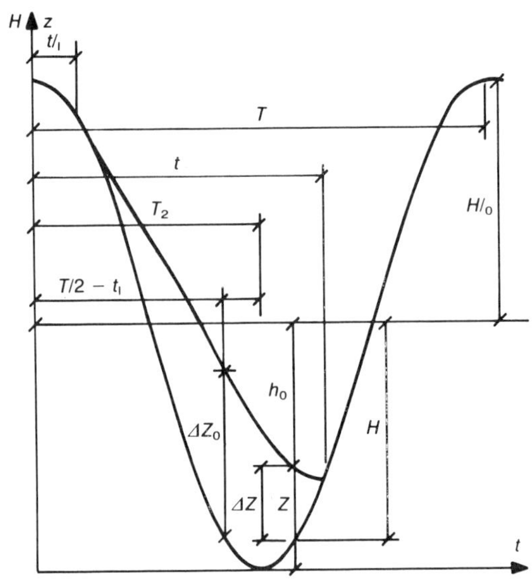

Figure 2 Definition of Terms for Calculation of Movement of the Outcrop Point

68. The unsaturated flow region was found to occur primarily between the mean water level and the front slope. It was concluded that the presence of two phase flow (unsaturated) had a significant influence on the hydraulic conductivity of the porous medium. Better agreement between numerical and experimental results was achieved when the effect of air entrainment was considered (Hannoura and McCorquodale (1979)).

MOVEMENT OF THE OUTCROP POINT

69. When water flows into a porous medium such as a breakwater, the phreatic line and the external free surface are usually adjacent to each other along the outer face of the breakwater. When water flows out from the porous medium a seepage face may be interposed between the external free surface under certain conditions. The movement of the outcrop point is required as a boundary condition for internal flow calculations. Dracos (1969) reported the results of an investigation of unsteady flow in a porous hydraulic structure in which movement of the outcrop point was determined as a function of time and maximum vertical drop rate.

70. The maximum vertical drop rate of the outcrop point, v_o, was:

$$v_o = \frac{-k}{m} \sin^2 \beta \qquad (27)$$

where

k = permeability coefficient
m = porosity
ß = angle the structures forms with the horizontal

71. Dracos defined the drop of the outcrop point as a fast or slow drop, depending upon the maximum vertical drop rate of the outcrop point and the ensuing vertical drop of the external free surface, v_v. If $v_v \leq \frac{-k}{m} \sin^2 \beta$ then the drop was slow. If $v_v \geq \frac{-k}{m} \sin^2 \beta$ then the drop was fast. The external vertical velocity, v_v, depended on the forcing function of the external water surface. Typically with wave-structure interaction, the fall of the outcrop point does not keep in phase with the drop of the free water surface. Figure 2 shows the typical lag found by Dracos.

72. In a rising phase or a slow drop phase, Dracos assumed that the movement of the outcrop point coincided with the movement of the external free surface at the interface. In the fast drop case, a seepage face develops and the movement of the outcrop point becomes more complex. However, Nasser and McCorquodale (1974) reported that results from analytical studies undertaken by Nasser (1974) and McCorquodale (1970) showed that a seepage face developed, even in the slow drop case. This was attributed to the fact that the movement of the phreatic line follows non-Darcy flow laws and that the phreatic line is not tangential to the rock face but meets it at an angle of about 45 degrees.

73. Nasser (1974) determined the motion of the outcrop point as follows:

For $0 < t < t^*$

where

t^* = time at which the internal and external movement starts to diverge

$$Y_c = Y_F = h_o + A_o \sin\left(\frac{2\pi}{T}\right) t \qquad (28)$$

where

Y_c = elevation of the outcrop point
Y_F = elevation of the free surface
A_o = wave amplitude
h_o = initial water depth (to the mean water level)
T = wave period

For $t^* < t < 3T/4 - t^1$ where $t^1 = t^* - T/4$

$$Y_c = \left(h_o + A_o \sin\left(\left(\frac{2\pi}{T}\right)\left(\frac{T}{4} + t^1\right)\right)\right) - V_* \left(t - \left(T/4 - t^1\right)\right) \qquad (29)$$

(The outcrop point lags behind the external wave and falls with a maximum internal velocity, V_*.)

For $3T/4 - t^1 < t < t_c$

$$\ddot{Y}_c + gm\left(a + mb\left|\dot{Y}_c\right|\right)\dot{Y}_c + \frac{g}{\Delta Y_p} Y_c = \frac{g}{\Delta Y_p}\left(h_o + A_o \sin \frac{2\pi t}{T}\right) \qquad (30)$$

where

$$\Delta Y_p = 2A_o \sin 2\pi \frac{t^*}{T} - V_* \left(T/2 - 2t^1\right)$$

(In this case the two curves gradually converge until they rejoin at $t = t_c$. The motion of the outcrop point is governed by a damped non-linear equation.)
74. The effect of the outer slope on the internal damping depends upon the horizontal velocity of the uprushing external wave, v_h, and the internal seepage rate, v_*. Nasser (1974) found that when $v_h < v_*$, the resistance of the flow increased in the vicinity of the interface due to air entrainment.
75. Nasser proposed classifying the waves running up a breakwater slope as fast or slow, depending on their uprush velocity relative to the internal seepage rate.
76. For a fast rising wave, Nasser (1979) proposed the following relationship to describe a modified non-Darcy friction term, F^1, where

$$F^1 = \left(\frac{v_h}{v_*}\right)^{0.5} F \tag{31}$$

and

$$v_h = \frac{\pi}{T} A_o \cot\theta \tag{32}$$

where

θ = angle of the embankment to horizonal

The modified resistance term, F^1, was applied to the first grid spacing adjacent to the interface in Nasser's model.

SUMMARY AND CONCLUSION
77. It is evident that numerical solutions to describe wave structure interaction do indeed have an important role to play in the design of rubblemound structures.
78. It is recognised that numerical modelling of internal breakwater flow kinematics is a very complex procedure, which has been approached from a variety of theoretical treatments. Several of the more recent modelling methods have been discussed and the relevant governing equations presented.
79. Methods for incorporating the effects of two phase flow and inertia were presented.

80. Many of the models presented are limited by the validity of the assumptions required in order to enable a numerical solution. This paper provides a summary of the major characteristics of each of the model types and indicates the applicability of the solution to internal flow in rubblemound breakwaters. Actual comparisons of the predictive accuracy of the models are not possible in most cases as each model type calculates different parameters which cannot be directly related. In addition, various models are placed in sub-groups and all members of the sub-group may not have exactly the same restrictions and limitations.

81. The models of McCorquodale, Hannoura and Nasser and that of Austin and Schleuter provide the most flexibility and allow the effects of inertia, two phase flow and deformation of the matrix to be included.

82. Several recommendations can be made for improving the predictive accuracy of these models. The major limitation of the McCorquodale type models results from utilisation of the Forchheimer equation. Equations of this type have been developed for slowly varying unidirectional unsteady flow. Clearly the flow within a breakwater is rapidly varying, cyclic flow. Hall (1987) has shown that this equation is not valid for most rubblemound breakwater configurations. The Forchheimer coefficients were shown to vary with flow regime and the model underestimated turbulent losses in the media. Modification of the Forchheimer equation or the development of an unsteady cyclic flow relationship would improve the predictive accuracy of the McCorquodale models. Additionally, large scale models and prototype verification of the model is required.

83. The numerical model of Austin and Schleuter should be advanced to include:

- Removing the restriction of total separation of units (no contact).
- To allow the use of additional shapes of media other than square blocks.
- To include three dimensional flow.
- To allow the use of multi-layered structures, more representative of rubblemound breakwaters.

REFERENCES

1. ABBOT, MB, PETERSON, HM and SKOVGAARD, O. "Numerical modelling of short waves in shallow water". Journal of Hydraulics Research, Volume 16, Number 3, 1978. pp 173-204.

2. AUSTIN, DI and SCHLEUTER, R. "A numerical model of wave breakwater interaction". 18th International Conference on Coastal Engineering, Cape Town 1982. pp 2079-2096.

3. BIESEL, F. "Equations de l'ecoutement non lent en milieunpermeable, application au calcul des filtres a houle". La Houille Blanch, Number 2, March 1950.

4. DRACOS, T. "Calculation of the movement of the Outcrop Point". 13th International Association for Hydraulic Research Congress, Volume 4, Japan 1969. pp 9-16.
5. GODA, Y and IPPEN, AT. "Theoretical and experimental investigations of wave energy dissipators composed of wire mesh screens". Massachusetts Institute of Technology, Hydrodynamics Laboratory, Report Number 60, 1963.
6. GOPALAKRISHNAN, T and TUNG, C. "Run-up of non-breaking waves. A finite element approach". Journal of Coastal Engineering, Volume 4, Number 1, 1980.
7. HALL, KR. "A study of the stability of rubblemound breakwaters". PhD Thesis, University of New South Wales, 1987. pp 600.
8. HANNOURA, AA. "Numerical and experimental modelling of unsteady flow in rockfill embankments". PhD Thesis, University of Windsor, Windsor, Canada, 1978.
9. HANNOURA, AA and McCORQUODALE, JA. "Air-water flow in coarse granular material". Journal of Hydraulics Division, American Society of Civil Engineers, Volume 104, Number HY7, July 1978. pp 1001-1010.
10. HANNOURA, AA and McCORQUODALE, JA. "Environmental wave forces on rubblemound breakwaters". Proceedings Civil Engineers in Oceans IV, San Francisco, August 1979. pp 367-379.
11. HANNOURRA, AA and BARENDS, F. "Non-Darcy flow: A state of the art. Flow and transport in porous media". Editors: Verruijt, A., Barrneds F, Balkema, AA, Rotterdam, The Netherlands, 1981.
12. HANNOURA, AA and McCORQUODALE, JA. "Rubblemounds: hydraulic conductivity equation". Journal of Waterway, Port, Coastal and Ocean Engineering, Volume III, Number 5, September 1985. pp 763-799.
13. HANNOURA, AA and McCORQUODALE, JA. "Rubblemounds: numerical modelling of wave motion". Journal of Waterway, Port, Coastal and Ocean Engineering, Volume III, Number 5, September 1985. pp 800-816.
14. HETTIARACHCHI, SSL. "The influence of geometry on the performance of breakwater armour units". PhD Thesis, Imperial College of Science and Technology, 1987. pp 480.
15. KOBAYASHI, No. and JACOBS, BK. "Riprap stability under wave action". Coastal Engineering, Volume III, Number 3, May 1985. pp 552-566.
16. KONDO, H. "An analytical approach to wave transmission through permeable structures". Coastal Engineering in Japan, Volume 13, 1970. pp 31-42.
17. KOUTITAS, C. "A numerical model for rubblemound breakwater stability". Proceedings Symposium on Engineering in Marine Environment, Brugge, 1982. pp 21-26.
18. KUELEGAN, GH. "Wave transmission through rock structures - hydraulic model investigation". US Army Waterways Experiment Station, Research Report H-73-1, June 1973.

19. LE MEHAUTE, B. "Permeabilitie des digues en enrochements aux ondes de gravite periodiques". University of Grenoble, 1957.
20. MADSEN, OS. "Wave transmission through porous structures". Journal of the Waterways, Harbours and Coastal Engineering Division, Volume 100, Number WW3, August 1974. p 167-188.
21. MADSEN, OS and WHITE, SM. "Reflection and transmission characteristics of porous rubblemound breakwaters". Technical Report 207, RM Parsons Laboratory, Massachusetts Institute of Technology, 1975.
22. MADSEN, OS and WHITE SM. "Energy dissipation of a rough slope". Journal of Waterways, Harbours and Coastal Engineering, Volume 102, WW1, February 1976. pp 31-48.
23. MADSEN, PA and WARREN, IR. "Performance of a numerical short wave model". Journal of Coastal Engineering, Volume 8, 1984. pp 73-93.
24. MASSEL, SR and MEI, CG. "Transmission of random wind waves through perforated and porous breakwaters". Coastal Engineering, Volume 1, Number 1, 1977. pp 63-78.
25. MASSEL, SR and BUTOWSKI, P. "Wind wave transmission through porous breakwater". Procedures 17th International Coastal Engineering Conference, Australia, 1980, Volume 1. pp 333-346.
26. McCORQUODALE, JA. "Finite element analysis of Non-Darcy flow". PhD thesis, University of Windsor, Windsor, Canada, 1970.
27. McCORQUODALE, JA and NASSER, SM. "Numerical methods for unsteady Non-Darcy flow". Proceedings International Symposium on Finite Element Methods in Flow Problems, University of Wales, United Kingdom, 7-11 January 1974. pp 545-551.
28. MIZURMA, K. "Flow analysis in idealised rubblemound breakwaters". Journal of the Waterways, Port, Coastal and Ocean Engineering Division, Volume 110, Number 3, August 1985. pp 344-355.
29. NASSER, MS. "Theoretical and experimental analysis of wave motion in rockfill structures". PhD Thesis, University of Windsor, Windsor, Canada, 1974.
30. NASSER, MS. "Experimental analysis of wave interaction with rockfill embankments with impervious cores". Proceedings Ports and Oceans Under Artic conditions, Norway, 1979, Volume II. pp 1351-1364.
31. NICHOLAS, BD and HIRT, CW. "Numerical calculation of wave forces on structures". Proceedings 15th International Conference on Coastal Engineering, Hawaii, 1976. pp 2254-2270.
32. PACKWOOD, AR. "The influence of beach porosity on wave uprush and backwash". Coastal Engineering, Volume 7, Number 1, February 1983. pp 29-40.

33. SOLIT, CK and CROSS, RH. "Wave reflection and transmission at permeable breakwaters". Massachusetts Institute of Technology, RM Parsons Laboratory Technical Report Number 147, March 1972.

34. STRAUB, LG and HERBICH, JB. "Experimental studies of wave filters and absorbers". Project Report Number 44, St Anthony Falls Hydraulic Laboratory, University of Minnesota, January 1956.

8. The influence of armour stone gradation on dynamically stable breakwaters

K. R. HALL and J. KAO, Queen's University, Ontario, and
M. MULCAHY, Department of Fisheries and Oceans, Canada

ABSTRACT

1. Dynamically stable or berm breakwaters are structures in which the armour layer is comprised of a wide range of stone sizes and undergoes reshaping in response to wave action. As a result of this wave action, a stable profile is achieved.

2. The effect of the gradation of armour stones and the amount of rounded stones in the armour on dynamically stable breakwaters was assessed in a two dimensional wave flume. A total of 52 series of tests were undertaken at the Coastal Engineering Research Laboratory of Queen's University, Kingston, Canada using irregular waves. Profiles of the structure during the various stages of reshaping were measured using a profiler developed for this study. Four gradations of armour stones were used giving a range in uniformity coefficient (D_{85}/D_{15}) of 1.35 to 5.4. The core of the breakwaters was relatively permeable.

3. The volume of stones and the initial berm width required for development of a stable profile along with the extent to which the toe of the structure progressed seaward were chosen as representative parameters of the reshaped breakwater. The results indicated that the toe width formed as a result of reshaping and the volume of stones required for reshaping were dependent on the gradation of the armour stone. The initial berm width required for reshaping was found to be dependent on both the gradation and the percentage of rounded stones in the armour.

Test Details

4. The studies were undertaken in a 1.4 m wide channel in a 2 m wide wave flume in which a rubble mound breakwater with a permeable core was subjected to irregular wave attack.

5. The rubble mound structure consisted of a core having a D_{50} of 1.2 cm and an armour layer having a D_{50} of 1.9 cm, where D_{50} is the diameter exceeded by 50% (by mass) of the gradation. Four different gradations of the armour stones were used with D_{85}/D_{15} ratio ranging from 1.35 to 5.4 where D_{85} is the diameter of stone, that is larger by mass, than 85% of the sample and D_{15} is the diameter of stone, that is larger by weight, than 15% of the sample. These gradations

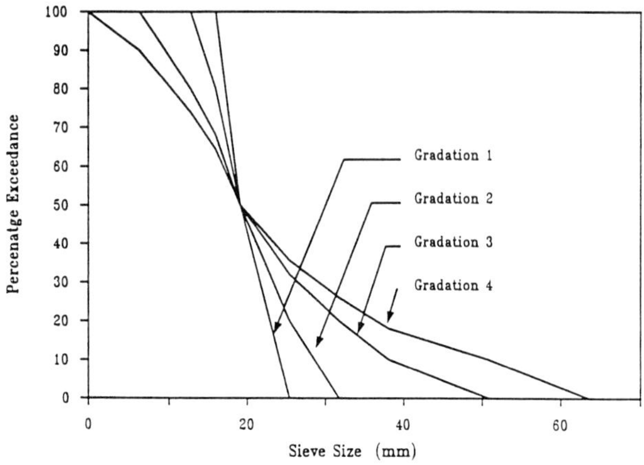

FIGURE 1 GRADATION CURVES

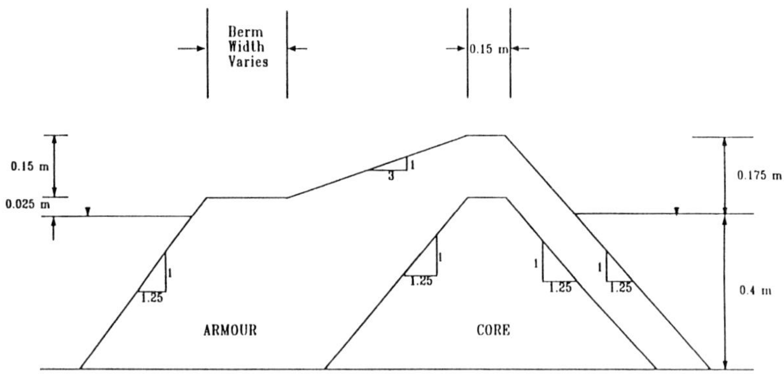

FIGURE 2 BASIC GEOMETRY OF THE TEST SECTION

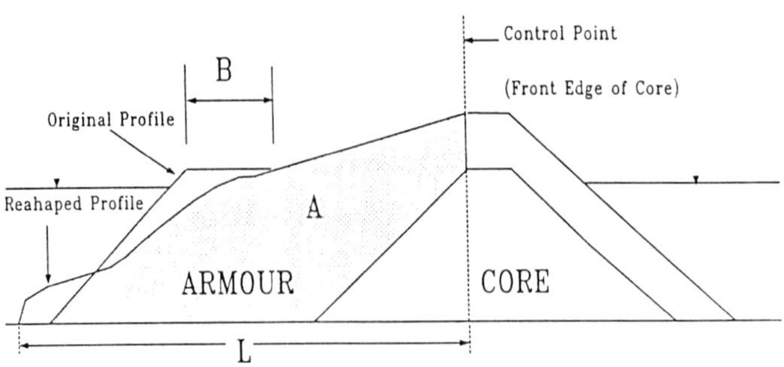

FIGURE 3 DEFINITION OF CHARACTERISTICS

are summarized in Table 1 and Figure 1. These gradations were selected based on a review of prototype quarry yields and reflect what may be realistically produced in a quarry. In particular, gradation 2 has characteristics of quarries used for many previous breakwater projects (Hall et al. (1983), Baird and Hall (1984), and Kao, (1990)). The in-place porosity of the armour ranged from 30 to 37% depending on the gradation. The specific gravity of the stones was 2.65.

6. Two other gradations of armour stones having a D_{50} of 1.9 cm and D_{85}/D_{15} ratio of 1.9 similar to gradation 2 but with different percentage of rounded stones, P_R, were also used in the tests, so that the influence of the degree of roundness was evaluated. These two gradations (2A and 2B) had 0 and 15 percent rounded stones respectively. The base gradations 1, 2, 3 and 4 contained 30% rounded stones.

7. Tests were undertaken on a dynamically stable breakwater whose initial configuration is shown in figure 2. The basic geometry of the test structure was selected based on previous experience and taking into consideration practical aspects which make construction simpler. Five capacitance type water level transducers were placed just in front of the breakwater to measure the dynamic water level fluctuations and enable determination of the extent of reflection off the breakwater. An additional water level transducer was placed immediately in front of the paddle to measure the deep water wave conditions.

8. A multi profiler system for automated, accurate profile acquisition, was mounted on the flume to measure the profile of the breakwater at predetermined intervals across the flume. The profiler basically consisted of a mechanical profiling arm connected to a potentiometer which was in turn linked to the computer via an amplifier and Analog to Digital converter. The horizontal position was determined using a photosensing device which could read the relative position along a carriage-way. A small circular disk (dia = D_{50} of the armour) was placed on the bottom of the profiling arm to eliminate profiling of "voids" in the armour.

9. The test structures were subjected to a design storm comprising of several segments of certain wave height - wave period combinations (significant wave height ranging from 6 to 14 cm and peak wave period ranging from 1 to 2 seconds). Each segment of each storm was run for a time equivalent to 3000 waves attacking the structure. Profiles were measured before the test and after each segment of test. For certain tests, several intermediate profiles were measured after 500, 1000, 1500 and 2000 waves. To assess the long term reshaping process, some tests were carried out for a duration of 36000 waves with profiles measured every 3000 waves. The dynamically stable breakwater was considered to have failed when erosion of the berm reached the toe of the 1:3 slope (see figure 2).

Parameters Studied

10. Three parameters describing various aspects of the berm breakwater were chosen as dependent parameters for the purpose of analysis. The aim of this study was not to evaluate profile formation, but to determine what initial conditions would lead to the development of a final stable profile, given initial design constraints. This approach provides useful information to the designer. The three parameters, illustrated in figure 3, are the bulk volume per unit length of armour stone on the front slope required for reshaping, (A), the width of berm required (B), and the distance from the control point to the outer extent of the toe following reshaping, (L).

11. In calculating the volume, A, only armour-stones on the front slope were considered; the volume of armour on the back slope was not included in the calculation since wave overtopping was not allowed during the tests, and thus there was no reshaping of the backslope. A good idea of the quantity of stones required at an early stage in the design process allows the designer and the contractor to determine the adequacy of potential quarries. As a result of compaction by waves, the volume, A, following reshaping was slightly smaller than the volume originally placed. From the present series of tests, compaction due to waves was found to be insignificant (less than 2%) thus the initial placed and subsequent reshaped volumes can be considered equal.

12. The toe width, L, is of importance in that it allows the designer to determine the extent to which the toe will reshape and depart from its original position. Suitable navigational buoys would need to be placed to mark the extent of the reshaped breakwater so that vessels would not become grounded. In the planning stage, such information is necessary to determine the area where vessels can be moored and where the fairway will be. It is also useful in assessing whether the reshaped breakwater will encroach onto adjacent property.

13. The width of berm eroded is important since it gives the designer an indication of the width of the berm required to ensure stability. Thus the designer need not provide a wider berm than necessary nor incur the undesirable consequence of providing an insufficient width of berm which would result in the crest of the structure being severely attacked by waves and potentially being damaged. In general, the width of berm eroded is related to the area following reshaping, and within construction error and possible variations of the angle of repose of the armour under water, the relationship is unique.

14. In subsequent analyses, A, B and L will be made dimensionless by dividing by D_{50} or D_{50}^2 as appropriate. Unless otherwise specified, the values used will be those corresponding to the attack of 3000 waves. It was found that at least 95% of all reshaping had occurred by this time in all tests. The parameters used in subsequent discussions

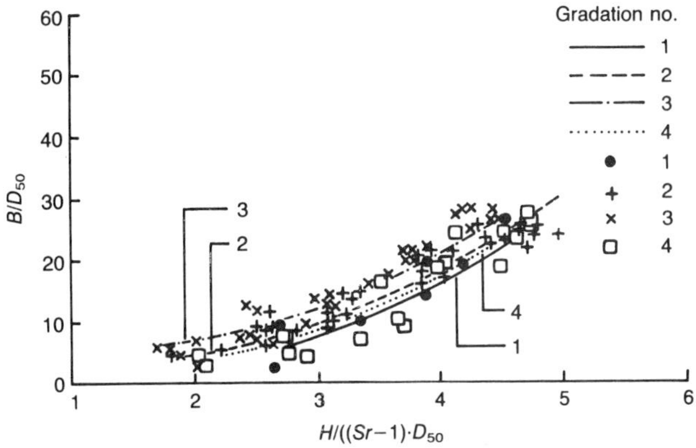

FIGURE 4 D_{50} VS. STABILITY NUMBER

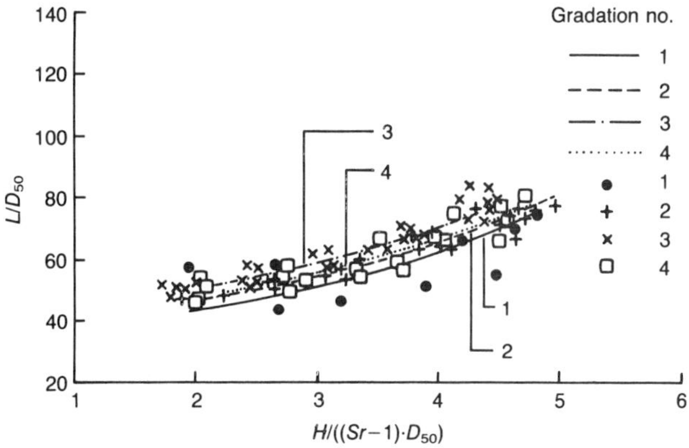

FIGURE 5 A / D_{50} VS. STABILITY NUMBER

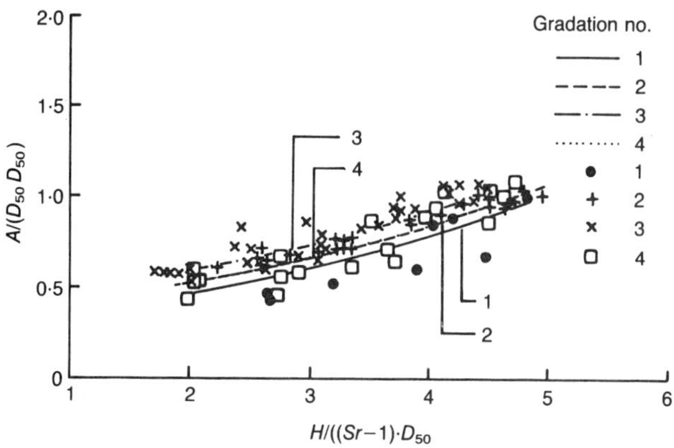

FIGURE 6 B / D_{50} VS. STABILITY NUMBER

will therefore be L/D_{50}, A/D_{50}^2 and B/D_{50}.

Empirical Model

14. A multiregression analysis was undertaken using minimization techniques in order to determine the relationship between the three dependent parameters and characteristics of the stone gradations including the D_{50}, the uniformity coefficient and the percentage of rounded stones (P_R).

15. The relationship between the dimensionless toe width and stone gradation was found to be

$$\frac{L}{D_{50}} = K_L + 11.05\frac{D_{85}}{D_{15}} - 1.47\left(\frac{D_{85}}{D_{15}}\right)^2 \qquad (1)$$

where K_L is a parameter dependent on the wave characteristics and is equal to $26.3 + 1.15\left(\frac{H_s}{(S_r-1)D_{50}}\right)^{2.2}$. Figure 4 shows the relationship between the experimental data and equation (1).

16. The model describing the volume of reshaping A/D_{50}^2 as a function of D_{85}/D_{15} was

$$\frac{A}{D_{50}^2} = K_A + 194.2\frac{D_{85}}{D_{15}} - 26.8\left(\frac{D_{85}}{D_{15}}\right)^2 \qquad (2)$$

where K_A is a parameter dependent on the wave characteristics and is equal to $148 + 30\left(\frac{H_s}{(S_r-1)D_{50}}\right)^{1.9}$. Figure 5 shows the relationship between the experimental data and equation (2).

17. The model to predict the value of B/D_{50} as a function of D_{85}/D_{15} , and p_R was

$$\frac{B}{D_{50}} = K_B + 7.52\frac{D_{85}}{D_{15}} - 1.07\left(\frac{D_{85}}{D_{15}}\right)^2 + 6.12\,P_R \qquad (3)$$

where K_B is a parameter dependent on the wave characteristics and is equal to $-10.4 + 0.51\left(\frac{H_s}{(S_r-1)D_{50}}\right)^{2.5}$. Figure 6 shows the relationship between the experimental data and equation (3).

CONCLUSIONS

18. The present series of tests indicate that for stones with the same D_{50} values both the relative gradation and the fraction of heavy stones in a gradation influence the reshaping of dynamically stable breakwaters. It appears that when the uniformity of the stones is small ($D_{85}/D_{15} < 3$), narrowly graded armour stones are less prone to reshaping. However, when the gradation of the armour stones is

sufficiently wide, the effect of the very large stones start to dominate over the effect of the voids inside the armour stone layer. The percentage of rounded stones was found to have an influence only on the width of berm eroded. The other dependent parameters, A and L, are dominated by the submerged section of the structure, where it appears stone mass not shape is dominant.

ACKNOWLEDGEMENTS
19. The authors would like to acknowledge the financial support provided by Small Craft Harbours Branch, Department of Fisheries and Oceans, Canada.

REFERENCES
1. BAIRD, W.F. and HALL, K.R., The Design of Breakwaters Using Quarried Stones. Proceedings 19th Int. Coastal Engineering Conference, pp. 1024-1036.
2. HALL, K.R., RAUW, C.I. and BAIRD, W.F. (1983). Development of a Wave Protection Scheme for a Proposed Offshore Runway Extension, Unalaska, Alaska. Proceedings - Coastal Structure 83, ASCE, pp. 157-170.
3. KAO, J.S.F. (1990). An Evaluation of the Factors Affecting the Stability of Dynamic Reshaping Breakwaters. Ph.D. dissertation, Queen's University.

Table 1
Relative Stone Grading

Gradation No.	D_{50} (mm)	$\frac{D_{85}}{D_{15}}$	P_r
1	19	1.35	30
2	19	1.9	30
2A	19	1.9	0
2B	19	1.9	15
3	19	3.0	30
4	19	5.4	30

10. A numerical simulation of some interaction aspects of a caisson vertical breakwater and its marine foundation

C. L. STERE, Bucharest Institute of Engineering

SYNOPSIS. The vertical wall breakwaters represent an engineer efficient solution for wave protective constructions of maritime harbours. However, when referring to the concrete caisson breakwater type and a sandy sea bottom, the structure-foundation dynamic interaction would generally bring in the most significant contribution to the structure particular behaviour and its stability. Some interaction aspects have been numerically studied, particularly related to the pore water pressure generation into the breakwater sandy foundation, possibly due to the cyclic action of the incident storm waves.

THE PHYSICAL DESCRIPTION

1. The real case-study was provided by the 20 m depth laid vertical wall caisson breakwater, belonging to a masterplan design alternative of the new Constantza-south harbour (Fig. 1), the construction of which started in 1980. The involved precast concrete caisson structure had a total height of 31.5 m and its submarine foundation consisted mainly of quite recent sea fine and medium-fine sand thick deposits. The initial design of the coarse sand ballasted caisson structure was carried out according to 50 years return period wave conditions and current prescriptions for static stability. Additional design measures had to be implemented through, in order to particularly incorporate some significant findings of subsequent studies upon the dynamic structure-foundation interaction and the structural safety (ref. 1,2,3).

2. Due especially to increased depths and sea environmental conditions, the caisson breakwater will be loaded - beside the gravitational loads and uplifts - by quite great horizontal forces, of a cyclic alternating nature, as given by the incident wave impacts with the breakwater structure. Following its quite stiff constructive features, this caisson type breakwater will likely undergo no overall structural deformations but mostly significant rigid swaying like displacements. They will further turn into cyclic pressing-depressing loadings onto the submarine foundation, which occur at its contact with the caisson basis and mostly under the caisson toe of the harbour side. Under such resulted cyclic alternating shear stress conditions, the loosely packed sandy foundation will be quite sensitive to the build-up

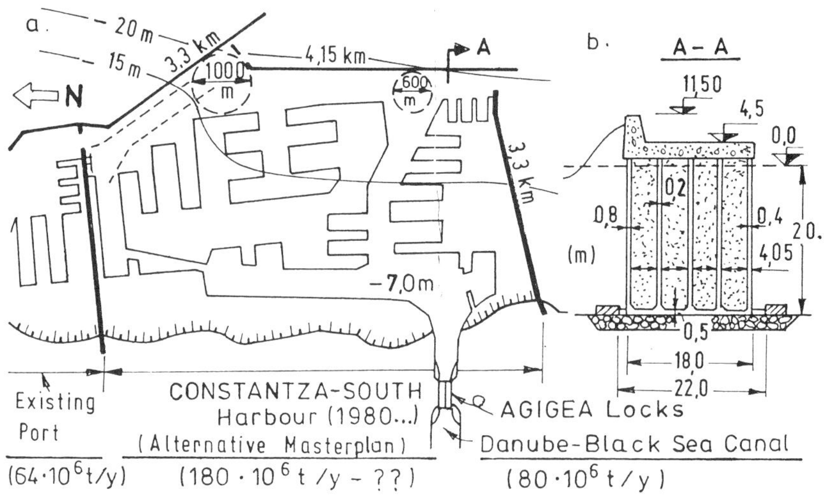

Fig. 1. The Romanian Constantza-south harbour: a- layout of a harbour alternative design; b- the caisson breakwater

of excess pore water pressure fields, high enough to bring in significant reductions in grain friction and hence a reduction of the sliding resistence of the sand granular structure(Fig.2)

3. The incident wave action upon the breakwater has been correlated to the still water level at its harbour side. For each wave crest and wave trough of a given regular or nonregular incident wave train,classical hydrodynamic pressure distributions and clapotis conditions have been considered,with refference to a Sainflou evaluated raise of the wave propagating medium level (Fig. 2-b).

NUMERICAL APPROACH

4. The consolidation phenomenon,which taking place within some deformable and saturated porous media,is mathematically governed by equations of the solis skeleton deformability (Navier), of pore water flow (Darcy) and of continuity,the latter expressing the physical connection between the pore water discharge and the in time subsoil skeleton voids ratio changes. As well known, the Biot's theory does include all the above mentioned aspects and, therefore, it seems to provide - in comparison to the Terzaghi's theory - a more reliable physical description of the nonsteady consolidation process and related possible excess pore water pressure generation in a sandy subsoil,due to cyclic wave loadings upon the breakwater and under the two-dimensional schematization.

5. Elements of the math model. The above mentioned analytical concept and the numerical program were worked out within the framework of the finite element approach (Fig. 1-b), by including the two-phase material of the foundation and linear elastic properties of the soil skeleton (ref. 1). The performed

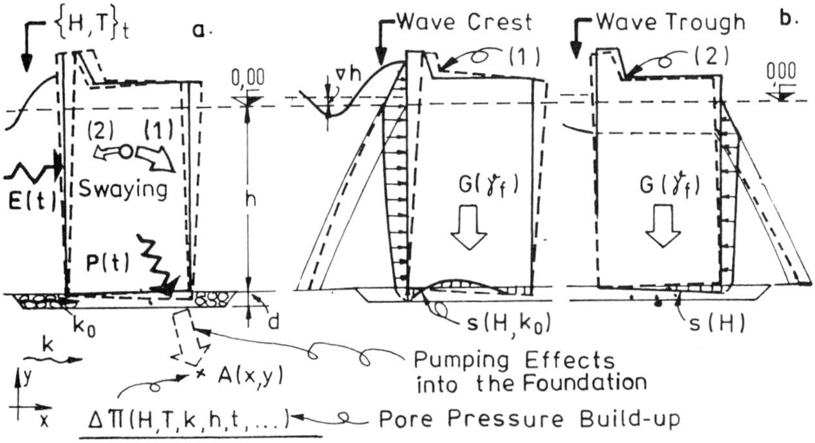

Fig. 2. Basic sketch about the caisson-foundation interaction: a- the mecanism of the pore pressure build-up; b- the considered two successive main caisson stages.

space discretization has accepted as time dependent main nodal unknowns the displacements $\{\delta(x,y)\}$ and the excess pore water pressures $\{\Pi(x,y)\}$.

6. As for the temporal discretization of any b(t) magnitudes, over a computing time step t, their convolution integral would be transformed into the new workable form of

$$[*b(t)]\Big|_{t_{n-1}}^{t_n} = \int_0^{\tau} b(\tau)\,d\tau = \Delta t[(1-\alpha)b(t_{n-1}) + \alpha b(t_n)] \qquad (1)$$

with the coefficient α=0.6667, as reccomended by Sandhu,R. for the parabolic type of the time interpolating scheme, actually selected (ref. 1).

7. The obtained system of linear algebraic equations had finally the following matrix form (ref. 1)

$$\begin{vmatrix} K & C \\ C^T & -\alpha\cdot\Delta t\bar{K} \end{vmatrix} \begin{Bmatrix} \delta \\ \Pi \end{Bmatrix}_{t_n} = \begin{vmatrix} 0 & 0 \\ C^T & (1-\alpha)\Delta t\bar{K} \end{vmatrix} \begin{Bmatrix} \delta \\ \Pi \end{Bmatrix}_{t_{n-1}} + \begin{Bmatrix} R \\ 0 \end{Bmatrix} \qquad (2)$$

whereat one may recognize the soil solid skeleton's stiffness matrix [K], the displacement-pore water pressure coupled matrix [C], the Darcy seepage matrix $[\bar{K}]$ and the external load vector {R}.

8. <u>The solution algorithm</u>. As the herein studied practical case did involve a loosely-packed fine/medium sand foundation with only possible excess pore water pressures $\{\Pi\}$, some specific provisions had to be incorporated with the solution algorithm, under the form of an unconditionally stable corrective iterative process. Basically, this means successive solvings - at the same time level (i) - of the system (2), provided its external load term $\{R\}_i$ is being amended through an error based redistribution technique, which involving actually some new nodal forces {r} of selected elements from the foundation (ref.1).

THE PORE WATER PRESSURE BUILD-UP

9. According to field wave measurements and their specific processings, proper wave conditions at the breakwater site were selected to run both the numerical model and a checking-up experimental model (ref. 1,2). The herein evidences of excess pore water pressure generation are related to an input of regular wave trains, which were mostly used for laboratory experiments.

10. The breakwater sandy foundation is particularly associated to a mollusk shell genesis, being averagelly characterized by elastic deformation moduli of 220 GPa, the Poisson ratio of .32, a slight cohesion of 400 MPa and internal friction angles of 28°. The measured permeability coefficients are in the range of 2.5·10 ...4·10 m/s. The discretization has covered a foundation zone extended laterally with 5xB and 3xB respectively, and in the bottom profoundness with about 8xh. These boundary limits were decided on the basis of a numerical sensitiveness study.

11. Isolines of excess pore water pressures, as the time response of the subsoil sands to different storm conditions, are presented in Fig. 3. One can notice, for seepage isotropy and anisotropy conditions as well, that the excess pore pressure build-up is particularly initiated - due to dominating moment loading - in the bottom zone situated under the caisson harbour toe. During the storm, due to the consolidation process which taking place in the sandy foundation and to the drainage effects of the much more permeable (k_0) breakwater quarry-run stone bedlayer as well, this pore pressure field will migrate around, with its peak value core systematically situated in the bottom profoundness at about (0.8...1.2)xh and an apparent shifting tendency towards the harbour direction. In principle, it seems the seepage anisotropy will give rise to lower pore water pressures, as particularly depending on the anisotropy ratio (n) & direction (θ) and the level of permeabilities.

12. Some influences are brought in by the dimensions (d,l) of the caisson artificial bedlayer (Fig. 4), which plays a significant role in the foundation consolidation process. Thus, the increase of the layer thickness (d) will encourage a diminution of the pore pressure field core peak values and, concomitantly, a slight increase of pore pressure values in some field peripherical points. As for the bedlayer extension (l) at the caisson backside, it seems to promote - as it being larger - general reductions of pore water pressures and a significant hastenning of the consolidation process of the subsoil. This is rather important, as a larger layer extension (l) will not raise so many technological constraints as a layer greater thickness (d), the dredging of which could not be driven in the sandy subsoil beyond some depth practical limits.

PARTIAL STABILITY AND SAFETY

13. As well known, the excess pore water pressure generation in the sandy grain skeleton will possibly lead to a temporary reduction in the bearing capacity of the soil, the actual extent of it depending on the wave induced loads pattern, the subsoil geotechnical structure and the consolidation conditions

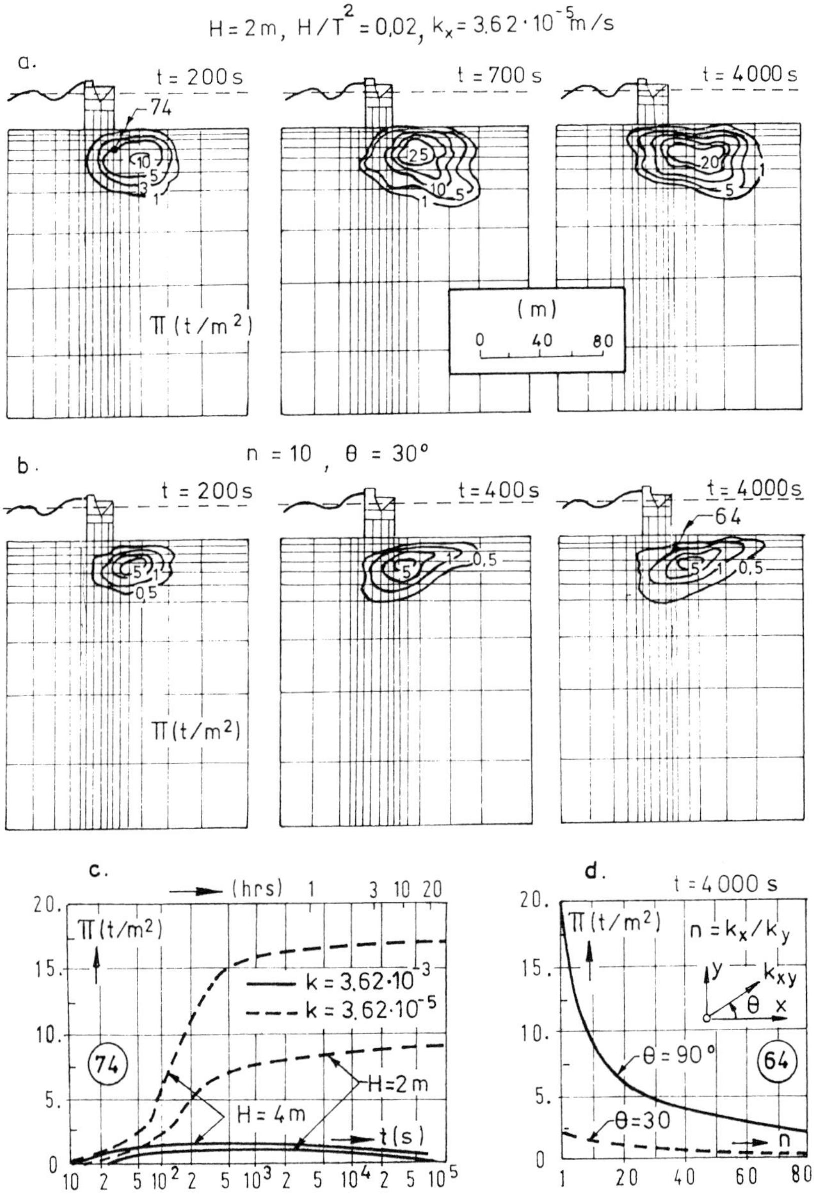

Fig. 3. Pore pressure fields in the breakwater sandy foundation: a- seepage isotropy; b- seepage anisotropy; c- average pore pressure histories for seepage isotropy conditions; d- influences of seepage anisotropy parameters

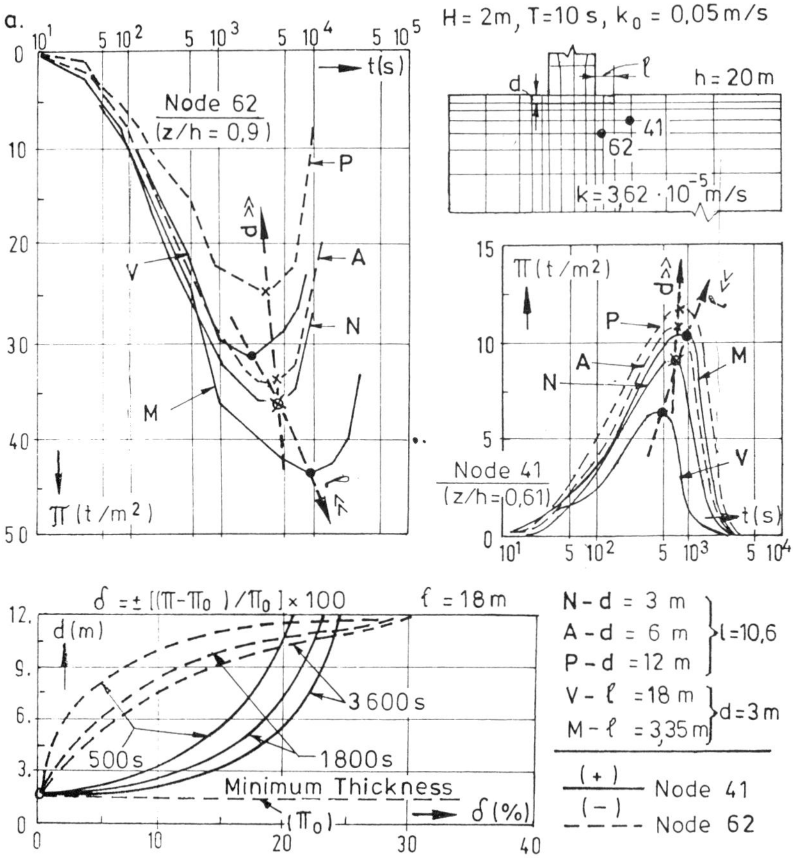

Fig. 4. The influence of rockfill bedlayer dimensions upon the pore pressure build-up: a- time evolution; b- the percentage influence of bedlayer thickness.

14. Under the circumstances, the particular failure mechanism of shallow circular surface sliding could be encouraged. The actual design criteria should properly include those cycling wave loading effects upon this kind of soil stability and the structure corresponding margin of safety (ref. 2,4). Thus, in comparison to a classical static approach (F_0) of the circle sliding stability, the taking into account of the dynamically generated excess pore water pressures will lead to actual safety ratios (F), the storm time histories of which do particularly bring up reductions of 25%...60% (Fig. 5). For quite slow consolidation conditions, the most affected circular sliding potential surfaces are likely situated in the foundation subsoil - in relation to storm intensities - at depths of z/h= 0.6...1.3, with an upward movement tendency during the storm duration (Fig. 5-c).

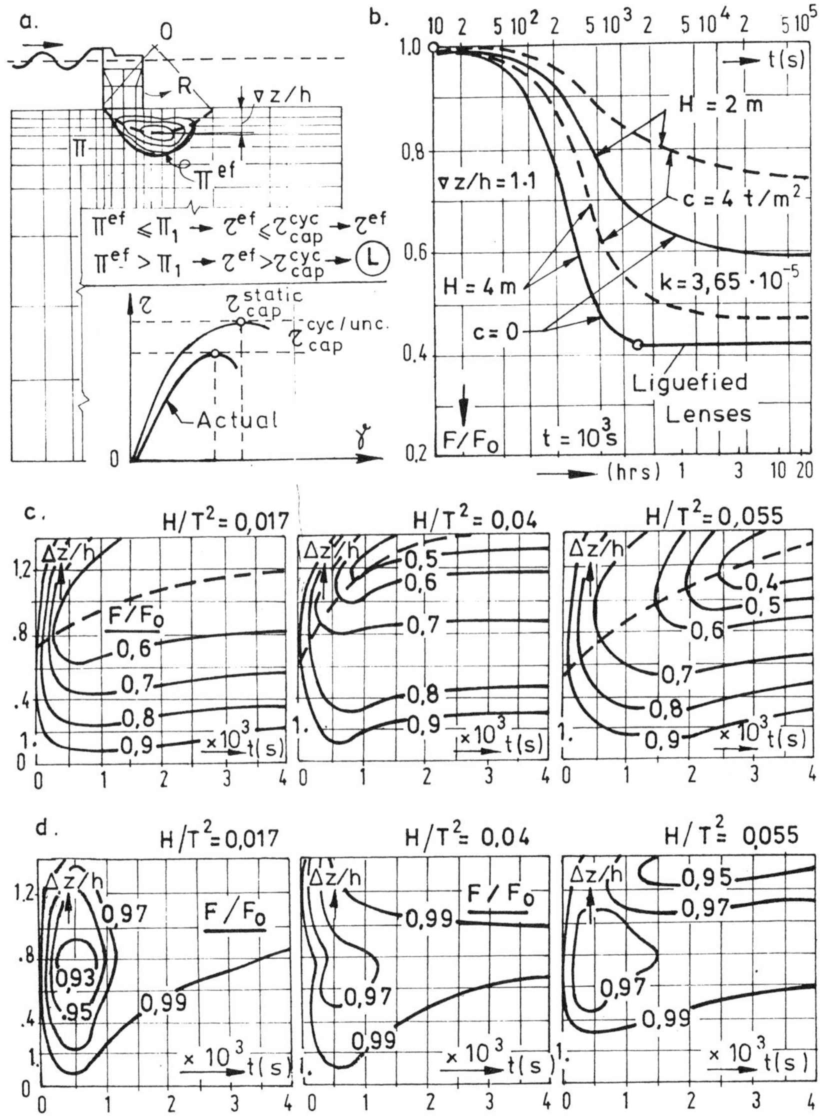

Fig. 5. Caisson circle sliding stability as affected by pore water pressures during the storm time: a- stability sketch; b,c,d- the ratio between dynamic (F) and static (F_0) stability coefficients in time, for seepage isotropy (c- k_x=.0000362 m/s, d- k_x=.00362 m/s)

15. As for ascertaining the real safety margins of the caisson breakwater, a risk analysis should be carried out about the above mentioned failure mechanism. Through an apprpriate discrete probabilistic approach (ref. 3), the actual domain of partial failure probabilities (P_f) has been evaluated, by taking into account main physical uncertainties, which characterizing the design parameters during different structure lifetimes (like wave heights & periods H-T, foundation & rockfill bedlayer permeabilities k-k_0, weight densities of the caisson ballast fill γ_f).

16. Under the more realistic conditions of irregular incident wave trains, a nonclapotis state and a harbourside nonstill water level, the pore pressure fields in the sandy subsoil will exhibit - in comparison to those ones assigned to monochromatic wave conditions - slower generation rithms and decreased pressure peak values with about 15%...20%.

REFERENCES

1. STEMATIU D. and STERE C. A finite element modelling of the dynamic interaction between the marine foundation and a vertical caisson breakwater. Hydrotechnics,no.11,12,1982 (Romanian)
2. STERE C. Upon the dynamic stability of vertical wall caisson breakwater of Constantza-south harbour extension. Institute of Civil Engineering, Division of Inland Waterways,Port & Coastal Engineering,1985,Report 283-HS.
3. STERE C. Safety analysis of vertical marine breakwaters. Scientifical Bulletin of Civ.Engrng.Instit.Bucharest,no.2, p.22-41,1983 (in Romanian).
4. HETEREN J.van et al. Verification of geotechnical design criteria for Eastern Scheldt storm surge barrier. Publications on hydraulic and geotechnical problems,Rijkswaterstaat,p.295-303,Delft,1989.

11. Port of Zeebrugge breakwater monitoring

L. V. D. VAN DAMME, Department of Environment and Infrastructure, Ostend, and D. J. O. VANDENBOSSCHE, HAECON N. V., Ghent

SYNOPSIS. The paper gives an overview of the general inspection and monitoring programme, applied on the breakwaters of the outer harbour at Zeebrugge. Special attention is paid to the instrumentation of a section of the NW breakwater. Measurement results are evaluated.

INTRODUCTION

1. In 1989 the construction of the breakwaters of the Zeebrugge Outer Harbour were completed. The breakwaters are of the rubble mound type without a concrete parapet. Fig. 1 shows a typical cross-section (ref. 1)

2. In Belgium the contractor-designer is responsible, during a period of ten years, for the good performance of his construction, according to the design criteria. To cover his possible risks the contractor needs an insurance on, in this case, the breakwaters. Since the unusual character and the very high value of the construction, the insurers are only willing to underwrite a policy on condition that the principal organized an inspection, monitoring and maintenance programme on the new breakwaters.

MONITORING AND INSPECTION PROGRAMME

3. The programme is based on a complete design file, which not only reflects the design philosophy, but also gathers all data used for the design of the breakwaters and the as-built drawings.

4. With this design-synthesis it is possible for any engineer to understand the complete design and to conduct the necessary maintenance activities.

5. The monitoring and inspection programme is composed of the following parts :

(a) visual control of the armour layer, the crest of the breakwater and the filter construction next to the road

(b) topographic measurements.

(c) bathymetric soundings of the different zones around and in the harbour.

(d) aerial remote sensing of the armour units.

(e) side scan sonar recordings of the underwater armour layer and the wave breaking carpet and berms.

(f) evaluation of the hydraulic design conditions on the basis of data collected by five measurement stations and seven wave-buoys in the surroundings of the harbour.

(g) evaluation of some design hypothesis by means of an instrumentation of a section of the NW-breakwater.

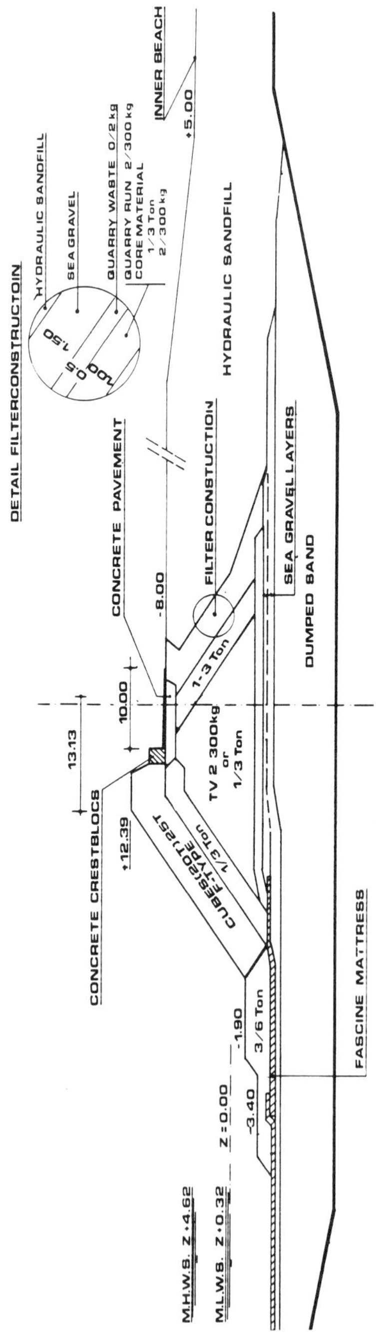

Fig. 1. Typical cross-section Southern West breakwater

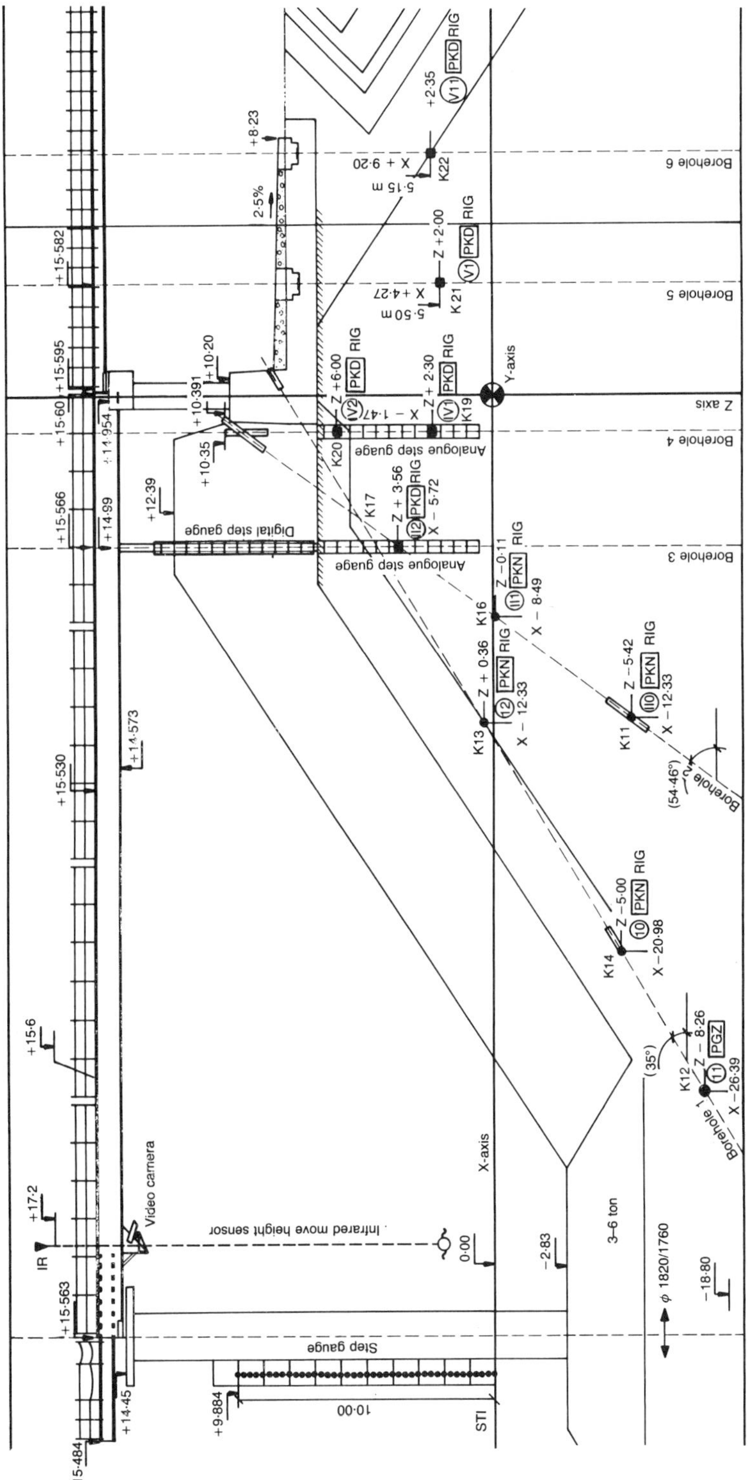

Fig. 2. Monitoring infrastructure

6. All inspections and measurements are carefully planned, prepared, carried out and reported in a standard prearranged format and using a comprehensive predefined vocabulary.

7. All the results of the measurements and the data collected in the inspection programme are issued in a yearly volume, which is an appendix to the design file.

INSTRUMENTATION OF THE NW-BREAKWATER

8. The monitoring programme started in 1987. It includes theoretical studies, laboratory investigations and field measurements. This paper gives the objectives, the set-up, the measurements, the collected data, data analysis and results.

AIM OF THE PROGRAMME

9. The objectives of this instrumentation programme are :

(a) better understanding of the geo-hydraulic phenomena affecting the overall stability and behaviour of the armour layer of rubble-mound breakwaters subjected to wave loads.

(b) improvement of the simulation techniques of these phenomena in physical models, leading to higher reliability of model testing.

(c) the determination of the real forces acting on the armour units and the real hydraulic pressure set-up in the core and foundation layers.

INFRASTRUCTURE

10. As shown on fig. 2 a measurement jetty of ca. 70 m long has been built. Six borings have been carried out into the breakwater core. Galvanized steel casings are placed in these boreholes. These casings are perforated in order not to disturb the overall permeability. In these casings the instruments are placed. A data centre is mobilized in a 20' container, placed on the leeside of the jetty.

MEASURING EQUIPMENT

11. The main considerations kept in mind during the selection of instrument layout were :

- good reliability to marine environment and wave action ;
- good accessibility for instrument installation and maintenance ;
- provision of reliable power supply.

12. The following equipment has been installed :

- waverider buoys :	2
- infra-red wave height sensor :	1
- pressure transducers :	11
- analog step-gauge :	3
- digital step-gauge :	1
- video camera :	2
- 3D ultrasonic current meter :	1
- temperature :	1

13. The data centre is equiped with all necessary interface units, multiprogrammer and computer unit which allows a continuous measuring during 17 hours with sampling rates of 10 Hz. A watch-dog program is continuously running on the computer system to check all sensors.

14. Fig. 3 is presenting a global overview of the data-acquisition system and off-line data processing.

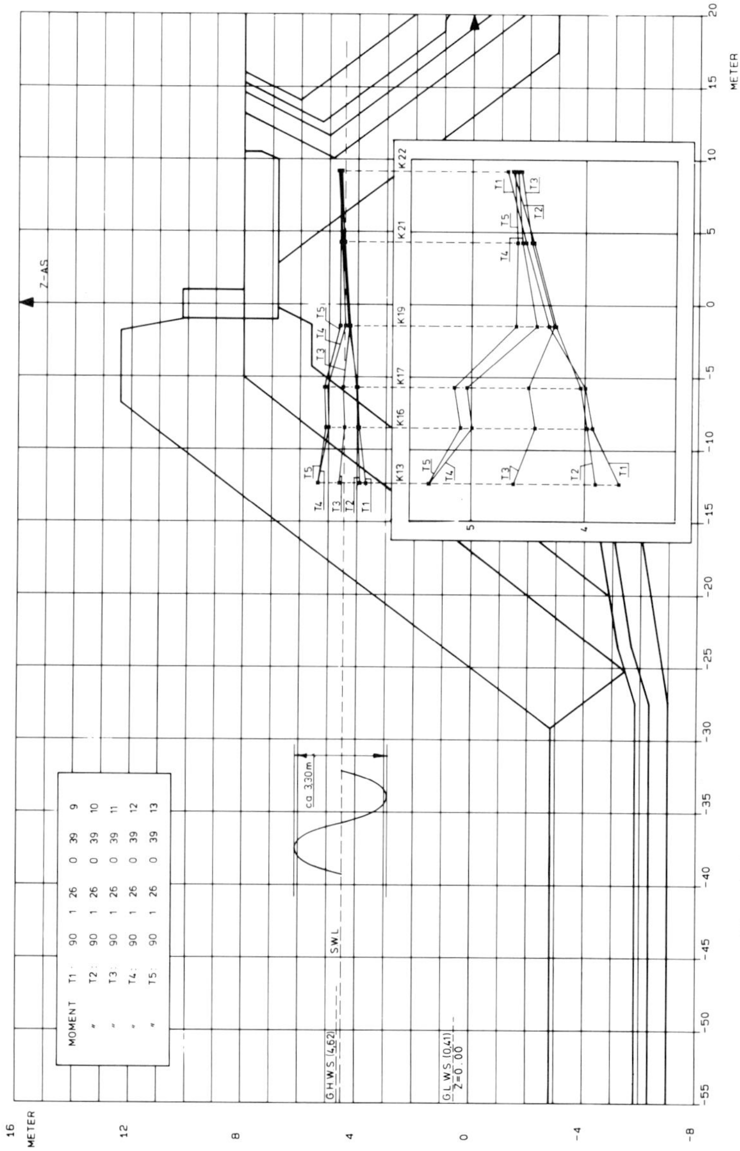

Fig. 4. Wave propagation of an incident wave with H = ca. 3,30 m

15. A detailed description of the measuring equipment, the hardware and software configuration and the basic software for off-line data processing is given in ref. 2.

RESULTS

16. Measurements have been performed during the storm periods of March 1988, March-April 1989 and January-February 1990. The collected data are processed. As an example fig. 4 shows the wave propagation of an incident wave with H = ca. 3,30 m. Pressure meter K13 is placed on the L.W. line and approximately 1 m below the interface between the 1-3 ton filter layer and the core. At the location of K13 the amplitude has decreased to ca. 1,90 m.

17. One also remarks :

- the clear damping of the amplitude when moving into the core of the breakwater ;
- the set-up due to wave action in the breakwater core.

18. The large scale tests conducted in the Groszen Wellen Kanal at Hannover show the same tendences.

CONCLUSIONS AND PERSPECTIVES

19. The inspection and monitoring programme, as part of the insurance policy, is a necessary tool to keep the breakwaters in perfect condition.

20. In the Zeebrugge case special efforts are done to get a better understanding in the phenomena acting in the breakwater core and in the armour layer during storm periods.

21. Together with the results of large scale tests, the prototype measurements will hopefully lead the breakwater engineering community to a better insight in the geo-hydraulic behaviour of these structures.

REFERENCES

1. VAN DAMME L., VANDENBOSSCHE D., Zeebrugge Main Breakwaters. Proc. ICCE 82, Am. Soc. Civ. Engrs.
2. VAN DAMME L., WENS F., de SAINT AUBAIN T., Zeebrugge Monitoring Programme of the NW breakwater, Proc. 9th Int. Harb. Congress, Antwerp, June 1988.

Discussion

P. LACEY, Ove Arup & Partners

This contribution is about the patented Carfer breakwater system. The breakwater is essentially a substantial pile with fin elements placed on the top. Both pile and elements are constructed in concrete.

The pile is driven into the sea bed, and the mixed elements are slipped onto the pile and separated by spacers. The piles are then arranged in several rows, distanced and positioned according to the sea conditions.

The pilings are placed at regular predetermined intervals and in rows parallel to each other which, in

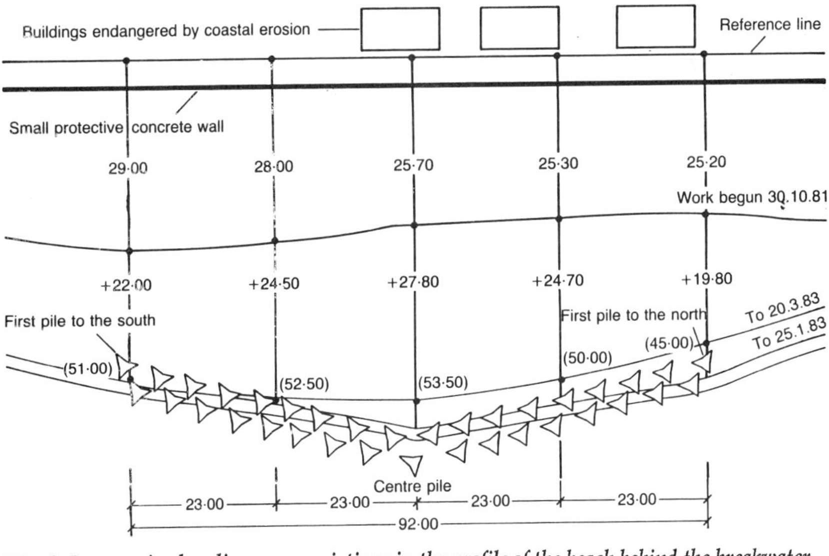

Fig. 1. Increase in shoreline area: variations in the profile of the beach behind the breakwater system (in parentheses, total distances up to 20. 3. 83)

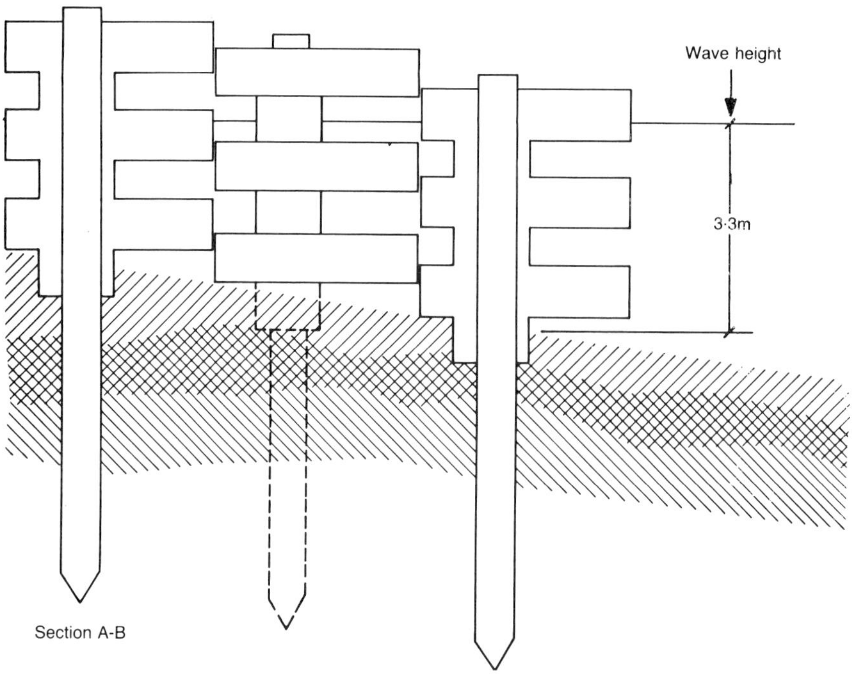

Fig. 2. Three-element system for sea-bottom for depths to 3.3m

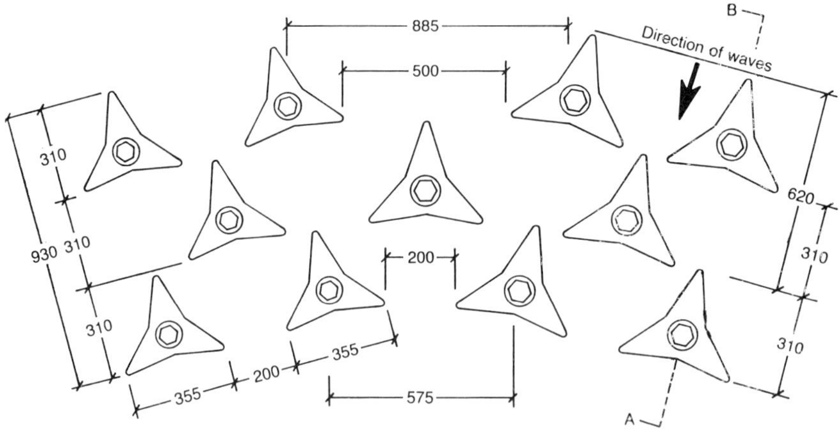

Fig. 3. Plan of barrier with three parallel rows of the Carfer breakwater system (measurements in m)

relation to the prevailing wind and tide may form a convex angle, as shown.

In fact, the Carfer breakwater is an energy dissipator capable of greatly reducing wave force, and trapping suspended material from the longshore or undertow current. This barrier is substantially different to the traditional breakwater because the breaking can be regulated in proportion to protective requirements, by the density of the single elements making up the entire structure.

Essentially, the structure lets the sea flow past, and slows it down, which is very different to the traditional closed breakwater type. The deposits which form behind the barrier do not cause water stagnation or leave decomposing organic material.

The new breakwater is also very flexible since individual piles and elements and the density pattern of the array can be installed as required, should the prevailing tidal conditions change or original positionings are found not to have been suitable to start with.

The breakwater has another important advantage for maintaining beach equilibrium, which concerns the longshore transport which remains unaltered. In fact, wave motion is only reduced and not cancelled under the lee of the protective work itself. This sediment distribution trend allows, in time, a uniform shoreline advancing, avoiding the presence of accentuated regression of the coast that is usually found in the beaches adjacent to those protected with traditional barriers.

Construction of prefabricated parts is relatively simple and once installed, does not require any maintenance or attention, and is therefore much more economical than solid rock breakwaters.

A Carfer breakwater was installed in 1981 on the Adriatic Coast (just south of Ancona) to protect a row of houses called Lido del Nazione. The beach, which had eroded back to the house foundations, has steadily accreted since the breakwater was installed, and has gained over 40 m in width.

The system is now being constructed at three new positions near the town of Ravena in Italy, as part of a large scheme of coastal protection.

P. LACEY, Ove Arup & Partners

I would like to ask Professor Bucharth about monitoring. Should we not be making out a very strong case for monitoring to be carried out as part of the specification?

M. A. LOSADA, University of Cantabria

For the pressure under the crown, it appears that

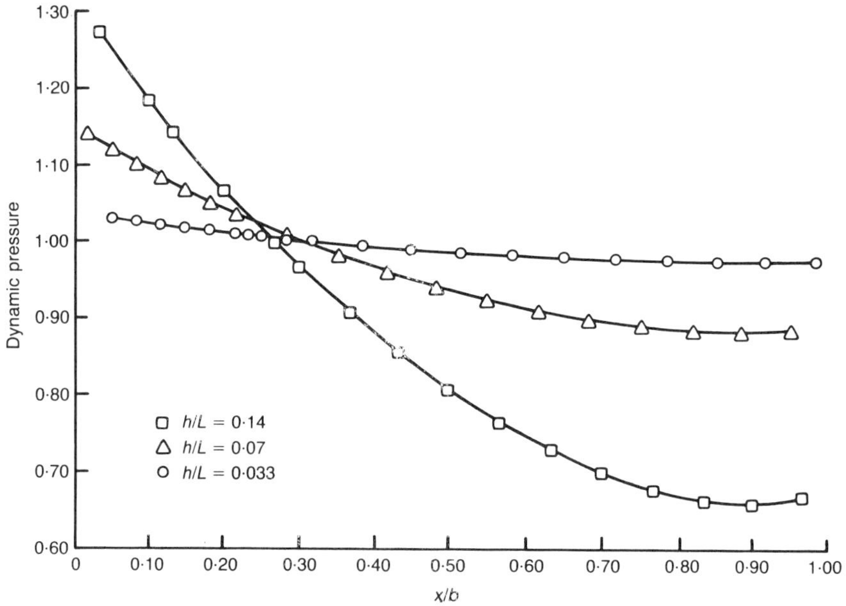
Dynamic pressure
x/b
h/L = 0·14
h/L = 0·07
h/L = 0·033

Fig. 4

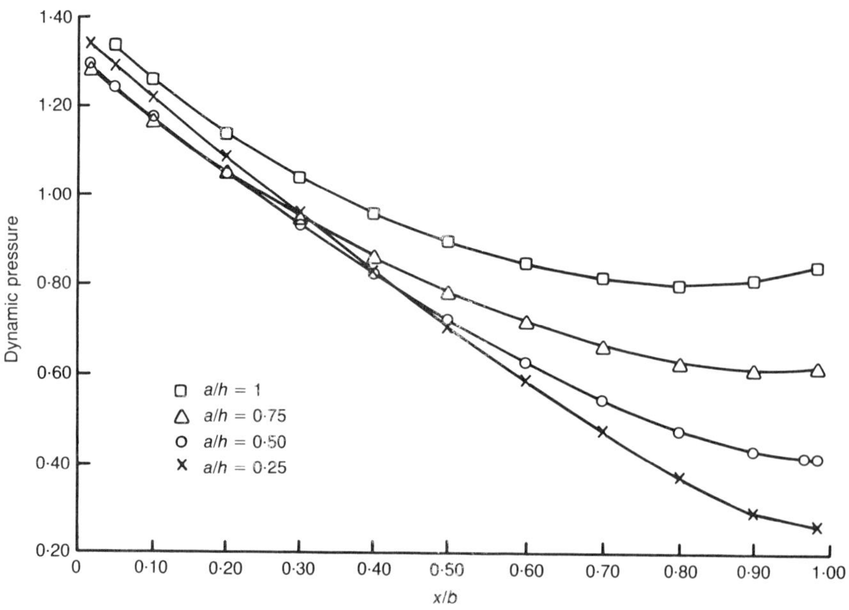
Dynamic pressure
x/b
a/h = 1
a/h = 0·75
a/h = 0·50
a/h = 0·25

Fig. 5

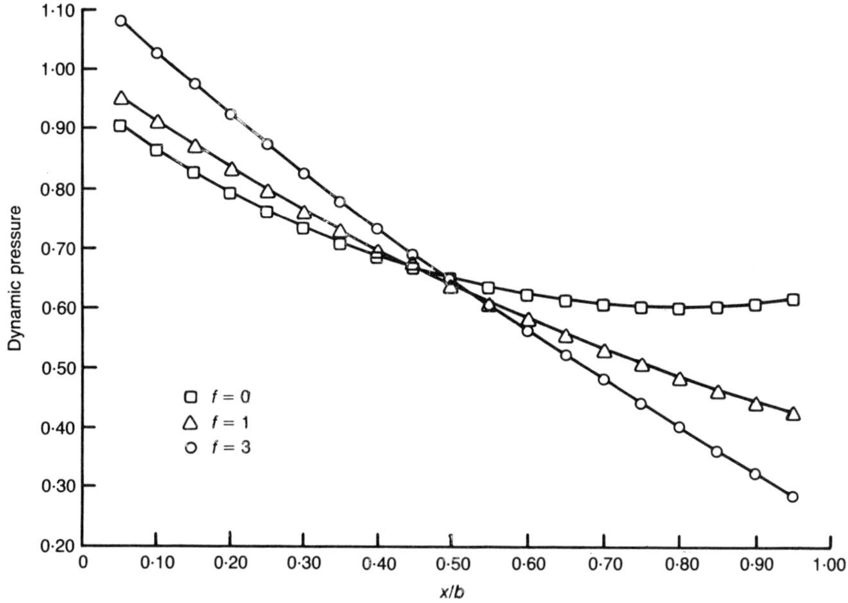

1·10
1·00
0·90
0·80
0·70
0·60
0·50
0·40
0·30
0·20
Dynamic pressure
0
0·10
0·20
0·30
0·40
0·50
0·60
0·70
0·80
0·90
1·00
x/b
f = 0
f = 1
f = 3

Fig. 6

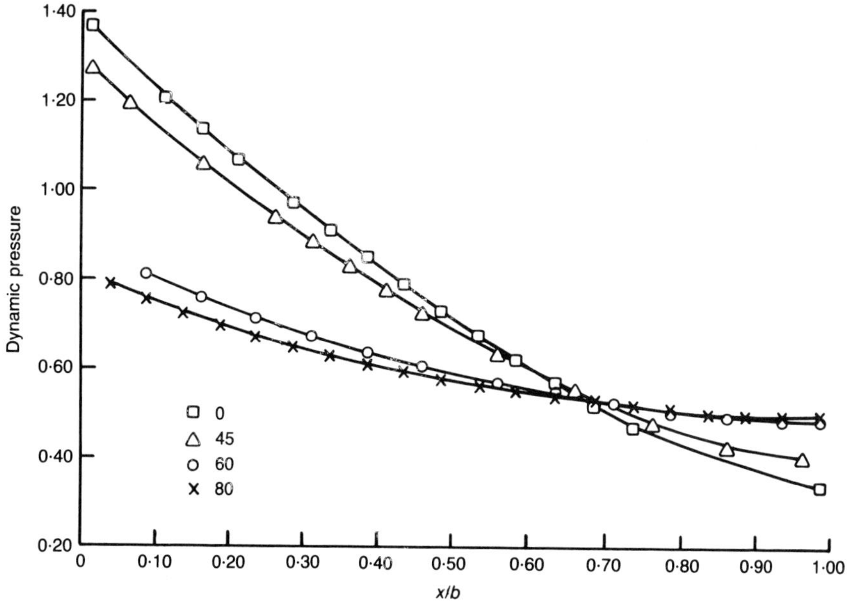

1·40
1·20
1·00
0·80
0·60
0·40
0·20
Dynamic pressure
0
0·10
0·20
0·30
0·40
0·50
0·60
0·70
0·80
0·90
1·00
x/b
0
45
60
80

Fig. 7

Professor Bucharth is saying that, as a safe solution, a linear distribution can be assumed between the measurable pressures at the front and the rear edges of the plate.

Let me show some results of the pressure under a crown obtained in a recent paper still under review: 'Water waves on crown breakwaters', by J. J. Losada et al., to be published in the ASCE, WW Journal.

We consider the interaction of an obliquely incident gravity wave train with a single homogeneous isotropic vertical porous structure of width b and height a crowned by an impervious element of the same width, between two semi-infinite fluid regions of constant depth h. Characteristics of the porous medium are the porosity ε, the linear friction factor f and the inertial term s. For all computations s=1, f=1, ε=0.45 and a/h=0.75; otherwise it is indicated in each figure.

Fig. 4 shows the influence of the wave period on the dynamic pressure under the crown. Figs 5-7 show, respectively, the influence of the porous height, the friction factor and the angle of incidence.

These results show that the linear pressure gradient assumption usually applied may be, in certain circumstances, unsafe.

C. T. BROWN, Seabee Developments

In developing his design formula, Hudson used both a conceptual model and an experiment. His conceptual model was framed in terms of the stability number H/d(Sr-1). There was a discrepancy between the model and theory in terms of wave angle, due to his use of a fixed direction for incident water flow.

This was addressed by Brown (Hamburg 1978) who considered rotating flow and found a close similarity between theory and experiment.

The crucial point of the model study is that stability is dependent on the slope normal dimension of the armour unit. The cube root of the volume is only an approximation to this. Also of importance is the shape and orientation of the unit, as these affect the cone of departure of the unit. This development of Hudson's conceptional momentum model, called 'Blanket (Linear) Theory', leads to important conclusions on the design of artificial units and the potential for improved performance of natural materials. This is perhaps best demonstrated by the high stabilities achieved at the Umpqua Jetties in Washington State, where stability coefficients of the order of 20 have been reported.

As regards the stability of arranged units, it is true that the transition from stability to continuing loss is of the order of 15%. However, most

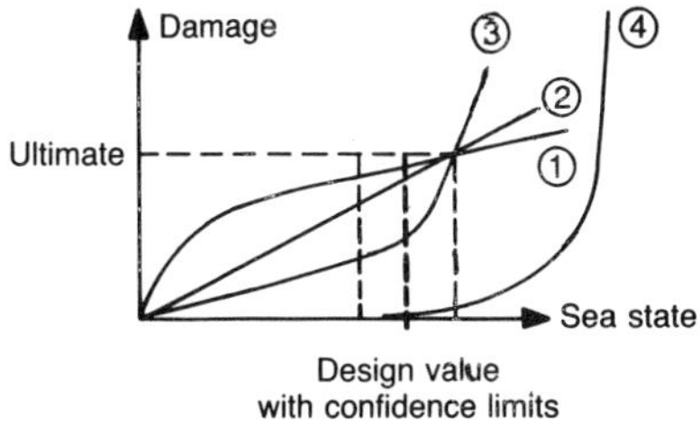

Suggested Fig. 12 of Paper 4

practitioners use load factors in their analyses, and we also carried out checks for extreme wave and depth-limited waves. These systems show marked economies compared with random and semi-random units.

J. W. VAN DER MEER, Delft Hydraulics
For progressive failure you need a safety factor for design. For Accropodes (4) this is about 1.5 on $H_s/\Delta D_n$. The advantage is then that a higher wave height than the design wave height does give no increase in damage.

Therefore Fig. 12 in Paper 4 is not giving the right figure for design. It should be as seen here.

L. FRANCO, Politecnico di Milano
I have some comments on Mr Lacey's presentation. I was pleased to hear about a recent example of the Italian fantasy and inventiveness) in design of coastal structures, although I was surprised to see it presented (without notice) by an English Consultant, and to see the new star-shaped poles named differently from the original Italian patent (called 'Ferr-An' by the inventor).

I do not agree with the claimed protection efficiency and environmental friendliness of this system. In fact I visited the site near Ancona (Adriatic Sea) with my students last May and we could see that another similar application at Scossicci, just a few kilometres north of the one shown here (S. Maria di Potenza) has not proved to be at all successful! Installed on a slightly steeper beach, many piles collapsed and the beach behind the barrier actually eroded (especially by underwater scour). I find that these isolated R.C. elements with sharp corners and corroded steel bars jutting out are surely not friendly for the beach users and the currents through the pile gaps during mild wave conditions may be very dangerous. This system may be promising if properly tested and improved, with special regard to the structural stability and durability, also considering the heavy and 'unpleasant' torsional stresses on the horizontal fingers. But I have to warn the audience that, as with many other coast protection works, it may be effective only in a limited range of conditions, and there is already an example of failure.

H. LIGTERINGEN, Frederic R. Harris BV
Professor Bucharth refers to the recent research in the Netherlands on the strength of concrete armour units. He mentions the tests on impact loads, but the five-year programme was much more comprehensive and dealt with the kinematics of rocking armour units, loads and stresses at impact and concrete technology. The studies resulted in a simulation model for the behaviour of cubes and tetrapods on a breakwater slope, which calculates for given storm conditions the number of rocking units and the number of broken units. Although the model is still based on a strong schematization of reality, it is the first tool which allows the designers to make a more quantitative assessment of the damage including breakage.

The simulation model has meanwhile been applied in the rehabilitation of a breakwater at Ponta Delgada, in the Azores. On the basis of the results of these computations, it was possible to apply large tetrapods. These would otherwise not have been selected because of the high and undefined risk of breakage.

K. TANIMOTO, Saitama University
I am very interested in Mr Bouchet's new project of breakwaters on which people can live. I have a short question about the durability, the lifetime of structures. How many years of life do you expect for the structures in La Condamine Project?

J. D. METTAM, Scott Bertlin
I have three questions for Mr Bouchet. There has been research on the horizontal plate breakwater, which has been published, at least since 1974 (list of references below). Is it then possible for you to patent the system?

Looking at Fig. 13 in Paper 5, I wonder whether the pontoon can be made to float upright, or at all? The horizontal forces on the pontoon will be very considerable. How will these be resisted?

References:
1. PATARAPANICH M. Maximum and zero reflection from submerged plate. J. WatWays Harb. Coast. Oce. Engn Div., 1984, 110, 171-181.
2. PATARAPANICH M. Forces and moment on a horizontal plate due to wave scattering. J. Coast. Engn, 1984, 8, 279-301.

3. PATARAPANICH M. and CHEONG H. F. Reflection and transmission characteristics of regular and random waves from a submerged horizontal plate. J. Coast. Engn, 1989, 13, 161-182.
4. SILVESTER R. Coastal engineering. Elsevier, 1974, 341-343.
5. PATRAPANICH M. and SILVESTER R. Optimum conditions for a submerged platform breakwater. Proceedings of the 4th Congress of the Asian and Pacific Division of the Incorporated Association for Hydraulic Research. Chiengmai, Thailand, 1984, Vol. 1, 539-550.
6. IJIMA T. et al. Breakwater and quay wall by horizontal plates and permeable materials. Coastal Engineering Research Laboratory, Kyushu University, Japan. Technical Report No. 2, 1971.
7. PATARAPANICH M. Wave reflection from a fixed horizontal plate. Proceedings of the International Conference on Water Resources Engineering. Asian Institute of Technology, Bangkok, 1978, 427-446.
8. SIEW P. F. and HURLEY D. G. Long surface waves incident on a submerged horizontal plate. J. Fluid Mech., 1977, 83, 141-151.
9. DURGIN W. W. Wave induced pressures on submerged plates. Wat. Harb. Coast. Engn. Div., 1975, 101, No. WW1.
10. DATTATRI J. et al. Laboratory investigations of submerged plate breakwaters. Proceedings of the 17th Congress of IAHR. Baden-Baden, 1977, Vol. 4, 88-89.

L. FRANCO, Politecnico di Milano
I realize that the wave-attenuating capacity of the immersed slabs of any assigned size is very sensitive to the wave period, being effective for a limited range of wave/water level conditions. Will the system be effective in real sea applications where random multidirectional waves are present? Has Mr Bouchet tested the three-dimensional performance of the 'oscillating seawall'?

Could he also provide some information on the behaviour of the experimental application at Barcelona after more than a year's monitoring; are there plans for the use of a system of these original barriers for beach protection in the near future?

H. LIGTERINGEN, Frederic R. Harris BV
I have a question for Mr Bouchet related to the breakwater for Port de la Condamine, as described in Section 3.1 of Paper 5. From Figs 12 and 13 the impression is given that these pontoons are floating. Can Mr Bouchet clarify this?

A second question is related to the wave forces on the pontoons (whether floating or fixed on piles); have short- and long-period wave forces been taken into account?

J. W. VAN DER MEER, Delft Hydraulics
Referring to para. 36 in Paper 7, the restriction that this method cannot describe a braking wave is not valid any more. Within the European research programme MAST this VOF-method is used to simulate breaking waves on and also in rubble structures. It looks to be a very promising method for describing wave motion in two dimensions.

F. BARENDS, Delft Geotechnics
Tests on the turbulent porous flow law, i.e. Forehheimer, are done in one dimension. They provide the value of the empirical parameter a for the linear part and b for the turbulent part. The turbulent part is proportional to the square of the bulk velocity q. The resistance is therefore expressed as F=a+bq. For the numerical simulation a generalization to two dimensions is required. In the present state of the art the generalization is achieved by making a scalar of the resistance: F=a+bq . This is convenient because it satisfied the principle of independence to the (chosen) coordinate system, but there is no physical justification. Since the resistance is dependent on the actual bulk velocity, it may be that the resistance is still dependent on the bulk velocity component in the corresponding direction. Therefore, by the choice F=a+bq one makes a constraint and eliminates induced anisotropy. Does Professor Hall have any information on this aspect?

H. OUMERACI, University of Hannover
In connection with Paper 7, I would like to give a warning to those who are developing numerical codes to simulate wave-induced internal flow in rubble mound structures, as well as to those who are using the results of these models. My opinion is that there is still a long way to go before we will have a reliable numerical prediction tool. The reason is simply that we know too little about the physical processes involved and their mathematical formulation. Therefore, I try to identify below a number of problems which should first be solved before one embarks on numerical considerations.

External flow field (seaward). Present knowledge related to the description of wave kinematics in front of a rough permeable structure is almost nil (partial clapotis). The first step is to perform detailed measurements in a large-scale model in order to obtain a mathematical formulation for the inflow boundary condition needed for the numerical model.

Transition and interaction between external and

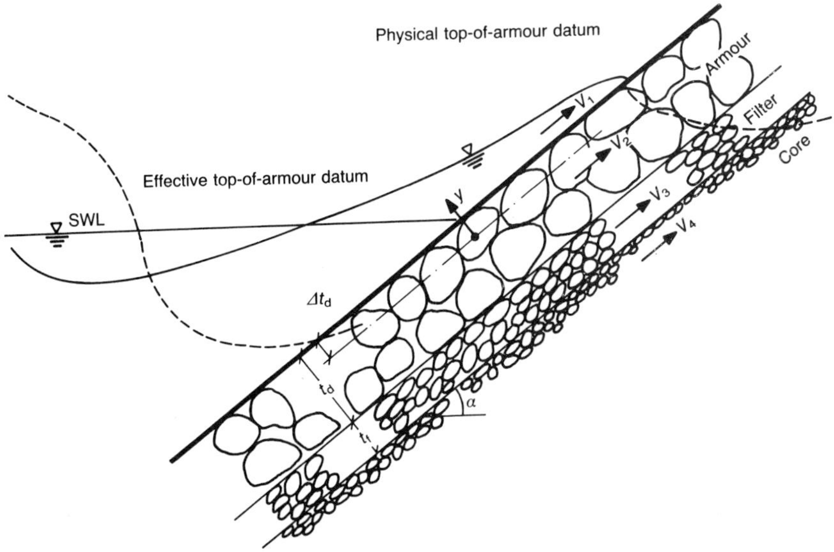

Fig. 9. Transition between external and internal flow

internal flow. Since the model describing the external flow should implicitly assume a flow within the armour layer, the effective datum or the external flow must be beneath the physical top-of-armour datum (Fig. 9). The question is then how to obtain the thickness Δt_a. A possible answer may be obtained by taking the flux on the slope V_1 (external flow) as equal to the flux V_2 for $y=\Delta t_a$ (internal flow). The same procedure must be valid for the transition of the flow between the different layers ($V_2=V_3$ and $V_3=V_4$).

A further problem yet unresolved is how to account for the energy dissipation on the slope in the case of breaking waves. (In the case of non-breaking waves, a formulation has been already proposed by Oladsen 1976).

Furthermore, nothing is yet available on the interaction between the external flow at (and on) the slope and the flow induced by wave action within the structure. The first step should thus consist of the performance of systematic large-scale model tests in order to obtain enough data for the understanding of the physical processes involved; the formulation of the coupling problem; and the validation of the numerical model.

Internal flow field induced by wave action. Because of the following difficulties, my opinion is that we have to formulate a porous flow model starting from first principles by writing the force balance on a solid body in unsteady flow, the force balance on the pore fluid and by spatially averaging the motion of the solid skeleton (GV, 1990). In fact, the questions which arise in dealing with internal flows are: does a Furchheimer-type equation apply for the flow in the armour layer; what is the relative importance of the different forces involved?

Furthermore, most of the existing porous flow models contain 1, 2 or 3 terms:

(a) linear velocity term
(b) quadratic velocity term
(c) term proportional to flow acceleration

For the last two terms, however, no agreement exists in the definitions by the various porous flow models.

In addition to the three aforementioned problems there are also the three following problems which should be solved in order to be able to develop a reliable numerical code.

(a) External flow field at and in the shorewards slope of the structure and interaction with the internal flow (coupling problem).

(b) Air entrainment: the effect of air entrainment (essentially by breaking!) on the internal flow is known to consist of the reduction of the effective hydraulic conductivity of the porous medium (Oumeraci and Partenscky 1990) and should necessarily be allowed for in the constitutive equation to be used in the numerical model.

(c) Internal stability of the porous medium.

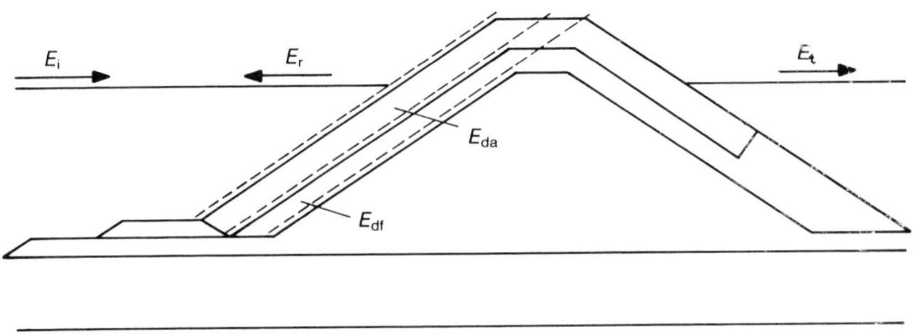

$$E_i = E_d + E_r + E_t$$

$$K_R = (E_r/E_i)^{1/2}$$

$$E_{dt} = E_d + E_t = (1 - K_R^2)\, E_i$$

$$E_d = E_{da} + E_{df} + E_{da}$$

Fig. 10

Generally, the porous matrix is assumed to be fixed; this may be valid in a first step. Actually, however, this is not the case (Oumeraci and Kohlhase 1988) and should be accounted for in the final version of the numerical code.

As a conclusion, the only way to simulate numerically the wave-induced internal flow in a rubble mound breakwater is to develop an integrated numerical code which is capable of simultaneously describing the flow in front of the structure, behind the structure, on, at and in the structure, by accounting for the incident reflected, dissipated and transmitted wave energy components involved (see Fig. 10).

R. J. MADDRELL, Sir William Halcrow & Partners

The design of a breakwater which would prevent oil ingress, even under storm conditions, while allowing water to flow to submerged intakes, was based on mathematical and physical model test results from Hydraulics Research Wallingford. The two models complemented each other and the results were in overall agreement.

The flows described above were simpler than those described by Professor Hall and Dr Hettiarachchi and we can have some confidence in the results. What further research is required before a significant degree of confidence can be placed on mathematically modelling flows in rubble mound structures?

D. MCDOWELL, Brighton Polytechnic

Dynamically stable breakwaters, by definition, depend on re-working of the armour layer by wave action. The experiments described in Paper 8 can only apply to normally incident waves. In all other cases, armour material is transported along the breakwater, leading to depletion at the updrift end. There must be an adequate supply of armour at the updrift end such that there is time for replenishment between major storm events. This is particularly important at breakwater ends or round-heads where wave action is inevitably tangential at some point.

Tests using normally incident waves have limited value, though the work described in Paper 8 does give useful qualitative information about the effect of grading the armour material.

The effects of grading can be seen on many natural beaches. There are some which appear to be very stable despite having a small maximum stone size. The presence of widely graded material, from fine sand to small pebbles, has produced a compacted armoured layer which resists erosion. Movement of coarser material takes place over the armoured layer. Is there a use for such material in breakwaters? Not, perhaps, as armour because it has low porosity, but as core material which would resist erosion in the event of failure of the primary armour.

J. D. SIMM, Robert West & Partners

In response to Professor D. M. McDowell's question on the usefulness of the results in Paper 8, in the light of longshore transport problems in the three-dimensional case, I would comment that the work by Burcharth and Frigaard[2] suggests some useful preliminary guidance, as noted in the CIRIA/CUR rock manual.[2] For berm breakwaters $H_s/\Delta D_{n50}$ will normally be less than 5 or 6, and Burcharth and Frigaard suggested the following design limits to keep below the incipient level for longshore motion. For trunks exposed to steep oblique waves, $H_s/\Delta D_{n50}<4.5$. For trunks exposed to long oblique waves, $H_s/\Delta D_{n50}<3.5$. For roundheads, $H_s/\Delta D_{n50}<3$.

Given these limits, the results of Hall et al. give a very useful first approximation to the cross-sectional design. In practice, of course, any preliminary design concept should always be checked in a three-dimensional physical model, at least until firmer design guidance can be provided.

Use of the formulae presented here by Hall et al. must take account of the fact that the D_{50} size presented here is not the same as the D_{n50} used by other authors in consideration of armour stability and proposed as the standard in definition of rock.[2] However, extensive tests in The Netherlands carried out in connection with the Eastern Scheldt Storm Surge Barrier project have established a relationship between the seive size D and the nominal diameter D_n which is generally found to hold, namely $D/D_n=0.84$. When including this work by Hall et al. in the CIRIA/CUR rock manual[2] this conversion was applied.

References:
1. BURCHARTH H. F. and FRIGAARD P. On the stability of berm breakwater roundheads and trunk erosion in oblique waves. Seminar on Unconventional Rubble mound Breakwaters, Ottawa, 1987.
2. CIRIA/CUR. Manual on the use of rock in coastal and shoreline engineering, 1991.

C. T. BROWN, Peter Fraenkel BMT

I was interested in Mr van Damme's use of sonar. Since March this year my company has been endeavouring to put in train a long-term monitoring programme for the Peterhead Breakwaters (Scotland).

These breakwaters (see Paper 14) were built out into nearly 20 m of water, and are subject to waves of the order of 10 m.

The North Breakwater was built straight on the rocky seabed, while the Southern Breakwater was built on an extensive rubble mound of 3-6 tonne stones, armoured by **a terrace of apron slabs, 15 ft square and 5 ft thick.** This mound does not appear to be stable, with some slabs upturned and others completely flipped over on top of others.

We found that the side-scan sonar was capable of resolving the 200 m^2 lifting eyes of the blocks.

Mr van Damme is fortunate in having such a programme in place. What is its annual cost, and how does this compare with the capital value of the breakwaters? This information would assist us all in arguing the case for the competent monitoring of breakwaters.

H. OUMERACI, University of Hannover
What kind of pressure transducers did Mr van Damme use, i.e. what is the natural frequency of these transducers? He mentioned that he is using a sampling frequency of 10Hz for the pore pressure records. According to our experience at the Large Wave Flume of Hannover (GWK), you need at least a sampling frequency of 40Hz.

Did Mr van Damme notice any time lag between the pore pressure records at different locations in the direction of wave propagation?

J. D. SIMM, Robert West & Partners
With regard to the question of monitoring programmes an the level of monitoring required, delegates may find Fig. 11 of interest. It sets our various levels of monitoring and was prepared by myself for the CIRIA/CUR Rock Manual[1] based on a concept originally developed by Bradbury and Allsop.[2] In defining a monitoring programme, clients and engineers need to be aware of the objectives of the output monitoring which will be, having compared structure states at various time, to check design assumptions and performance and identify maintenance needs. Comparisons will for example permit the simple matter of checking that the breakwater is still where it was originally placed at level 1. At the more advanced level 2, comparisons will enable definition of damage parameters such as S in van der Meer's stability formula.[1,3]

References

1. CIRIA/CUR. Manual on the use of rock in coastal and shoreline engineering. 1991, Chap. 7.
2. BRADBURY A. and ALLSOP N. W. H.. Monitoring technique for armoured coastal structures. Report IT 343, Hydraulics Research Ltd, Wallingford, 1989.
3. VAN DER MEER J. W. Stability of breakwater armour layers. Coastal Engng, 1987, II, 219-239.

Measures of the state of a rock structure

Aspect of structure state	Survey technique
Level I: Location Two to 10 points on the structure measured in relation to a well-established grid and datum levels. Redundant survey points need if major movement possible.	Conventional survey techniques (any settlement markers should be installed during construction)
Level II: Geometry Outer surface description, related to level I survey points.	Conventional survey, using profiling techniques recommended in Appendix 1 of Ref. 1 but with profiles at wider spacings (say, 20–30 m). For underwater surveying, bathymetric techniques may give useful information
Level III: Composition Position and attitude of each piece of rock armour, including unstable pieces. Position and size of major voids and exposures of core or underlayers.	Armour-degradation inspection techniques Comparative photography Photogrammetry For underwater surveying, side-scan sonar techniques may be used in co-ordination with diver surveys to identify features on sonar traces
Level IV: Element composition Shape and size of armour rock, including any fractures.	Armour-degradation inspection techniques

Note: Levels III and IV armour degradation inspection techniques are difficult to implement on wide gradings and/or gradings with D_{15} less than about 0·3m. The techniques, however, work extremely well on large 'single-sized' rock armour ($D_{15} > 0{\cdot}3$ m; D_{85}/D_{15} in the range 1·2 – 1·4).

Fig. 11. Measures of the state of a rock structure

J. N. WOLFE-BARRY, Sir William Halcrow & Partners Ltd

I congratulate Mr van Damme on working for an organization prepared to set up a data collection programme on the scale described. It is only by studying the prototype breakwaters that we can finally come to understand and interpret the results of the model testing.

I have recently been involved with a project in West Africa which included repairs to two breakwaters. One of these was a rock rubble mound with no crest wall or roadway. We had as-constructed drawings but the original survey monuments had all been destroyed. Without these and without accurate records of the construction, it was virtually impossible to make sense of the drawings to assess the changes that had taken place in the 30-odd years since the breakwater had been built. I would appeal to anyone constructing a breakwater to ensure that at the very minimum accurate records are kept of the construction, that robust survey monuments are constructed, and that the client is urged to inspect and maintain them throughout the life of the structure.

D. J. VANDENBOSSCHE, Haecon NV Consulting Engineers

Delegates have raised a question about the costs to be

incurred by a proper breakwater monitoring programme with the aim of giving basic maintenance data. The programme set up for the Zeebrugge main breakwaters totalling about 8.4 km covers

(a) visual control of the armour layer, the crest and the filter structures next to the crest structure

(b) topographic measurements of the movements of 'witness' armour units

(c) bathometric soundings of the toe, berm and wave carpet structures

(d) measurements of some witness profiles

(e) analysis of the data and reporting to the executive bodies

The whole process could cost, in our experience, roughly yearly 0.025-0.05% of the capital cost incurred.

A proper monitoring programme is a prerequisite for proper maintenance of the breakwater structure. The people responsible for this, who have often not been involved in the design, have to acquire a 'feeling' for the structure.

H . LIGTERINGEN, Frederic R. Harris BV

I am puzzled by the high percentage indicated by Mr Vandenbossche as an annual cost of monitoring the Zeebrugge breakwater (0.025-0.05%). Although it may be small compared with the construction cost, it is a large sum in the port's budget for maintenance of the structure.

I would promote the definition of a kind of minimum requirement for monitoring to be meaningful, which would not include the rather extensive measurements made in Zeebrugge, in their attempt to generate more general useful insight.

P. LACEY, Ove Arup & Partners

In reply to Professor Franco, of course, with a breakwater such as the Carfer type there will be problems if not installed properly. Although there are three technical reports prepared it is not yet clear exactly how the breakwater functions, or how it reacts to large tidal ranges and large waves. However, what is clear is that it has worked near Ancona by attracting sand accretion. It has saved property and enhanced a leisure facility. It does need more study.

H. F. BURCHARTH, Paper 4

In reply to Mr Lacey, I certainly agree with the importance of monitoring, and I would like to refer to the section in my paper about the monitoring of prototype structures, in which I stress the importance of documentation of the as-built structure plus systematic monitoring during structural life.

In reply to Professor Losada, the linear distribution

of pore pressure under a superstructure base plate is chosen as a safe engineering solution for all cases where the base plate is still resting on the porous foundation, i.e. a failure state with an open wedge between the foundation and the base plate has not been reached. The linear distribution, between measurable pressures at the front and the rear edges, can be verified, by both theory and experiments, to be a safe solution. In fact the calculations presented in the paper that Professor Losada is referring to, do all show - as should be the case - that the linear approximation is a safe solution. Consequently I disagree with the comments of Professor Losada, the reason for which must be a misinterpretation of the otherwise correct figures which were shown by him.

In reply to Mr Brown, I agree that both the shape and the orientation (e.g. in the case of elongated units) influence the hydraulic stability of an armour layer. It is well known that placing elongated stones or parallelopipedic blocks with the long axis perpendicular to the slope improves the hydraulic stability.

The Hudson formula cannot in general describe the hydraulic stability of any kind of hollow blocks (like SHED and SEABEE) when placed in close contact with neighbour units. As far as I know Mr Hudson has never claimed this. I pointed out in my paper that to my knowledge no general hydraulic stability criteria for hollow cubes have been presented so far. Unfortunately, I am not familiar with the 'blanket theory' presented by Mr Brown, and so I cannot comment on it.

I agree with Mr Brown on the importance of load factors (safety factors) to be used in the design of armour which fails suddenly without a foregoing sign of significant damage (brittle failures). Fig. 12 in Paper 4 illustrates this point.

In reply to Dr van de Meer, I absolutely agree on the necessity of a safety factor. This must be in proportion to the brittleness of the armour failure. How big it should be depends on many factors, including the uncertainty on the wave climate estimate. Consequently a fixed value of the safety factor cannot be given, but it is certain that armour exhibiting brittle failures demands a bigger factor than armour with gentle failure development. In practice it means that brittle armour, like Accropods on the steep slope of 1 : 1.33 (recommended by SOGREAH), must be designed for a lower degree of damage (or, expressed in another way, with a bigger safety factor on the block size) than, for example, conventional rock armoured slopes. This is illustrated clearly in Fig. 12 of Paper 4. The same figure does not contain any information on the relative block size of the four types of armouring, and consequently no interpretation about the relative cost of the armour can or should be made from the figure. Dr van der Meer's last comment might be based on this kind of misinterpretation of the figure.

In reply to Mr Ligteringen, I am sorry that I did not include more details about the contents of the Dutch armour unit research project in my presentation. However, the key points are briefly mentioned in Paper 4, except that cubes were also investigated. The research represents a very useful contribution to breakwater technology. However, some important parts of the results have never been published, for example, how the results on impact speeds can be used to estimate the stresses in the armour units, and thereby the risk of breakage. Without information about this important - but also very difficult step - it is impossible to use the results in a quantitative way. Moreover, no information, to my knowledge, has been published about stresses from static loads (gravity, wedging) and from wave slamming, although such loads can be the dominating ones for large Tetrapods as well as for other larger slender interlocking concrete units. As far as I am concerned research on these aspects has not been undertaken in this programme. I hope that the very useful results of the Dutch research programme can be supplemented by further research in the near future, so that a comprehensive design method for Tetrapods can be produced.

R. B. BOUCHET, Paper 5

In reply to Professor Tanimoto, the probabilistic approach of our calculations takes into account a life time of 100 years. We expect the structures of La Condamine harbour will last a little longer.

To Mr Mettam, our works on horizontal plates were triggered by the reflections of Mr Piot and the tests he did at Chatou Laboratory in 1967. We made clear the phenomenon of the oscillating sea wall in a mathematical model in 1984, and we confirmed these results in the same year with hydraulic tests done at Sogreah Laboratory.

The pontoons are settled on fixed caissons on driven piles. The distance between two caissons is 60 m and the pontoons are 120 m long. We oppose horizontal forces with vertical forces, so we have smaller forces in both pontoons and caissons.

In reply to Professor Franco, we made many tests in flume, and also recently in basin, to check the effectiveness of slabs in 3D, but all the tests were done with regular waves. The slabs set in Barcelona are monitored to verify the forces acting on them, especially on piles. Their effectiveness seemed quite good, but we had no wave recorder in front or behind to quantify it.

As stated in Paper 5, the performance depends on the width and the immersion of the slab. The width is a function of the wavelength, and its immersion must be as low as possible. The tests we performed show that the effectiveness decreases rapidly with the immersion. In Barcelona the upper face of the slab is 2 m below the mean sea level. Many plans are underway for beach protection.

To Mr Ligteringen, we carried out many tests in flume and also in basin to assess the wave forces acting on pontoons, both with regular and random waves. An important point is to determine the dynamic behaviour of the structure struck by random waves, especially the waves whose period is close to the natural period of the structure, and also the response to seismic actions.

The two main factors for design are stability under seismic actions and comfort conditions for future dwellers; at the top of the building the acceleration must be less than 1/200 g during normal storms. In that case, it is surprisingly easier to cope with 8 m 10 s waves, than with 2 m 4 s ones.

K. HALL, Papers 7 and 8

Drs Oumeraci and Barends have made several valued points with which the Authors fully agree. Given the present state-of-the-art, it will be many years before the design engineer is given a useful reliable design procedure involving numerical modelling of breakwater flow (provided a good research effort is maintained).

Delineation of the external pressure field, subsequent coupling to the internal flow field and an examination of the actual physical processes involved in cyclic unsteady flow in porous structures, including air entrainment and inertial effects, are extremely critical areas. Studies are currently underway at Queen's University Coastal Engineering Research Laboratory (in addition to other research units) to examine each of these areas.

The inclusion of the internal stability of the porous medium (granular matrix deformation) is extremely important, as pointed out by Dr Oumeraci.

In response to Mr Maddrell, the above research (and that mentioned by Dr Oumeraci) is required before confidence can be placed in the mathematical modelling of flows. Additionally, verification using firstly scale models and secondly prototype data must be undertaken. The latter is extremely important, since mathematical modelling of breakwater flows relies on techniques developed in most cases for quite different materials (size, permeability, etc.) and configurations for a rubble mound breakwater. Only on completion of these studies can we be confident in the numerical techniques.

With respect to Paper 8, dealing with dynamically stable breakwaters, the comments made by Mr McDowell, while valid for a cobble beach system, do not apply to the 'berm breakwater' system as defined in the Paper. These structures have been tested at normal and oblique incidence and are designed so that no 'alongshore' movement occurs. All profile formation is in the onshore-offshore direction. The exception is at the head, where one of three solutions is used:

(a) increasing berm width
(b) increasing berm elevation
(c) increasing material size at the head.

All three solution techniques achieve the same purpose: to eliminate alongshore movement.

Although the 'berm family' can be extended to include materials sizes (relative to design conditions) which will move in the alongshore direction, no such structures have been designed so far. It is not recommended that this practice be utilized until a better understanding of the process has been obtained.

The Authors would like to thank Mr Simm for his comments addressing Mr McDowell's questions. The limits set by Burcharth and Frigaard to safeguard against longshore motion are very useful and very practical from an engineering perspective.

L. VAN DAMME, Paper 11

I do not agree that the measurements at Zeebrugge are 'extensive'! The given percentage covers the measurements I stated. In our experience the measurements would be rather the minimum requirement. This statement may be underwritten by T. N. Wolfe-Barry's West African experience. Compared with the contribution of J. S. Simm, the level 1 monitoring can be defined as the minimum requirement, but that may cost 0.025% of the incurred capital cost. Even compared with the maintenance costs, this amount can not be described as 'large'.

Description of the pressure transducers

Type: PTX 160/D - DRUCK
Dimension: 17.5 mm diameter x 220 mm length
Range: (-0.125 bar) - (+ 3.375 bar)
Accuracy: 0.1% FS
Resolution: 0.0005% FS
Linearity: 0.05% FS
Temperature influency: +/- 0.3% total error bound between -2°C +30°C.

In response to Dr Oumeraci's question about sampling rate, the sampling rate of the sensors is always 50 Hz. Depending on the data processing that is used, we can reduce the information to produce a sampling rate between 0.5 and 50 Hz.

12. Historical development of breakwater structures in the world

K. TANIMOTO, Port and Harbour Research Institute, and
Y. GODA, Yokohama National University

SYNOPSIS. Historical development of breakwater structures in the world is reviewed on the basis of a literature survey. Recent development of breakwater structures in Japan is also introduced. Various sequences of the evolution of breakwater structures can be observed from the primitive natural stone breakwaters to the modern structures with greater stability against waves. Recent topics of interest are the revival of the primitive breakwater (berm breakwater) in one direction and the development of sophisticated caisson breakwaters in the other direction.

HISTORICAL EVOLUTION OF BREAKWATER STRUCTURES

1. History of breakwaters can be traced to the ancient ages of Egypt and Mycenae. In the present paper, however, the historical evolution after the 18th century will be summarized to provide a common understanding of modern breakwater design in the world.

A treatise on historical development of breakwater design by Y.Ito

2. Many discussions on breakwater structures have been made after the construction of the first modern breakwater at Cherbourg (France) around the end of the 18th century. Y.Ito published a paper entitled "A history of the development of breakwater structures", which reviewed those discussions up to 1969 by referring to 160 papers (ref.1).

3. Figure 1 shows the flow of historical development of breakwater structures which was originally prepared by Ito in 1972 and rearranged and modified slightly by K.Tanimoto in 1989. Beginning from the primitive dike of natural stones, various sequences are noticed in the structural development toward the increase in the stability against waves or saving in the construction cost. Some sequences will be explained with typical breakwater sections.

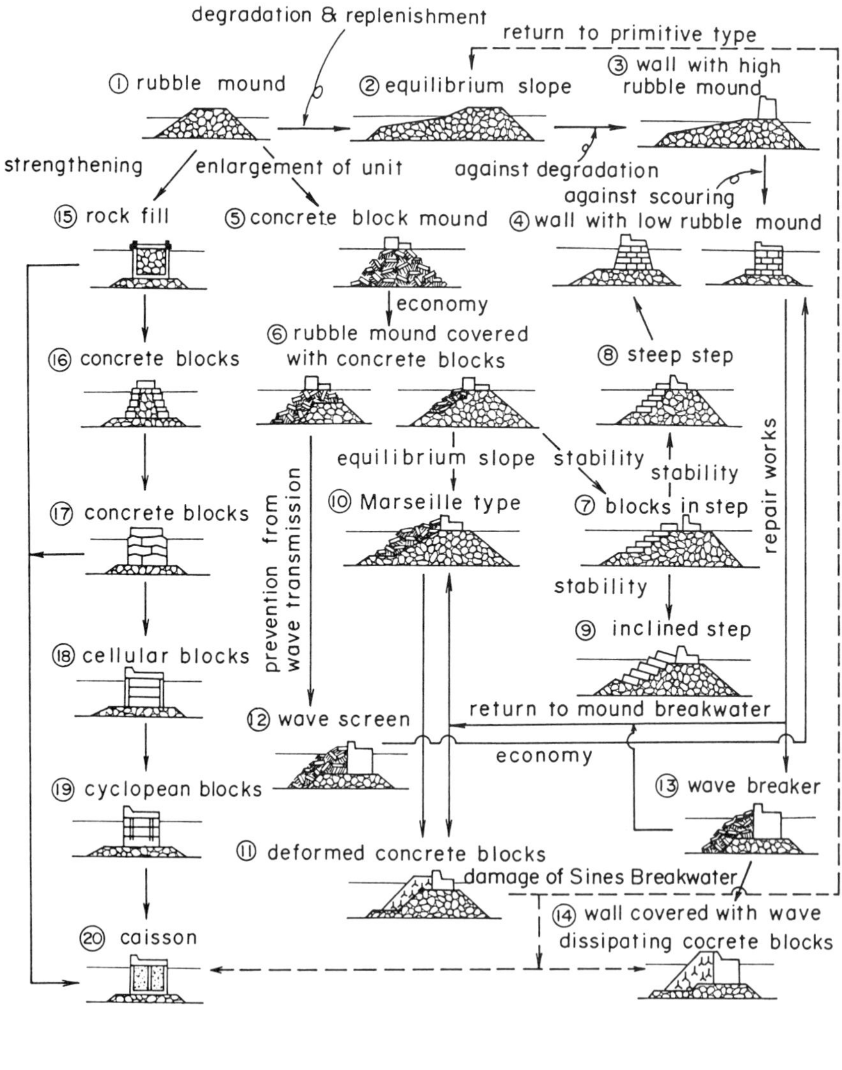

Fig.1 Evolution of breakwater structures

Evolution from rubble mound breakwaters to wall breakwaters

4. Introduction of upright superstructures. It is said that the modern breakwater construction began at Cherbourg in 1784. Figure 2 shows the cross section after the completion in 1853. The initial design was that of wooden frames in the cone shape filled with stones, but the cone breakwaters were abandoned in 1789 due to heavy damage, and a quarry stone breakwater having the seaside slope of 1 on 3 adopted for reconstruction. By the end of 1790, stones of 2.7 million cubic metres had been placed in position for the extension of 3,850 metres.

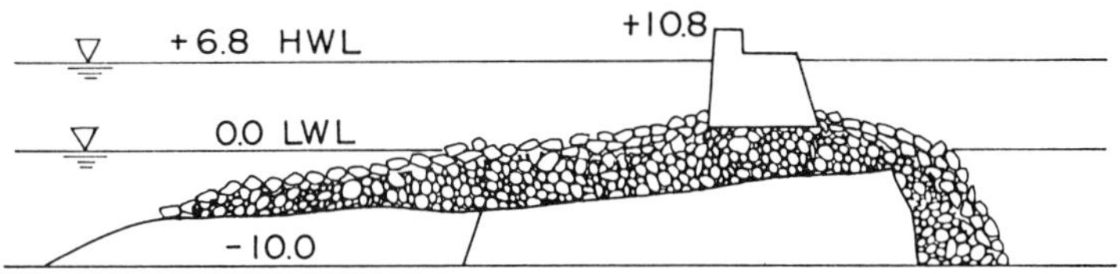

Fig.2 Cross section of Cherbourg breakwater

5. The quarry stone breakwater, however, had never been accomplished because of degradation of the top level work by storm waves. In 1828, by considering the damage caused to the main breakwater body, it was decided to build the present breastwork with granite blocks. The job continued for 23 years from 1830 to 1853.

6. Because the structure was of a mixed type composed of a vertical wall and a high rubble mound foundation, impulsive wave forces due to breaking waves were exerted upon the vertical wall. During the construction of the superstructures, the vertical wall and the rubble mound foundation were frequently damaged and the repair work is still continued at present. The present profile having a seaside slope of about 1 on 7 is the result of erosion of armour stones in a large quantity, which have been replenished since the beginning of the 19th century.

7. According to the report by J. Boulard, the total amount of replenished materials is approximately 6 million cubic metres of stoneblocks and 0.5 million cubic metres of masonry (ref.2). The work of Cherbourg breakwater is compared with that of a recent breakwater such as that of the Antifer oil port, which is 3,500 metres long in water with the depth of 8 to 25 metres (average of 20 metres). It is concluded that the material volume of the Cherbourg breakwater is 3 times that of Antifer, the construction term 20 times, and the cost 4 times, if both works are converted to comparable dimensions in length and height.

8. Deepening of the bottom level of wall. The bottom level of vertical wall in the Cherbourg breakwater was on the

low water level. One of the solutions to decrease the scouring of rubble mound as well as wave forces on the wall is to deepen the bottom level of wall. This solution was pursued mainly in the United Kingdom. The bottom level of wall reached to -3.5 metres in the Alderney breakwater in 1849, and to -13.1 metres in the Peterhead breakwater in 1898 as shown in Fig.3. Prior to these, the construction of a moundless vertical breakwater had been undertaken at Dover in 1847, in consideration of the difficulties in the constructions of rubble mound breakwaters at Cherbourg and Plymouth.

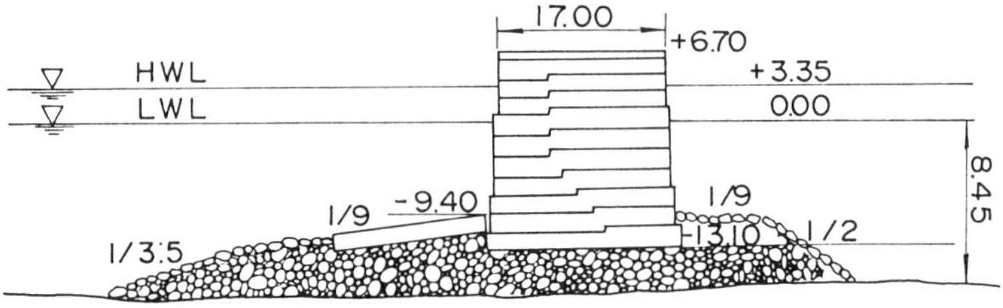

Fig.3 Cross section of Peterhead breakwater

Evolution through concrete block mound breakwaters to wall breakwaters.

9. From quarry stones to concrete blocks. Another solution to save the amount of materials for mound breakwaters, which had to become a gentle slope structure in the equilibrium state, was to use artificial concrete blocks of large size instead of quarry stones so that a steep slope mound could be formed. A mound breakwater built with concrete blocks only was first constructed at Alger (Algeria) in 1830's. Then, the section was improved as shown in Fig.4 in 1840's, in which concrete blocks were used as an armour layer for quarry stones to save the construction cost.

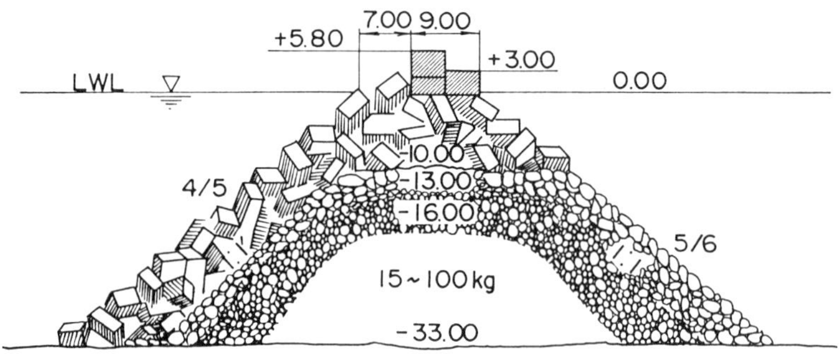

Fig.4 Cross section of Alger breakwater

10. <u>From pell-mell to stepwise placement.</u> An idea to increase the stability of concrete blocks was to place them in steps having an overlap. Such a breakwater as shown in Fig.5 appeared at Genoa (Italy) in 1879.

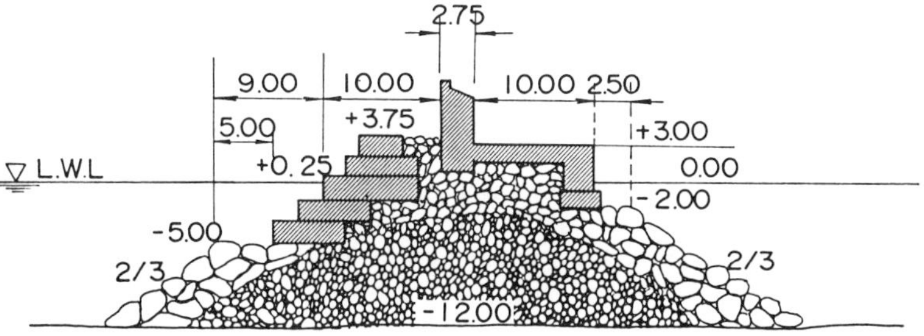

Fig.5 Cross section of Genoa breakwater

11. <u>From overlapping blocks to a vertical wall.</u> As blocks were overlapped more and more to increase the stability, the structure eventually became a vertical wall. In this sequence, a vertical breakwater with a low rubble mound appeared in Naples (Italy) as shown in Fig.6 in 1900. Since then, a vertical breakwater with a low rubble mound became a predominant structure of breakwaters in Italy. The structure of vertical walls was developed from concrete blocks weighing several tens of tons to cyclopean blocks weighing hundreds of tons through cellular blocks during the early decades of the 20th century.

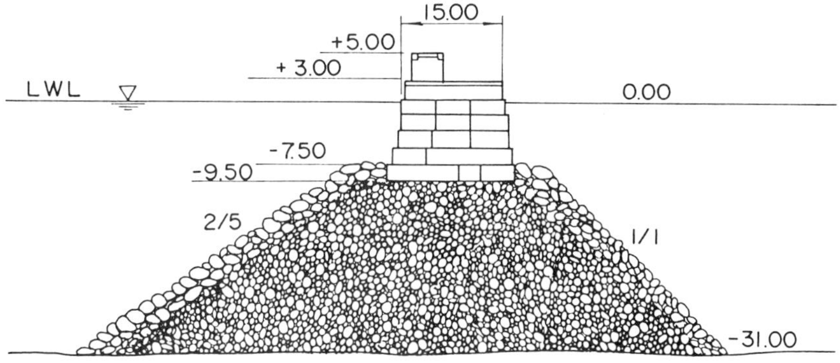

Fig.6 Cross section of Naples breakwater

<u>The first and second shocks in the history of breakwaters</u>

12. <u>The first shock</u>. In 1930's, damage to vertical wall breakwaters occurred one after another, for examples at Catania (Italy) in 1933, at Alger and at Leixões (Portugal) in 1934. Figure 7 shows an example of damage in Catania. Most of the damaged breakwaters were successfully

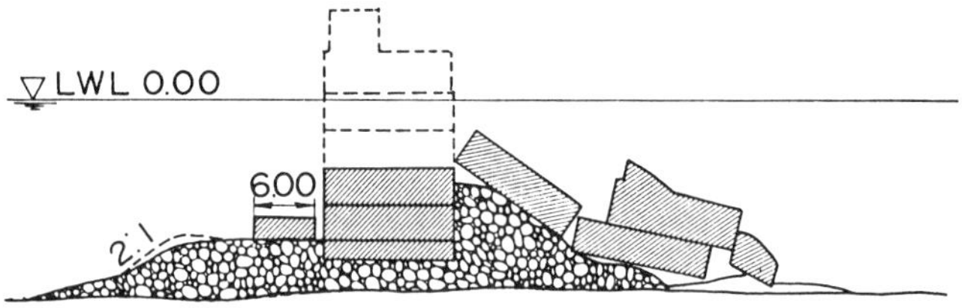

Fig.7 Damage of Catania breakwater

reconstructed as mound breakwaters of the Marseille type. Namely, the breakwater structure partly returned to the mound type from vertical wall type. These incidences could be considered as one of the reasons that rubble mound breakwaters are still predominant in the world, although there are various other reasons for the damage of vertical breakwaters. The authors call these incidences as the first shock in the history of breakwater construction.

13. The second shock. The second shock happened unfortunately at Sines (Portugal) in 1978/79 after about a half century from the first shock. The heavy damage on Sines breakwater is too well known among coastal and harbour engineers to need any explanation. It can be said that the damage has greatly stimulated discussions on structures and design philosophy of breakwaters in the world. One of the consequences from a viewpont of the structural aspect is the highlight upon a primitive berm breakwater. Figure 8 shows an example of berm breakwaters in Brazil (ref.3).

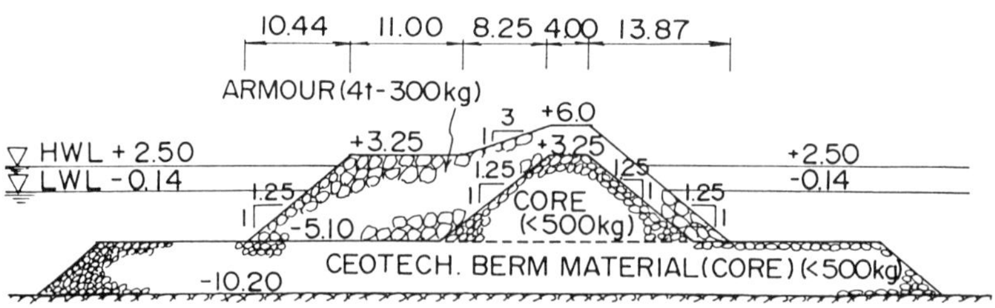

Fig.8 Cross section of Sergipe breakwater (ref.3)

HISTORICAL DEVELOPMENT OF BREAKWATERS IN JAPAN (ref.4)

14. The first modern breakwater construction in Japan was undertaken at Yokohama Port in 1891, which were designed by H.S. Palmer (a retired British Major-General). One of the sections is shown in Fig.9. The breakwaters having a total length of 3,630 metres were first completed in 1896. They were redesigned however with solid concrete blocks for the whole upright section after damage by a storm in 1902.

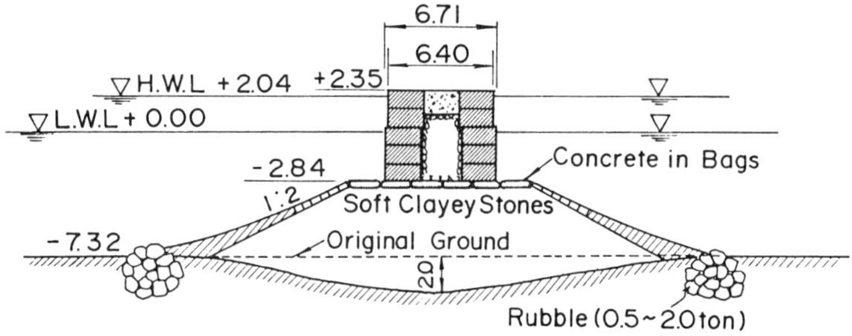

Fig.9 Cross section of Yokohama breakwater (1)

15. A vertical breakwater consisting of solid concrete blocks was first built at Otaru Port by I.Hiroi from 1897 to 1907. He adopted the breakwaters at Karachi, Madras, and Colombo as the prototype of Otaru breakwater. Breakwaters with reinforced concrete caissons were introduced at Kobe Port for the first time in 1911, and then utilized for the island breakwater at Otaru Port in 1912. Because the wave conditions are much severer at Otaru than Kobe, the breakwater was quite sturdy as shown in Fig.10. The caissons weighing 883 tons were filled with poured concrete. Caisson breakwaters soon became the major structural type of breakwater for use in withstanding rough seas.

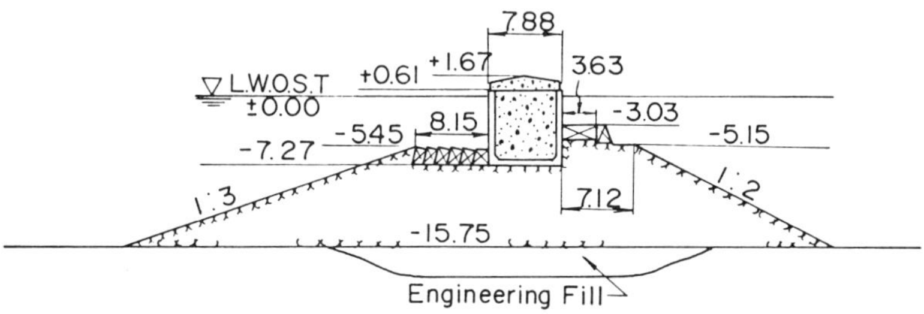

Fig.10 Cross section of Otaru breakwater

16. The filler material for the caisson cells has gradually evolved from concrete to gravel and then to sand. One of the early examples of a sand-filled breakwater is the outer breakwater at Yokohama Port, which was designed by S.Samejima and built from 1928 to 1943. As shown in Fig.11, the caissons were very sturdy, with an outer wall 60 centimetres thick. The interiors of the caissons were filled with a mixture of sand and cobbles to give the greatest possible density. Filling of caissons with sand was first done in water areas where the wave conditions were relatively mild, but sand filler soon became popular in areas of rough seas as well. This was particularly so in the period shortly after World War II,

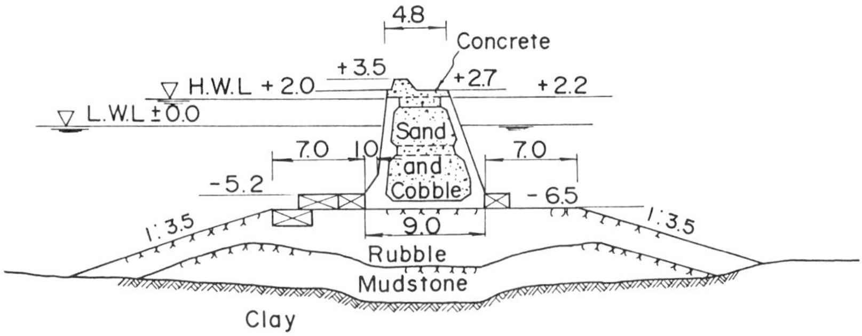

Fig.11 Cross section of Yokohama breakwater (2)

when cement was scarce because of damage to production facilities.

17. Breakwaters are also constructed as barriers against tsunamis and storm surges. Storm surge protection breakwaters, of which the total length is 8,250 metres, were built at Nagoya Port from 1962 to 1964 after the catastrophe caused by Isewan Typhoon in 1959. In succession, a tsunami protection breakwater as shown in Fig.12 was built at Ofunato Port from 1963 to 1967 after the damage due to Chilean earthquake tsunami in 1960. The breakwater is located at the bay mouth, where the maximum water depth is 38 metres. At the central harbour entrance with the opening of 200 metres, a submerged dike up to the level -16.3 metres was built to reduce the opening as much as possible.

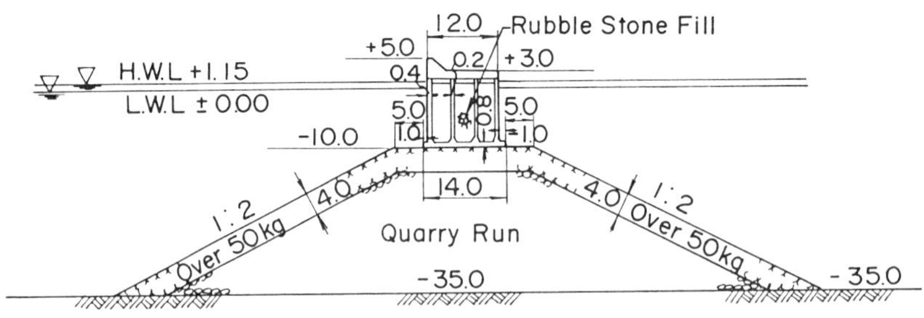

Fig.12 Cross section of Ofunato breakwater

18. The widest caisson breakwater is found at Hedono Port in a south-west island. The caisson width is 38 metres as shown in Fig.13 against design waves of 9.7 metres significant wave height, 17.5 metres maximum wave height, and 13.2 seconds significant wave period.

19. The appearance of artificial concrete blocks called tetrapods, which was proposed by P.Danel in 1949 (ref.5), greatly affected the design of breakwater structures in Japan. Similar blocks were first used at Yagi Port in 1955. Then a caisson breakwater at Rumoi Port, which was built

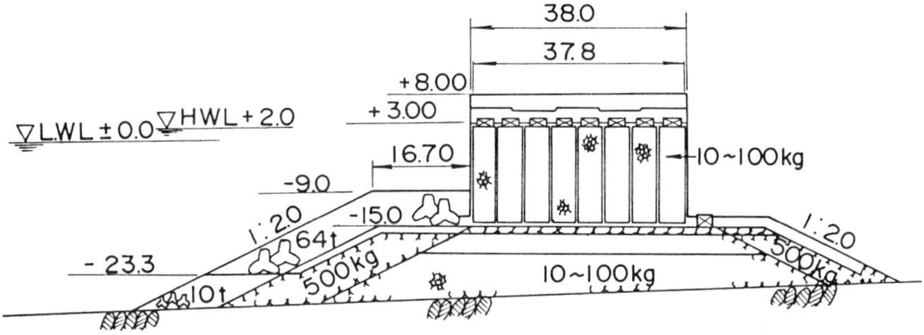

Fig.13 Cross section of Hedono breakwater

from 1911 to 1933, was reinforced with artificial concrete blocks as shown in Fig.14. This type of breakwaters became very popular as a breakwater in relatively shallow water areas where waves are broken. The function of artificial concrete blocks in front of a vertical wall is not the same as armour units for mound breakwaters but to reduce wave forces, wave overtopping and wave reflection by dissipating waves. Consequently, they are called wave dissipating concrete blocks in Japan.

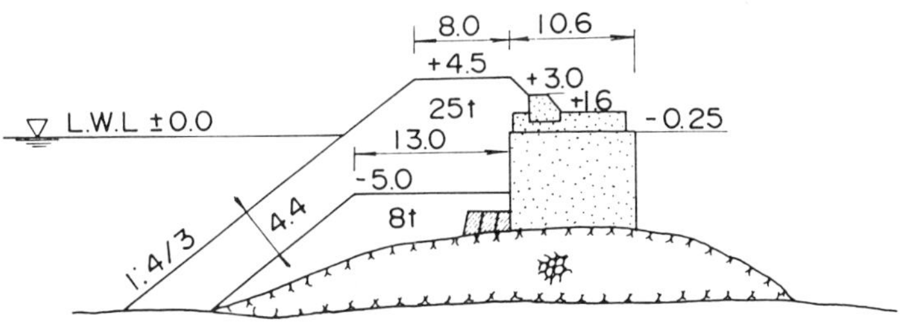

Fig.14 Cross section of Rumoi breakwater

RECENT DEVELOPMENT OF CAISSON BREAKWATERS (ref.6)

Caisson breakwaters utilizing wave forces to aid stability

20. Basic shapes of wave barrier wall. The principal function of breakwaters is to stop the wave propagation there, and the simplest method is to intercept the water by a solid wall. The shape of this solid wall can be considered arbitrary, but the basic four shapes as shown in Fig.15 are selected for the discussions. They are a vertical wall, a sloped wall, a concave wall and a convex wall.

21. A vertical wall is often applied as a vertical breakwater. The length of the wall is the shortest, but a horizontal wave force acts on the wall almost with the same phase from the top to the bottom. On the other hand, the

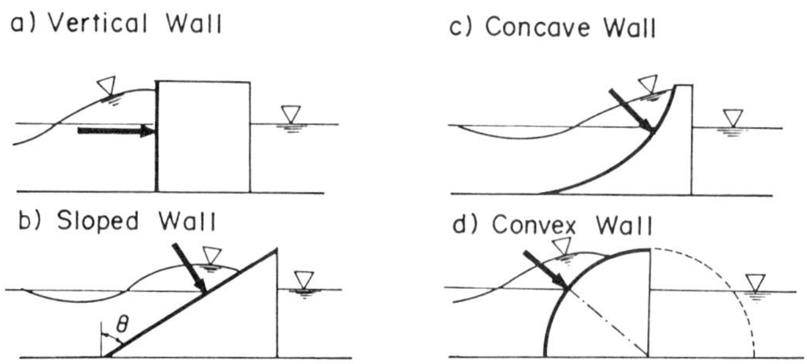

Fig.15 Selected shapes of wave barrier wall

others have a longer wall, but a wave force acts with a downward component. Since the downward component works as a weight to resist sliding, breakwaters having these walls can be said to be breakwaters utilizing the wave forces for the stability. Some of them are introduced.

22. Sloping top caisson. A triangular caisson is ideal for the stability against sliding, because it is automatically stable under the action of a wave crest when the slope is selected appropriately. It is difficult, however, to build a sloped face caisson breakwater for the whole height. The actual application of this concept becomes feasible by making a sloping face only at the upper part of the front wall. A typical example of such a breakwater can be found in so called the Hanstholm type breakwaters (ref.7).

23. Multi-cellular caisson. As an application of a concave wall, a multi-cellular caisson as illustrated in Fig.16 has been developed for breakwaters in relatively deep water with rough seas in Japan. The idea originated from a sloped caisson. A concave wall is then conceived to

Fig.16 Artist image of multi-cellular caisson breakwater

reduce the width. Finally, the concave wall is divided into three segments and piled up vertically to make a compact structure. After laboratory studies, a full size unit of the multi-cellular caisson was fabricated in 1985, and was tested successfully at Wakayama Port in 1986/87 (ref.8).

24. Semi-circular caisson. A distinguished merit of a convex wall appears when the section is a semi-circle. Wave pressures on the semi-circle face act towards the centre. Consequently, the resultant force acts towards the centre and makes no rotational moment. It results in a quasi-uniform distribution of the reaction at the bottom slab. This merit can be used as a breakwater on very soft ground. Because of the high stability against waves and the soft feature with a round top, it is expected that a semi-circular caisson is very favourable for offshore breakwaters to protect a beach from erosion particularly in a marine recreational area. Figure 17 shows an example of cross section by a possible design for extremely high breaking waves.

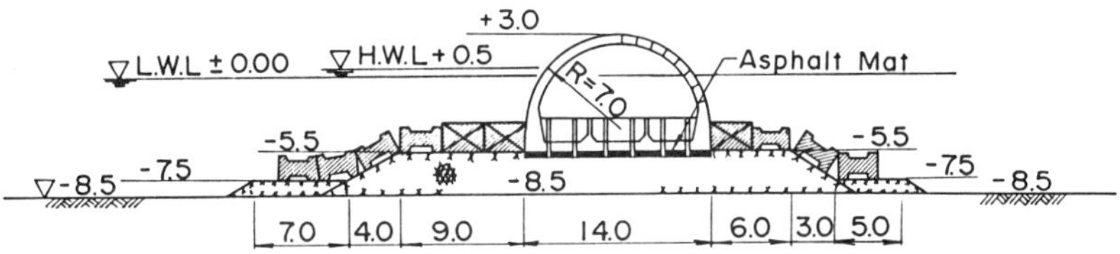

Fig.17 Possible cross section for extremely high breakers

Wave absorbing caisson breakwaters

25. Wave dissipating caisson. One of disadvantages of a vertical breakwater is its high wave reflectivity. G.E. Jarlan proposed a perforated caisson to cope with this disadvantage in 1961 (ref.9). Any shape of perforation is acceptable with an appropriate opening ratio, although the original one by G.E.Jarlan was a circular hole. In Japan, a wave dissipating caisson having a vertical slit wall became gradually popular in a relatively calm water area, since the first one was used at Takamatsu Port in 1970. Perforated caisson breakwaters are also found at several ports in Italy (ref.10).

26. Curved slit caisson. A curved slit caisson was proposed to apply the wave dissipating caisson to rough seas in 1976 by K.Tanimoto. After laboratory and field tests, a curved slit caisson breakwater of 150 metres long was completed in the construction at Funakawa Port in 1984. Figure 18 shows its cross section. The curved slit members were prefabricated with prestressed concrete and fixed to the main body of caisson by a dry joint method.

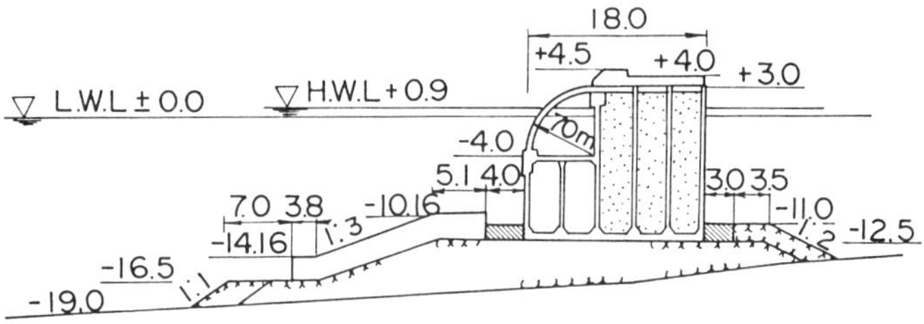

Fig.18 Cross section of Funakawa breakwater

27. Dual cylindrical caisson. According to a calculation, a trapezoidal caisson and a cylindrical caisson become more economical than a rectangular caisson, as the water depth becomes deeper and deeper in a rough sea. The deepest breakwater in Japan, probably so in the world, is under construction at Kamaishi Port with a trapezoidal caisson. As it is introduced by a separate paper in the present Conference, a cylindrical caisson is described here.

28. Even in deep water breakwaters, the construction of low reflective structures and permeable structures is requested sometimes to minimize the environmental impact. Responding to this request, a dual cylindrical caisson as shown in Fig.19 has been developed, although field tests are still in operation at Sakai Port from 1989. The outer cylinder has a perforation and form a wave chamber of doughnut shape bounded by an impermeable inner cylinder.

Fig.19 Artist image of dual cylindrical caisson breakwater

29. Wave power extracting caisson. Although the wave energy is just dissipated by the wave dissipating caisson, the same effect to reduce wave reflection can be achieved by converting it into usable energy, for example electricity. A combination of the wave power converter and

the caisson breakwater is particularly attractive because the construction cost can be jointly borne by the accounts for power generation and harbour protection. A series of field tests of a wave power extracting caisson breakwater is being conducted at Sakata Port from 1987 (ref.11).

ACCESSIBLE BREAKWATER WITH AMENITY

30. In Japan, breakwaters are generally kept unaccessible for people to diminish the danger by overtopping waves. Nevertheless, many people like to stay on breakwaters, for instance, for fishing. A recent important movement is to make breakwaters accessible to the public. Some of the breakwaters have been equipped with handrails and been opened to fishing people. Besides them, a new breakwater with a promenade to which people can have free access like a park is under construction at Wakayama Marina City. Figure 20 shows its image made by Prof.Tsuyoshi Sasada, Environmental Engineering, Osaka University, with computer graphic. The design of promenade has been carried out by taking an architectural aspect into consideration, and a green plantation on the breakwater is being planned to increase the amenity. Such a breakwater with promenade was built at Brighton Marina in 1978 (ref.12).

Fig.20 Accessible breakwater with amenity

CONCLUDING REMARKS

31. The history of breakwater construction was a series of struggles against powerful waves. Various structures appeared through various sequences from a natural stone breakwater to increase the stability and/or to save the construction and maintenance cost.

32. Recent topics of interest are the revival of primitive berm breakwaters for a relatively calm sea and

the development of sophisticated caisson breakwaters for a rough sea. Various new caisson structures, depending on water depth and wave condition, have been developed greatly in Japan. They are no more breakwaters struggling against waves but breakwaters harmonizing with waves.

33. In addition, demands of an accessible breakwater are increased recently together with popularization of marine recreations. In such a breakwater, a human sense should be taken into consideration in the design as well as natural conditions.

REFERENCES

1. ITO,Y. A history of the development of breakwater structures. Tech. Note of Port and Harbour Res. Inst., No.69, 1969. (in Japanese)
2. BOURARD,J. The port of Cherbourgh—The outer breakwater— Cherbourg Naval, Private memo, 1980.
3. MURRAY,M. and SAYAO,O.J. Offshore breakwater for the Sergipe Marine Terminal, Brazil. Book of Abstracts, ICCE'90, 1990.
4. GODA,Y. Random seas and design of maritime structures, University of Tokyo Press, 1985.
5. DANEL,P. Tetrapods. Proc. of Fourth Conference on Coastal Engineering, 1953.
6. TANIMOTO,K., TAKAHASHI,S. and KIMURA,K. Structures and hydraulic characteristics of breakwaters—The state of art of breakwater design in Japan —, Rept. of Port and Harbour Res. Inst., Vol.26, No.5, 1987.
7. AGERSCHOU,H., LUNDGREN,H. and SØRENSEN,T. Planning and design of ports and marine terminals, John Wieley and Sons.
8. OKADA,Y., WATANABE,T., SUGAWARA,T. and TANIMOTO,K. Recent developments of new type breakwaters in Japan, 27th International Navigation Congress, Osaka, 1990.
9. JARLAN,G.E. A perforated vertical breakwater, The Dock and Harbour Authority, Vol.41, No.488, 1961.
10. ROMITI,G., NOLI,A. and FRANCO,L. The Italian experience in composite breakwaters, Breakwaters '85, Thomas Telford Ltd, 1985.
11. TAKAHASHI,S. Hydrodynamic characteristics of wave-power extracting caisson breakwater, 21st ICCE, 1988.
12. TERRETT,F.L., GABLY,P. and STUBBS,S.B. Harbour works at Brighton Marina: Investigtion and design. Proc. Inst. of Civil Engineer, 66(1), 1980.

13. Large-scale model investigation: a contribution to the revival of vertical breakwaters

H. OUMERACI and H. W. PARTENSCKY, University of Hanover, E. TAUTENHAIN, Large Wave Flume, and H. NICKELS, University of Hanover

SYNOPSIS. A research programme for the investigation of caisson breakwaters subject to breaking wave impact forces is outlined as a contribution to the revival of vertical breakwaters. Since the large-scale model study constitutes the key part of the programme, the first results of this study related to the impact loading and dynamic response of a caisson breakwater model are presented and discussed.

INTRODUCTION

1. A number of cases of serious damage were experienced by large vertical breakwaters before and after World War II. Since then, caisson breakwaters have been widely used only in such countries like JAPAN and TAIWAN where it is difficult to quarry sufficient quantities of sound large-sized rock units. Elsewhere, this breakwater type has been abandoned in favour of rubble mound breakwaters. However, potential possibilities for the revival of vertical breakwaters are appearing to emerge, essentially due to the catastrophic failures of a series of large rubble mound breakwaters in the last 15 years.

This may be encouraged by the fact that, despite extensive research programmes on rubble mound breakwaters which followed the aforementioned failures, only very limited results related to the most urgent aspects (structural strength of armour, geotechnical stability, crown-wall stability etc.) have been obtained.

In addition, this revival could be promoted by the following relatively recent developments:

a) Considerable improvement of the technology of caisson structures and their foundation, particularly in the offshore oil industry. The largest caisson which has been built to date has a volume of more than 160,000 m^3 (70m·45m·51m) with construction costs of about $486/$m^3$ (ref.7). Potential possibilities for further developments exist with respect to reducing wave impact forces, construction time, foundation work and costs.

b) An increase of the need for protective structures in deep water, essentially due to the increase of ship sizes in the last decades. The economic advantage and the considerable savings in construction time which would result from the use of composite breakwaters is thus obvious.

c) An improvement of model techniques and the availability of large-scale testing facilities ("Super wave tanks") in which the dynamic, hydraulic and hydro-geotechnical aspects can be simultaneously investigated.

d) The availability of more reliable wave records and wave hindcast models for the evaluation of extreme wave conditions.

e) Experience in the numerical modelling of wave-structure-soil interaction, particularly in the offshore industry.

2. Considering these developments and the state of the art on the breaking wave impact loading of vertical structures, the FRANZIUS-Institut has started an extensive research programme on caisson breakwaters which is supported by the German Research Council (DFG) within the Special Research Unit SFB 205 in HANOVER. The main objective of this programme is to improve the understanding of the physical processes involved and to elaborate more reliable design methods for vertical structures subject to breaking wave forces.

3. The underlying philosophy of this paper is to question the fundamental requirements and conditions for caisson breakwaters to become more competitive than the rubble mound type. This is illustrated by showing the research programme and strategy and discussing some results of the ongoing investigations at the University of HANOVER.

BRIEF REVIEW OF VERTICAL BREAKWATER FAILURES

4. Most of the disasters occurred before World War II, with the exception of the failure of the GENOA breakwaters and of some more recent less known failures in JAPAN like the breakwater collapse in NIIGATA Harbour.

The most catastrophic failures are summarized in Tab.1.

5. A detailed analysis of these failures and a series of further more or less serious cases of damage in JAPAN (ref. 5) and elsewhere is being performed and will be the subject of a forthcoming paper.

Table 1- Relevant vertical breakwater failures

Breakwater (year)	Country	Design Wave H(m)/ L(m)	Actual Wave H(m)/ L(m)	Major causes of failure
Byzerte (1905) (1915)	Tunisia	-	-	Too high rubble found. Breaking waves Overtopping
Valencia (1926)	Spain		7/150	Waves underestimated Scouring, soft soil Overtopping Breaking waves
Antofagasta (1928) (1929)	Chile	6/90	9/250 8/800	Waves underestimated No connect.betw.blocks Breaking waves
Catane (1930) (1933)	Italy	- -	7/200 7.5/250	Waves underestimated No connect betw.blocks Too high rubble found. Breaking waves
Algiers (1934)	Algeria	5/80	9/300	Waves underestimated Scour, soft soil Overtopping Breaking waves
Genoa (1955)	Italy		7/200	Waves underestimated No connect betw.blocks Breaking waves
Niigata (1976)	Japan	7/260	7/280	Soft soil Breaking waves

However, some of the relevant results which have already been obtained may be summarized as follows:

a) During the storms which damaged these breakwaters, breaking waves or breaking partial clapotis rather than pure clapotis were generally observed in front of or at the structure; i. e. breaking was not solely caused by shallowness. It might be also attributed to the irregularity of the waves (frequency modulated wave groups, storm wave -long wave interaction), to the phase interference of incident and reflected waves as well as to the effect of wind on waves.

b) Almost all these structures were heavily overtopped by the waves which caused the collapse.

c) Although the design wave climate was largely underestimated for almost all the breakwaters which were damaged prior to the Sixties, this was rarely the case for the more recent structures. In addition, the type of structure which consisted of vertically non-connected blocks was abandoned since the Thirties in favour of monolithic structures. Nevertheless, sliding and tilting of the new structures continued to occur, not only in the case of soft ground.

d) None of the collapsed breakwaters was designed to withstand the loading induced by breaking waves or/and overtopping. It was thought that the fundamental characteristic of these structures was to reflect the waves without the occurrence of shocks. It is also surprising that this concept is still implicitly present in the existing standard design formulae and textbooks (ref.4).

e) Single waves caused breaches of up to 150m width, which were then widened by the next waves. The most striking feature of these gaps, however, was their distribution along the breakwater. The first important breach generally occurred adjacent to the breakwater head.

f) One of the further striking failure modes was the seaward tilting of a series of breakwaters (VALENCIA, ALGIERS, NIIGATA etc.). Although this failure mode is generally attributed to scouring or/and liquefaction(refs. 11 & 15), the mode of transfer of the dynamic loading to the foundation, and thus the actual loss of stability are still not understood.

5. These results allow us to draw the following requirements for an eventual revival of vertical breakwaters:

a) The existing design methods which are still based on static calculation concepts must be rejected. In fact, the design of caisson break-waters is a purely dynamic problem involving the wave-caisson-foundation interaction. Therefore, the wave breaking load history as well as the dynamic response of the whole system are needed.

b) Since the breaking wave phenomenon and the mode of transfer of the dynamic loading to the structure have a three-dimensional character, the final design methods should be checked under 3-D conditions of wave attack.

c) The development of methods for the evaluation of the impact loading und dynamic response of the structure and its foundation is not sufficient, however. Therefore, it should necessarily be accompanied by a further development of caisson breakwater technology towards reducing impact loading and dynamic response, and facilitating construction and foundation work (ref.6).

RESEARCH OBJECTIVES, STRATEGY AND PROGRAMME

6. Based on the aforementioned findings and according to the research facilities available at the FRANZIUS-Institut, the following objectives have been derived:

- Improvement of the understanding of the breaking wave impact generation and mechanisms (qualitative investigations);
- Evaluation of impact loading and dynamic response of caisson breakwaters;
- Development of design methods and guidelines;
- Evaluation of counter measures to increase the stability of the structure and its foundation.

7. In order to achieve these objectives the research strategy and programme shown in Tab.2. have been established.

Table 2- Research programme and strategy

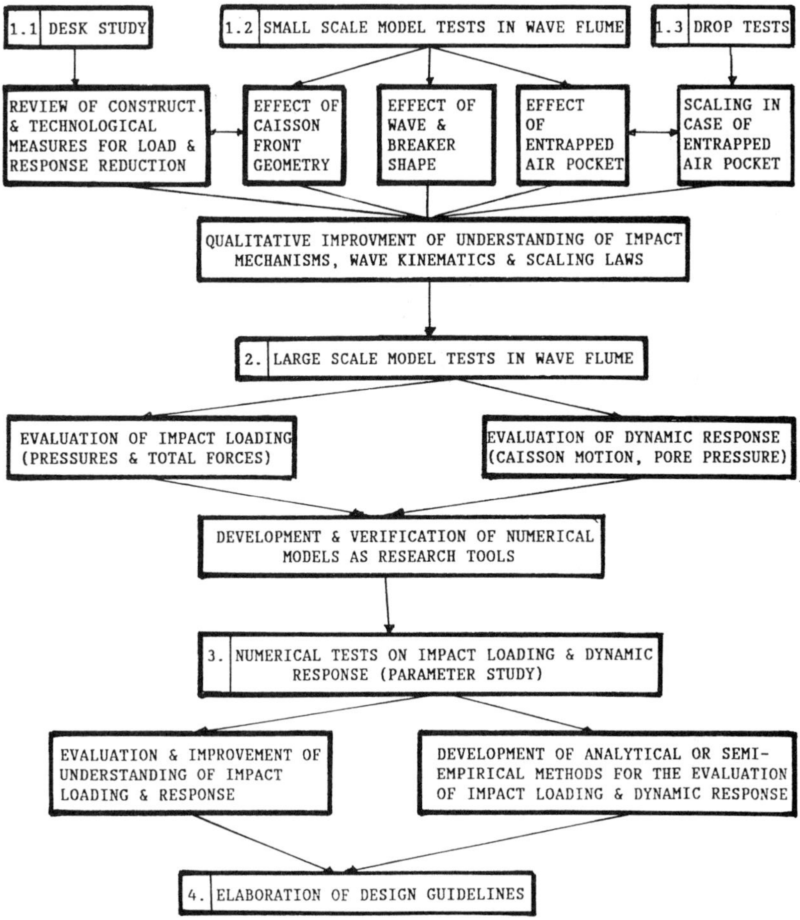

EXPERIMENTAL SET-UP AND TEST PROCEDURE

8. Description of the breakwater model. A front view of the caisson breakwater tested is given in Fig. 1. It is seen that the model consists of two independent parts.

On the first one, for which a cross section is shown in Fig. 2, total vertical and horizontal forces are measured.

For the vertical force measurement three 500 kN-force transducers of type C2 from HOTTINGER BALDWIN MESSTECHNIK have been used. Their natural frequency is 10.5 kHz.

For the horizontal force measurement two 200 kN-force transducers of type U2A from HOTTINGER BALDWIN MESSTECHNIK are used. Their natural frequency is 5.6 kHz.

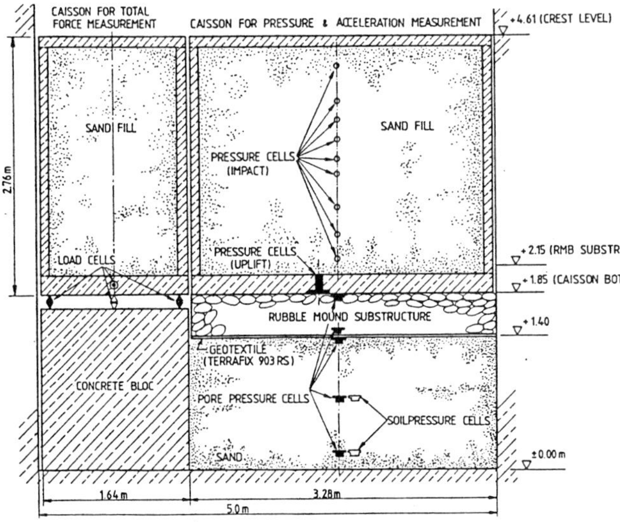

Fig. 1 - Front view of the caisson model

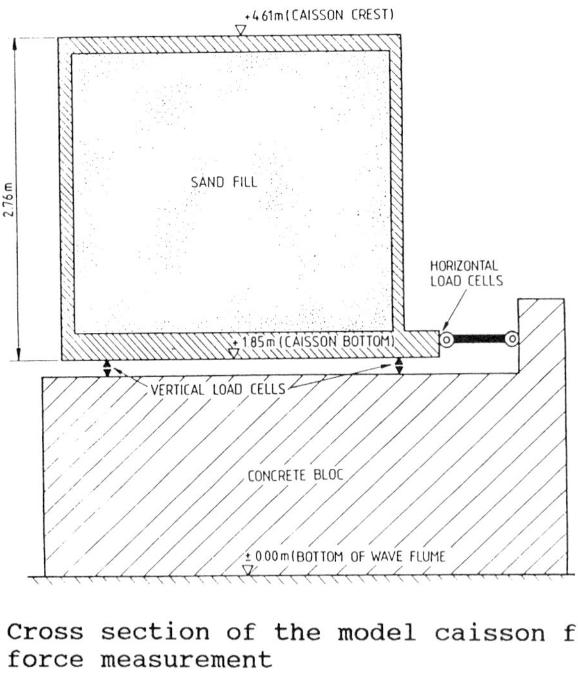

Fig. 2- Cross section of the model caisson for force measurement

On the second caisson part, for which a cross section is shown in Fig.3, pressures and accelerations are measured.
Piezo-electric pressure transducers of type PDCR 830 from NATEC SCHULTHEISS are used for the measurement of impact pressure on the caisson front, the uplift pressure at the bottom of the caisson and the pore pressure within the rubble mound foundation. Their natural frequency varies from 28 kHz to 360 kHz for the pressure range of 0.35-35 bar.
The total wave-induced stress variations in the sand layer beneath the rubble mound foundation were measured by PHILIPS pore water pressure transducers of type KS 2150 and KULITE soil pressure cells of type O234.

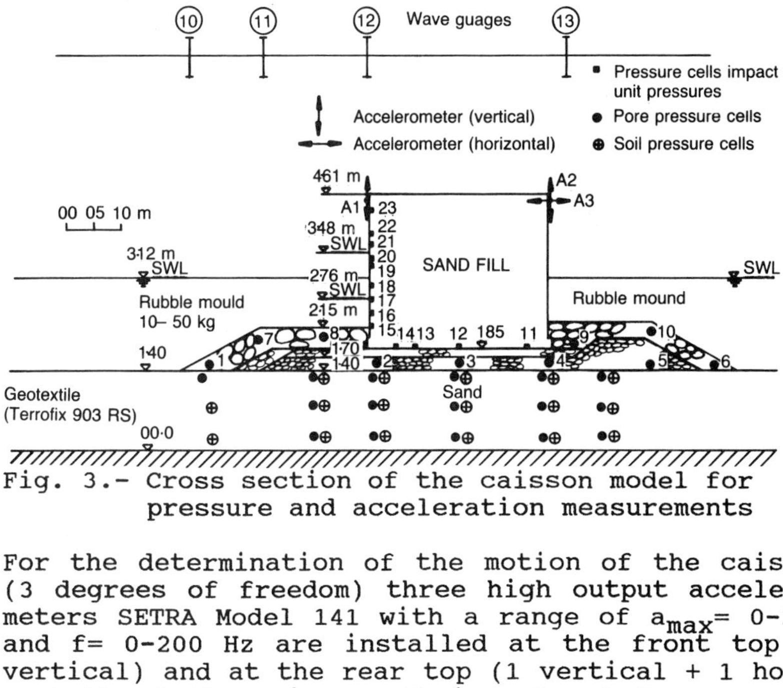

Fig. 3.- Cross section of the caisson model for pressure and acceleration measurements

For the determination of the motion of the caisson (3 degrees of freedom) three high output accelerometers SETRA Model 141 with a range of a_{max}= 0-2 g and f= 0-200 Hz are installed at the front top (1 vertical) and at the rear top (1 vertical + 1 horizontal) of the caisson. Their natural frequency is 275 Hz. These accelerometers have been selected, because of the quite low range of response frequency (f= 5-20 Hz) and acceleration (a= 0.2g) to be expected. Since $x_{max}= a_{max}/4\ \pi^2\ f^2$, the corresponding maximum amplitudes x_{max} of the caisson motions are in the range of millimeters.

9. Test programme and procedure. Monochromatic waves with heights H= 0.60-1.20 m and periods T= 4-7 s, as well as irregular waves with the same range of maximum heights have been used.
The water depths, wave heights and periods were selected in such a way that breaking, broken and

standing waves could be obtained. Tests under overtopping wave conditions have also been performed.
Two alternatives of the vertical caisson front have been tested for the same conditions: plane and cylindrical fronts.
Impact pressures and forces, uplift, pore water and soil pressures as well as accelerations (motions) of the structure have been simultaneously measured. Detailed measurements and observations (video) of the waves at the structure have also been performed.
After completion of the hydraulic model tests, pendulum tests were run on the same model in order to determine the dynamic characteristics of the structure and its foundation.

FIRST EXPERIMENTAL RESULTS

10. Classification of impact loading cases.
Impact pressure and force histories, together with more than 500 corresponding simultaneously recorded pictures (high-speed video) of breaking wave shapes have been analysed on the basis of defined criteria related to water depths and velocities.
Based on the results of this analysis, four main breaking wave impact loading cases have been identified (Fig. 4). The detailed description of these loading cases as well as the procedure which led to their determination are given in ref. 9.

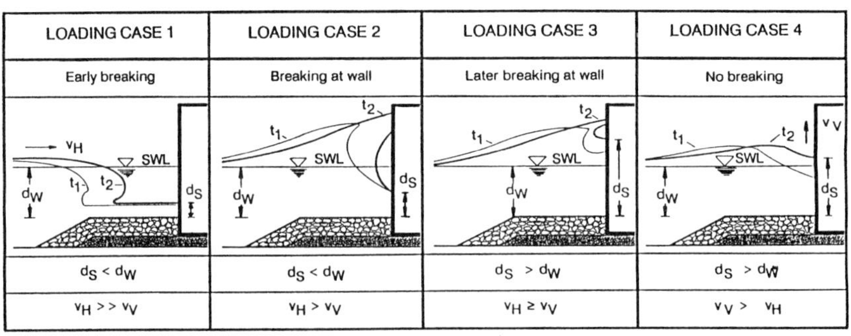

Fig.4- Classification of impact loading cases

11. Impact pressure distribution on the caisson front. Detailed spatial and temporal impact pressure distribution and the resulting horizontal force can be easily determined from the test results for time steps as small as Δt = 0.11 milliseconds. An example of such pressure distributions is shown in Fig. 5 for time steps Δt= 5.9 ms. They are caused by a plunging breaker of about 1 m height (loading case 2).

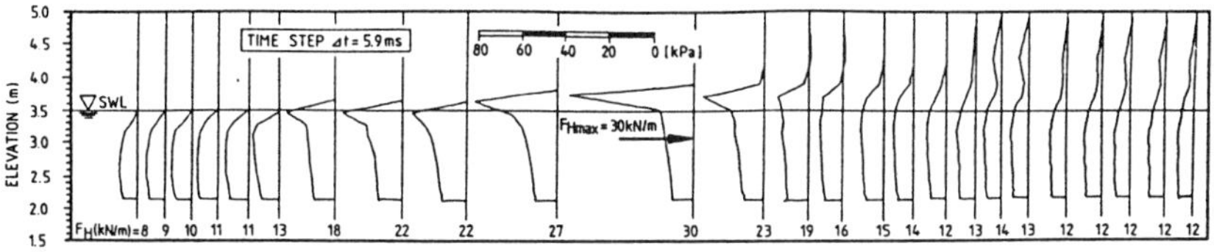

Fig.5 - Spatial and temporal impact pressure distribution

This kind of result, together with the aforementioned findings related to loading case classification, not only permit improvement of the understanding of the breaking wave impact mechanisms, but are also of considerable importance for the verification and calibration of numerical models simulating breaking wave impact loads.

12. Uplift pressure distribution. The uplift pressure distribution under the caisson base can be determined in the same manner as the impact pressure on the front. However, the uplift pressure is subject to much less variation than the pressure on the caisson front (Fig.6).

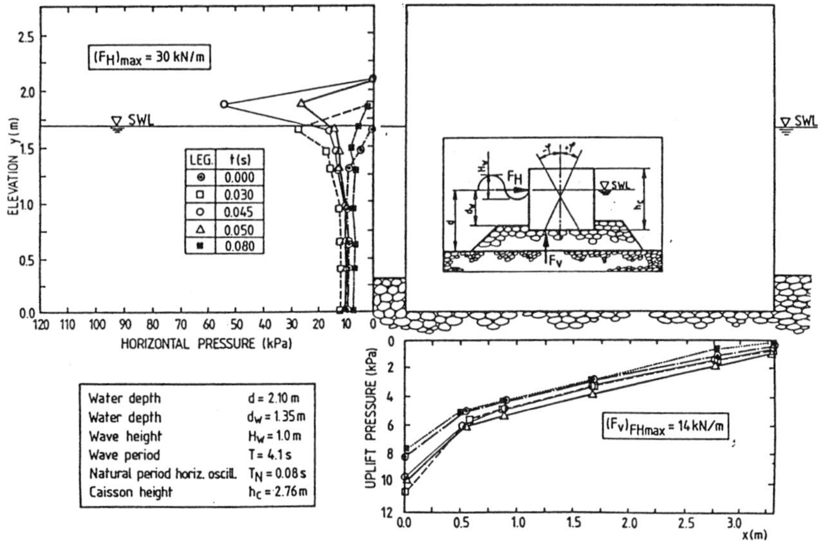

Fig.6 - Uplift and impact pressure

It is also seen from Fig.6 that the actual uplift pressures are not linearly distributed as usually assumed in standards and textbooks. In addition, the time lag between the occurrence of the maximum

impact load on the caisson front and that of the maximum uplift force is relatively short (in the in the range of 50 milliseconds).
Further, it is expected that the contribution of the uplift force to the total overturning moment will be relatively important).

Dynamic response of the structure

13. Position of the problem. The impulsive load induced by a wave breaking on the caisson front will cause swaying and rocking motions. These motions may be superimposed to give a rocking motion around a certain unknown point along the vertical axis going through the centre of gravity of the structure (Fig.7).

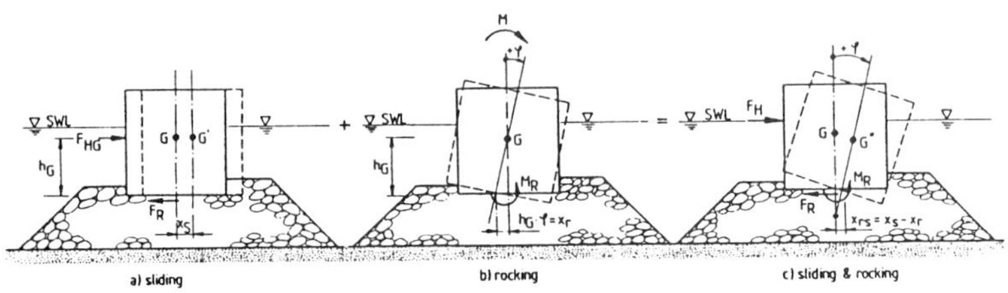

Fig. 7- Motions of a caisson subject to impulsive loading

The simplified equation of the free motion of the caisson, taken as a single linear oscillator is:

$$I_\phi \cdot d^2\phi/dt^2 + C_\phi \cdot d\phi/dt + K_\phi \cdot \phi = 0 \qquad (1)$$

Where:
I_ϕ = Mass moment of inertia of the caisson with the added moments due to the hydrodynamic and geodynamic mass (kg m²).
C_ϕ , K_ϕ = Rotation damping (kg m²/s²)and stiffness (kg m²/s²) of the system, respectively.
ϕ = Rotational degree of freedom (rad.)

At small damping, the natural period T_n of the oscillations is determined solely by the inertia I_ϕ and the elasticity K_ϕ of the vibrating system:

$$T_n = 2\pi \sqrt{I_\phi/K_\phi} \qquad (2)$$

Based on Eqs.(1) & (2), response curves like the one shown in the principle sketch in Fig. 8 may be drawn. The dynamic coefficient K_d can be assumed to

be equal to the ratio of the static equivalent force F_{sta} causing the same response and the actual dynamic force $(F_{dyn})_{max}$.
Actually, there is a distinct similar curve for each ratio of rise time and force duration T_e/T_t. However, for simplification, only the curve for $T_e/T_t = 0.5$ has been drawn to highlight the effect of loading duration, inertia and stiffness on the response of the structure.
Even based on a very simplified analysis, Fig. 8 clearly shows the necessity of a dynamic analysis for the design of caisson breakwaters when subject to impulsive loading. Here, the most relevant difficulties, however, still consist in the evaluaion of the damping, stiffness and added mass coefficients (water and foundation masses which are constrained to vibrate with the caisson). In fact, considering the hydrodynamic mass and the geodynamic mass will lead to an increase of the natural period T_n in the order of 25% and 10%, respectively (ref. 3).

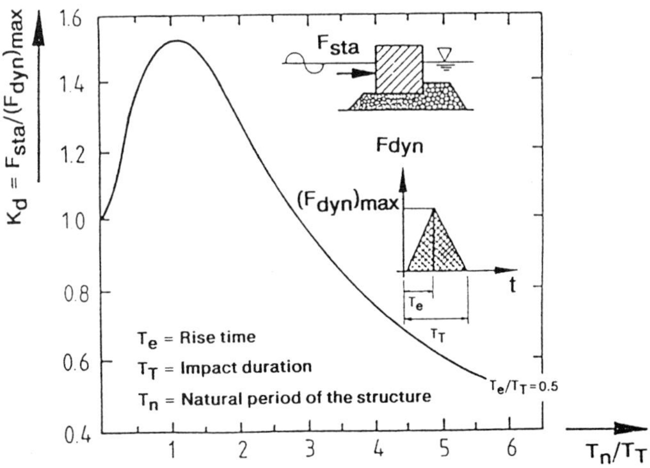

Fig. 8- Evaluation of the static equivalent force F_{sta} (Principle sketch)

In addition, it can be shown that for impulsive loading, i.e. for $T_n/T_t > 1$, the response is primarily controlled by inertia, whereas for longer-lasting quasi-static forces, i.e. for $T_n/T_t < 1$, it is rather governed by stiffness.

14. Impact force and dynamic response. An example from the results of the simultaneous records of horizontal forces and accelerations of the caisson is shown in Fig. 9.

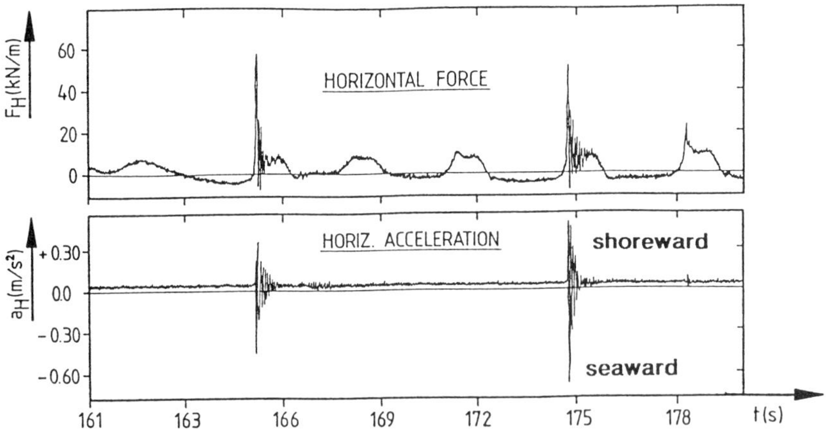

Fig. 9- Simultaneous records of horizontal forces and accelerations

According to Fig.9 and to the relationship $x_{max}=a_{max}/4\,\pi^2\,f^2$ it is seen that the motion of the caisson measured at the top is in the range of ± 0.10 mm.

The most striking result in Fig. 9, however, is that, even with a slightly smaller peak value, the 2nd impact force induces a much more significant motion than the second one. A detailed analysis of the reasons for this phenomenon has clearly shown that this is primarily caused by the more pronounced force oscillations after the force spike, and that these oscillations are induced by the entrapped air pocket expanding and being compressed cyclically under the highly transient pressure field of the plunging wave (ref.10). In addition, it is known that a number of load cycles are necessary for resonant amplification to build up. This is to show that entrapped air not only has a damping effect on the impact pressure, but may also considerably amplify the response of the structure. Furthermore, it makes clear that the evaluation of the whole load history is required for the final answer of the design problem.

CONCLUDING REMARKS AND PERSPECTIVES

15. Based on the analysis of past failures and the first results of the large-scale model tests which have been analysed so far, the following concluding remarks can be drawn:

- The suggestion commonly found in the literature that impact pressures are not significant, and hence should not be used for the design, could not be confirmed by the simultaneously measured impact forces and response of the caisson breakwater. The only evidence so far is that the whole load history - and especially the force oscillations after the peak value - is determinant.

- It is unwise to assume that impulsive loading due to breaking waves will not occur in deep water. Most breakwaters may be subject during their lifetime to all breaking wave conditions, and hence to the loading cases mentioned in Fig.4.
- The type and intensity of the impact loading is principally governed by the breaker shape at the structure front. However, the maximum peak value of the impact force does not necessarily induce the maximum dynamic response. The force oscillations with much lower amplitudes and periods in the range of the natural period of the structure appear to be much more relevant; i.e. plunging breakers with an entrapped air pocket may constitute the most critical loading case. This finding is reinforced by the fact that such breakers always induce a double-peaked impact force. The second spike generally occurs whilst the structure is still responding to the first spike. This may lead to a resulting cumulative response which is far more significant than that due to a single peak.
- The dynamic response of caisson breakwaters subject to impulsive loading cannot be properly evaluated by the commonly used simplified methods of dynamic analysis assuming triangular impulses. More sophisticated methods of simulation of the dynamic response by considering the actual whole load history are needed.

16. The present efforts within SFB 205 are concentrated on the following tasks which should be solved in the near future:

- Analysis of the data related to the response of the foundation (rubble mound and sand layer);
- Scaling problems of the two relevant effects related to the entrapped air pocket (wave energy cumulation and pulsation periods);
- Development and calibration of numerical models for the impulsive loading and the dynamic response.

ACKNOWLEDGEMENTS

This study is supported by the German Research Council (DFG). Additional support is provided by the European Community within the Research Project MAST G6-S: Research on Coastal Structures.

REFERENCES

1. ARAMI, A.; HATTORI, M.: Experimental study on shock wave pressure - Chuo University, Dept. Civil Eng. Report No 32 (1989),pp. 37-63 (in Japanese).

2. DE ROUVILLE, M.A.; BESSON, P.; PETRY, P.: Etat actuel des études internationales sur les efforts dus aux lames. Annales des Ponts et Chaussees Vol. 108, 1938, pp. 5-113.

3. GERSCHUNOV, E.M.; SMIRNOV, G. N.: Some questions concerning the dynamic analysis of coastal structures. Trudy koordinacionnych po gidrotechnike, Vyp. 20, 1965, pp.17-28 (in Russian).

4. GODA, Y.: Random sea waves and engineering applications. Tokyo University Press, Tokyo, 1985.

5. HATTORI, C. et al.: Disasters of breakwaters by wave action. P.H.R.I. Techn. Note No 485, 1985, pp.1-281 (in Japanese).

6. LUNDGREN, L.: Development of caisson breakwater technology. PIANC Bulletin No 50, 1985 pp. 19-27.

7. MAEDA,K.; SHIRAISHI, S.: World's largest caisson. ASCE, Civil Engineering, August 1990,pp. 47-49.

8. MITSUYASU, H. : shock pressure of breaking waves. ASCE, Proc. 10th ICCE, Tokyo 1966, pp. 268-283.

9. OUMERACI, H.; KLAMMER, P.; PARTENSCKY, H. W.: Classification of breaking wave impact loads on vertical structures. ASCE, J. Waterway, Port, Coastal & Ocean Eng. Div. (Paper submitted).

10. OUMERACI, H.; PARTENSCKY, H.W.: Breaking wave impacts on vertical fronted structures (in preparation).

11. RENAUD, P.J.M.: La jetee Mustapha au Port d'Alger. Annales des Ponts et Chaussees, Vol. 105; 1935, pp.553-828.

12. RUNDGREN, H.: Water wave forces- A theoretical and laboratory study- Trans. Roy. Inst. of Techn.No 54, Stockholm 1958.

13. TANIMOTO, K.: ; TAKAYASHI, S.; KIMURA, K.: Structure and hydraulic characteristics of breakwaters- The state of the art of breakwater design in Japan- Report P.H.R.I Vol.26,1987, pp. 11-55.

14. WITTE, H.H.: Stochastische & deterministische Vorgänge bei der Druckschlagbelastung senkrechter Wände. HANSA, Vol. 126. No 19/20, 1989, pp. 1378-1384.

15. ZEN, K.; UMEHARA, Y.: Analysis on wave-induced pore water pressure in sand layers under breakwaters. Proc. Int. Symp. Ocean Utilization'85, Tokyo, June 1985.

14. Stability of deep water caisson breakwater against random waves

K. TANIMOTO, Port and Harbour Research Institute, and
Y. GODA, Yokohama National University

SYNOPSIS. The deepest caisson breakwater in Japan is under construction at the water depth of 60 m in Kamaishi Bay. The most remarkable feature of wave pressures due to standing waves in a deep water is that the negative peak is larger than the positive peak in the absolute value. In the present paper, this fact is demonstrated by both results of irregular wave tests and calculations with the fourth order approximate theory of finite amplitude standing waves. Sliding distances of the caisson are also discussed with the results by experiments and theoretical calculations.

TSUNAMI BREAKWATER AT KAMAISHI PORT

1. The most common facility to prevent or reduce tsunami run-up on the land is seawalls which are built along the waterline. In a port area, however, seawalls with a high crest causes hindrance and inconvenience to the daily port operation. Reduction of tsunami height by means of offshore breakwaters provides a solution for the protection of the port area from tsunamis as well as storm waves. Tsunami breakwaters restrict the inflow of tsunami waves into a harbour by narrowing the entrance and make change the mode of harbour reasonance.

2. A typical example of tsunami breakwaters is found in Ofunato Bay along the Sanriku Coast, where the tsunami breakwater was completed at the bay mouth in 1967 after the Chilean earthquake tsunamis in 1960. A new tsunami breakwater is under construction in Kamaishi Bay which is located at the distance of 30 km in the north from Ofunato Bay.

3. Figure 1 shows the plan of Kamaishi Bay. The site of breakwaters is the bay mouth of 2,100 m width. The breakwaters consist of the north breakwater 990 m long and the south breakwater 670 m long, and have a central opening 300 m wide as well as two side narrow openings.

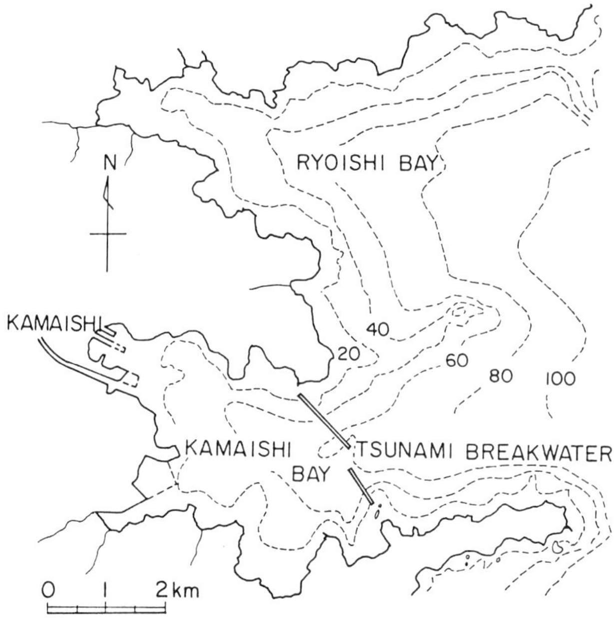

Fig.1 Plan of Kamaishi bay

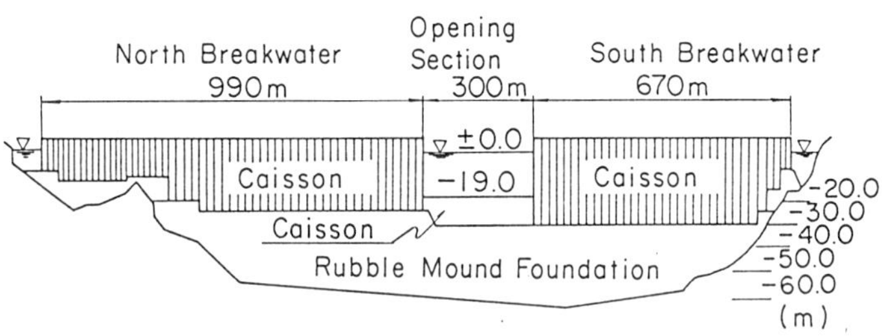

Fig.2 Longitudinal section of tsunami breakwater

4. Figure 2 shows the longitudinal section of the tsunami breakwaters. As seen in this figure, 70% of the total length is located at the water depth deeper than 50 m and the deepest one is 63 m. At the central opening, a submerged dike with the crest level of -19 m is planned to be built to reduce the opening section as much as possible.

5. Figure 3 shows the cross section at the deep water part. The upright section is shaped in a trapezoid at the lower part to have a wide stable bottom slab and is a wave dissipating type with double horizontal slit walls at the upper part. Three units of caisson have been already placed at the site by the end of 1990.

6. A lot of studies for the breakwater design has been

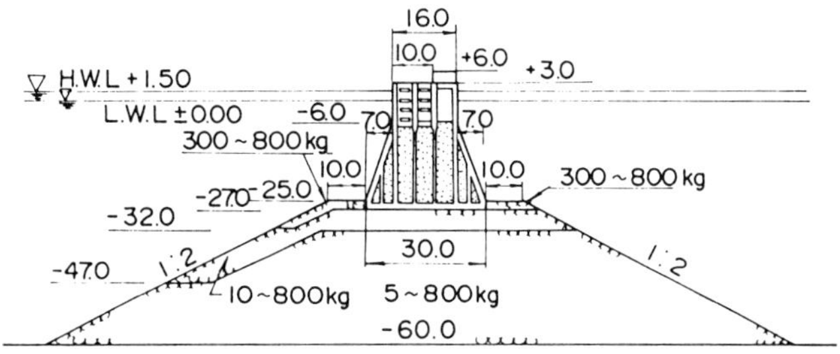

Fig.3 Cross section of Kamaishi breakwater (deep section)

conducted against storm waves, tsunamis and earthquakes. As for storm waves, two and three dimensional experiments have been carried out to investigate wave transformation and the stability. In the present paper, discussion is focused only to an aspect of the stability of deep water caisson against sliding (ref.1).

LABORATORY EXPERIMENTS

7. The laboratory experiments were carried out in a large wave flume with the scale of 1/40. The model of upright section is slightly different from the executed prototype in the dimension, because the caisson width was reduced from originally proposed one after the laboratory experiments. Figure 4 shows the dimension of upright section together with the locations of pressure transducers as P1, P2, P3···.

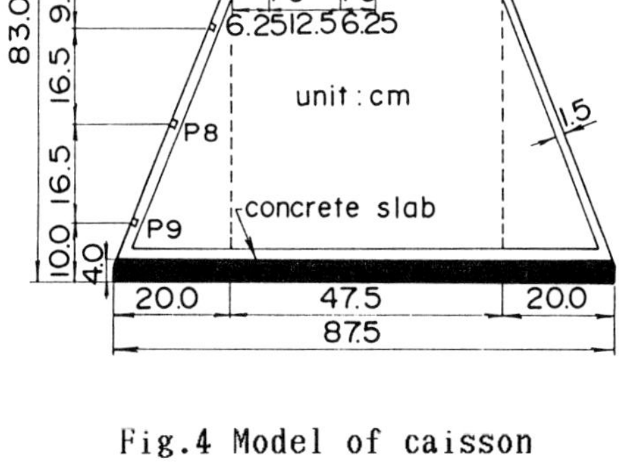

Fig.4 Model of caisson

8. In the experiments, two alternative sections were tested in order to verify the effectiveness of the wave energy dissipation devices. One was the proposed section and the other was a section with the solid wall in the front. The latter was realized by placing the rear side of the caisson to face the offshore side. These two sections are called the absorbing type and the reflective type, respectively. In the present paper, however, the results for the reflective type only are described.

9. Irregular waves were generated so as to realize the wave spectrum of Bretschneider • Mitsuyasu type (ref.2). All measurements with the breakwater model were made in a stationary state to consider the influence of multi-reflection between a breakwater model and a wave paddle. Incident wave heights were corrected from ones measured in advance without the breakwater model by multiplying the amplification factor due to the multi-reflection. The amplification factor, however, was 1.1 at most in those experiments, because only the width of 28% to the whole flume width with a partition wall was partially reflective by the breakwater model.

10. The water depth including a tide level, h, at the breakwater site is 153.8cm and the significant wave period, $T_{1/3}$, is 1.90 s. The significant wave height, $H_{1/3}$, was changed to five ranks from 7 to 24 cm. The effective duration for the analysis was about 200 waves for a train of waves and three different wave trains were tested for each rank of the significant wave height.

11. In order to investigate the stability against sliding, the weight of model caisson was changed to three ranks by controlling the quantity of filling material. Their weights are expressed by the notation as W_{100}, W_{80}, W_{70}. The notation of W_{100} indicates the weight corresponding to the design weight, and others are corresponding to the weights of 80% and 70% of the design weight in air, respectively. The weight corresponding to W_{100} is 5.5 kg/cm in the model.

12. The displacement of caisson under the action of waves was measured around the top in the lee side with a displacement metre. Wave pressures were also measured on the front wall.

ESTIMATION METHOD OF INCIDENT WAVE HEIGHT FOR INDIVIDUAL WAVES

13. Figure 5 shows an example of records of a water surface elevation in front of the caisson and wave pressures on the front wall. The water surface elevation, η_S, was measured at the location right over the front edge of the bottom slab of caisson, and consequently

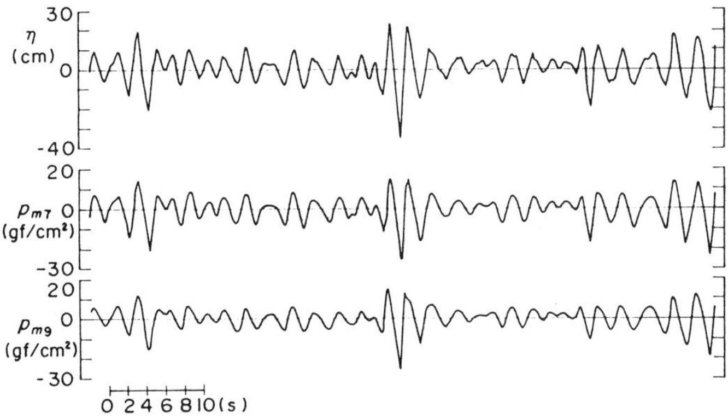

Fig.5 Records of water surface elevation and wave pressures

includes the reflected waves from the breakwater. The wave pressures, p_{m7} and p_{m9}, were measured at the points of P7 and P9 corresponding to the depths of 28.3 and 61.3 cm below the still water level, respectively. Peak pressure intensities for individual waves were analyzed for both positive and negative.

14. The incident wave height for individual waves was estimated from the relation between the representative wave heights like H_{max} , $H_{1/10}$, $H_{1/3}$ and the corresponding representative peak intensities of wave pressure measured at the point of P9 on the wall, where H_{max} is the largest wave height, and $H_{1/10}$ is the largest one-tenth wave height. The representative wave heights are those corrected for the influence of multi-reflection from the wave heights measured without the breakwater model. This method to estimate the incident wave heights was selected as a practical one among several alternatives.

15. Figure 6 shows the relations between the

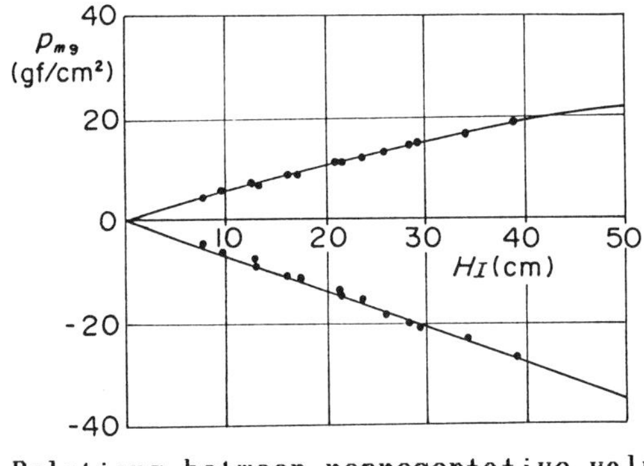

Fig.6 Relations between representative values of incident wave height and wave pressure

representative wave heights and the corresponding representative wave pressure intensities. The relations can be expressed by the following regression equations:

$$p_{m9}^{+} = 0.428\ H_I + 0.000506 H_I^{\ 2} \qquad (1)$$

$$p_{m9}^{-} = -0.481 H_I - 0.00559\ H_I^{\ 2} \qquad (2)$$

where, H_I is the incident wave height, and the notations of + and - attached to p_{m9} indicate the positive and negative peak values, respectively.

WAVE PRESSURES ON THE WALL

16. In 1966, Y.Goda pointed out according to calculations with the fourth order approximate theory of finite standing waves that the negative peak of wave force on a wall becomes larger than the positive peak in the absolute value when the relative water depth, h/L, is larger than 0.25 (ref.3). The present condition of h/L=0.28, of which the wavelength L is calculated for the significant wave period, falls into this range.

17. Figures 7 and 8 shows the pressure intensity measured at the points of P7 and P9 for the reflective type. In the figures, the relations calculated by the Goda formula (ref.2) for the positive peak and by the finite amplitude standing wave theory of the fourth order approximation for both peaks are also indicated. In the calculations, the wave period is assumued as a constant equivalent to the significant wave period and the upright section is replaced by a vertical wall. In addition, the effect of the rubble mound foundation is disregarded for the finite amplitude standing wave theory.

18. The measured and calculated results demonstrate that the absolute value of the negative peak intensity is larger than the positive peak intensity. The constant value of p_{m7} in the region of large wave height is caused by an exposure of the pressure transducer to the air.

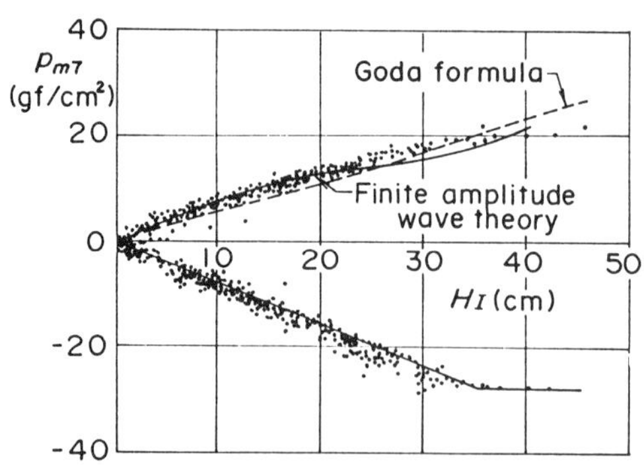

Fig.7 Wave pressure at the depth of 28.3cm below the still water level

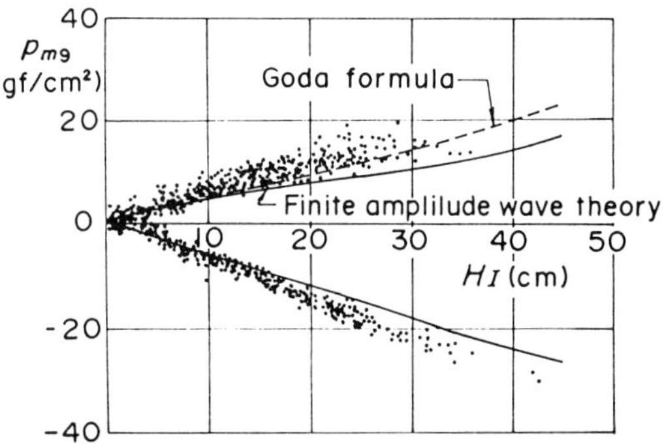

Fig.8 Wave pressure at the depth of 61.3cm below the still water level

19. It is observed in the results that the finite amplitude standing wave theory predicts slightly smaller values than measured wave pressures for both positive and negative peaks. For the positive peak, the Goda formula predicts slightly larger value than that by the finite amplitude standing wave theory.

STABILITY AGAINST SLIDING

20. Figure 9 shows an example of records of the horizontal displacement of caisson, s' , the water surface elevation, η_S , and the wave pressure at the point of P9, p_{m9}. It is observed that the caisson moves to both directions towards the shore and towards the offshore. In the records, the shoreward displacement is defined as the positive direction.

21. The displacement by individual waves was analyzed for both shoreward (positive) and offshoreward (negative)

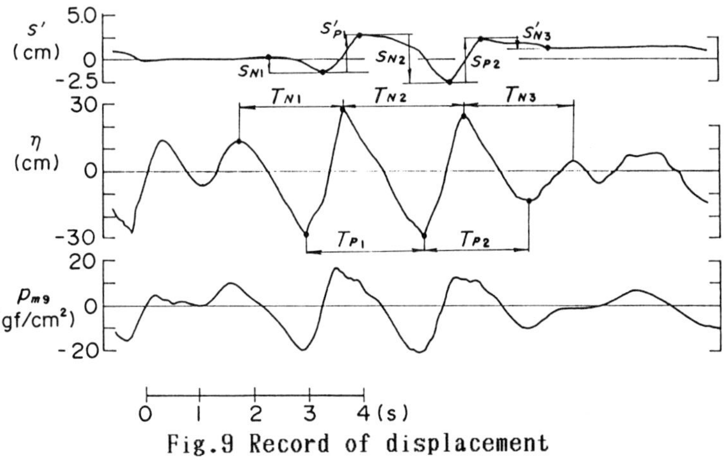

Fig.9 Record of displacement

directions by the following defintions:
a) shoreward displacement; the difference between the maximum and the minimum in the record of displacement during the time between two successive negative peaks in the record of water surface elevation,
b) offshoreward displacement; the difference between the minimum and the maximum in the record of displacement during the time between two successive positive peaks in the record of water surface elevation.

22. An example of plotting of the shoreward and offshoreward displacements against the incident wave height is shown in Fig.10. The data are widely scattered because of their inherent variability and errors caused from various assumptions in the data analysis. The broken lines in the figure indicate the regressive relations by the least square error method. It is noticed that both shoreward and offshoreward displacements are almost the same. This is also confirmed by the fact that the residual displacement after the wave action is not large and the direction depends on the last large displacement of shoreward or offshoreward.

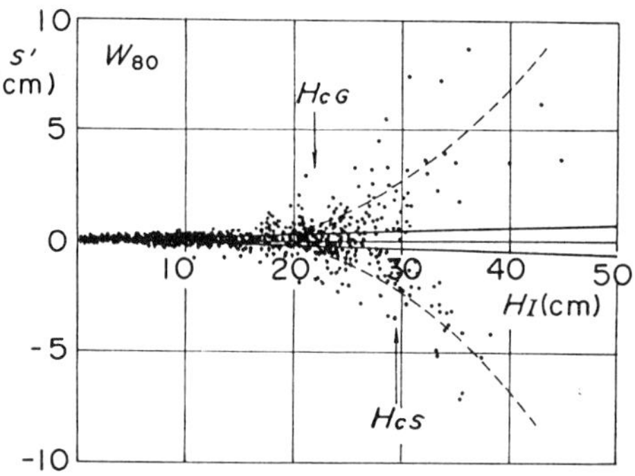

Fig.10 Shoreward and offshoreward displacements

23. In the displacement, the influence of rocking motion is included, because it was measured around the top of caisson. Consequently, the displacement due to rocking must be subtracted from the measured displacement to get a sliding distance. The displacement due to rocking was estimated by a linear relation with the incident wave height as shown by solid lines in the figure, which was determined by a least square error method for the data in the region of small wave height.

24. In the figure, threshold wave heights against sliding for both shoreward and offshoreward directions are also indicated according to the theoretical calculation. The threshold wave height against the shoreward sliding, H_{cG}, is calculated by the Goda formula and that against the

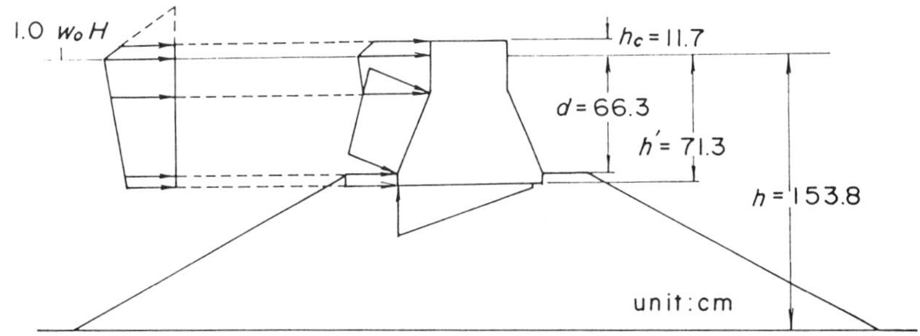

Fig.11 Wave pressure distribution by the Goda formula at the wave crest

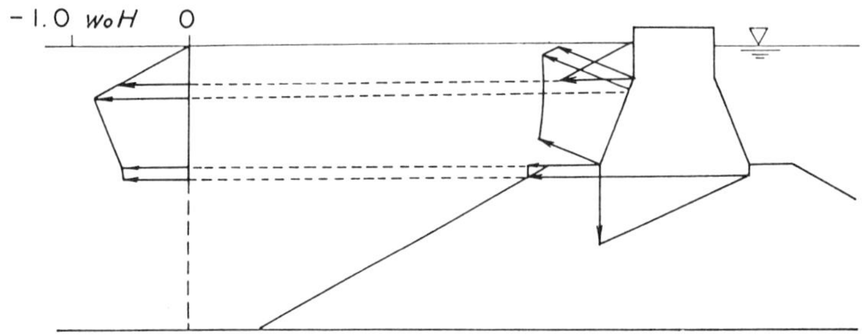

Fig.12 Wave pressure distribution by the finite amplitude standing wave theory at the wave trough

offshoreward sliding, H_{cs}, is calculated by the finite amplitude standing wave theory of the fourth order approximation. Figure 11 and 12 show the wave pressure distribution calculated for the threshold states.

25. The calculated threshold wave heights against shoreward and offshoreward sliding exhibit a different tendency from the experimental results. Namely, the threshold wave height against offshoreward sliding is apparently larger than that against shoreward sliding in the calculation, while both are almost the same in the experiment. The reason of this difference will be considered later on.

26. Figure 13 shows all data of the sliding distance, s, for the reflective type with the weight of W_{70} and W_{80} in a non-dimensional form. No sliding was observed for the caisson weighing W_{100} in the experiments.

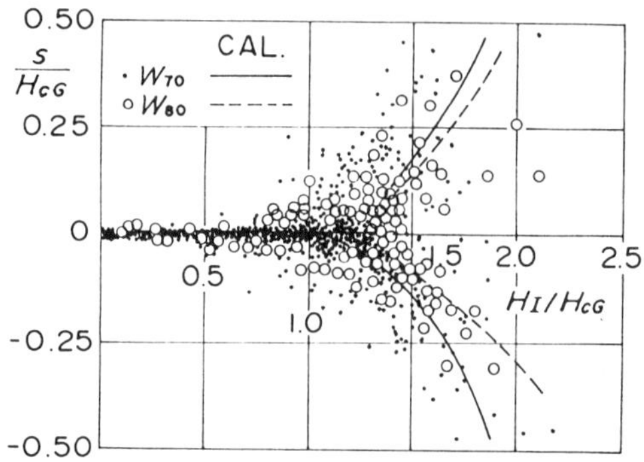

Fig.13 Shoreward and offshore ward sliding distances

In the figure, calculated relations are also shown, and they will be explained in the next Chapter.

THEORETICAL CONSIDERATIONS FOR SLIDING DISTANCE

27. Sliding model. A simple theoretical model for sliding of a vertical wall is considered to investigate the reason arising in the paragraph 25. Because wave pressures are comparatively static due to standing waves, the rubble mound foundation can be supposed to be a solid mound.

28. The equation of motion of the wall during sliding is expressed,

$$(W/g+M_a)\ddot{x}=F_H\mp\mu(W_O-F_V) \tag{3}$$

where, W is the wall weight in the air, W_0 is the wall weight in the still water, M_a is the added mass, $\ddot{x}$ is the acceleration of the wall motion, F_H and F_V are horizontal and vertical components of wave force respectively, g is the gravitational accerelation, and μ is the friction factor. The sign $\mp$ in the right hand side of Eq.(3) corresponds to the sign $\pm$ of F_H .

29. The added mass must be considered for both sides of wall and the water surface elevation at the front wall, η_s , is preferable to be considered. Then, the added mass is supposed as the following relation similarly to the piston motion in a wave maker theory:

$$M_a=C_m\{(h'+\eta')^2+h'^2\}W_O/g \tag{4}$$

$$\left.\begin{aligned}\eta'&=\min\{\eta_S,h_c\} \quad ; \ \eta_S\geq 0\\ &=\max\{\eta_S,-h'\} ; \ \eta_S<0\end{aligned}\right\} \tag{5}$$

where, h' is the water depth at the bottom of wall, h_c is the crest height of wall above the still water level, and C_m is the added mass coefficient. The coefficient C_m is supposed to be 0.54 according to the small amplitude wave theory.

30. The wave pressure must be considered for the whole

circumference of the wall, because considerable wave overtopping is caused by high waves. In the calculation, however, the wave pressure at the front wall, at the bottom slab and at the crest face is considered. For the front wall and bottom slab, the wave force is calculated on the basis of the following assumptions:

a) The time history of horizontal and vertical wave force components varies proportionally to that of the water surface elevation calculated by the finite standing wave theory of the fourth order approximation by a different way for the positive and negative wave pressure distributions.

b) The positive peak wave pressure distribution is given by the Goda formula as shown in Fig.11.

c) The negative peak wave pressure distribution is given by the finite amplitude standing wave theory of the fourth order approximation as shown in Fig.12. Then, the resultant horizontal wave force at the negative peak is increased by 1.1 times, because it is noticed that the absolute intensity of wave pressure on the front wall is slightly larger than that predicted by the finite amplitude standing wave theory.

31. The wave force acting on the crest surface of wall due to overtopping wave, F_{VS}, is calculated by the following relations:

$$F_{vS}=w_0 Q(t) \tag{6}$$

$$Q(t)=(2/3)(2g)^{1/2}K\int^{t}(\eta_S-h_c)^{3/2}\,dt;\quad \eta_S\geq h_c \tag{7}$$

where, w_0 is the specific weight of fluid, K is the discharge coefficient and is assumued to be 0.32. The amount of overtopping, Q, becomes its peak, when the water surface elevation, η_S, falls to the level of the crest height, h_c. Consequently, the force, F_{VS}, becomes its peak at this moment. It is assumed that the force decreases from the peak symmetrically in the time. This can be considered so as to include the influence of transmitted waves at the rear wall.

32. Time histories of wave force and sliding motion. The equation of motion, Eq.(3), is integrated numerically from the following initial conditions at $t=0$:

$$\dot{x} = x = 0 \tag{8}$$

where, $\dot{x}$ is the velocity of motion and x is the displacement. The sliding ends when the velocity of motion becomes zero.

33. An example of calculated results is shown in Fig.14 with distorted ordinates for different variables. The weight of caisson corresponds to W_{70} and the wave period and height are 1.9s and 30cm, respectively. In the figure, the notation of F_{VO} indicates the vertical component of wave forces on the front wall and the bottom slab. Consequently, the total vertical component of wave force,

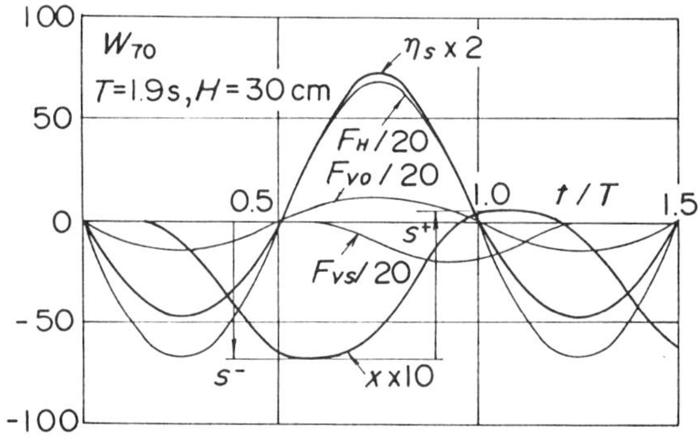

Fig.14 Time histories of wave forces and displacement (1)

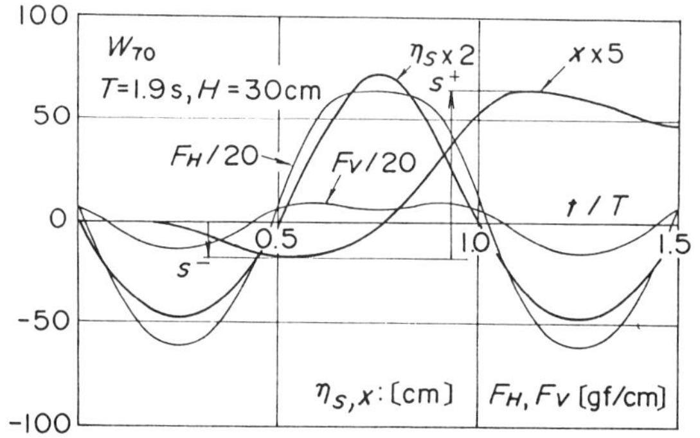

Fig.15 Time histories of wave forces and displacement (2)

F_V is expressed by the following relation:

$$F_V = F_{VO} - F_{VS} \quad (9).$$

In this example, the offshoreward sliding distance x^- is 6.8cm and the shoreward sliding distance x^+ is 7.1cm. The residual displacement after one cycle of wave is only 0.3cm towards the shore.

34. On the other hand, Fig.15 shows the result calculated by the different assumptions for the wave force as follows:

a) the intensity and its time history of wave pressure on the front wall and the bottom slab are given by the finite amplitude standing wave theory,

b) the wave force on the crest surface of caisson is disregarded.

35. These assumptions are probably more usual than those in the paragraphs 29 and 30. However, the shoreward sliding distance is 16.5cm and much larger than the offshoreward one of 3.5cm in this calculation. The major reason why the shoreward sliding distance is so large is explained as follows:

a) the positive peak of horizontal wave force, F_H , is

flat and lasts longer, although the peak value is almost the same as that in Fig.14,
b) the wave force on the top surface, which works downward so as to decrease the sliding, is disregarded.
36. Comparizon with experimental results. The calculated sliding distance, which is shown already in Fig.13, represents an average relation of experimental data which are scattered widely.

CONCLUDING REMARKS

37. The offshoreward wave force at wave trough should be considered carefully as well as the shoreward wave force at wave crest in the design of deep water caisson breakwater.
38. It is expected that the caisson of Kamaishi breakwater will slide in both shoreward and offshoreward directions during one cycle of individual waves, when waves much higher than the design waves attack the breakwater. It may result in relatively small residual displacement of caisson after the storm. This is greatly different from a case of a shallow water caisson breakwater, where the caisson will slide in one side towards the shore so as to accumulate the displacement.
39. Secondary phenomena like a wave force on the crest surface of a caisson due to overtopping are also important to predict the sliding distance. A sophisticated investigation on wave forces and behaviour of caisson in a large wave flume is expected to bring a further development for the design of caisson breakwaters (ref.4).
40. Unit conversion. $1gf/cm^2 = 98.1Pa$.

RFERENCES

1. TANIMOTO,K., KIMURA,K. and MIYAZAKI,K. Study on stability of deep water breakwater against waves, Rept. of Port and Harbour Res. Inst., Vol.27, No.1, 1988 (in Japanese).
2. GODA,Y. Random seas and design of maritime structures, University of Tokyo Press, 1985.
3. GODA,Y. and KAKIZAKI,S. Study on finite amplitude standing waves and their pressures upon a verticalwall, Rept. of Port and Harbour Res. Inst., Vol.5, No.10, 1966 (in Japanese) or Coastal Engineering in Japan, Vol.10, 1967.
4. PARTENSCKY,H. New investigations on impact pressures on vertical coastal structures, Third European Workshop on Coastal Zones, 1990.

15. Cellular coffer dams as breakwaters and coastal structures

J. H. WOTTON, Crouch Hogg Waterman

SYNOPSIS. This paper discusses the use of cellular cofferdams as breakwaters and coastal structures in particular locations. The factors affecting the decision to adopt this form of construction at different sites are discussed, together with particular design problems encountered. Various construction methods which have been adopted are described and compared. Plant and equipment used and rates of progress achieved are discussed.

INTRODUCTION

1. Straight web cellular cofferdams are used for both permanent and temporary structures. This paper is concerned with the former category which includes:-

- breakwaters
- bulkhead walls
- training walls
- groynes
- jetties
- dolphins

2. There are many examples of the use of cellular cofferdams as breakwaters and coastal structures, both in the UK and abroad. This paper describes works by the Author's firm, Crouch Hogg Waterman, formerly Crouch & Hogg at:-

Mallaig Harbour	Reclamation
Mallaig Harbour	Breakwater
Lower Clyde	Reclamation
Lochmaddy	Jetty
Brodick	Jetty
Port Glasgow	Slipway

and discusses other locations in the UK where cellular cofferdams have been used as permanent coastal structures, e.g.,

Papa Westray	Breakwater
Bristol	Royal Portbury Dock Training Wall
Leith	Harbour Entrance Lock

The location of these various sites is indicated on Fig. 1.

Overseas some of the best examples of this type of construction are to be found on the Great Lakes in the U.S.A. These are referred to below.

Fig. 1

3. Cellular cofferdams are mass gravity structures which derive their stability by the development of hoop tension forces from the retention of fill material contained within the cell and/or connecting arc. They occur either as circular, diaphragm or clover leaf, Fig. 2. Circular construction is most commonly used as there are problems associated with the latter two types, particularly in the filling cycle of the operation. Clover leaf construction is relatively uncommon and only tends to be used for particularly high structures.

All structures, regardless of type, have to be stable against the following failure modes under the worst combination of loading:-

. Sliding on the base
. Overturning

. Interlock failure
. Overstress of pile webs
. Horizontal shear failure of the infill
. Vertical shear failure of the infill
. Bearing capacity failure
. Circular slip

Minimum factors of safety of 2 are generally the norm, although this will always be a matter of engineering judgement taking all factors into account for a particular location and duty.

Additional consideration should be given to the likely deformations which occur during filling and/or backfilling where these may result in unacceptable service conditions. As a guide, permanent deformations of 1 to 1½% of the overall cell height are generally acceptable.

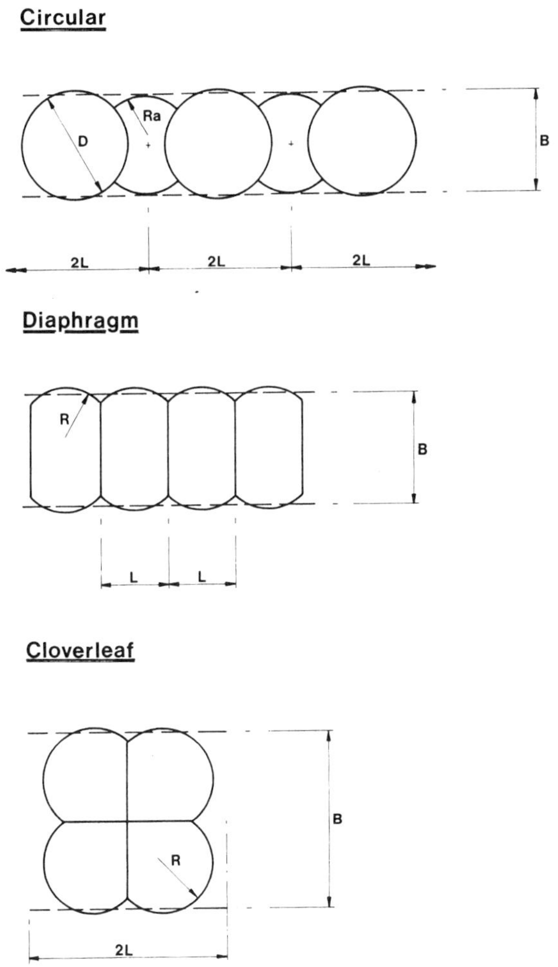

FORMS OF CONSTRUCTION

Fig. 2

4. During construction, care needs to be taken to ensure the adequacy of the temporary works template which can either be internal or external. Internal templates are more commonly used. External templates are generally restricted to particularly high structures.

Pitching of piles takes place between the "fixed" junctions. Any binding should be shaken out before driving commences. The first two or three piles of the adjoining arc should be pitched and driven with the main cell. This avoids any problems in closing the arcs which would otherwise result from barrelling of the cell during filling.

Driving normally proceeds on alternate passes, working around the cofferdam keeping the lead between adjacent piles to an acceptable limit, dependent on ground conditions. On completion of driving, and before filling, a positive head of water should be maintained within the cell in order to promote the development of ring tension forces in the cell and thus enhance temporary stability.

Filling should be hard inert granular material placed centrally to the cell, either hydraulically or in the dry.

The temporary works should not be removed until the cell is approximately two-thirds filled.

On removal of the temporary works and completion of filling, the piling plant can move forward to build the next cell, following which the arc is then completed and the cycle repeats.

MALLAIG RECLAMATION

5. In 1973, in response to increasing demands of the local fishing fleet, Mallaig Harbour Authority commissioned Crouch & Hogg to prepare a long term plan for the development of the fishing port to provide additional land, shelter and berthage. The plan, Fig. 3, provided for major reclamation of the tidal foreshore, new inner and outer breakwaters, marginal quays and fish handling facilities.

6. By 1974, Mallaig had become the premier herring port in Europe with annual landings of 70,000 tonnes. However, the growth of the port was severely constrained by a lack of developable land. Accordingly in 1974 the first element of the plan was implemented to provide a new lorry marshalling area by the reclamation of approximately 0.5 hectare of tidal foreshore. The area was virtually square in plan 70m x 70m, Fig. 4.

7. Rock outcropped over virtually the whole site, other than along the eastern margin where a thin layer of boulder clay up to 3 metres in thickness overlay mica schist bedrock.

8. A number of alternative forms of construction were considered for the north and east seaward perimeters of the reclamation. As there was no requirement to provide berthing facilities alongside the reclamation it was concluded that a rock armour slope would provide the most economic solution. Based on theoretical wave predictions, 2 to 4 tonne armour stone was selected. Investigations were undertaken to establish the likely yield of armour stone of this size local to Mallaig. These indicated that the local rock was highly foliated, and was therefore unlikely to yield a sufficient quantity of armour stone for such an operation to be economic. More remote sources were also investigated but rejected because of the high costs involved in obtaining and transporting relatively small quantities of rock armour to the remote Mallaig location.

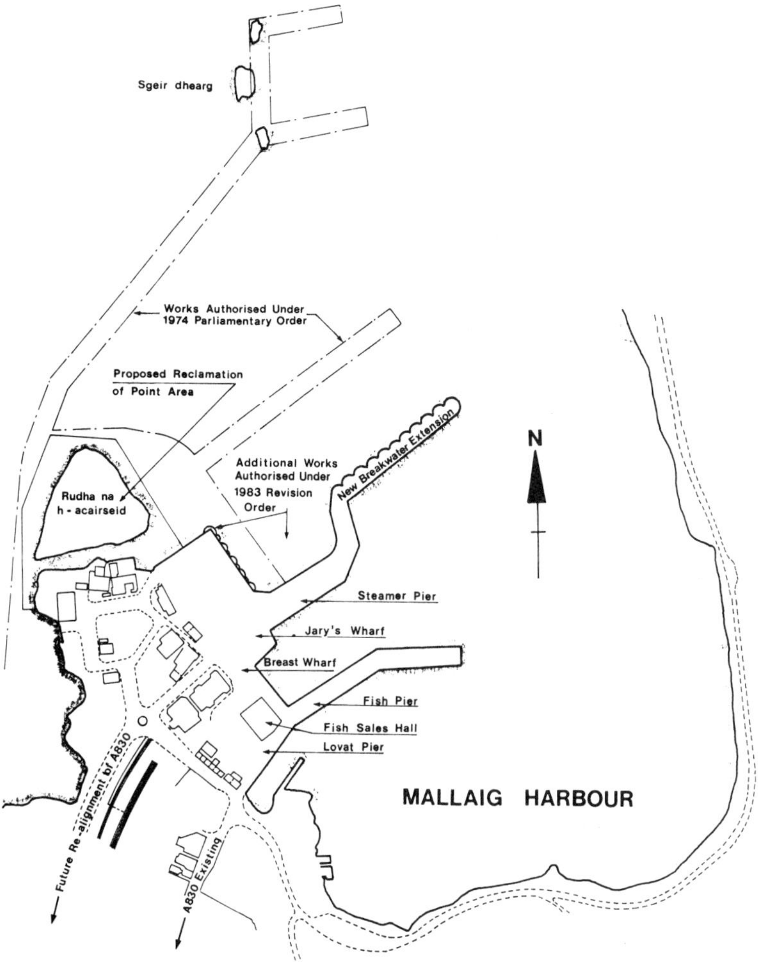

MALLAIG HARBOUR
DEVELOPMENT PLAN

Fig. 3

9. Alternative methods of construction for the seaward margins to the reclamation were therefore investigated and it was concluded that 11.5 metre diameter cells should be used to form the main east bulkhead over water whilst the north return over the foreshore was designed as a mass concrete gravity retaining wall.

10. There were no unusual aspects to the design of the cells. Factors of safety ranged from 2 to in excess of 5, the limiting failure mode being sliding on the base. Infill and backfill were specified as either dredged or land won hard inert granular material.

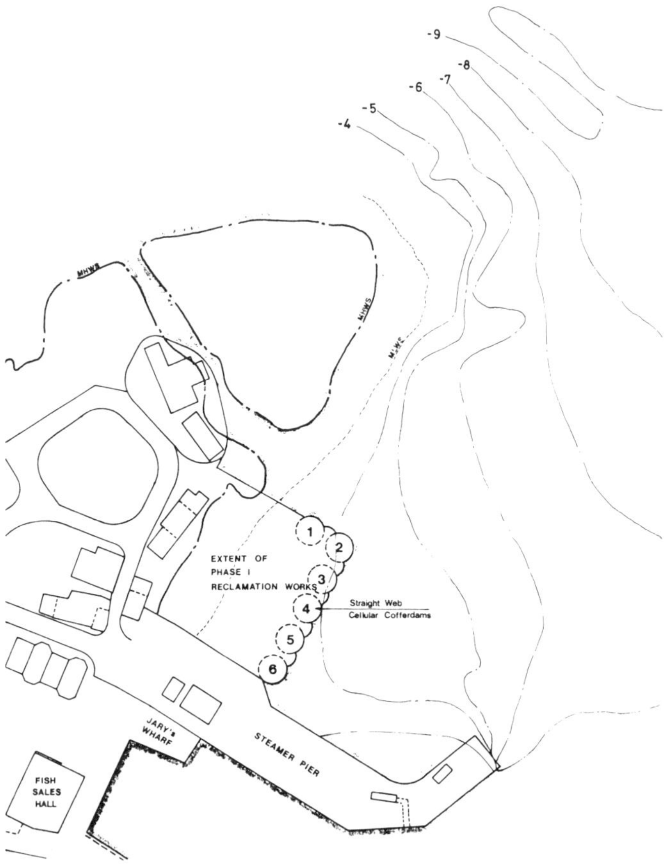

MALLAIG HARBOUR RECLAMATION

Fig. 4

11. The Contractor opted to build the cells using a fixed 10 Ton Scotch derrick crane located on the foreshore within the main reclamation area. The derrick was located such that all six cells could be covered from the one position. The cells were to be built around an inner template consisting of a space frame supported on four adjustable spud legs with frames at two levels at approximately three metres vertical pitch. The frame which was fabricated from square hollow sections was sized to be handled by the derrick. Prior to the commencement of cell construction, the sea bed was prepared by removing all soft and loose material overlying the boulder clay. These superficial deposits which varied up to a maximum thickness of 1 metre were removed by the derrick rigged with a grab.

12. Work progressed from north to south commencing with cell No. 1 which formed a return on the north face. Cells were to be filled by dredged sand from the harbour for which purpose the Contractor had employed a specialist subcontractor who had deployed a small cutter section dredger. This method was used to fill and secure cell No. 1.

13. During the construction of the corner cell No. 2 and when it was at its most vulnerable prior to filling, a storm

developed with sustained winds up to force 5/6. A steep sea with waves of short periodicity developed in a short space of time and gave rise to local wave effects around and between the two cells.

The external wave forces on the partially built cell caused a temporary loss of positive head which, together with the local venturi effect between the two cells, resulted in the collapse of cell No. 2. Cell No. 1 was in turn pulled over by wire stays which had been installed to stabilise cell No. 2 during the critical stages of its construction.

14. An inspection immediately after the collapse indicated that much of the steel piling had suffered extensive damage and could not be re-used. Cell No. 1 was therefore abandoned and replaced by an extension to the mass concrete gravity wall.

15. Cells 2 to 6 were completed satisfactorily despite problems with the dredger which had been badly damaged during another storm. As this occurred at a critical time during cell building, the Contractor was forced to import crushed rock from local quarries for cell and arc filling. The bulk reclamation was however completed using dredged material.

16. The piles were installed using a BSP 700N air hammer and were driven to a penetration of between 1 and 2 metres into the boulder clay. The maximum pile length handled was 12.5 metres and each cell had a bulk volume of $850m^3$. Table 1 summarises these principal parameters.

Filling was the critical item on the cell production cycle which became prolonged as a result of the reversion to land won material as this had to be stockpiled off site and trucked to the works as required. The overall cell production cycle was 15 days.

MALLAIG BREAKWATER

17. In 1978, Mallaig Harbour Authority, through Crouch & Hogg, commissioned the Hydraulics Research Station, Wallingford, to undertake model testing of the overall harbour development layout. The model testing confirmed the significant wave height as 1.3 metres with a zero crossing period of about 5 seconds. The test also showed that acceptable shelter conditions would exist within the harbour as proposed in the development plan, Fig 3.

18. The model included an inner breakwater, some 100 metres long, forming an extension to the existing steamer pier. A number of construction options were considered, including rubble mound, concrete caissons, tied sheet piled walls and cellular cofferdams.

19. In terms of the shelter provision alone, the rubble mound structure was cheapest but had significant disadvantages. Among these were the substantial additional cost of providing the lee side berthage required by the client and the reduction in navigable water space resulting from the extensive breakwater side slopes.

Concrete caissons were more expensive than either of the two sheet piled options.

20. Cellular cofferdams were adopted as the final design because they offered the following advantage over the tied sheet pile wall alternative.

- flexibility and ease of construction using reasonably sized plant working end over.

- reduced risk during construction, only one cell at risk at any one time whereas in double wall construction long lengths would be at risk.

. economy, cells are cheaper.

One disadvantage of cellular construction is the higher cost of providing a continuous berthing face on the lee side because of the scalloped profile of the cells. The final layout of the breakwater is indicated on Fig. 5.

21. Cells of 15.24m diameter were adopted to meet the design requirement for a breakwater of 13 metres overall height. The design loading considered the combination of Clapotis wave loading, together with bollard pulls on the lee-side.

Passive pressure was ignored in the calculations which indicated that factors of safety varied from 2 to 4. The critical failure mode was overturning.

Particular attention was paid to settlement and bearing capacity failure given the massive nature of the structure. Site investigation had indicated the presence of loose to medium dense sands varying in thickness from 6 to 10 metres overlying dense gravels. It was therefore concluded that settlement would be relatively instantaneous. There were no apparent bearing capacity problems.

The cells were designed with a penetration of 3 metres increased to 5 metres on the last cell, No. 6 to provide additional protection against scour.

Frodingham SW1A piles in mild steel were specified. The longest piles were 20.5 metres overall.

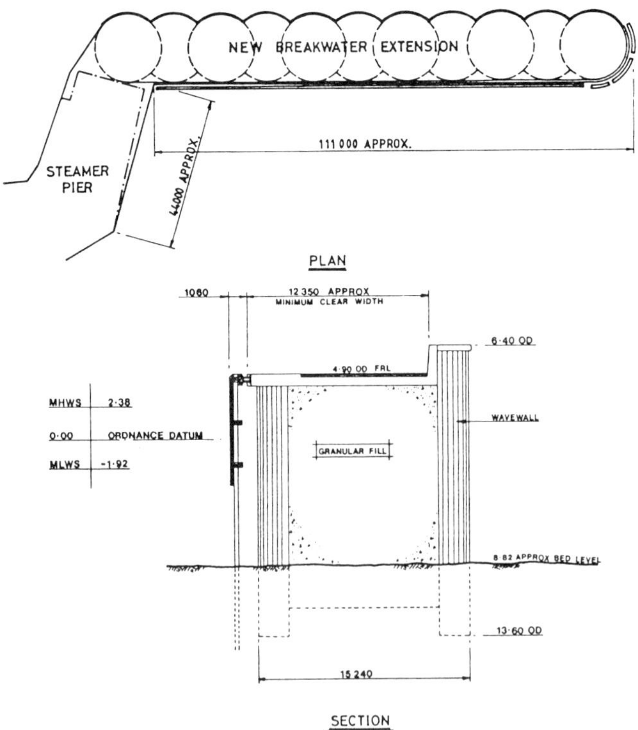

MALLAIG BREAKWATER
Fig. 5

22. The Contractor opted to build the cells end over, working from the existing steamer pier using a 100 ton crawler crane. The piling template was constructed in prefabricated quadrant sections supported off a secondary internal frame. Ground support to the template was provided by twelve steel H piles driven to level. The vertical pitch between the pile frames was four metres.

23. Given the exposed nature of the site each cell was only commenced when a settled period of weather was predicted. This proved a wise precaution with no significant down time being experienced on cell construction, albeit storm conditions up to force eight were experienced during the construction of cell 5 when the temporary works proved their worth.

24. Following completion of the temporary works, the Contractor then pitched and drove successive quadrants of the cell. Whilst some concern was expressed at the merits of this proposal at first, given the preference to pitch the whole cell at the one time, no real problems were encountered.

25. During construction of cells 1 to 5, the ground conditions were found to be somewhat looser than had been anticipated form the site investigation. This gave rise to problems in holding the temporary works piles during driving. These were however offset to some extent by the easier conditions experienced in driving the permanent piles.

26. No particular bed preparation was considered necessary, as sea bed cross falls within any given cell were nowhere greater than 10%. However, problems did occur during filling. The cross fall within the cell resulted in unequal pressures from the filling being imposed on the cell piling resulting in some lozenging and out of shape. This in turn gave problems in closing the interconnecting arcs.

27. Whilst installing the temporary piles in cell 6, the Contractor reported very soft ground conditions, so much so that the H piles were reported as "falling-in". A series of eight CPT's were therefore carried out which revealed the presence of a very soft clay layer coincidentally just at the founding level of cell 6 i.e. 5 metres below sea bed. The soft clay layer was 2 to 3 metres thick. The piling in cell 6 was therefore extended by five metres to 26 metres overall length and the piles driven through the soft layer to found in the dense gravel below. Long term time related settlements at cell 6 have subsequently been recorded since construction completion.

28. Piling was carried out using a 700N air hammer. The average pile length was 18 metres, increasing to 26 metres on cell 6. Cells were filled with beach won sand and gravel placed centrally by conveyor through water. The material was imported by sea and offloaded to a stockpile located on the vehicle marshalling area. Filling took place in two stages, stage one to the level of the lower piling frame and stage two following removal of the temporary works. The unit bulk volume per cell was 2,500m^3.

29. The overall production cycle was 14 days. Output rates are illustrated on Fig. 6.

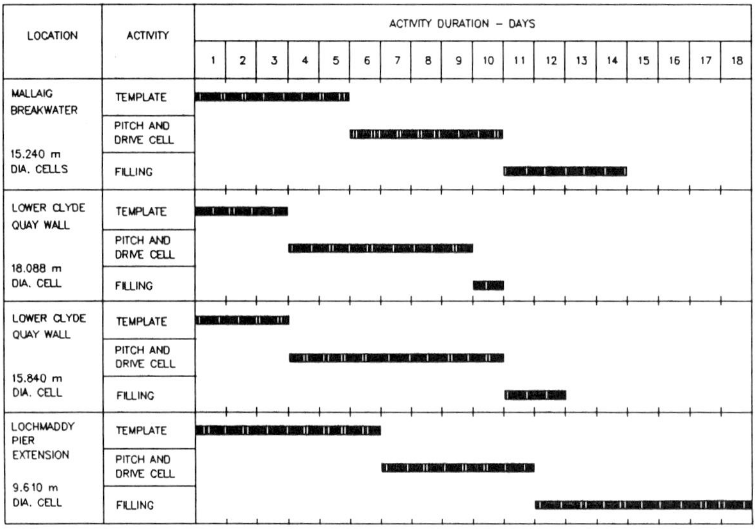

COMPARATIVE RATES OF PRODUCTION

Fig. 6

LOWER CLYDE

30. In 1985 Crouch & Hogg were commissioned as consulting civil engineers for a major development on the lower Clyde. A major element of the works was the reclamation of approximately 1½ hectares of tidal land.

31. Feasibility studies were undertaken to determine the most appropriate form of construction for the seaward boundary to the reclamation.

The depth of water and proximity to existing jetty structures precluded the use of an armoured embankment. Whilst the lack of overburden to provide passive resistance ruled out a conventional tied sheet pile wall. Cellular cofferdams of 12 and 15 metre diameter retaining 11 and 16 metres respectively were adopted as the most appropriate design solution.

32. The geology of the site was confirmed by a comprehensive site investigation. This revealed chlorite schist bedrock at depths ranging from 1 to 6 metres on the line of the cells dipping across the cell by as much as 15 degrees. The bedrock was overlain by a thin layer, 1 to 3 metres, of glacial till described as a stiff silt underlying soft alluvial silts having no appreciable strength.

33. The design required the removal of the soft alluvial silts and the subsequent benching of the relict till material to form a stepped level formation. Prior to cell construction, the specification required the placing of granular fill material to form a predetermined level bed platform upon which cells were to be constructed. After

driving the cells were to be filled to full height and the area then backfilled. A reinforced concrete service tunnel supported on steel bearing piles driven through the cell infill formed the closure structure to the rear of the existing jetties. The void behind the tunnel was subsequently to be filled to final level. The form of construction is indicated on Fig. 7.

34. The limiting condition was sliding of the cells on the inclined plane of the bedrock for which a factor of safety of approximately 2 was calculated.

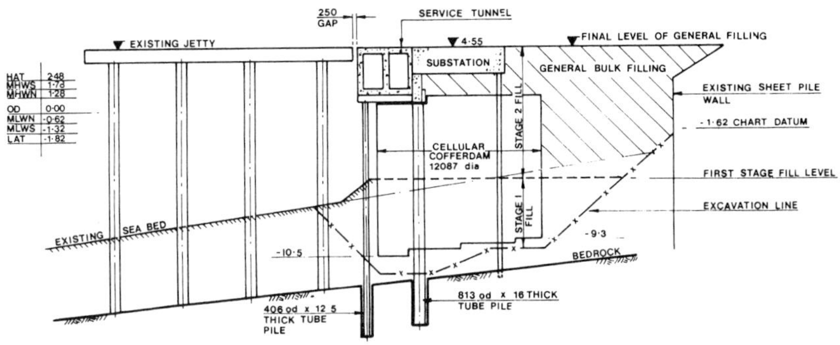

TYPICAL SECTION 12m CELL

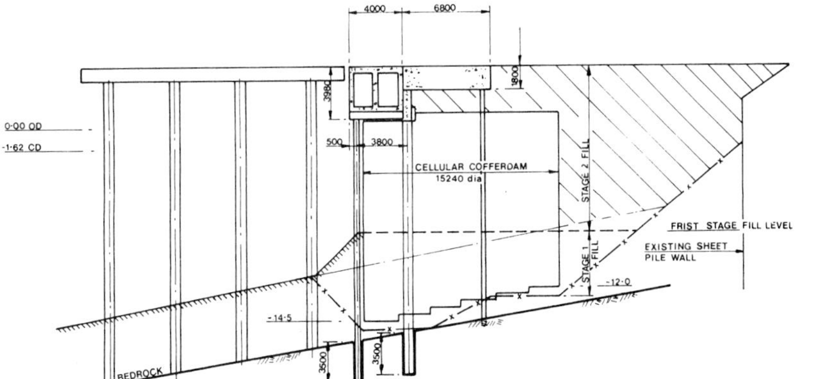

TYPICAL SECTION 15m CELL

LOWER CLYDE

Fig. 7

35. The cells subsequently had to be checked against more onerous imposed loadings. This work concluded that modifications were required to enhance the various factors of safety to acceptable levels, sliding being particularly critical.

36. The following design modifications were implemented:-

- the cells were to be founded on material having an insitu strength equivalent to an angle of internal friction of 40° or a cohesive strength of 150kN/m², dependent on the nature of the founding stratum. Further site investigation was carried out to determine the design formation level on a cell by cell basis. Deeper excavation resulted.

- all filling materials were specified to have an angle of internal friction of 40°. Crushed granite was used and inclinometer measurements taken in the stock piles to demonstrate compliance with this requirement.

- the earth retaining face of the cofferdams was roughened by the addition of welded attachments to promote the generation of wall friction.

- the service tunnel support piles were modified to drilled and socketed 813 diameter steel tubular piles designed to act as a deeply embedded passive anchors.

37. Cell construction was carried out from north to south using a combination of shore based and floating plant. The lead activity was cell bed preparation, using an Ajax crane rigged for grabbing working from the shore during the early stages and latterly from a Uniflote barge. The early parts of the works were affected by the presence of uncharted buried timber pile stobs which had to be removed. After a number of trials, the Contractor evolved an efficient vibrating pile extractor and attachment which became known as "the green gizmo"! As pile stobs were encountered the heads were exposed by divers and the extractor attached. Pile stobs of 3 metres in length were generally extracted in under fifteen minutes. Following cell bed preparation, stage one filling was placed by a follow-up crane working from completed cells. The stage one filling was levelled by divers.

38. On completion of the stage one filling the cofferdam template was floated into position on a Uniflote pontoon and located on four tubular spud piles. Following the installation of the spud piles the template was lifted clear of the pontoon using wires and tirfor winches working over pulleys located at the top frame level and the pontoon was floated clear. The template was a single unit fabricated from rolled hollow sections.

39. With the template located and secured, the complete circle of cofferdam piles was pitched and driven to final level. Filling was then placed by conveyor up to the underside of the template and continued to final level on removal of the temporary works.

40. The completed cells were then backfilled to tunnel formation level and the tunnel support piling installed some time thereafter. This was to allow take-up of cofferdam displacements caused by backfilling. Displacements of 60mm equal to ½% of the retained height were recorded.

41. The Contractor used an Eiger crane for the stage one filling, template location and cofferdam pile pitching and driving. Driving was by a 700N air hammer with all piles being driven to a specified level.

42. The crushed granite fill materials came from Furnace quarry on Loch Fyne and were brought to site by sea and offloaded to a temporary stock pile. At all times the Contractor was required to maintain a buffer stockpile equal to two cell volumes to guard against the risk of breakdown in supply. The unit bulk volume of the 12 metre cell was 750 cubic metres and of the 15 metre cell was 1,650 cubic metres.

43. Typical cycle time for construction was 10 and 12 days for the 12 and 15 metres cells respectively. Fig. 6 illustrates the various activities and durations.

LOCHMADDY, BRODICK AND PORT GLASGOW

44. Cellular cofferdams were utilised at each of the above locations, albeit not specifically as breakwater structures. At Lochmaddy and Brodick cells were used as part of extensions/conversions to existing jetty structures. Whilst at Port Glasgow cells were used to form the ground support to a ship building berth mat. Brief descriptions of each of these installations follow.

<u>Lochmaddy</u>

45. In 1986 Lochmaddy pier in North Uist was converted to a roll on-off ferry berth. The works included a 45 metre long extension to the existing pier, which incorporated a 9.5 metre diameter cellular cofferdam forming the new roundhead.

46. The ground conditions local to the cell comprised steeply sloping bedrock outcropping at the sea bed but partially infilled with sand. Cell bed preparation involved the removal of the sand, the levelling of the Lewissian gneiss bedrock and partial replacement with crushed rock to form a level bed for cell construction. Following completion of the bed preparation, cell construction proceeded conventionally without any problems and the completed cell was filled with crushed rock from a local quarry.

47. All work was carried out by an Ajax crane working from the new pier extension.

48. Continental pile sections were used, with the maximum pile length being 16 metres. Experience with this particular pile at Lochmaddy suggests it is a more robust section in handling than the SW1 & 1A sections manufactured in this country.

49. The unit bulk volume of the completed cell was 900 cubic metres and the construction cycle was 18 days. Figure 6 illustrates the durations of individual activities.

<u>Brodick</u>

50. Brodick Pier on the Island of Arran was converted to a roll on-off ferry berth in 1970. The works included the construction of five cellular cofferdams; two formed an extension to the existing berthing face; two acted as combined dolphins and bridge lifting points whilst the fifth supported the bridge hinge and formed the end point to a new approach causeway.

51. The ground conditions comprised medium dense sands of substantial depth which gave rise to no particular design or construction problems.

52. Cells were built from a crane mounted on a Uniflote barge. Piles were driven using an air hammer.

53. The unit bulk volume of the larger cells on the lifting dolphins was 700 cubic metres and 550 cubic metres on the pier extension and hinge cells. No data is now available on rate of progress.

Port Glasgow

54. In 1969, Lithgows Ltd., commissioned Crouch & Hogg to prepare the designs for a new ship building berth in Port Glasgow for 250,000 ton dwt half tankers. The design requirement was to provide a ground stiffness capability of 25 Ton/m² in the static build areas increasing to 100 T/m² at the launch way ends.

The ground conditions consisted of very soft alluvial silts having a cohesive strength of less than 25kN/m² overlying stiff boulder clay at depths to 10 to 15 metres.

55. The initial design solution incorporated cellular cofferdams driven through the soft silts to found in the stiff boulder clay at depth. The cofferdams were then capped with a heavily reinforced concrete slab.

56. Only five cofferdams were installed as problems were experienced during driving because of the presence of buried shipyard rubbish such as timber, concrete, steel and the like. The design was therefore modified to a scheme incorporating steel bearing piles driven to refusal on rock and capped with a reinforced concrete slab.

57. The cells were driven around an inner template using 7½ Ton Scotch derrick cranes mounted on rails. Piles were driven using a combination of air and diesel hammers. The longest pile length handled was 16 metres.

LOCATION	TYPE OF STRUCTURE	DATE	N° OF CELLS	CELL DIA. m	MAX. PILE LENGTH m	UNIT CELL VOLUME cu.m.
PORT GLASGOW	SHIPBUILDING BERTH MAT	1970	5	11.561	18	N.A.
BRODICK	PIER EXTENSION	1970	5	8.408	12.5	700
MALLAIG	QUAY WALL	1975	6	11.561	12.5	850
MALLAIG	BREAKWATER & PIER EXTENSION	1984	6	15.240	24.5	2,500
LOCHMADDY N. UIST	PIER EXTENSION	1986	1	9.610	16	900
LOWER CLYDE	QUAY WALLS	1990	22	12.088 & 15.240	11.5 16.5	750 1,650

PRINCIPAL PARAMETERS

Table 1

OTHER UK INSTALLATIONS

Bristol, Royal Portbury Dock

58. As part of the construction of the Royal Portbury Dock in Bristol 35 No. 12 metre diameter cells were constructed to provide an aligning structure for the entrance lock to the dock and to act as a training wall to prevent the build-up of silt in the approach channel to the lock, Ref. 1.

59. SW1A piles were used in the construction. The cells were built around an inner template and the piles driven to found in the red marl. Cells were filled using locally dredged sand and gravel.

Papa Westray

60. Papa Westray is one of the northerly group of islands in the Orkney archipelago. In the early 1980's the harbour at Papa Westray was developed by the provision of a new breakwater/pier constructed using straight web cellular cofferdams.

Leith, Entrance Lock

61. In the late 1960's the entrance lock to Leith Harbour was reconstructed using cellular cofferdams.

EXAMPLES OVERSEAS

62. There are a number of examples on the use of cellular cofferdams as breakwaters or coastal structures in the USA, Ref. 2. The following are the most notable examples.

Presque Isle, Pennsylvania	- Groyne
Grand Marais Harbour, Michigan	- Breakwater
Port Sanilac, Michigan	- Breakwater
Fairport Harbour, Ohio	- Breakwater
Wilmington Harbour, Delaware	- Groyne

CONCLUSIONS

63. Cellular cofferdams are a versatile form of construction which have been used extensively as breakwaters and other coastal structures. Given that they can be built using relatively modest plant and that once filled they are immediately stable they represent a flexible and economic form of construction for this type of duty.

ACKNOWLEDGEMENTS

64. The assistance and co-operation in the preparation of the paper of the following Authorities and Companies is gratefully acknowledged.

Mallaig Harbour Authority
Property Services Agency
Western Isles Islands Council
Caledonian MacBrayne Ltd.
Christiani & Nielsen Ltd.
Edmund Nuttall Ltd.

REFERENCES

1. IRWIN - CHILDS .F. et.al The Royal Portbury Dock, Bristol. Proceedings of the Institution of Civil Engineers, February 1978.

2. US ARMY COASTAL ENGINEERING RESEARCH CENTRE, Shore Protection Manual, Volume 2, Chapter 6.

Discussion

H. D. OSBORN, Haka UK Ltd

Vertical walls subject to wave action, as we have seen, invariably suffer overtopping, particularly when the wind is behind the waves, helping them over the wall. To eliminate this it is necessary to absorb the energy

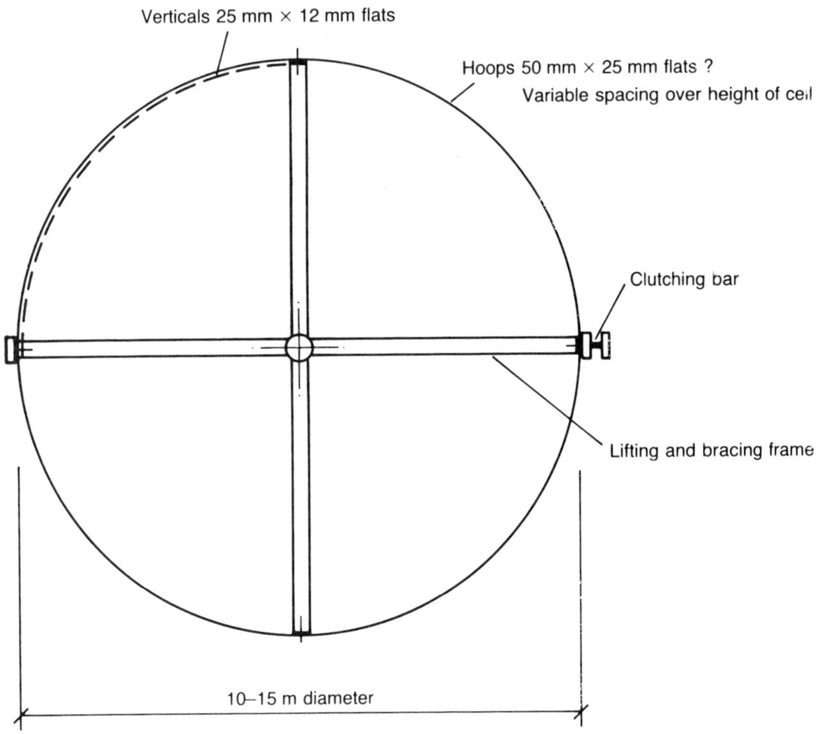

Fig. 1

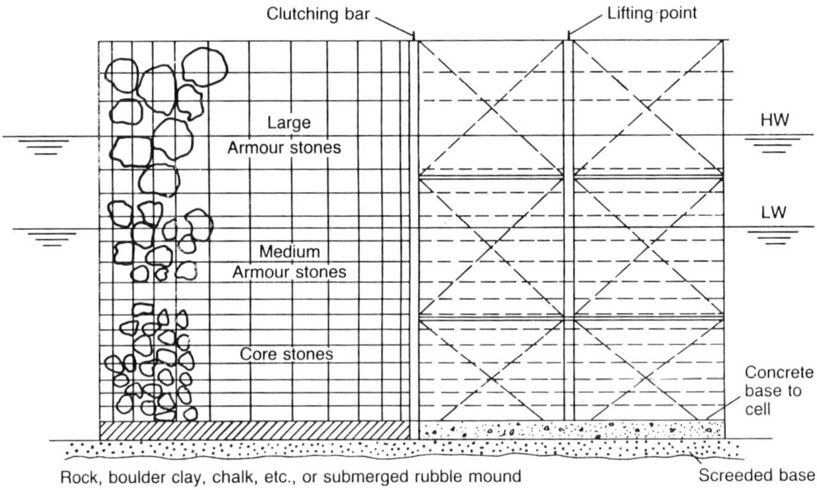

Fig. 2

of the waves as they reach the wall.

I offer the conference a possible solution to this, still using vertical-faced walls. Fig. 1 shows a plan sketch of a cell formed by steel hoops, using heavy flat bars at intervals held together with vertical flat bars welded to the hoop bars. Fig. 2 shows an elevation of the cells with reinforced concrete bases, founded on a screeded stone base, on a rock or rubble foundation. The cells would be clutched together as indicated, and would be filled with armour stone, thus allowing the wave action to be absorbed in the structure.

The cells, complete with the concrete base, could be prefabricated onshore and placed in position by a suitable heavy-lift crane.

Wave energy absorption methods have been used on a number of reinforced concrete structures, such as the Jarlan walls on the Ekofisk platform, and our Japanese friends have indicated similar wave absorption structures in their papers.

K. BERT, Zinkon Contractors UK
The very advanced development of caisson structures in Japan, as constructed already, seems in contradiction to the experimental studies performed in Europe. Does Japan work on a trial and error basis, or is Europe inventing the wheel again?

A. BEZUIJEN, Delft Geotechnics
I have some questions for Dr Oumeraci and Professor Partenscky.

The natural response time of the vertical breakwater appeared to have an influence on the results. What was the response time in the model? Did the authors study what variation is possible for different conditions (a breakwater on a subsoil of stiff sand or weak clay). Or, to ask it in another way, how generally applicable are the results?

Concerning the air content in the pore water of the subsoil, this can be quite different in model from reality; normally it is less in the model. Did the authors take special measures to lower the air content in the subsoil?

The authors have shown that there are soil pressure transducers in the model. Are results available? Measuring dynamic fluctuations, should one expect reflections from the walls of the flume to disturb the measurements?

H. F. BURCHARTH, Aalborg University

In the context of Paper 13, as to pressure distributions under the base plate the usual engineering assumption is that a straight distribution is on the safe side. This is a correct assumption in the case of quasi-static flow in the porous media. The authors' measurements from the large-scale model tests show that for highly non-stationary pressures caused by breaking waves the simple assumption will also be on the safe side.

J. METTAM, Scott Bertlin

Referring to the previous comment by Professor Burcharth about pressure under a caisson-type breakwater, the new BS 6439:Part 7 suggests that one should not just use a triangular shape for uplift pressure. Where there is a tendency for overturning, there is also a tendency to open up a gap under the base. In this space the uplift pressure will tend to equal the pressure at the seaward face. This concept has also been considered in design of dams.

Mention has been made by Dr Oumeraci (and Professor Burcharth) of the difference between air bubble formation in sea water and fresh water. I do not think any laboratory will be keen to use salt water in contact with all the expensive equipment in a wave flume. Perhaps some other solution can be found. I was once asked why it is that the bubbles in champagne persist so long. Perhaps this is a related problem indicating, if not an alternative fluid for use, at least that some suitable alternative can perhaps be found.

H. F. BURCHARTH, University of Aalborg

The proposal from Mr Mettam belongs to a state of

failure where the caisson has been lifted free of the foundation. This is an 'after limit state' situation which is not usually taken as the design limit state situation. It would be similar to the situation where a block in a block breakwater is on its way away from the rest of the structure.

D. M. MCDOWELL, Brighton Polytechnic

Attention should be directed to the work of R. Bagnold, who conducted research into impact pressures on vertical breakwaters in 1936 while working as a research fellow funded by the Institution of Civil Engineers. He demonstrated the mechanism that gave rise to high impact pressures and expressed the pressure as a function of wave dimensions and the thin-ness of a layer of entrapped air. One general conclusion was that high pressures depend on particular wave conditions. The effect on a structure during its lifetime needs to be studied in terms of probability of occurrence of potentially damaging waves.

N. W. ALLSOP, Hydraulics Research, Wallingford

The discussion in the context of Paper 13 on air entrainment and bubble persistence gives me the opportunity as technical co-ordinator for my EC colleagues in the MAST research programme (G6) to note some of the collaborative work on salt water effects being undertaken at Polytechnic South West by Drs Peter Hewson and Joan Griffiths. Their involvement in wave impacts in salt water has been supported by the Single Layer Armour Research Club, and by MAST G6 Coastal Structures, as well as SERC.

Early observations have noted that bubbles in salt water are smaller than in fresh water, and persist many times longer. With the observations that even the most severe wave motion in front of a vertical wall seldom leads to much bulk displacement of water, we have the possibility of modelling this volume of water numerically as a zone of (uniformly) 'soft' water.

Naturally these simplifying ideas can only give part of the truth, so research continues. In discussion with our colleagues, it has been suggested that modifications to the fluid surface tension by the addition of detergent may give a more acceptable approach than using salt water in these expensive large flumes.

H. LIGTERINGEN, Frederic R. Harris BV

It looks like the discussions in the new PIANC Working Group on vertical wall breakwaters will become interesting. I wish to make two comments regarding Dr Oumeraci's paper.

He questions the validity of Goda's formula for wave

pressures on a vertical wall. From my own experience at Delft Hydraulics, where we compared the available measurement results from various studies with the predicted pressures, we came to the conclusion that Goda's formula does cover conditions with breaking waves, excluding impulsive wave pressures.

Regarding the aspect of force increase due to dynamic response of the caisson system, I would like to refer to theoretical work at Delft University some 10 years ago. When taking into account the added mass of the foundations and the surrounding water, one finds for normal soil conditions (sand) natural frequencies of oscillation of the structure, which do not coincide with the frequency range of the exciting force. Only on rock foundations could the frequencies become high enough to cause resonance, but then the foundation problem does not exist.

F. BARENDS, Delft Geotechnics

The present-day state of the art is capable of predicting the moment of instability, in some cases rather well, in other cases rather poorly. In reality, not the moment of failure but the amount of deformation is essential, as it defines the actual degree of damage or malfunction. One should therefore put emphasis on the deformation of structures while failing. This requires physical tests at full scale and sophisticated mathematical simulation models. One achievement in the progress of numerical modelling is worth mentioning. It concerns the double sliding mechanism, which is important in fictional materials. It was suggested by de Tosselin and Tang in 1957, and it has been recently included in a non-associative elasto-plastic computer model. Applied to a caisson-type structure, the result shows an anticlockwise failure, which is in the first instance not according to intuition. However, usually failure does show this behaviour, as for example at the Niigata breakwater. If one could include a simple Mohr-Coulomb failure mechanism, a clockwise failure would occur. Therefore, in order to assess the deformation at failure, it is essential to apply appropriate mechanical models. The example discussed shows that it is possible.

H. F. BURCHARTH, Aalborg University

In Paper 14, equation (3) shows that the rubble foundation is taken as a rigid structure. This is a simplification because elastic-plastic deformation of the mound takes place. It might be a reasonable assumption for the actual performed model test where larger horizontal displacements of the caisson are studied; but for a prototype situation with a caisson placed on a high foundation, the model is probably too simple.

H. OUMERACI, University of Hannover
Professors Tanimoto and Goda assume that breaking waves are only caused by shallowness, but the experience of breakwater failures and the recent knowledge on wave instability and breaking in deep water have shown that breaking can also occur in deep water for other reasons. In my opinion it is too risky not to account for breaking waves in the design.

M. A. LOSADA, University of Cantabria
Peak forces by short breaking waves in deep water are punctual and very rare. Although these forces may be locally very high, they are not important for the stability of the whole breakwater, in which the huge mass concrete beam distributes these forces well along.

C. T. BROWN, Peter Fraenkel BMT
In Paper 14 the authors describe horizontal sliding of the breakwater caisson in both directions. The Peterhead Breakwater mentioned in Paper 12 is somewhat similar. It is constructed from blocks about 1.6 m high, 2.5 m wide and 4 m deep, weighs approximately 45 tonnes, and has about 1-3% voids due to vertical joints. The terrace slabs forming the scour protection to the seaward toe have in places been moved away from the breakwater and the parallel joints have opened up and we had considered that this was due to the downflow and strong horizontal pressures in the water column. However, the displacements observed are similar to those reported by the authors. Could the authors comment on the behaviour of the rubble mound and toe protection in their experience, and on whether the movement of the breakwater could be the driving force for such displacement? Could they also comment on the effect of the vertical slots in venting high-pressure spray and air from the entrapped air?

N. W. H. ALLSOP, Hydraulics Research, Wallingford
The structures shown in Paper 15 are vertical and, in relation to probable wave lengths, are relatively smooth. Up to the point that incident waves start to overtop the structure, waves will therefore be reflected back at a reflection coefficient C_r close to 100%. Wave conditions immediately seaward of such a structure under storm attack will be confused, and instantaneous water velocities much greater than in front of a rubble structure. Wave-induced scour will probably be greater, although this will depend on many other factors.

At Hydraulics Research, Wallingford we are particularly concerned with the safety aspects of wave reflections from these vertical walls, particularly at harbour entrances. Such reflections frequently lead to waves which for a brief moment greatly exceed the limiting steepness that would be calculated by simple or complex wave theories. These very steep waves pose a major safety hazard to small boats, and may well be the direct cause of a number of vessel sinkings in various English, Scottish and Irish harbours. Our initial considerations suggest that these safety hazards may be much reduced by modifying the vertical walls to improve their reflection performance, by perhaps reducing C_r to values nearer 40-60%. Perforated wave screens or voided concrete blocks might be attached to the front of the present walls to give this improvement.

Is Mr Wotton aware of this type of safety hazard at any of the harbours mentioned in Paper 15, and would he agree on the desirability of improving the hydraulic reflections and overtopping?

I. MIRCHANDANI, W. S. Atkins

Steel sheet piled cofferdams have not been much used in the UK for permanent structures but Mr Wotton appears to have been concerned with a good proportion of the total. W. S. Atkins has recently designed a cellular cofferdam quay wall for port development at Londonderry for which construction has just commenced, so this paper is of particular interest.

The paper contains much useful information on construction methods and progress rates. I have three specific questions.

The structures were designed at various times from about 1970 to the present. Has there been any change in the design approach over the years regarding cell stability (horizontal and vertical shear failure of the infill)? Which of the various design theories (Terzaghi, Cummings, etc.) does he currently use?

Section 27 mentions the unexpected presence of 2-3 m of soft clay which was left in place in the cell. Granular infill is usually considered essential for cell stability. How was this reconciled relative to the original design, and have there been any adverse long term consequences?

Have there been any recent surveys of the condition of the sheet piling of the older structures, e.g. residual wall thickness? If so, what were the findings?

R. D. DAWSON, W. A. Dawson Ltd

In relation to Paper 15, I would like to refer to the cellular pile breakwater under construction at Westray in Orkney. The straight web pile cell form of construction is an appropriate form of construction in isolated and partially exposed areas where quality concrete work/quality welding of steelwork/ quality paintwork cannot be carried out economically because of a lack of skilled labour, quality material resources or exposed weather conditions.

H. LIGTERINGEN, Frederic R. Harris, BV
Mr Wotton has presented very interesting information about the construçtion and performance of cellular cofferdams, but did not mention if there had been evidence of scour effects seaward of these vertical structures. As this is always considered to be an important design aspect, I am curious to know if the structures had scour protection and if there have been any problems in this area.

K. TANIMOTO, Paper 12
In response to Mr Bert, it could be said that the design techniques of caisson breakwaters in Japan have developed on the basis of abundant experience in the field and practical research work. Impulsive pressures on a vertical wall due to breaking waves and the behaviour of the wall under the action of impulsive wave pressures have also been exciting subjects of research in Japan. However, the research had not developed so as to be applicable to practical design procedures. On the other hand, there has been an urgent request to improve the design techniques of caisson breakwaters in order to avoid possible damage. Responding to this urgent request, the following three approaches have been made

(a) improvement of the calculation method of design wave forces, including engineering experiences
(b) presentation of the conditions by which the action of dangerous impulsive wave forces can be avoided
(c) development of smart caisson structures which overcome the powerful wave forces.

These approaches are reasonable and practicable for an urgent solution. ·In fact, caisson breakwaters with great lengths have been constructed, and new caisson structures are becoming popular in Japan. Of course, a more scientific approach is preferable to make clear the various phenomena. Such an effort is continuing partly in Japan, and we expect great progress in Europe. In this sense, I am very interested in the research programme at Hannover University.

I agree with Professor Burcharth's view. Actually, the rubble mound in the scaled model is considerably more rigid than that in the prototype. This is a reason why we focused our attention only on the sliding motion of the caisson in the model experiment. The elastic-plastic deformation of the mound could be investigated by numerical calculations with the finite-element method of the prototype, if necessary.

The principal purpose of the theoretical considerations in Paper 12 is to explain the fact that the caisson in deep water slides in both the shoreward and offshore directions, owing to the action of unexpectedly high waves, and to find what factor is important for the sliding. For this purpose, I think the simplified theoretical model is reasonably applied, because the wave forces are relatively static in this case.

In reply to Dr Oumeraci, we know that breaking waves exist in deep water. In the design of breakwaters, they are called partial breaking waves, and it was standard to consider the partial breaking wave force in addition to the standing wave force in Japan. At present, however, the partial breaking wave force is not considered explicitly, because it is of the same order as the standing wave force when the maximum wave height is applied to the design calculations. We are using the maximum wave height corresponding to $H_{1/250}$ for calculations of the design wave force.

In addition to Professor Losada's comment, I can say that breaking waves in deep water are important for floating structures rather than gravity-type vertical structures.

In reply to Mr Brown, when the caisson wall slides in the offshore direction, the toe protection block is undoubtedly moved in the same direction; this is observed in our experiment. Consequently, it can be said that the offshore movement of the caisson wall becomes a driving force for the toe protection block. However, I do not think the situation in the Peterhead breakwater is the same, because it is a relatively shallow water breakwater. In shallow water sliding of the vertical wall can only be caused in the shoreward direction. The offshore displacement of toe protection blocks is generally observed even if no displacement is caused in the vertical wall; this is due to the downflow when the uplift force becomes large. The uplift force is very important for the stability of toe protection blocks. Experimental results indicate that the stability of toe protection blocks with vertical holes is greatly improved, because the uplift force is reduced.

As to the other question, I suppose it concerns the breaking wave forces on a vertical slot caisson breakwater. If so, I can say that the vertical slots are very effective in reducing impulsive wave pressures on the breakwater, if the wave chamber behind the vertical slots is opened upwards. When the wave chamber is closed by a ceiling slab, weak impulsive pressures are caused in the wave chamber by the compression of the air trapped at the upper part.

H. OUMERACI, Paper 13

Even without any additional wind effect, overtopping still constitutes one of the reasons for the failure of relatively low-crested vertical breakwaters.[1] On the other hand, the high reflective properties of traditional vertical structures generally result in inconvenience to navigation, or in seabed scour and erosion of the toe berm which may be detrimental to the overall stability of the structure. Therefore, any innovation towards the reduction of reflected waves and overtopping is welcome. However, the principal of the solution presented by Mr Osborn is not new; it is the same concept as for crib structures that have been used in the past in the USA and Scandinavia, but which still have a number of technical and economic advantages (reflection coefficients of 0.3 - 0.6, small uprush, relatively small stones).

Reviewing and analysing the performance of crib structures[2] will certainly help improve the solution suggested by Mr Osborn, particularly with respect to the distribution of the stone sizes in the cell. In addition, there are a number of problems (related to the construction and function of the system) to be solved before such a solution will definitely be feasible.

In reply to Mr Bert, neither have we the intention to re-invent the wheel, nor would we like to question the Japanese caisson breakwater technology. Essentially due to the difficulties in quarrying sufficient amounts of large rocks, Japan has become the country with the longest stretch of vertical breakwaters (several hundred of kilometres), and undoubtedly also the country with the most advanced caisson breakwater technology.

On the other hand, there are on average more than six disasters per year related to caisson breakwaters in Japan. For instance, more than 100 breakwaters collapsed between 1965 and 1982, not only due to caisson sliding but also to further failure mechanisms of hydraulic, structural and geotechnical natures. The most striking finding by examining these failures is that the design wave conditions were rarely exceeded.[1]

One of the reasons for these failures was nicely demonstrated[3,4] by Goda who showed that the standard wave pressure formulae used in the past for the design could not explain the sliding experienced by the damaged breakwaters. The new pressure formulae (which are known as Goda's formulae and which became standard formulae in Japan and the UK), even though they failed in some cases, could indeed better predict the occurrence of the actual sliding failures than the formulae of Hiroi et al.

Furthermore, there are still a number of important

modes of failure which cannot be explained by the existing design tools and empirical formulae, clearly showing that the problem of the design of vertical breakwaters has not yet been satisfactorily solved.[1] For instance, most of the caisson breakwaters in Japan and elsewhere are built in such a way that waves are invited to plunge on the wall, while the problem of impact loads induced by breaking waves (e.g. plunging breakers) is 'solved' by stating that 'the occurrence of impulsive pressure should be avoided'.

Based on the results of the review and analysis of the failures experienced worldwide by vertical breakwaters, the research strategy shown in Table 2 of Paper 13 has been developed. The design of such structures is indeed an integrated problem involving hydraulic, structural and soil mechanics aspects, and the research strategy now adopted by the MAST Research Programme (MAST G6-S) represents an efficient solution to account for all these aspects and elaborate a sound physical and technical basis for the development of rational design tools.

To the question of Mr Bezuijen, I have the following replies.

The natural period of oscillation of the breakwater in the model is mentioned in Fig. 6 of Paper 13 (T_N = 0.08 s). The transfer of the model results to a prototype is a difficult task which is partly dealt with in a further paper.[5] Nevertheless, the strategy adopted in this research is to develop a numerical model which will be validated by the test results for various model conditions, so that a reliable prediction tool is obtained. The latter can then be used to predict the dynamic response of prototype structures.

The air content in the pore water of the subsoil (sand) constitutes the most important difficulty when measuring pore water pressure within the sand bed in the wave flume. The usual measures were taken to reduce the air content, but it was observed that even a small amount of air considerably affects the results. Therefore, additional investigations of this effect were systematically performed in the soil mechanics laboratory (Department of Professor Richwien) in order to interpret better the results obtained from the large-scale model tests.

Some results of the soil pressure measurements are available,[6,7] suggesting that there are two components of the total stress: one high-frequency component which is probably induced directly by the shock wave, and a lower-frequency component induced by the motion of the rocking structure. It is obvious that the former is affected by the solid boundaries of the flume (walls and bottom).

To Professor Burcharth: if the uplift pressure values

directly measured at both the seaward and shoreward edges are joined by a straight line, then you are certainly on the safe side. However, one should bear in mind that if for some reason the rubble mound foundation should become less permeable at the rear side, the uplift pressure on the shore side will tend to increase considerably. This actually occurred and was measured at a prototype breakwater in Italy.[8]

I have the following replies for Mr Mettam.

The occurrence of a very different uplift pattern when there is a tendency of the caisson structure to overturn is obvious. However, the caisson structure is generally designed in such a way that an overturning will not be allowed to occur. Nevertheless, this case is interesting when investigating the development of the failure beyond the limit state.

It is not our intention to recommend the use of salt water for hydraulic testing in wave flumes. We only wanted to stress that

(a) air content is a determinant influencing factor for the measured impact pressure

(b) air bubbles' sizes and persistence are very different in fresh water and salt water, and this may affect the impact pressure obtained form model tests using fresh water

(c) something should be undertaken to investigate this effect and eventually to provide the corrective factors when using fresh water for model testing. In this respect, Polytechnic South West (Plymouth) has already started a research programme within the MAST G6-S/ Project 2.

To the comment of Professor McDowell, in fact it is not the highest wave which may cause the most critical loading. There are some waves and wave sequences which are potentially more damaging than others. I would therefore plead for replacing the 'design wave heights' concept by the concept of 'design wave loads'.

Now for my replies to Mr Ligteringen. The validity of Goda's formulae has not been questioned in general, only for breaking waves which are associated with high shock forces. The specification commonly found in the literature that these formulae are valid for 'breaking and non-breaking waves' is misleading and should probably be replaced by 'broken, non-breaking and partially breaking waves'. If - as admitted by Professor Goda and now confirmed by Mr Ligteringen - Goda's formulae are not valid for 'impulsive pressure', two questions then arise.

(a) Why have these formulae been applied (apparently with success) to the breakwaters investigated by Goda,[3] even though all these structures have been built in such a way that plunging breakers - and thus 'impulsive pressure' - were invited to occur?

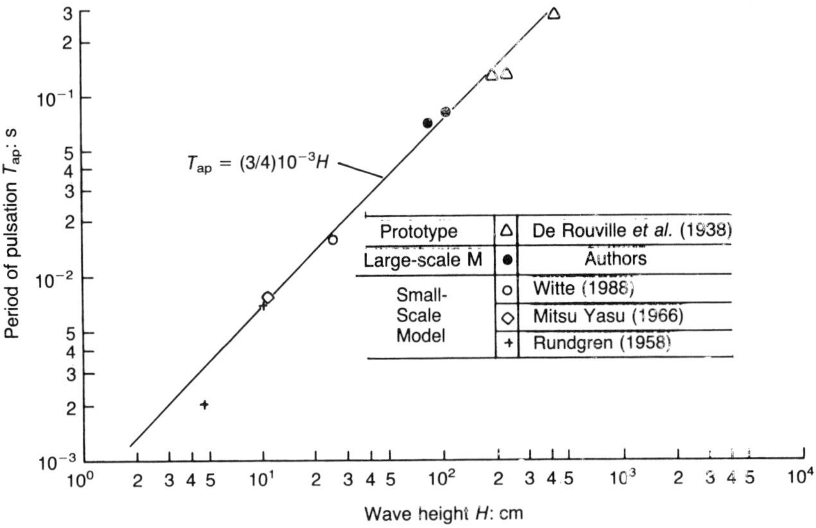

Fig. 3

(b) Are we missing formulae for the prediction of impulsive pressure induced by breaking waves?

Theoretical considerations (hydrodynamics and thermodynamics) as well as the results of prototype measurements and hydraulic model tests have shown that if an air pocket is entrapped between the vertical wall and a plunging wave (with height H) impinging on it, an impact force is induced which exhibits oscillations with the period T_{ap}.

$T_{ap} = 3 \times 10^{-3} H/4$ (T_{ap} in s and H in cm)

For plunging waves of the height H = 1-10 m, periods of the force oscillations of T_{ap} - 0.075-0.75 s will result.

Considering that prototype vertical breakwaters built on normal soil (sand) have natural periods of oscillations of T_N = 0.05 - 0.5 s,[1] it is seen that the frequency range of the exciting force may well coincide with the natural frequency of oscillations of the structure-foundation-water system.

Professor Barends presented the very interesting concept 'double sliding' which might contribute to the explanation of why a number of vertical breakwaters failed by tilting seaward and not shoreward.

Since all the breakwaters which tilted seaward were subject to breaking waves and overtopping, the following possible explanation has been suggested.[1] Under the repetitive action of the wave loads the structure exhibits rocking motions. Since the maximum oscillation amplitudes are located at the seaward and shoreward edges of the monolithic structure, repetitive percussions of the foundation at these locations generally result, leading to the initiation of shear failure under both edges (two slip surfaces in opposite directions). However, the suction action of the receding waves eroding the toe berm, together with the long-lasting pressure behind the breakwater under overtopping conditions and probably with some waves breaking on the front side of the caisson deck, have caused the structure to fail seaward rather than shoreward. These effects are probably reinforced by some 'abnormal' forces induced by heavy overtopping and which are still not understood. There are also a number of cases where seabed scour has clearly contributed to seaward tilt.

I also fully agree with the suggestion of Professor Barends that it is the extent of deformation rather than the moment of failure which is essential for the definition of the actual performance of the structure. Adopting this suggestion, however, would require suitable prediction tools and threshold values which are not yet readily available.

References

1. OUMERACI H. Review and analysis of vertical breakwater failures - Lessons learned. Preliminary Res. Report MAST G6-S/Project 2, 1992.
2. GAILLARD D. D. Wave action in relation to engineering structures. 1904, Professional Papers No. 31, US Army Corps of Engineers.
3. GODA Y. A new method of wave pressure calculation for the design of composite breakwater. Rep. Port & Harbour Res. Inst. (PHRI), 1973, 12, No. 3 (in Japanese).
4. GODA Y. Random seas and design of maritime structures. University of Tokyo Press, 1985.
5. OUMERACI H. and PARTENSCKY H. W. Breaking wave impact loads on vertical structures. Proceedings of the First Project Workshop: MAST G6-S/Project 2. Hannover, 1991.
6. WOLTER H. Stress induced beneath a caisson breakwater under horizontal impact loads. Dipl.- Ing. Thesis, Inst. Grundbau, Bodenmech. & Energiewaasserbau, University of Hannover, 1991 (in German).
7. OUMERACI H. Dynamic loading and response of caisson breakwaters - Results of large-scale model tests. Res. Report MAST G6-S/Project 2, 1991.
8. MARCHI E. et al. Storm wave pressure on the breakwater of Genoa harbour. Proceedings of the 16th IAHR Congress, Sao Paulo, 1975, Vol. 1, 246-253.

J. H. WOTTON, Paper 15

In response to Mr Allsop's point regarding wave

reflections from vertical smooth faced structures I would acknowledge that there is clearly a risk of such a phenomenon occurring. Indeed, in storm conditions at Mallaig very disturbed seas can be observed on the outer side of the new breakwater. I would agree that it would be desirable to incorporate measures to improve the hydraulic reflections of such structures; however, this will always have to be considered in relation to the additional costs involved.

As regards Mr Mirchandani's questions, we design cellular cofferdams against a wide range of failure modes which include sliding, overturning, bursting both in the ring and in the junction piles, horizontal shear failure of the fill (Cummings) vertical shear failure of the fill (Terzaghi), foundation failure, circular slip failure where appropriate, log spiral failure mechanism after Brinch-Hansen. All of these have been standard checks over many years. The soft clay layer was encountered at depth and the piles were driven further to contain the material. Granular sea bed deposits and the cell infill were sufficient to ensure stability of the cell. There has however been ongoing long-term settlement of the cell in question, which has been continuously monitored. This now appears to be approaching its 90% value. As regards residual thickness of metal no specific measurements have been undertaken.

Mr Dawson's contribution is noted with interest. I would endorse his views.

Mr Ligteringen's reference to scour seaward of the structure is noted. No specific scour protection methods were incorporated. However, at Mallaig regular surveys are undertaken to assess the change in sea bed. To date no adverse problems have been revealed.

J. D. METTAM, Scott Bertlin & Partners

The method proposed in BS 6439 : Para. 7 for treatment of uplift pressure under a caisson-type breakwater may be considered analogous to the classic Euler analysis of buckling of long columns. In that analysis a long column, which does not buckle if it is straight and axially loaded, is analysed assuming that it has a slight bow in it. The test of safety against buckling is whether or not the column, once it has some initial deflexion, then continues to deflect. The introduction of an initial bowing creates what Professor Burcharth describes as an 'after limit state' case. This approach is justified by the increasing stresses which result once the initial limit state is passed.

If a caisson begins to overturn and the front toe starts to lift off the foundation, the uplift pressure will quickly increase from the triangular distribution towards a rectangular distribution. This could lead to an unexpected failure. It is therefore prudent to use the rectangular distribution when analysing overturning.

For this reason BS 6439 allow the triangular distribution to be used when checking factors of safety against sliding, but requires the more severe rectangular distribution to be used when checking factors of safety against overturning.

16. Standardizing the approach to specifying quarried rock in coastal structures and breakwaters

J. D. SIMM, Robert West & Partners, J.-P. LATHAM, Queen Mary & Westfield College, and C. ORBELL-DURRANT, May Gurney & Co. Ltd

SYNOPSIS

Quarried rock, widely used in coastal structures and breakwaters, is often ill defined and specified. Economy is thereby often lost and payment disputes are frequent. This paper summarises the main problems encountered and sets out a workable and coherent set of defined parameters together with guidance values for these parameters. Aspects covered include density, durability, nominal diameter, shape, grading, profile measurement and payment procedure. The paper concludes with reference to standard gradings and methods of optimising quarry output and to the European standardisation issue.

INTRODUCTION

1. Quarried rock is widely used in coastal structures and breakwaters and can be employed in a wide range of structure types. The growing trend towards "softer" options for coastal defence, including maintaining sand and shingle beaches, often dictates that the wave climate at the shoreline be suitably controlled. Rock can be used here to significant effect in structures such as detached breakwaters, groynes, fishtail breakwaters, cills, etc. It can also be used as a protection to existing vertical coast defence structures either in the form of a sloping revetment or an anti-scour blanket. The former provides a physical protection but both assist in dissipating wave energy and hence mitigate the worst effects of local scour. Quarried rock is also, of course, a major component in conventional rubble mound breakwaters. The berm breakwater concept also permits the use of quarried rock in armour layers for large breakwaters in deep water exposed to significant wave climates where previously only very large artificial concrete armour units would have been considered.

2. In the United Kingdom, the above considerations, combined with an appreciation of the futility of "chasing down the toe" of hard coastal defences and concerns about rising sea levels, has led to a significantly increased interest in rock structures. A number of schemes have already been constructed around the U.K. coast and a considerable number more are planned. Approximately 750,000 tonnes of armour was used in the

U.K. in 1989, an annual figure which is expected to double by 1994. This will impose considerable demands on the quarry resources within the U.K. and overseas. For the east coast of England, continental sources of rock are often competitive, but for west coast schemes shipment costs from the mainland Europe are high and U.K. quarries therefore need to plan and stockpile to meet the likely demand.

3. With these increasing demands, and with the amount of information now available on rock material and designing with it, it would be foolish indeed to continue with the present situation where specifications from one contract to another impose widely varying demands for rock quality, shape, grading, layer thickness and packing density, construction technique, and (perhaps most controversially) methods of measuring the quantity of rock placed for payment. For too long, rock projects have suffered from the assumption by engineers that "anyone can design a pile of rocks to go in the sea." In some senses, "anyone can", and indeed they have, and the resulting confusion is only too obvious. When preparing specifications and bills of quantities for payment, the resulting documents and drawings often reflect "plan chest engineering" (use of historical precedent even if incorrect or unduly onerous) or even worse, indescriminate use of manuals, such as Shore Protection Manual. These problems have been compounded by a lack of properly defined methods of testing for rock quality and geometry, tests for aggregates being applied even though some of these are hardly appropriate.

4. This paper describes work which has been carried out as part of the preparation by CIRIA in the U.K. and C.U.R. in the Netherlands of a joint Manual on the Use of Rock in Coastal and Shoreline Engineering (CIRIA/CUR, 1991), which has set out to resolve some of these problems by providing inter alia detailed guidance on the materials, hydraulics and geotechnical design of rock structures, and suggested standard specifications and methods of measurement for payment. The paper therefore commences with a discussion of some of the problems mentioned above in a little more detail, followed by a description of three levels of standardisation or standard approaches:

i. A first level of standardisation comprising a set of common definitions and methods of approach, an essential foundation for emerging from the problems in this field.

ii. A second level of standardisation in which guidance values are described for the parameters defined at the first level of standardisation.

iii. A description, at a third level, of methods of achieving more economic use of available rock. These methods include (a) use in the European context of suggested standard gradings, which permanent quarries can produce generally and stockpile, (b) methods of adjusting rock weight and density to suit the available resources and the aggressiveness of the environment, and (c) a description of the approach to the use of temporary dedicated quarries for large breakwaters and in remote sites, where the key to economic design is matching the quarry yield to the quantities of rock required in the structure.

HISTORICAL PROBLEMS IN SPECIFICATION AND MEASUREMENT

5. The properties of quarried rock may be seen in three distinct categories:

i. Intrinsic properties of the rock before winning from the quarry face.

ii. Manufactured properties which are those of the quarried rock after winning (generally by blasting), production in the quarry and transport to site.

iii. Execution-induced properties, which are those that the rock has once it has been placed in the coastal structure or breakwater. These properties are summarised in Figure 1, where they are related to functional requirements of a rock structure.

6. For specification of the intrinsic properties of the rock to be used designers have normally sought guidance in the limited literature available. Before publication of the CIRIA/CUR rock Manual (Ref 1), published guidance was restricted to papers (e.g Ref 2) and brief references in codes of practice (Ref 3). Recommendations available in such sources ran the risk of being taken out of context and being used without proper understanding of the implications of the specified test result ranges, a situation exacerbated by the use of tests originally designed for aggregates.

7. One particular intrinsic property - that of density - is of particular interest to coastal engineers and has been the source of much confusion. Minimum densities are often specified from durability considerations. Density is also critical to hydraulic stability, all armour stability formulae using this property, converted to the relative buoyant density when submerged in water ($\Delta = (\rho_i / \rho_w) - 1$). Designs are typically made assuming and specifying a minimum density of 2.6 tonnes/m^3 although lower oven dry densities, ρ_r, may occur in some useable sedimentary rocks and some igneous rocks on the market at present have densities as high as 3.1 tonnes/m^3. However, in computations of Δ and assessment of hydraulic stability, the figure that should be used for the density, ρ_i, of the rock is saturated surface dry density, $\rho_{ssd} = \rho_r (1 + 0.01 W_{ab})$ where W_{ab} is the water absorption, the latter being the more important parameter in assessing durability (see Table 1).

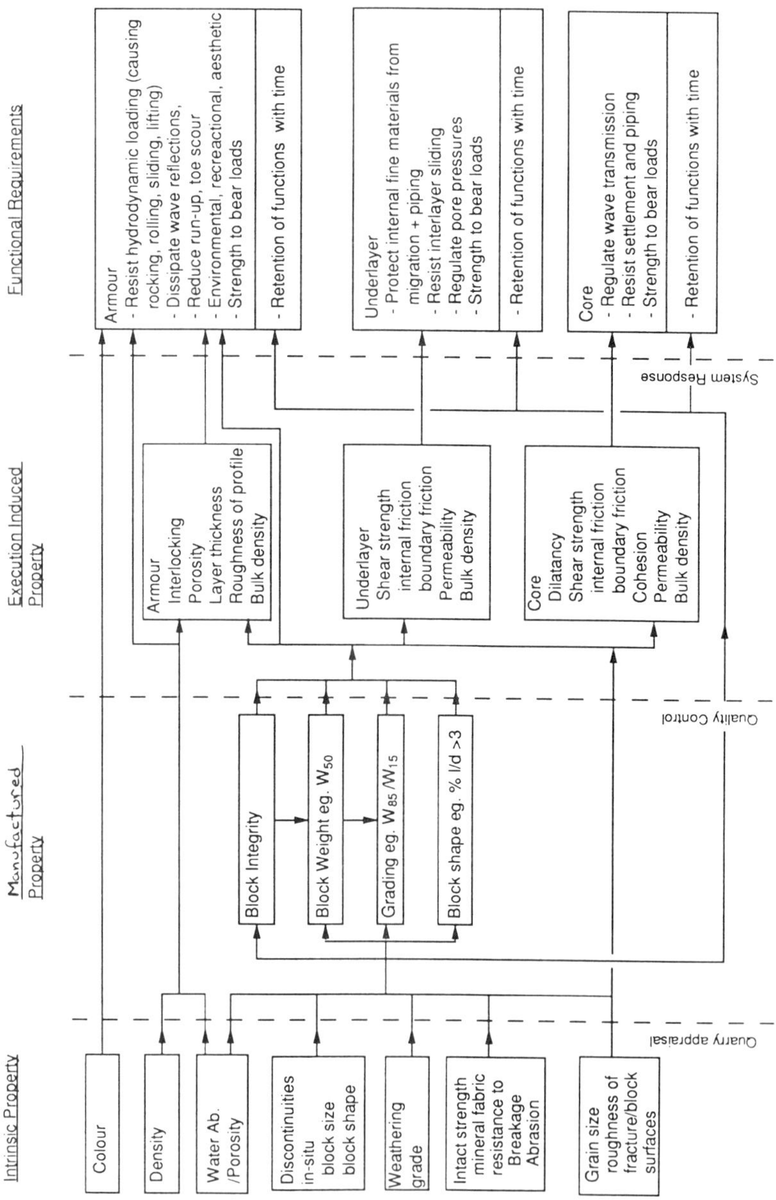

Fig 1 Flow diagram for rock properties and functions (from Ref 1)

	Excellent	Good	Marginal	Poor	Comments
Oven Dry Rock Density Ap. 2.6 ρ_r (t/m³)	>2.9	2.6-2.9	2.3-2.6	<2.3	Physical property affecting hydraulic stability via. Good indicator of durability except dense but weathered basic rocks
Water Absorption Ap. 2.7 W_{ab} (%)	<0.5	0.5-2.0	2.0-6.0	>6.0	Single most important indicator of resistance to degradation. Good indicator of weathering resistance. Often misleading for porous limestones with large free draining pores.
Magnesium Sulphate Soundness BS 812/BS 6349 MSS (%)	<2	2-12	12-30	>30	Indicates resistance to weathering. Important test for porous sedimentary rocks for use in hot dry climate. Good correlation with water absorption.
Freeze/Thaw Ap. 2.4 FT (%)	<0.1	0.1-0.5	0.5-2.0	>2.0	Important test for freezing winter climates (especially reservoir dams). Good correlation with water absorption.
Methylene Blue Absorption Ap. 2.10 MBA (g/100g)	<0.4	0.4-0.7	0.7-1.0	>1.0	Indicates presence of deleterious clay minerals
Fracture Toughness ISRM (1988) K_{IC} ($MP_a.m^{½}$)	>2.2	1.4-2.2	0.8-1.4	<0.8	Indicates resistance to Type 2 breakage, degree of weathering (geological). Good correlation with abrasion resistance. Can be misleading for impact resistance of large blocks
Point Load Index ISRM (1985) $I_{s(50)}$ (MP_a)	>8.0	4.0-8.0	1.5-4.0	<1.5	Indicates resistance to Type 2 breakage. Quick test with high test variability Can be misleading for impact resistance of large blocks
Wet Dynamic Crushing Value Ap. 2.5 WDCV (%)	<12.0	12-20	20-30	>30	Indicates resistance to Type 2 breakage. Quick test
Mill Abrasion Resistance Index Ap. 2.9 k_s (loss per 1000 revs)	<0.002	0.002-0.004	0.004-0.015	>0.015	Indicates resistance to abrasion by mutual grinding of saturated rock surfaces. Test result may be used with rock degradation model. Section 3.3.4.2
Block Integrity Drop Test Ap. 2.11 I_d (%)	<2	2-5	5-15	>15	Indicates resistance to Type 1 breakage of large blocks

Table 1 Guide to rock durability from test results (from Ref 1)

8. When considering manufactured properties, confusion here relates both to poor definition of terms and lack of knowledge about acceptable values of such terms.

9. Confusion commences with things as simple as definition of the nominal diameter of a piece of rock. A range of values are available in the literature from which the unsuspecting may choose. Van der Meer (Ref 4) uses the now usual definition of the side of the equivalent cube i.e. $D_n = (W / \gamma_r)^{1/3}$, where γ_r is the unit weight of the rock, this definition being built into his new but now widely used formulae. Serious errors in the application of van der Meer's formulae will arise if the diameter of the equivalent sphere is used instead of the side of the equivalent cube, because the weight of rock calculated thereby from the nominal diameter will be 52% of that with the correct cubic definition. A further confusion is the difference between nominal cubic diameter and sieve or grizzly size. Extensive tests in Holland have established that nominal diameter is 84% of the equivalent square sieve size opening for the 50% passing value.

Grizzly spacings (horizontal parallel bars in one direction only) are, by contrast, generally smaller, the exact relationship between grizzly spacing and nominal diameter being dependent upon the efficiency of the grizzly.

10. The shape of rock is the second as manufactured property over which much confusion has been generated, both in regard to the descriptive terms used and in regard to the aspect ratio of pieces of rock. Confusing and vague terms such as "prismoidal" (generally taken to mean "cuboidal" or similar to a parallelepiped) abound in existing specifications. The majority of these shape descriptors do not have any rational scientific basis. Aspect ratio, the ratio of the maximum to minimum dimensions of a piece of rock, is again often crudely defined, even though typical British specifications require the aspect ratio not to exceed 2 for armourstone. In practice, two schemes are used. One is the enclosing cuboid system. Under this system the aspect ratio is the ratio of the longest to the shortest side of the enclosing cuboid and a maximum aspect ratio of 2 is often specified. The alternative system is that where the maximum dimension is taken as the maximum dimension which can be measured across diagonals. The difference between these schemes is illustrated in Figure 2. It is worth noting that a cube, which with the enclosing cuboid system obviously has an aspect ratio of 1, the aspect ratio of 1.73 on the alternative definition.

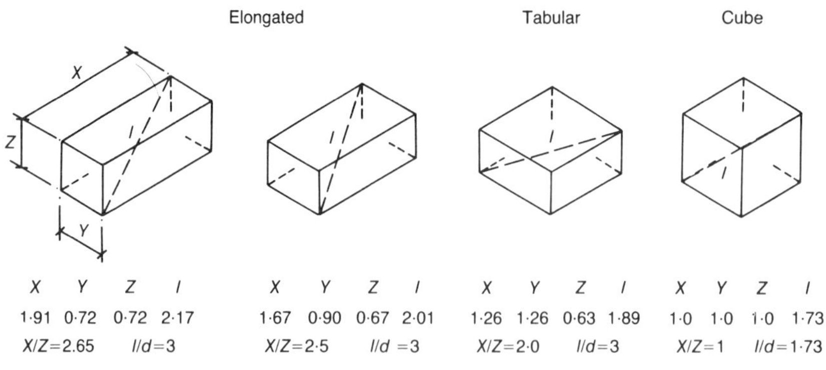

Fig 2 Illustration of blocks with equal volume and l/d=3, compared with cube with l/d = 1.73 (from Ref 1)

11. A third manufactured property which designers have specified in different ways is grading of armourstone. Often, the grading limits between 75% to 125% of the nominal W_{50} weight suggested in the Shore Protection Manual (Ref 5) are adopted. In fact, tests (Ref 4) have shown that this requirement is unneccessarily onerous and that conventional two stone thick armour layers perform quite satisfactorily up to D_{85}/D_{15} of about 2.5, equivalent to a value of W_{85}/W_{15} of 16, which vastly differs from the Shore Protection Manual's figure for W_{85}/W_{15} of 1.7. A further problem in the specification of grading is a lack of understanding of an

appropriate way to impose suitable grading limits, even when the approximate width of grading desired is understood.

12. As-constructed (or execution-induced) properties, even given the above comments on intrinsic and manufactured properties, have perhaps been the greatest source of contention. The principal areas of controversy have been final surface profile (including appropriate tolerances), and areas related to surface profile definition including layer thickness, porosity, and packing density. As all these matters have not only been of concern technically but also the basis on which a contractor has been paid for completed work, the controversy has been more than academic.

13. The surface profile measurement technique has, historically, rarely been specified. If levels are measured with a conventional level and staff system, the argument is always whereabouts on each piece of rock the staff should be placed. Even having measured the surface profile, defining whether or not this is within tolerance, has again been difficult, primarily because of vague tolerance specifications. The nominal straight line surface found on many drawings for the upper surface of an armour layer is variously interpreted (Ref 6). Clearly, one cannot take this line to mean that which is only just touched by the outermost points on the armourstone, and therefore the proportion of the armourstone which must protrude above this line has been the source of much contention.

14. Layer thickness and packing density are clearly related matters. Given porosity $n_v = 1 - (\gamma_b \div \gamma_a)$ then the respective expressions for layer thickness and number of units in area A are: $t_a = n\, k_t\, D_{n50}$ and $N_a = nA\, k_t\, (1-n_v) D_{n50}^{-2}$, where γ_b = bulk weight density as laid, n = number of layers making up the total thickness of armour and k_t = layer thickness coefficient. k_t and n_v have normally been selected from values tabulated in the Shore Protection Manual (Ref 5). These values, together with lack of clarity on profile survey methods already discussed, have been the root cause of later contractual disputes between designer and contractor. This is hardly surprising since k_t has not even been constant in the Shore Protection Manual itself (Ref 5), being reduced from 1.15 in the 1978 edition, to 1.00 in the 1984 edition.

15. Contractual disputes have arisen because of the normal practice of measuring as-placed armourstone for payment on a volumetric basis. Thus, the hands of both the contractor and the resident engineer are often tied to the nominal thickness shown on the drawings because of the payment implications. However, if the designer then wishes to impose further constraints in terms of packing arrangement, as well as maintaining a nominal layer thickness, the field is immediately open to disputes which may involve considerable sums of money.

FIRST LEVEL OF STANDARDISATION-COMMON DEFINITIONS AND APPROACHES

16. In specifying intrinsic rock properties, perhaps the most important consideration is to specify saturated surface dry density. However, since this is also related to the water absorption percentage w_{ab}, which in turn has a marked influence on durability, the designer is strongly recommended to quote a maximum value for water absorption as well (as will be discussed in paragraph 23). This will then give more flexibility in permitting a wider range of density.

17. The nominal diameter of armourstone (D_n) should always and only be specified as the side of the equivalent cube $(W / \gamma)^{1/3}$.

18. When describing the shape of armourstone, it is recommended that only objective terms (very round, semi-round, equant, irregular, elongate and tabular) be used. These shape descriptors can in turn be expressed in terms of one number , the Fourier Asperity Roughness coefficient, P_R. Typical rocks of each of these shape types, together with their P_R values, are given in Figure 3. A limited number of flume tests (Ref 7) suggests that a surprising bonus of using P_R is that it may then be used directly in van der Meer's hydraulic stability formulae (Ref 4), replacing the coefficient 6.2 in the plunging wave formula by C_{pl} and the coefficient 1.0 in the surging wave formula by C_{su} where $C_{pl} = 5.4 + 70\ P_R$ and $C_{su} = 0.6 + 40\ P_R$.

19. Aspect ratio should always be defined as l / d for consistency and uniformity, despite the fact that the authors recognise that for certain rock types the enclosing cuboid system may seem attractive. Measurement of aspect ratio is illustrated in Figure 4.

20. Rock gradings should be defined by a consistent scheme (Ref 1). The system is summarised in Figure 5. It essentially revolves around defining weights for a series of four class limits, together with a restriction on the range of the mean weight of the stone in the grading. The mean weight is being defined in a particular way in which stones smaller than the extreme lower class limit (ELCL) or the lower class limit (LCL) are excluded. The class limits are defined by weight for gradings heavier than 300 kg and by weight or by size for gradings in the range 10 - 300 kg, reflecting the way in which the rock will be produced and assessed in the quarry.

21. For surface profile measurement, a consistent scheme should be adopted in which a semi-smooth surface profile is determined using a staff or sounding line with a spherical end of diameter 0.5 D_{n50}, to measure levels at intervals of 0.75 x D_{n50} across the profile. This technique has the advantages of being extremely practical, and of smoothing out the worst of the peaks and troughs in the surface profile. It also matches closely the

(a) ELONGATE/TABULAR (ET)
$P_R > 0.015$

(b) IRREGULAR (IR)
$P_R = 0.013\text{-}0.015$

(c) EQUANT (EQ)
$P_R = 0.011\text{-}0.013$

(d) SEMI-ROUND (SR)
$P_R = 0.009\text{-}0.011$

(e) VERY ROUND (VR)
$P_R < 0.009$

Fig 3 Photographs for visual comparison of block shapes (courtesy HR Wallingford)

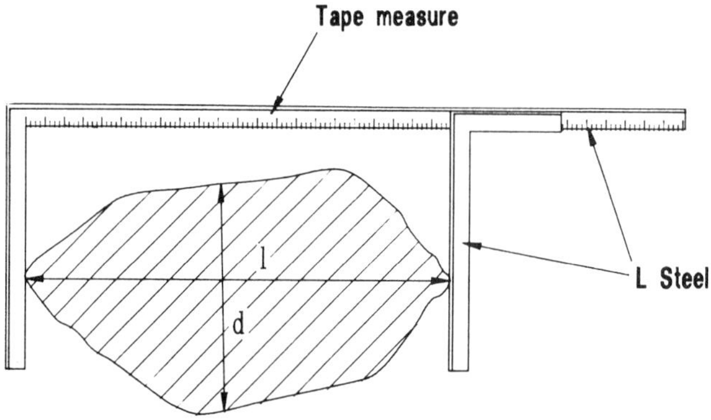

Fig 4 Measurement of l and d (photograph courtesy : G Laan, Rijkswaterstaat)

techniques commonly used in hydraulic laboratories for measuring surface profile, thereby ensuring that the layer thicknesses produced on site match those that have been used in the laboratory and for which the stability formulae have been derived.

22. Measurement of rock for payment should again be by a consistent scheme. The method given in Ref 1 is consistent with the British Civil Engineering Standard Method of Measurement (Ref 8) for earthworks. The major difference to previous procedures is that measurement of armourstone greater than 300 kg is by weight, a method strictly controlled by imposing on the contractor a target tonnage per m^3 or per m^2 of armour facing. The system assumes that the works will commence with construction of a trial area in which placement technique, packing arrangement and layer thickness can be established to the satisfaction of

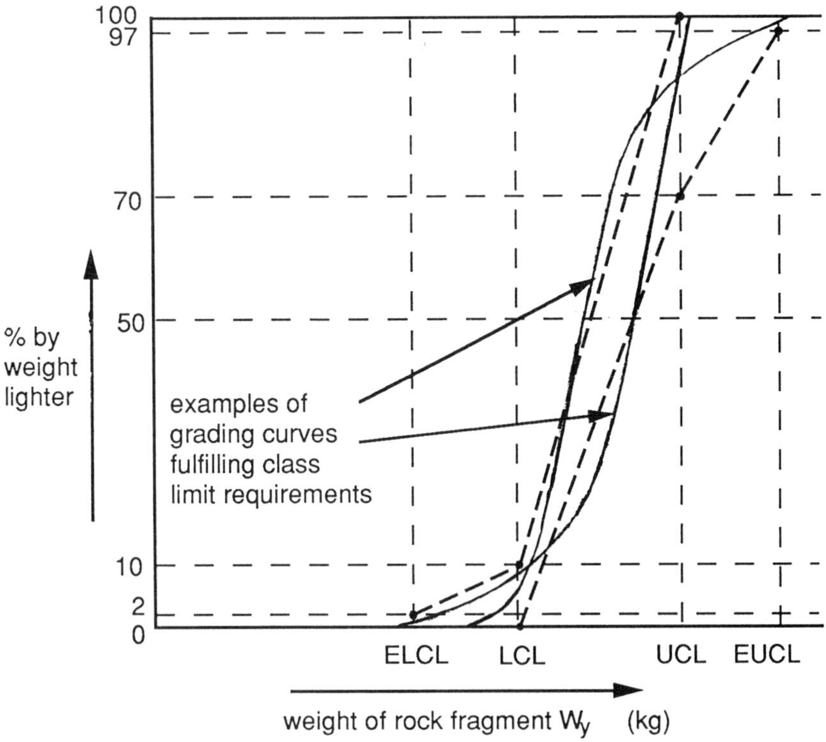

Fig 5 Explanation of the grading class limits

both contractor and designer, and then used as the basis for measurement for payment. The method slightly shifts the balance of risk away from the contractor to the client, but this is hardly unfair in view of the potential range of achievable layer thickness coefficients. The Method of Measurement also includes provision for adjusting the design density should a contractor wish to supply different density material, in a manner which the final price can be controlled for the client. It also permits design adjustments to be made where appropriate.

SECOND LEVEL OF STANDARDISATION: GUIDANCE VALUES FOR DEFINED PARAMETERS

23. In terms of intrinsic rock properties, rock density is constrained by hydraulic stability considerations. Otherwise, the requirements for intrinsic rock quality are largely dictated by durability considerations. A series of tests which may be used to assess rock quality are given in Table 1, in which the relevance of each property tested is given in relation to rock durability and suggested ranges for each of the test parameters are given in four quality categories. Generally, the designer should impose no more severe a requirement than that rock in armour layers be better than the lower limit of the good category. Marginal rocks may, however, be used

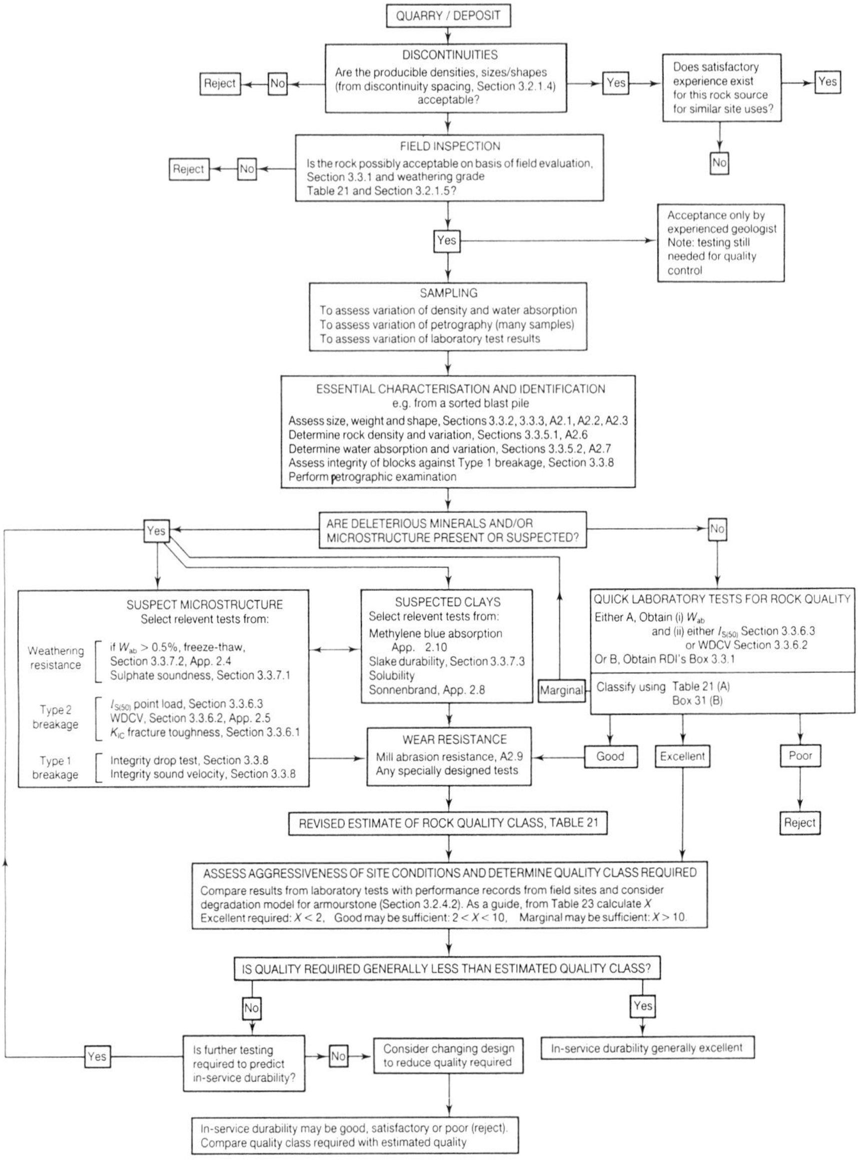

Fig 6 Flow diagram for in-service rock durability evaluation (from Ref 1)

for core and underlayer materials and, in addition, they may be used for armour stone where the application is in a relatively non-aggressive environment. In the latter case it is recommended that the QMW Rock Degradation Model, (Ref 1), be applied to assess the suitability of the rocks being considered. The overall appraisal of rock quality in any given quarry or deposit will normally follow the procedure described in the flow diagram in Figure 6.

24. Limits on aspect ratio should be set taking account of both the rock types envisaged for use and practical and construction considerations. Generally, equant rock is the least problematic to handle and place, exhibits a smaller number of breakages, and has a smaller range of layer thicknesses achievable in two layer armour construction. However, inclusion of only equant rock is economically and technically an unnecessarily severe restriction and irregular rock, at least, should also be permitted, a safeguard to the designer being always the construction of the test section of the works. There is no technical reason for restricting rock to an aspect ratio less than 2. Indeed, it should be noted that a previously commonly adopted criterion of $X/Z > 2$ in the enclosing cuboid system in fact represents a criterion of $l/d > 2.7$. The Rock Manual (Ref 1) based on this consideration and a large number of measurements made by the Rijkswaterstaat during the Eastern Scheldt Project, recommends that not more than 5% of stones should have l/d greater than 3 and not more than 50% of stones should have l/d greater than 2. This criterion will permit most quarry sources which are suitable from a quality point of view (see paragraph 23 above) in massive sedimentary and igneous together with some metamorphic rock to be included without recourse to excessive shape sorting.

25. Grading widths should not be unnecessarily restricted. Tests by van der Meer and others show no difference in stability between grading ranges for two layer armour up to D_{n85}/D_{n15} of 2.5 ($W_{85}/W_{15} = 16.0$), and all the standard gradings proposed by the Rock Manual (Ref 1) fall well within this limit. Where, in special circumstances, "single size" grading is imperative, D_{n85}/D_{n15} should be set at about 1.5. Further tests carried out as part of the preparation of the Rock Manual (Ref 1) suggest that very wide, "quarry run", gradings with D_{n85}/D_{n15} between 2.5 and 5 should not be used for two layer armour as bleeding of fines leads to progressive damage, although they are not necessarily excluded from structures such as berm breakwaters, where layers may be typically up to 10 stones thick, and which will normally be model-tested anyway.

26. Assuming the profile measurement technique described in paragraph 21 is adopted, Table 2 gives some guidance on achievable vertical placing tolerances for rock materials. These tolerances are based on what is known to be practical. Placing in the dry and in water depth less than 5m can be much more accurate and for armour layers can be tied to the mean nominal diameter, D_{n50}. In deeper water, the placing tolerance is more a function of construction technique and, therefore, is largely independent of rock size.

27. Guidance on layer coefficients and porosities is notoriously difficult to give, as the compilers of the Shore Protection Manual (Ref 5) have proved. For equant rock, depending on the degree of packing and placing technique, early results indicate layer coefficient k_t may vary from 0.8 to

Depth of placing below low water	Bulk placed rock of grading where		All armour layers and individually placed rock with gradings where W_{em} greater than 300kg	
	Effective Mean Weight (W_{em}) less than 300kg	W_{em} greater than 300kg (not armour layers)	On individual measurements (m)	Design profile to actual mean profile (m)
Dry, i.e. above low water	± 0.2 m	+ 0.4 m - 0.2 m	± 0.3 D_{n50}	+ 0.35 D_{n50} - 0.25 D_{n50}
Less than 5m	+ 0.5 m - 0.3 m	+ 0.8 m - 0.3 m	± 0.5 D_{n50}	+ 0.6 D_{n50} - 0.4 D_{n50}
5m to 15m	+ 1.2m - 0.4m			
Greater than 15m	+ 1.5m - 0.5m			

Notes 1. Tolerances apply even if gradings being placed have not been selected from the standard gradings in Clause 4.

2. All tolerances refer to the design profile to actual mean profile unless stated otherwise.

Table 2 Vertical placing tolerances for rock materials (from Ref 1)

1.15, and porosity from 38% to 37% (again using the profile measurement technique described in paragraph 21). For irregular rock, the range on layer coefficient is wider with k_t values from .75 to 1.20 being possible. Porosities for irregular rock tend to be higher, in the range 40% - 39%. Notwithstanding the provisional nature of these figures, they highlight again the importance of carrying out a trial section of the work in order to determine the actual values of k_t and n_v for the particular rock type and placement technique being adopted.

THIRD LEVEL OF STANDARDISATION: METHODS OF ACHIEVING MORE ECONOMIC USE OF AVAILABLE ROCK

28. For this final section of the paper, the term standardisation is used to indicate the need for consistent approaches in two common situations. In the developed world, use of standard gradings produced by permanent quarries is to be encouraged. In the developing world and for major projects where temporary dedicated quarries are envisaged, optimal use of total quarry output becomes the overriding consideration.

29. For projects in the developed world, including north west Europe, where temporary dedicated quarries are normally unnecessary, projects will depend largely on supplies from permanent quarries. These supplies could be greatly improved if the variation in existing rock specifications were

reduced, as discussed in the opening paragraphs of this paper. One very practical way in which to do this is to set up a system of standard gradings of rock. All quarries can then invest with confidence in the production equipment necessary to produce these gradings and quarries with adequate space can also stockpile them. Quarry producers and contractors alike (both within the European Community and elsewhere) who have been contacted during the course of the preparation of the Rock Manual (Ref 1) have greeted the concept of standard gradings with enthusiasm. The gradings proposed in the light and heavy ranges are summarised in Table 3 and relate to the class limit system shown earlier in Figure 5. The majority of the gradings reflect those already adopted by the Rijkswaterstaat in Holland and which have been successfully supplied to projects in Holland from quarries in places such as Belgium and Germany.

<table>
<tr><td rowspan="2"></td><td rowspan="2">Class Designation (kg)</td><td>ELCL ($y<2$)</td><td>LCL $0<y<10$</td><td>UCL $70<y<100$</td><td>EUCL $97<y$</td><td colspan="2" rowspan="2">Requirements for range of effective mean weight i.e. excluding pieces less than ELCL</td><td colspan="4">Additional Information
Expected range for compliance with standard gradings</td></tr>
<tr><td colspan="4">where y is the % by weight lighter on the cumulative plot</td><td colspan="2">W_{50}</td><td colspan="2">W_{85}/W_{15}</td></tr>
<tr><td></td><td></td><td colspan="4">Class Limit Definition by weight, W_y (kg)</td><td colspan="2"></td><td colspan="2"></td><td colspan="2"></td></tr>
<tr><td>HEAVY GRADINGS</td><td>300 - 1000 kg
1000 - 3000 kg
3000 - 6000 kg
6000 - 10000 kg</td><td>200
650
2000
4000</td><td>300
1000
3000
6000</td><td>1000
3000
6000
10000</td><td>1500
4500
9000
15000</td><td colspan="2">540 - 690 kg
1700 - 2100 kg
4200 - 4800 kg
7500 - 8500 kg</td><td colspan="2">595- 760 kg
1800-2200 kg
4200-4800 kg
7500-8500 kg</td><td colspan="2">2.3 - 3.8
2.2 - 3.6
1.6 - 2.2
1.4 - 1.8</td></tr>
<tr><td>LIGHT GRADINGS</td><td>10 - 60 kg
60 - 300 kg
10 - 200 kg(wide)</td><td>2
30
2</td><td>10
60
10</td><td>60
300
200</td><td>120
450
300</td><td colspan="2">20 - 35 kg
130 - 190 kg
30 - 90 kg</td><td colspan="2">26 - 46 kg
150 - 220 kg
70 - 130 kg</td><td colspan="2">3.2 - 7.7
2.8 - 6.0
5.0 - 11.4</td></tr>
<tr><td></td><td></td><td colspan="4" rowspan="3">Class Limit Definition by Square Hole (mm)</td><td colspan="6">Average weight retained on LCL hole, W_L (kg)</td></tr>
<tr><td></td><td></td><td colspan="6">for rock density, ρ_r (t/m³)</td></tr>
<tr><td></td><td></td><td colspan="2"><2.5</td><td colspan="2">2.5 - 2.9</td><td colspan="2">>2.9</td></tr>
<tr><td></td><td></td><td></td><td></td><td></td><td></td><td>min</td><td>max</td><td>min</td><td>max</td><td>min</td><td>max</td></tr>
<tr><td></td><td>200 - 350mm
350 - 650mm
200 - 550mm (wide)</td><td>100
200
100</td><td>200
350
200</td><td>Not def- -ined</td><td>350
650
550</td><td>20
115
45</td><td>40
180
80</td><td>25
130
50</td><td>45
200
90</td><td>25
145
55</td><td>50
240
100</td></tr>
</table>

Table 3.2.2 Requirements and supplementary information for standard gradings

Note: The 10-60, 60-300 and 10-200 kg classes are equivalent to the 200-350, 350-650 and 200-550mm classes respectively

Table 3 Requirements and supplementary information for standard gradings

30. In the light grading range, a wide grading of 10 - 200 kg has been included as well as the Dutch gradings to reflect production techniques for such rock commonly adopted in U.K. quarries. This would involve eye selection by a machine operator to remove individual stones heavier than the top end of the grading, together with passing the remaining blast products over a grizzly to remove the fines. Light gradings have been specified both in terms of weight and in terms of size. From the point of view of a specifier, the weight specification may seem more attractive, but on the other hand a size specification reflects more realistically the production techniques that are likely to be employed for such gradings. The authors are content to let time be the judge of which of the two specification techniques for the light gradings is more attractive.

31. When, however, a temporary dedicated quarry is opened for a particular project, the concept of standard gradings should firmly be shunned. Instead, the one-to-one correlation between the quarry and the project should encourage the designer to make maximum use of the yield of the quarry, taking account of not only any stone requirements, coastal structures and breakwaters, but also any requirements for aggregates for concrete or roadstone. It is now widely recognised that one of the most efficient ways to match quarry yield to stone demand in a breakwater is by use of the berm breakwater concept. This concept avoids waste of material between size gradations in a conventional multi-layered breakwater with a two stone thick armour layer by essentially splitting the quarry yield into two, using the heavier stone in an armour layer which is typically 10 - 15 stone diameters in thickness and the remainder in the core of the breakwater. The armour layer will reshape under the action of early storms, to a dynamically stable 'S' shaped profile. A number of breakwaters of this type have been built recently in places such as Iceland (Ref 10) and for initial design purposes, reference may be made to work by van der Meer (Ref 1) and Kao and Hall (Ref 12), the essence of which is again summarised in Ref 1.

EUROPEAN STANDARDISATION

32. Approaches have been made by British and Dutch representatives to CEN committee TC154 in connection with adoption of the proposals in the CIRIA/CUR manual, in particular Appendixes 1 and 2 of the manual (Model specification and testing standards for quarried rock applications in coastal and shoreline engineering). The authors will advise delegates of the latest position in this regard at the conference.

ACKNOWLEDGEMENTS

33. The authors acknowledge the assistance of CIRIA and CUR, through whom the work for the Rock Manual was initiated. Funding for the Manual project was provided by the Department of the Environment (Construction Industry Directorate) and the Ministry of Agriculture, Fisheries & Food (with responsibility for coastal defence) in the U.K. and by the Department of Public Works (Rijkswaterstaat), Ministry of Economic Affairs, and private companies in The Netherlands. The support of the various organisations from The Netherlands and the U.K. construction industries, and the British Aggregate Construction Materials Industries (BACMI) is acknowledged, as is the personal assistance of Mr. G. Laan of the Rijkswaterstaat who provided the authors with much information and stimulation during the course of the Manual preparation.

REFERENCES

1. CIRIA/CUR Manual on the use of rock in coastal and shoreline engineering 1991.

2. FOOKES, P.G. and POOLE, A.B. Some preliminary considerations on the selection and durability of rock and concrete materials for breakwaters and coast protection works. Q. Jnl. Engng. Geol., **14**, 97-128.

3. BRITISH STANDARDS INSTITUTION BS6349 British Standard Code of Practice for Maritime Structures. Part 1: General Criteria 1984.

4. VAN DER MEER, J.W. Stability of breakwater armour layers-design formulae. Coastal Engng., 1987 **11**, 219-239.

5. SHORE PROTECTION MANUAL U.S Army Corps of Engrs Coastal Engng. Res. Center. U.S. Govt. Printing Office, Washington, 1984 4th Edn.

6. LATHAM, J-P, POOLE, A.B. and MANNION, M.B. Developments in the analysis of armour layer profile data. Proc. Conf. Breakwaters'88 - Design of breakwaters. Thomas Telford, London, 1988, 343-361.

7. BRADBURY, A.P., LATHAM, J-P and ALLSOP, N.W.H. Rock armour stability formulae - influence of stone shape and layer thickness. Proc 22nd Int. Coastal Engng. Conf. Am. Soc. Civ. Engrs, 1980.

8. INSTITUTION OF CIVIL ENGINEERS Civil Engineering Standard Method of Measurement, 2nd Edition. Thomas Telford, London, 1985.

9. ALLSOP, N.W.H. Rock armouring for coastal and shoreline structures: hydraulic model studies on the effects of armour grading Report EX 1989, Hydraulics Research, Wallingford, 1990.

10. BAIRD, W.F. and WOODROW, K.W. The development of a design for a breakwater at Keflavik, Iceland. Proc. Conf. Breakwaters '88 - Design of breakwaters. Thomas Telford, London, 1988, 225-234.

11. VAN DER MEER, J.W. and KOSTER, M.J. 1988 Application of computational model on dynamic stability. Proc. Conf. Breakwaters '88 Design of breakwaters. Thomas Telford, London, 1988, 333-342.

12. KAO, J.S. and HALL, K.R. Trends in stability of dynamically stable (berm) breakwaters. Proc. 22nd Int. Coastal Engng. Conf. Am. Soc. Civ. Engrs, 1990.

NOTE: Section numbers and Appendix (Ap.) numbers quoted in Table 1 and Figure 6 refer to the relevant parts of Ref 1.

17. Wave-induced stresses in hollow cube armour units

D. M. HERBERT, Hydraulics Research, and P. WALDRON, University of Bristol

SYNOPSIS

A major collaborative research project is being undertaken to investigate the hydraulic and structural performance of hollow cube armour units. The research is aimed at improving the understanding of the forces that the units are expected to withstand. The project has used scale model tests and field studies to quantify the hydrodynamic forces. These results have then been used to compute estimated stresses induced in units in service.

INTRODUCTION

1. In the 1950's rubble mound breakwaters were traditionally protected from wave attack by two layers of randomly placed rock or bulky concrete armour unit such as the cube. There was little interlock between adjacent units which relied almost entirely on their own self weight for stability. Excess wave energy was dissipated within the voids in the armour layer, underlayer or core.

2. More complex armour units were then produced such as the Tetrapod, Dolos, Stabit and Accropode. The Tetrapod and Dolos were placed with random orientation in two layers, whereas the Stabit and Accorpode were placed similarly but in a single layer. These new units generated larger restraining forces through increased interblock friction, and hence smaller sized units were required for the same level of stability. However, the units were difficult to place accurately with controlled orientation. The resistance of these armour layers therefore varied stochastically.

3. Further increases in hydraulic efficiency were obtained by the introduction of the Cob hollow cube armour unit in the 1970's. The Cob was developed by Coode Blizard and Partners and was first used on La Collette breakwater at St Helier in Jersey in 1973/74 (Ref 1). A

second unit, the Shed, was developed by Shephard Hill in the early 1980's (Ref 2). At least 21 breakwaters or seawalls around the world have been armoured with Cob or Shed units.

4. Hollow cube armour differs from conventional armour in some distinct ways. Wave energy is dissipated in the voids in the centre of the units, and not in the voids between adjacent units. The armour units are placed in a regular close fitting single layer which, with a porosity of 60%, contributes to a very high hydraulic stability. Tests in four different laboratories have demonstrated that it is difficult to fail the armour layer by removing individual armour units. In fact, physical model tests have suggested that a slope armoured with hollow cube units could tolerate a significant wave height twice that for Dolos or Accropode units of the same weight. The use of these units thus offers potential cost savings in the quantities of armour required. Further economies can also be made because the units allow the use of steeper slopes than randomly placed armour.

5. Traditionally, the suitability of a particular armour layer has been assessed by physical model tests. Test results have concentrated on the hydraulic performance and stability of the armour layer. Little detailed consideration was generally given to the long term durability of individual armour units. This approach was acceptable for robust armour units such as cubes, but was likely to present problems when slender limbed units were being used. The fragility of slender armour units was dramatically illustrated by the failure of the Sines breakwater in Portugal in February 1978. The most significant factor in this failure was the breakage of the 42 tonne unreinforced Dolos units during a storm believed to be no greater than the design condition (Ref 3). Following this, many other failures involving breakage of armour units were reported, including breakwaters at Arzew el Djedid in Algeria, Tripoli in Libya and Diablo Canyon in California. Most of these failures involved Dolos or Tetrapod armour units.

6. These failures have led to concerns about the stresses that armour units are required to withstand in service. Researchers in the USA and Denmark have concentrated on the Dolos unit, whereas the design of the Tetrapod and the cube is being addressed by agencies in the Netherlands. In the UK the Single Layer Armour Unit Research Club was formed in April 1987 "to identify and develop techniques to allow the quantitative evaluation of hydrodynamic effects and loadings on single layer concrete

armour units". The Club includes civil engineering consultants, contractors and research institutions. The first phase of studies were completed in March 1991 and a second phase will continue until March 1994.

7. An analysis of cracked units on hollow cube armoured structures at La Collette, Jersey, and Pyrgos Marina, Limassol, Cyprus, indicated that damaged units were concentrated in the inter-tidal zone. There are virtually no cracked units near the crests of the structures where significant wave impact events are unlikely. Conversely, damage was greatest in areas that withstood the most severe wave attack. The lower units at La Collette, which are visible between tides, had suffered significant damage. However, no cracking was detected in the permanently submerged lower units at Pyrgos.

8. The surveys appeared to indicate that cracking might have been caused by large wave impacts on the slender unreinforced limbs of the units. A series of hydraulic model tests and fieldwork deployments were therefore devised to quantify wave impact pressures and, where possible, wave forces. In conjunction with these studies finite element stress models of Cob and Shed units were developed. The models have enabled hypothetical load cases, including that of wave impacts, to be investigated and potential stress concentrations identified.

HYDRAULIC MODEL TESTS

9. Two series of two-dimensional random wave physical model tests were undertaken to investigate the wave-induced forces and pressures on individual Cob and Shed armour units (Ref 4). Preliminary tests to measure regular wave forces on a Shed unit had previously been conducted by Hettiarachchi and Holmes (Ref 5). These tests gave an indication of the magnitude of the hydraulic loading and enabled a force transducer to be designed.

10. The tests on the Cob units were carried out at a scale of 1:31.25 and incorporated three force transducers and six pressure transducers. The force-gauged units were positioned at three different elevations on the armoured slope. They were isolated from adjacent units and the filter layer, thus enabling whole body wave-induced forces to be measured. Each instrumented unit was designed to measure forces parallel (drag) and perpendicular (lift) to the slope. Impact and quasi-hydrostatic pressures on the units were measured using long water filled hypodermic tubes connected to the pressure transducers buried deep in the rubble core. The hypodermic tubes were positioned so that pressures were recorded at three positions (the upper,

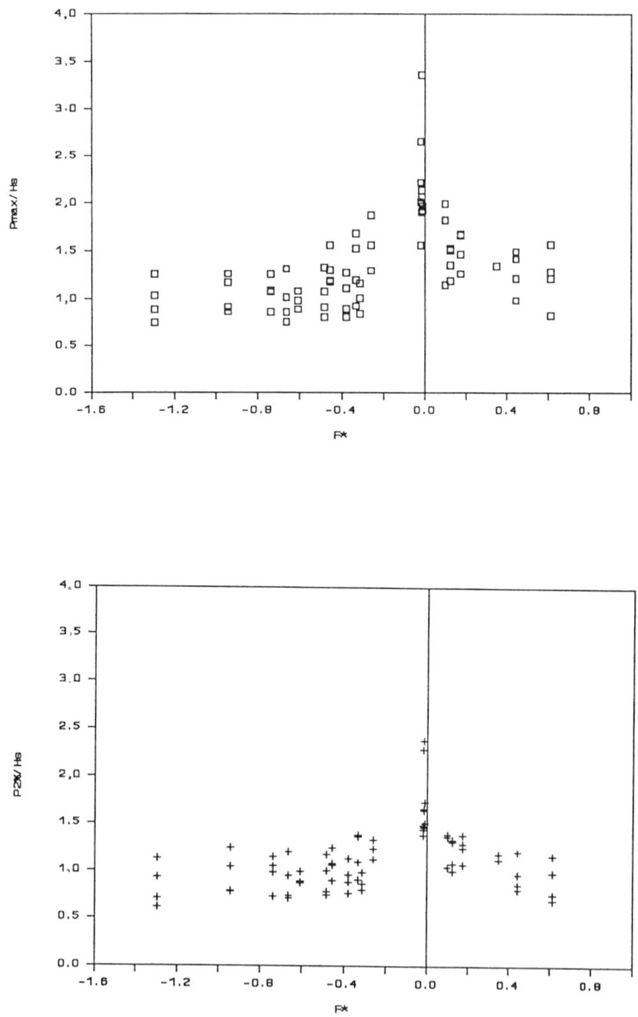

Fig 1 Variation of impact pressures with freeboard.

bottom and one of the side limbs) on the top face of each of two armour units.

11. Similar instrumentation was used in the Shed armour unit tests which were completed at a scale of 1:32.5. However, the force transducers were redesigned so that forces in all three planes could be measured, including forces parallel to the slope but perpendicular to the direction of wave travel (transverse).

12. The data obtained from these tests were analysed using a threshold crossing method which enabled the peak forces or pressures to be identified. An exceedance probability analysis was then completed on these peak values to quantify some of the more extreme values. The maximum impact pressures on the Cob units were found to be slightly larger than those on the Sheds. However, the longer duration wave-induced forces (the drag and lift forces) were similar in magnitude for both types of unit. Not only does the shape of a unit effect the size of the wave impact events, but it has a significant bearing on the way the imposed forces are distributed within the unit itself (Ref 6).

13. Wave impact pressures exhibited typical rise times in the order of ten to twenty milliseconds. The size and frequency of these impacts was found to be dependent on the relative elevation of water level to pressure transducer. This is illustrated in Figure 1, where the maximum pressure to significant wave height ratio, P_{max}/H_s, and the 2% pressure to significant wave height ratio, $P_{2\%}/H_s$, are plotted against the non-dimensionalised freeboard, F^*, for the Cob unit tests. The non-dimensionalised freeboard is defined as:-

$$F^* = F/H_s \qquad (1)$$

where F is the distance of the centre of the hypodermic tube above still water level.

14. Significant impact events were noticeable for $-0.2 < F^* < 0.2$. For the range of wave heights considered, this limits wave impact events to the three units situated around still water level. For $F^* < -0.2$ the units are almost constantly submerged and hence wave impact events are rare. Conversely, for $F^* > 0.2$, the waves have broken before they reach the measurement unit, which is loaded only quasi-hydrostatically by waves running-up the structure. The largest and most frequent wave impacts occurred on the upper and side limbs of the units with maximum impacts equivalent to approximately 3.5 times the significant wave height. Wave impact events on the

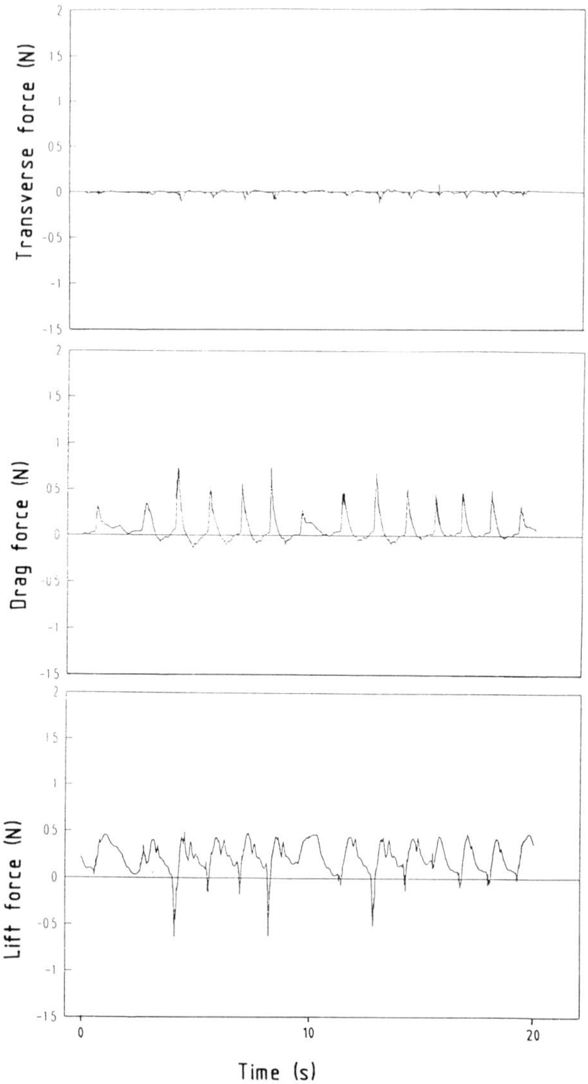

Fig 2 Example time series traces of the wave induced forces on a shed armour unit.

bottom limb, which faces away from the incoming wave front, were rare. The maximum impact on this limb was equivalent to twice the significant wave height.

15. The forces behaved in a similar manner to the pressures, in that they exhibited a dependence on the freeboard of the units. The dominant hydraulic force was the upslope drag force which was approximately twice the inslope lift force. These two forces were found to be greatest on units situated at or just above still water level. For a given wave the maxima of the two forces coincided, corresponding to the instance when the incoming wave hit the unit. The orientation of the wave determined the relative magnitude of the forces with plunging waves imposing large inslope lift forces, but small upslope drag forces. Surging waves reversed this effect.

16. The out of slope lift force, which is usually of overriding concern to the designer, was of sufficient magnitude to overcome the component of the buoyant weight of the unit acting in the lift direction. This force became particularly important during run-down, when water forced into the core during run-up, drained out of the structure. However, none of the units were plucked out of the slope thus indicating that significant inter-unit restraining forces were mobilised. In every recorded case the downslope drag force was significantly smaller than the out of slope lift force. Transverse wave forces, caused by vortex shedding as the wave travels up the slope, were extremely small. However, three dimensional physical model tests have indicated that for angled wave attack and at breakwater roundheads the transverse force may be greater than the upslope drag force. Example time series traces of the transverse, drag and lift forces imposed on a single unit are illustrated in Figure 2.

FIELD DATA ACQUISITION

17. Scale models are limited in their validity and accuracy by scale effects. The extent and importance of these limitations can often be identified by comparing results with other models or simulation techniques, but some comparisons require detailed information at full scale. It was considered particularly important to identify any scale effects inherent in the pressure measurements made during the physical model tests. The degree of aeration is likely to be underpredicted in the model, giving impacts of shorter duration and larger magnitude than occur at full scale.

18. Two fieldwork deployment exercises were therefore completed on La Collette breakwater at St Helier in Jersey.

The first deployment, from January to May 1988, involved instrumenting the upward facing limbs of a single Cob unit with five pressure transducers. The data were logged on to a manually operated analogue tape recorder. In general approximately 30 minutes worth of data were collected each day. This proved to be an expensive requirement and the second deployment, covering the winter of 1989/90, used a more sophisticated data logging procedure to overcome the need for manual operation. The new logger was controlled by an on board computer, which started recording when the water level sensed by a combined wave/tide gauge exceeded a set threshold. Data from an array of four pressure transducers, also positioned on the upward facing limbs of a single unit, were recorded on to an eight millimetre video cassette using a digital tape recorder. These tapes had a capacity of 2.3 Gbytes and lasted about four weeks. Over 20 Gbytes of data were collected during this deployment.

19. Resistance-type run-up gauges were deployed during both fieldwork exercises in order to give an indication of incident wave conditions. These instruments did not perform well. Problems were encountered with signal quality and the harsh surf zone environment caused the gauges to break soon after installation.

20. Both data sets were digitised at 500 Hz and processed on a Compaq 286 PC. The data were analysed with the same threshold crossing method used on the physical model data. Tidal levels in Jersey rise by as much as 0.4m/minute, so the data was analysed in five minute segments. This was the maximum time over which tidal level could be considered constant and yet still have a sufficient number of waves to investigate.

21. During the fieldwork exercises a maximum impact pressure of 8.0m was observed by the transducers on the upper limb, but generally impact pressures in excess of 4.0m were infrequent. Incident wave conditions at the structure were not measured directly in the deployments and the significant wave height can only be estimated to be in the order of 2.0-3.0m. This implies that peak pressures were equivalent to $2\text{-}4H_s$ which is in agreement with the physical model data. The impact rise times were generally less than twenty milliseconds and in isolated cases were as fast as five milliseconds. This was also in agreement with the physical model test results. However, it was unclear if this agreement was coincidence or due to the lack of scale effects in the model.

22. Impact events were both largest and most frequent when the tidal level was between 1.0m and 1.5m below the transducer. At these tidal levels, impacts during storms

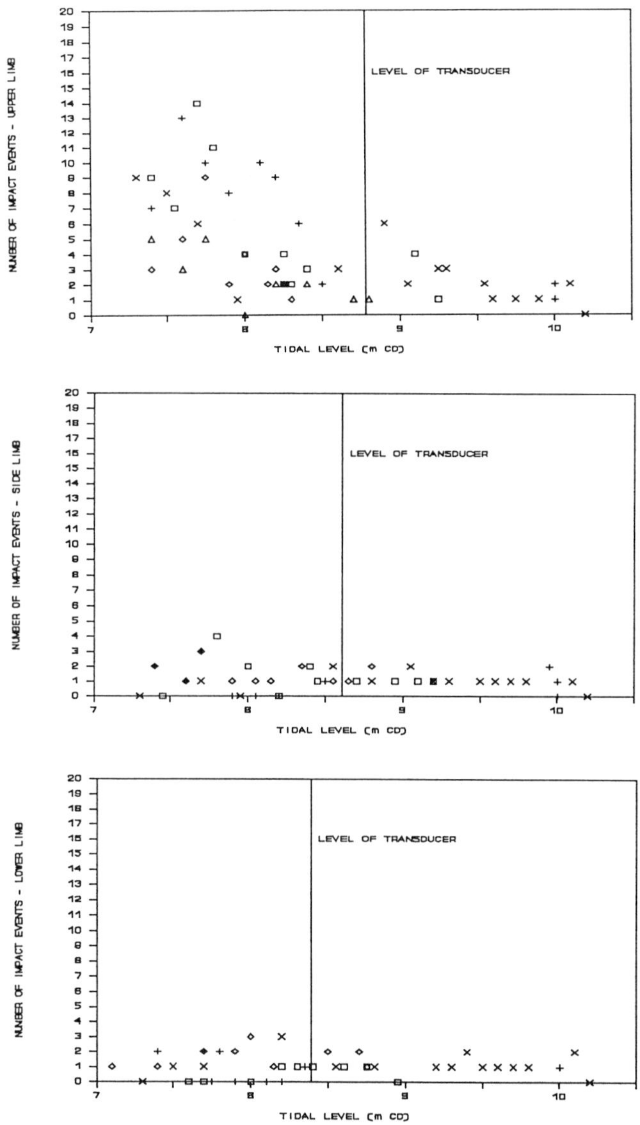

Fig 3 Variation in the number of impact events with tidal level.

occured at up to once every 20-60 seconds, corresponding to an impact every 3-8 waves. As many as 15 impacts were recorded in a five minute period but this number was markedly reduced as the water level approached that of the transducer. No significant impacting occurred when the tidal level approached that of the transducer. Figure 3 illustrates the variation, with tidal elevation, in the number of impact events recorded by the transducers in each five minute period.

23. Few impacts were observed on the side and lower limbs. This indicated that the orientation of the limb to the incoming wave front was extremely important in determining the frequency and magnitude of any wave impact events. There was also significant variation in the impact pressures measured on a single limb. Two transducers, situated approximately 0.5m apart on the upper limb, consistently recorded markedly different impact pressures.

WAVE INDUCED STRESSES IN ARMOUR UNITS

24. Having established that wave impacts are a potentially damaging load event, a preliminary assessment of the stresses induced in concrete units was concentrated in this area. Results from the field study indicated these effects to be very localised and the uppermost horizontal limb to be the most vulnerable. Accordingly, in the first instance, the effects of a wave impact acting on a single outward facing limb have been quantified. For the Cob unit this limb, identified in Figure 4, is particularly susceptible as it presents the largest possible surface area to the incoming wave. This then induces forces in the direction in which the limb has the shallowest depth and least bending resistance. The Shed unit is less susceptible in this regard as the individual limbs are not straight and vary in cross-section along their length.

25. In the analysis of slender armour units subjected to short duration loads, both the geometry of the unit and the duration of the load pulse are of particular importance. For the 2 tonne Cob unit, the effect of pulse velocity was investigated by performing a series of ten elastic finite element transient analyses for a range of different durations from 0.4-100 milliseconds. Because of the symmetry of the problem, only half of the unit had to be modelled, as shown in Figure 5. A total of 480 isoparametric brick elements were used, the solution being obtained by the Newmark method of implicit time integration with a time step of 2 microseconds. Young's modulus for the concrete was taken as 30 kN/mm^2 and Poisson's ratio as 0.2. An idealised triangular force - time history

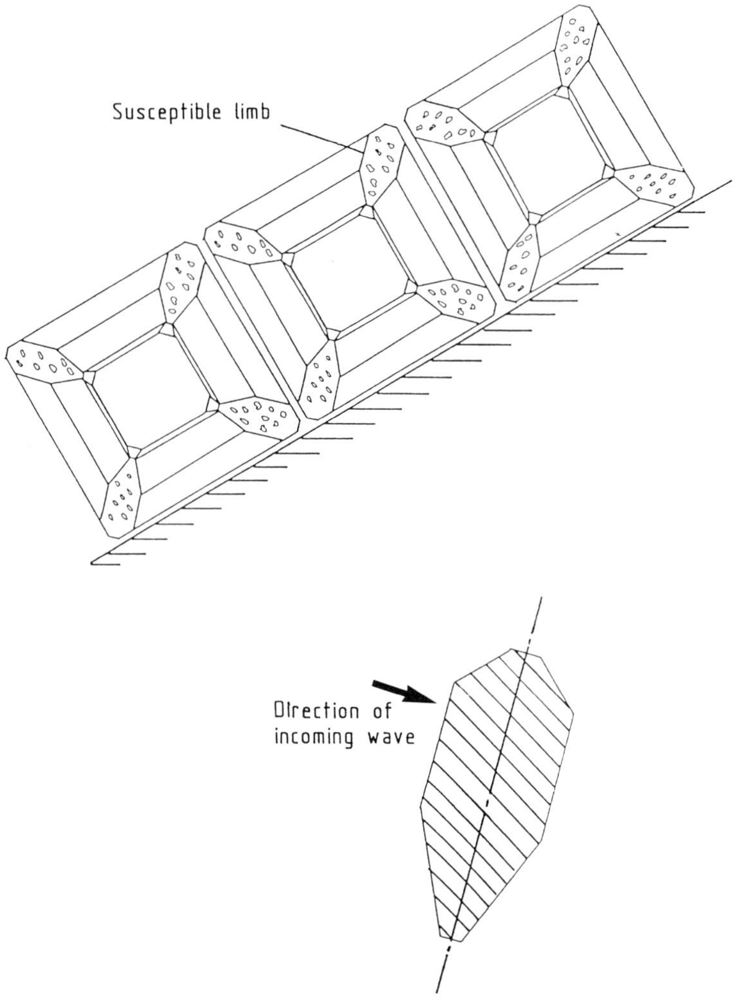

Fig 4 Most susceptible outward facing limb.

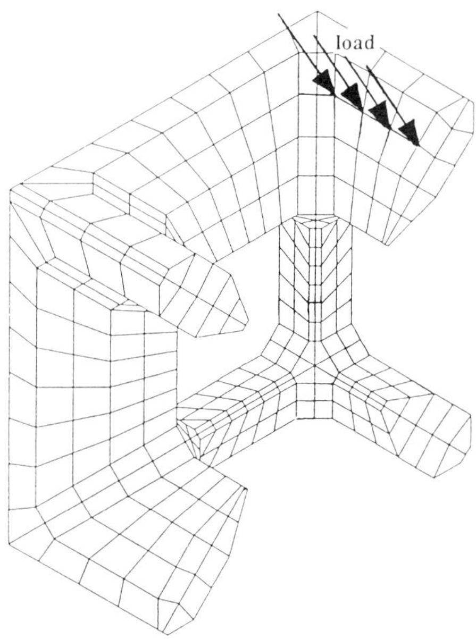

Fig 5 Finite element model of the Cob armour unit.

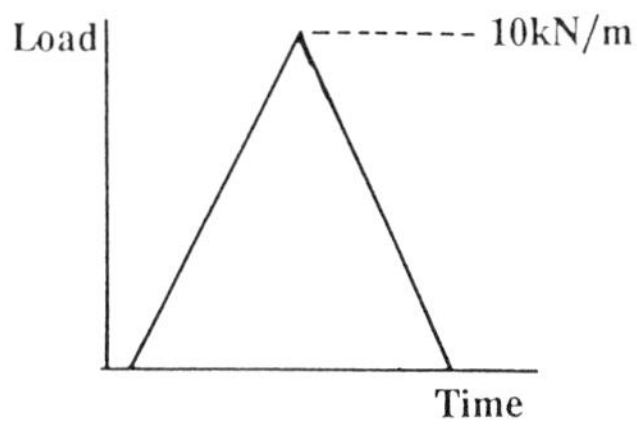

Fig 6 Idealised force-time history.

was assumed for the wave loading event, as shown in Figure 6, with a notional peak load of 10 KN/m run. This was applied to the model as a line load at the appropriate angle of incidence on one limb only (see Figure 5).

26. In all cases, the peak value of tensile principal stress occurred at the midpoint of the loaded limb on the unloaded face. Figure 7, in which this peak stress is plotted against pulse duration, indicates the three different modes of behaviour possible. For very short pulse durations (less than 1 millisecond) the applied load is removed before the limb has time to deflect fully under the peak load. In this case the total impulse represented by the area under the force-time history is the governing factor in determining peak stress, hence the direct relationship between pulse duration and peak tensile stress in the first part of Figure 7. At intermediate durations (1-6 milliseconds) dynamic amplification of load occurs, in this case by a factor of 1.15. The actual value of this amplification factor, which has a theoretical maximum value of 2 for a square wave pulse, depends on the force-time history particularly the rise time and the duration at maximum load. Finally, at long pulse durations, the unit behaves as if loaded with a static load, the peak stress being related to peak load only.

27. It can be seen from Figure 7 that for a typical impact pulse duration of 10-20 milliseconds, as measured in the field, the response of the Cob unit is in this final region. Hence the maximum stress can be evaluated by using a static analysis. Having established this, a linear elastic finite element analysis was subsequently performed. Apart from a considerable saving in computer time, this type of analysis has a number of advantages over a dynamic analysis. In particular, results may be scaled directly for other levels of loading or for different concrete properties without further analysis.

28. For the Cob unit the arrangement of the elements was the same as that adopted previously for the transient analyses. The boundary conditions shown in Figure 8 were chosen to be representative of those for a unit resting on the rubble mound on all four corners and prevented from tipping or moving up or down the slope. The load was applied to the model as a uniform pressure of 153 KN/m^2 over the exposed face of the loaded limb. This pressure intensity had been determined previously (Ref 6) from field and laboratory observations of the force-time histories associated with significant wave impact events. For the Shed unit similar considerations yielded a lesser value of 112 KN/m^2 for the equivalent uniform pressure.

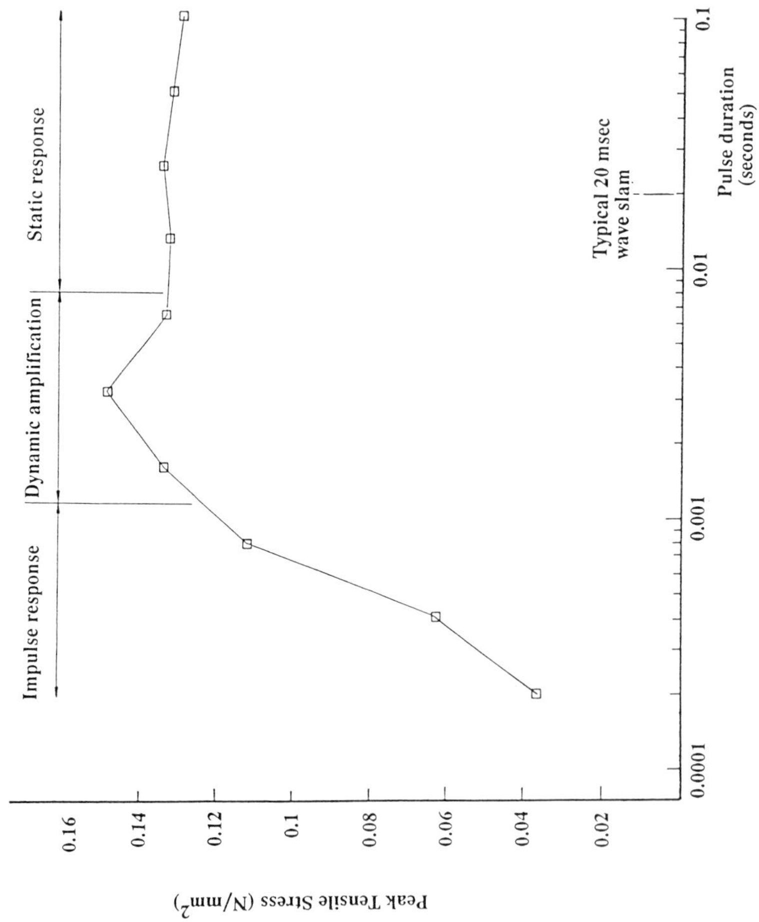

Fig 7 Peak tensile stress versus pulse duration.

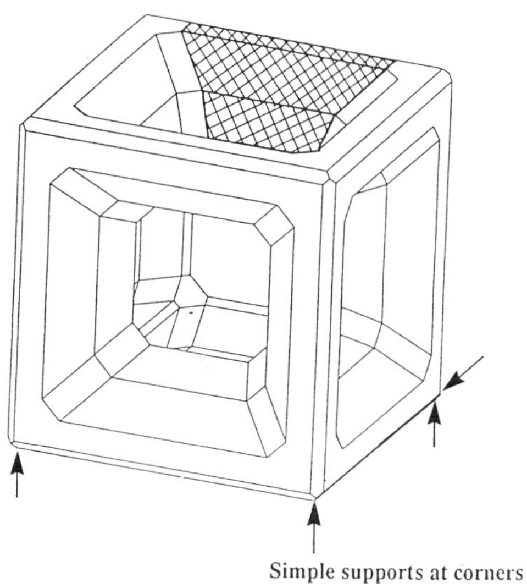

Fig 8 Assumed distribution of pressure and boundary conditions for the Cob unit.

29. Results of the linear elastic finite element analysis for both unit types were similar. Neglecting artificially high local stresses at the concentrated boundary supports, the maximum value of tensile stress occurred simultaneously on the unloaded face of the limb at midspan and at the internal corners at both ends of the loaded limb. For the Cob and Shed units, values of maximum tensile strength were found to be 1.18 N/mm^2 and 0.84 N/mm^2 respectively.

30. Since the tensile strength of concrete typically used in construction of the units is 3-5 N/mm^2, it would appear that the level of tensile stress developed by a single wave impact event is not, on its own, sufficient to initiate cracking. However, in the preliminary analysis reported here a number of assumptions have had to be made particularly with regard to the unknown effects of interference to the flow caused by adjacent limbs, the alignment of the limb with respect to the wave front, the boundary conditions, surface roughness of the concrete, air entrainment, etc. Given these assumptions, the resulting level of tensile stress may well indicate that wave impact events are a significant cause of damage, particularly when combined with other potentially damaging effects such as those caused by differential temperature, settlement and self-weight loading (Ref 7).

CONCLUSIONS

31. A series of laboratory and field deployment studies have identified much new data on the hydraulic performance of rubble slopes armoured with hollow cube armour units. These results have then been used in finite element models of hollow cube armour units to identify stress concentrations.

32. The wave pressure measurements exhibited considerable spatial variations on the surface of a unit, even along the same limb. The maximum wave impact pressures measured in the fieldwork deployments and model tests occurred on the upper limbs of units situated around still water level. These impacts were equivalent to 3-4 times the significant wave height.

33. The wave-induced forces were strongly dependent on relative water level. The dominant force was found to be the upslope drag force which was approximately twice the size of the next largest force, the inslope lift force. The out of slope lift force was larger than the component of the buoyant weight of the unit acting in the lift direction. However, due to the large restraining forces mobilised, no movement of the units was observed. The downslope drag force was significantly smaller than the out

of slope lift force. The transverse forces were extremely small although will increase considerably for angled wave attack and at breakwater roundheads.

34. The finite element analysis indicated that a single wave impact event was not, on its own, sufficient to initiate cracking in a unit. However, significant damage may be caused when wave impact events are combined with other loads.

FUTURE WORK

35. The second phase of the work began in April 1991 and will last for three years. This work will involve the instrumentation of a unit in the field so that wave-induced forces at prototype scale can be measured. It is planned to use this fieldwork deployment to identify scale effects present in the physical model during force measurements and provide further input data for the finite element models. Further development of these models will also take place enabling combinations of loads on a single unit and an array of units to be investigated.

ACKNOWLEDGEMENTS

36. The authors would like to thank Mr N W H Allsop and Dr M A Wastling for their help in the preparation of this paper. The work was funded jointly by the Science and Engineering Research Council, the Construction Industry Directorate of the Department of the Environment and the members of the Single Layer Armour Unit Research Club. The members of the research club include:-

Coode Blizard	G Maunsell & Partners
Kirk McClure & Morton	Posford Duvivier
Shephard Hill & Co	Soil Structures (to 9/90)
Hydraulics Research	University of Bristol
Polytechnic South-West	States of Jersey

REFERENCES

1. Wilkinson A R and Allsop N W H. "Hollow Block Breakwater Armour Units". Proc Conf Coastal Structures '83, ASCE, Arlington, Virginia, March 1983.

2. Dunster J A, Wilkinson A R and Allsop N W H. " Single Layer Armour Units". Proc Conf Breakwaters '88, ICE, Eastbourne, May 1988.

3. Port Sines Investigating Panel. "Failure of the Breakwater at Port Sines, Portugal". ASCE, New York, 1982.

4. Allsop N W H and Herbert D M. "Single Layer Armour Units for Breakwaters". Report SR 259, Hydraulics Research, Wallingford, March 1991.

5. Hettiarachchi S L and Holmes P. "Performance of Single Layer Hollow Block Armour Units". Proc Conf Breakwaters '88, ICE, Eastbourne, May 1988.

6. Wastling M A. "The Effect of Wave Forces on Individual Limbs of Single Layer Armour Units". Report number UBCE/C/90/8, Department of Civil Engineering, University of Bristol, March 1990.

7. Toner W L, Waldron P and Floyd, J A. "Structural Behaviour of Single Layer Concrete Breakwater Armour Units". Developments in Coastal Engineering Seminar, University of Bristol, March 1991.

18. Alderney breakwater: the quest for a final solution

N. W. H. ALLSOP, Hydraulics Research, M. G. BRIGGS, Coode Blizard, T. DENZILOE, States of Guernsey, and A. E. SKINNER, Imperial College

Abstract

This paper describes studies to identify reasons for the repeated damage to the Admiralty Breakwater at Alderney. The main causes of damage to the breakwater are identified. The main thesis is supported with results from a wide range of hydraulic, geotechnical and geological studies, as well as from diving and other surveys of the structure. The paper outlines possible engineering measures to reduce maintenance and repair costs.

Introduction

Background

1. Alderney is the northernmost of the Channel Islands 13 km off the Normandy coast of France, and is approximately 2.5 km by 5.5 km (see Fig 1). The Admiralty Breakwater provides shelter to Braye Harbour, moorings for fishing boats and visiting yachts, and protection to the shoreline around Braye Bay.

2. The harbour was built so that the British Navy could monitor the French Navy in Cherbourg. Construction was started in 1847, to a design by James Walker, the 2nd President of ICE. The initial breakwater design included a mound of stone to low water, surmounted by blockwork walls with rubble infill. The most of the stone for the mound and walls was quarried locally from the Mannez quarry on Alderney. Shortly into construction, the design was revised. The formation level was reduced to 3.5 - 4m below low water. The foundation stones, until then simply placed tightly, were laid using cement, and the batter of the wall was steepened. This construction, which was continued to a length of 823m by 1856, is shown in Figure 2. The design was then revised again. Construction of the outer section was completed in 1864, giving a total length of 1430m.

3. During construction, and after completion, the breakwater was damaged, on occasions being breached completely. The Board of Trade appointed engineers to visit the breakwater and to report on measures to secure the whole or part of it. Sir John Hawkshaw, 11th President of ICE, and Col. Sir Andrew Clarke visited Alderney in May 1870. Their report noted the instability of the mound and made recommendations. The government did not however consider that the costs were merited, and it seems that none of the recommendations were implemented.

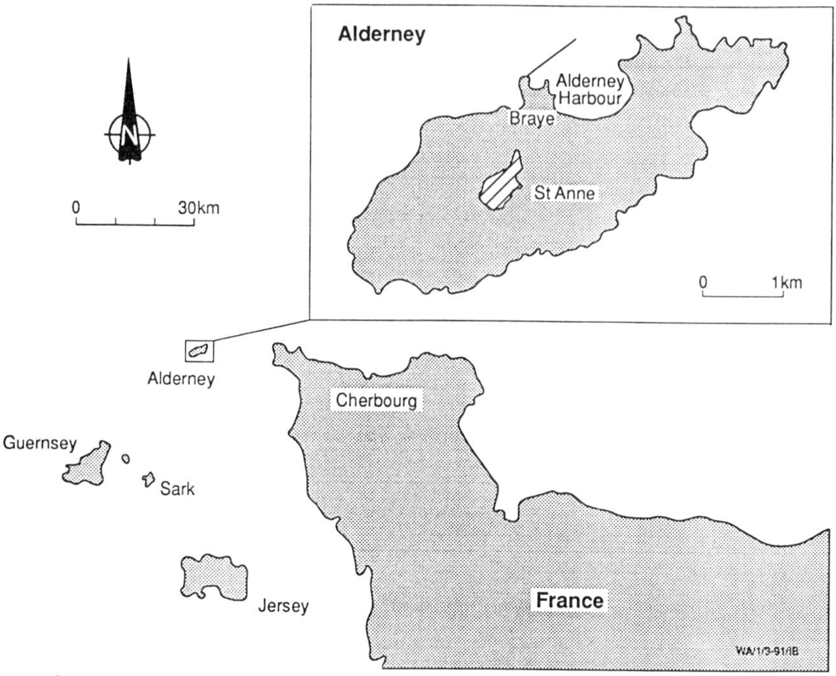

Fig 1 Location

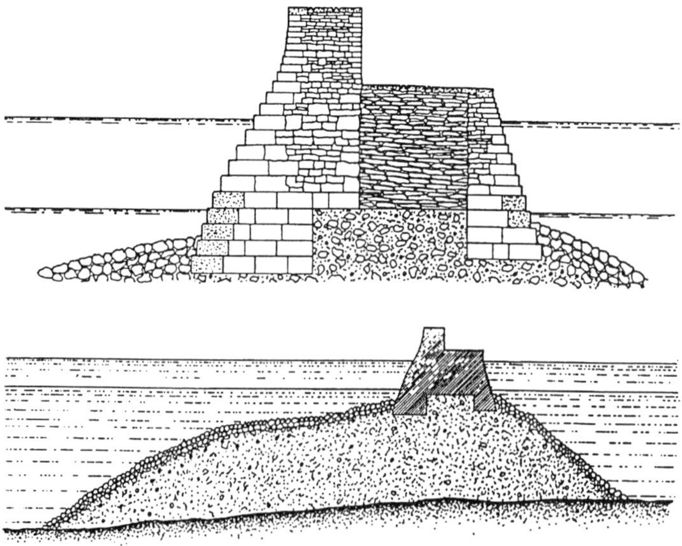

Fig 2 Breakwater cross-section

4. After the initial damage, the wall was protected by stone dumped to maintain the foreshore level. About 300,000 tons were tipped between 1864 and September 1871. From 1873, repair and maintenance work covered the inner length of 871m only. The outer portion was abandoned to the sea and the wall quickly collapsed, leaving a mound crest about 4m below low water. For the shortened section, approximately 20,000 tons of stone were dumped annually, and further work was still required to repair breaches in the superstructure. The dumping of rock was ceased in 1964.

Present situation

5. Waves at Alderney are frequently severe. Depths off the breakwater generally exceed 15 - 20m. Atlantic storms reach the breakwater with little reduction, with the 1 in 50 year storm condition of H_s=11.0m offshore corresponding to H_s=8.0 to 8.5m at the breakwater. The severity of wave impact is increased by the mound, causing waves to break directly onto the wall. Storms at Alderney usually persist for many hours, so the breakwater is exposed to the full range of possible wave and water level combinations, particularly those which allow waves to break directly against it.

6. Responsibility for the maintenance of Alderney breakwater was transferred to the States of Guernsey in 1987. The breakwater has suffered repeated damage since its construction, and maintenance costs for recent years have been estimated at around £500,000 per annum, excluding the costs of storm damage.

7. Damage takes two main forms. Direct wave impact on the wall shakes the breakwater, and cracks mortar joints. The impact pressures force water into the joints, and into voids behind. Loose rock from the mound is thrown against the wall, abrading the wall by a depth greater than 1m. The typical size of rubble on the mound has reduced, and the process has generated considerable quantities of sand.

8. Each summer a team of 8 men repoint the face of the wall above mid-tide level, fill cracks and replace damaged masonry. A team of 6 civil engineering divers carry out repair work at the toe, working both below and above water.

9. During 1989/90, storms battered the breakwater for six weeks. At its peak on 25/26 January 1990, the storm had a return period of about 1 in 25 years, with offshore conditions of H_s=10 to 10.5m. During the next six days the storm subsided slowly, then rose again to H_s above 7m. On 11 and 12 February, storm conditions again exceeded H_s = 9m. This continuous pounding cracked the masonry facing, and a large cavity was formed in the wall. Finally it was breached by an explosive failure clearly audible in and around Braye. Other sections of the structure also suffered damage.

10. An emergency procedure had previously been formulated, and permanent repair work was underway within 10 days. The cost of repair work occasioned by these storms has been estimated at £1.1 million.

Outline of the studies

11. On assuming responsibility for the breakwater, the States of Guernsey Board of Administration reviewed the maintenance required. In June 1988 they initiated studies to formulate a long term strategy, appointing Coode Blizard as consultants, supported by a team of specialist advisors led by HR Wallingford.

12. From the outset, it was agreed that a staged approach would be required. The first stage was principally a detailed review of survey and historical data, together with an inspection of the breakwater. Stage 2 studies involved detailed wave prediction and refraction modelling, supplemented by a programme of current measurements off the breakwater. At stage 3, the studies became more detailed. Further investigations on site addressed the origins and compositions of materials on the breakwater mound and adjoining beaches. A comprehensive programme of hydraulic and geotechnical studies examined the behaviour of different parts of the breakwater structure under a wide range of conditions.

13. The objectives for the studies were to identify the mechanisms by which the breakwater is damaged, and to propose measures to reduce the rate and occurrence of damage. It was not intended to produce a final design.

14. The studies described here have been described in more detail in a report to the States of Guernsey by Allsop, Shih & Skinner (Ref 1), on behalf of the study team.

Stage 1

15. The initial inspection showed that the breakwater walls were in surprisingly good condition, given the continual pounding that they were experiencing. Survey records showed small movements of the wall between 1984 and 1989, generally less than 10mm. At one location the blockwork walls seemed to have spread by about 50mm and lifted slightly, coinciding with a rock outcrop beneath the mound.

16. Many surveys of the mound were available. The best since the cessation of rock dumping covered 1970 to 1988. These were redrawn to constant scale and datum. The results were sometimes inconsistent, but it was clear that substantial volumes had been lost from the mound. As expected, the level of the upper foreshore had fallen. More surprising was the conclusion that the mound had also lost considerable volumes from the lower slopes, often in quite deep water. Even under Alderney's wave climate, wave and tide velocities are not sufficient to carry away other than small material from these depths. Direct hydraulic action was not therefore the cause of this erosion.

17. Diving surveys confirmed that the surface of the mound included a wide range of sizes, ranging from sand to boulders. Generally the lower slopes were well filled with sand. Shallow excavations in the upper surface of the mound showed that voids between the larger stones were nearly filled by sand, originating from the grinding down of the material in the mound.

18. It was now apparent that much of the mound behaves hydraulically as if impermeable to wave action, and geotechnically as if mostly sand. Conventional methods of modelling its performance might thus have encountered difficulties. The programme of work intended for Stage 3 was therefore substantially revised.

Stage 2

19. Work on Stage 2 was directed to identifying design wave conditions at the breakwater. This required further analysis of wave conditions calculated by the Meteorological Office Atlantic Ocean wave model, at grid points 20 -30km to the north-west of Alderney. These gave predictions for wave conditions in deep water for each of 8 direction sectors.

20. Wave conditions were calculated at 4 points close to the breakwater using a numerical refraction model. The tidal velocities here, particularly in the Swinge, are often substantial and can influence wave conditions at the inshore points. The refraction model therefore required data describing the strength and direction of currents over the area of the refraction grids. The Admiralty Tidal Stream Atlas gives current direction and magnitude for mean spring and neap tides. Calibration information was measured between Alderney and Burhou in June. The refraction modelling was completed in August 1989.

21. Waves, up to the 1 in 50 year return period, were derived for all 8 direction sectors. At the breakwater, the largest waves arrive from 320 - 330°N, very nearly normal to the structure. The most severe waves at the breakwater occur at mid-ebb, with the 1/50 year condition giving H_s=8.0 to 8.4m along the breakwater. Waves at the breakwater therefore vary with position along it, wind direction, and tide level, as well as return period. A simplified set of conditions close to the breakwater may be summarised:

Return period (years)	H_s (m)	T_m (s)
1	6.0	11.0
5	7.0	11.4
20	7.4	11.9
50	8.4	12.7
extreme	9.2	12.0

Stage 3

22. The stage 3 studies described further below included:
- a) Laboratory and desk studies on the material forming the mound;
- b) Numerical modelling of hydraulic processes on and within the mound;
- c) Calculations of geotechnical stability of the mound;
- d) Physical modelling to measure wave pressures acting on the mound, and hydraulic stability of possible protection schemes.

The breakwater mound

23. In cross-section the breakwater can be considered in two parts: the wall; and the mound. The wall is composed of blockwork on the sea and harbour sides with rubble infill. There is little data available on material in the mound, but a number of sources have been used to estimate a size grading curve.

Analysis of mound erosion

24. Data on the mound were available from surveys conducted between June 1970 and October 1988. In Stage 3, these were digitised, stored and processed by a Digital Ground Model (DGM). Successive DGMs were compared to indicate the main areas of erosion. The DGM software also calculated the volume of material over selected lengths, at each survey date.

25. This work was conducted in two parts. Early in the study period, levels were entered to the DGM for each of the 11 surveys listed above. Four masked areas were defined within the area digitised, the largest covering most of the seaward

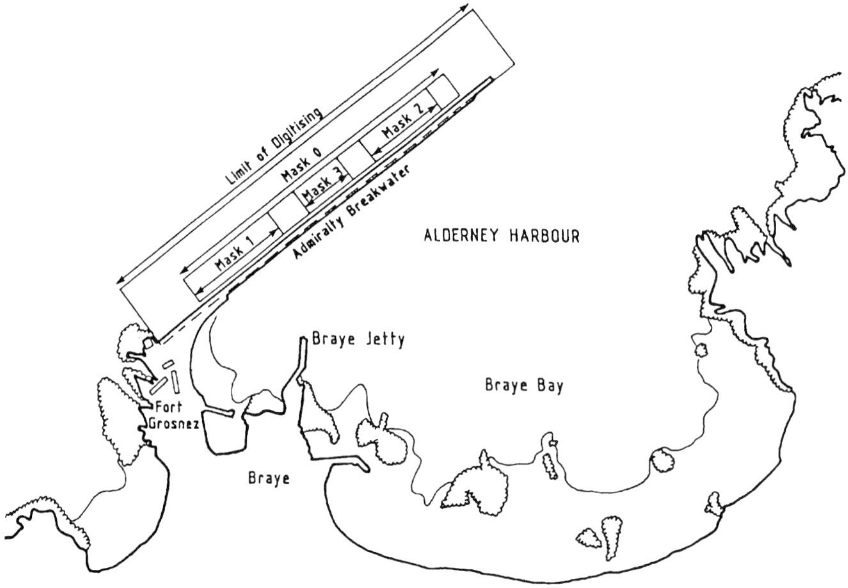

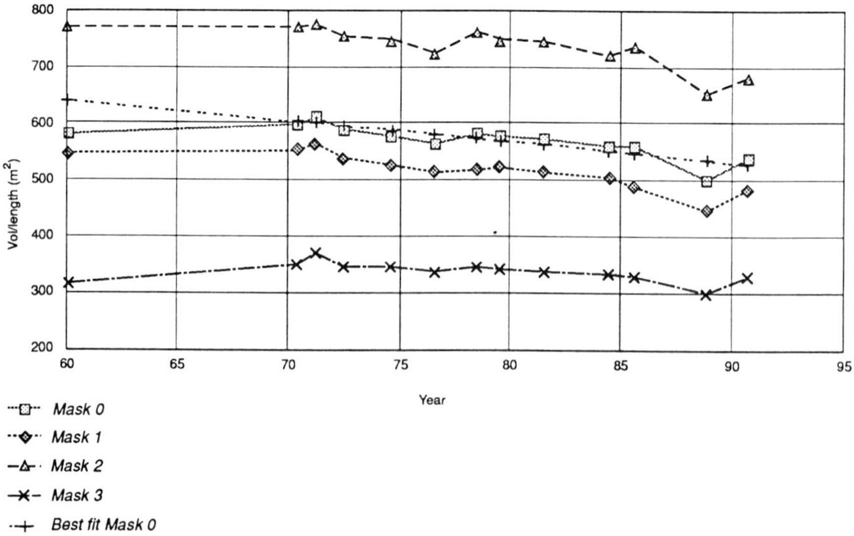

Fig 3 Changes to the mound volume

length of the mound. The mound volume was calculated within each mask area, at each survey date, giving a history of mound volume for each mask area. This suggested that the erosion process was relatively consistent over the different zones considered, but that it appeared to be accelerating.

26. A new survey was conducted in August 1990 and transferred to the DGM. Levels before construction in 1847-50 also became available, and the opportunity was taken to enter these levels into the DGM as well. The calculations of mound volume were repeated using the 1847/50 bed levels. Between 1970 and 1990 there was a general loss of material from each area analysed, see Figure 3. Erosion was least from the sheltered zone behind Braye Rocks covered by mask 3, and greatest from the outer end of the mound covered by mask area 2. This analysis also suggested that much of the material lost between 1985 and 1988 had returned, but it seems more likely that the 1988 survey contained an offset that was not identified.

27. Between 1970 and 1990, loss of material from Mask 0 averaged about 3.7 m^3/m per year. For the other mask areas the rate varied, with the most severe erosion from the outer section, Mask 2, and the least erosion from the area protected by Braye Rocks, Mask 3. The rates for each area may be derived from that for the full area by multiplying by a factor: Mask 1 (1.14); Mask 2 (1.19); and Mask 3 (0.54). Over the breakwater, the loss was equivalent to about 5,700 tonnes per annum. This is significantly lower than earlier estimates of the volume required to maintain the foreshore at rather higher levels, generally given as 20,000 - 40,000 tonnes/year. There was no evidence that the erosion was slowing. It was therefore accepted that the mound would continue to lose material, unless engineering measures were adopted to stabilise the mound, and/or reduce the forces acting on it.

The breakwater wall

28. The main failure mode of the breakwater wall is by removal of individual blocks, followed by erosion of rubble from behind the blockwork. The integrity of the blockwork depends upon the exclusion of wave pressures, and on frictional resistance between blocks. Where the foundation support offered by the mound is locally insufficient, blocks may be extracted. Experience has shown that a local failure spreads rapidly, and has often been followed by loss of support of the deck slabs and/or parts of the harbour wall. The damage caused in January and February 1990 was therefore typical. The primary requirements for the integrity of the breakwater wall are adequate foundation support to the sea wall blocks to obviate settlement cracking and loss of bond, and a reduction of wave pressures acting on the sea wall. The integrity of the breakwater wall therefore depends critically upon the stability of the mound.

Geological and geotechnical studies

29. Two failure modes have been considered for the breakwater mound: a slip of the outer face, occasioned by increased pore pressure; or the loss of fine material from the mound, prompting a local collapse of the foundation to the blockwork wall. These were addressed using empirical design methods, laboratory experiments and analytical calculations. Before studying these failure processes in detail, it was necessary to identify the principal material properties of the mound.

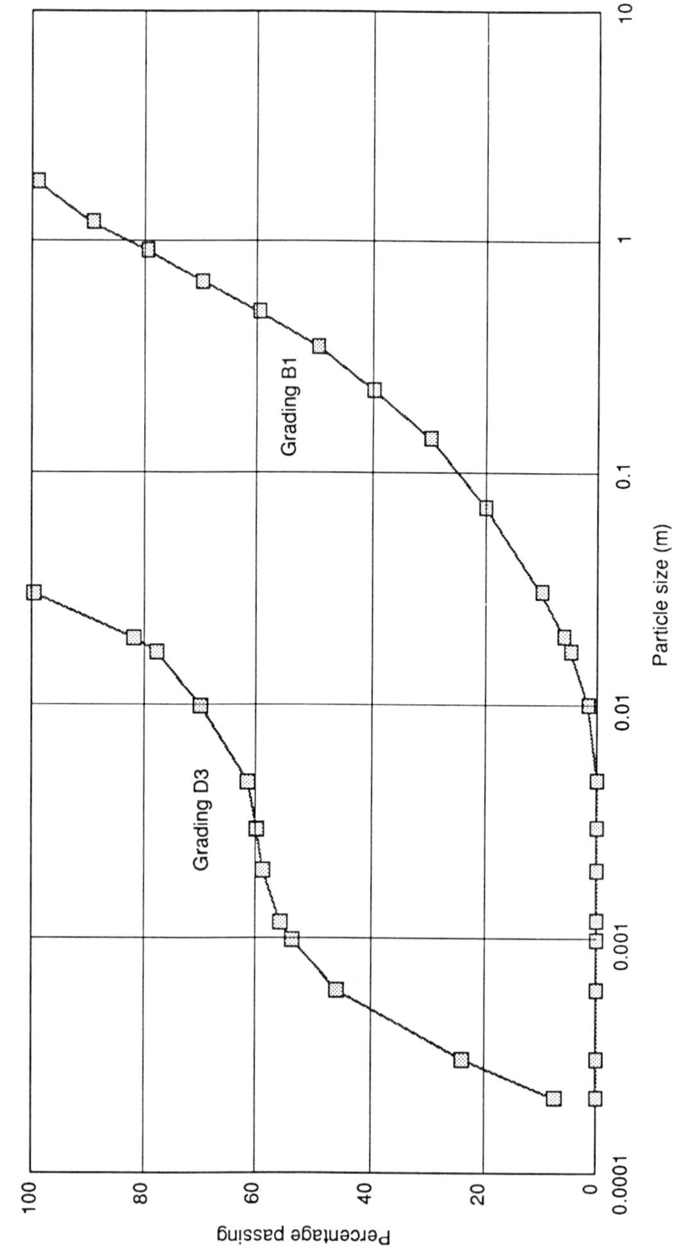

Fig 4 Size grading curves

Material properties

30. It has often been assumed that much of the sand and shingle around the breakwater, and in adjoining bays, was produced by abrasion of the breakwater. The Mannez quarry lies in sandstone, which outcrops only on the east and south of Alderney. This stone is variable in grain size and mineral proportions, but is distinct from igneous deposits elsewhere on the island. No sandstone outcrops on the coast around Platte Saline, Little Crabby, or Braye bay. Samples from the Mannez quarry, Little Crabby and Platte Saline beaches, and the upper slope of the breakwater were taken for analysis. This confirmed that much of the shingle on Little Crabby beach, and a proportion of the sand on Platte Saline, derive from the Mannez sandstone. Sand taken from the upper surfaces of the mound, again contained minerals from the sandstone, and could be from that source.

31. Historical data was used to compare rock sizes with a hypothetical blast curve derived by Drs Poole and Latham for the Mannez quarry. This was extended by assuming that 30% of the "as blasted" material would be finer, and 20% could have been coarser, see Figure 4.

32. Within the original cross-section, voids will have been filled by sand, as in Chesil Beach, where a distinct horizon of low permeability has been identified (Ref 7). The sand washed into the voids will be restricted by the size grading of the primary fabric, and by the local flow conditions. Samples of the sand from the mound were analysed for size grading and permeability.

33. After construction, more rock was added to the mound. Wave forces acting on the structure were now much changed. Wave velocities and accelerations across the upper slope of the mound will have been increased by reflections from the wall. Even during the initial maintenance period, it was noted that stones were quickly abraded, generating large quantities of sand. Some of this has built up Little Crabby and Platte Saline beaches. The material remaining on the mound will have built out from the original cross-section. Sand will have been included with recent material. This part of the mound therefore included material as blasted from the quarry, together with partially abraded material, and sand.

34. The upper slope includes material that is moved by direct wave action. This zone extends from the breakwater wall to about 50m out, and is up to 2m thick. Under storm conditions, most material here will be mobile, and will exhibit little geotechnical strength.

Geotechnical modelling

Stability of composite gradings

35. Under steady or non-steady flows, many wide or gap-graded mixtures can be unstable, depending on the grading curve, and the hydraulic forces, often given by the hydraulic gradient. Kovacs (Ref 5) gives conditions described simply by values of Hazen's uniformity coefficient, $U_c = D_{60}/D_{10}$, but this gives no information on the hydraulic conditions needed. A more realistic approach is based on work by Kenney & Lau (Ref 4), extended by den Adel et al (Ref 6). This uses the local gradient of the grading curve, over a size ratio of 4, to produce an F-H diagram. Low values of H indicate possible instability. Kenney argues that the limits of stability for the mixture are given by:

Stable	$H > 1.3F$
Semi-stable	$1.3F \geq H > 0.6F$
Unstable	$0.6F \geq H$

36. Den Adel et al extended this analysis using data on hydraulic gradients at the onset of movement. They suggest a tentative relationship between $(H/F)_{min}$, derived from the F-H curve, and the critical hydraulic gradient for the onset of suffusion, i_{cr}:

$$i_{cr} = 0.5\ (H/F)_{min}$$

37. This method was tested experimentally in this study. A sample was prepared for which the calculation method predicted a hydraulic gradient i_{cr}=0.06. The sample remained stable up to about i_{cr}=0.01, and superficial flow velocities of 10^{-6}m/s. Fine material started to wash out at flows from 10^{-6} to 10^{-3} m/s. During this process, the permeability increased. At higher flows the permeability was governed by the sizes in the primary fabric, and remained at about 0.025 m/s. In a repeat test the limiting hydraulic gradient for internal stability was about 0.03.

38. Previous work suggested that unsteady flows might cause instability at lower hydraulic gradients, so a brief set of tests was conducted with unsteady flows. Threee tests were run with a period of oscillation of about 10 seconds. The pressure loss and flow velocity were digitised as before. The critical hydraulic gradient at which suffusion occurred was similar to that measured in the steady state tests.

Slope stability

39. The mound would have had side slopes compatible with a safety factor of 1.0 during construction. The slope angles would then have been set by the maximum mobilisable strength of this material. The estimated size distribution for the original material is well graded, with a uniformity coefficient of $U_c = D_{60}/D_{10} =$ 16.5. For low stress states in the mound, a reasonable value of the strength parameter ϕ' would be near 40°.

40. This can be checked by considering the underwater slope with respect to slab sliding under drained conditions. The 1870 to 1872 profile shown by Vernon-Harcourt (Ref 3) indicates slopes of the lower face of 36° to 38°. The undisturbed original slope would have been deposited with a safety factor close to 1.0. A value of F = 1.1 gives a value of ϕ' of 41°, supporting the estimate of ϕ'= 40°.

41. Pore pressures would have been small in the mound as placed. Now most of the mound is filled with sand. Flows will therefore set up pore pressure gradients. The strength of the sand infilled rubble depends significantly on the proportion and properties of the sand. Bishop (Ref 2) indicates that a significant change in the mound slope angle had occurred by 1905, compared with the slope as built. By 1905 the outer face of the mound had flattened to about 28°. What change in material behaviour occurred between 1872 and 1905?

42. The sand infill is substantially derived from the abrasion products of the mobile stone on the mound foreshore. This sand would have built up within the mound and washed out of the base of the mound. At the same time rubble would have been included with the sand. The mixture of sand and stone would now have adopted a slope compatible with its strength, shallower than the original 36-38°. Flow through the outer layers of the mound would now be impeded, and pore pressures would increase under wave attack. The strength is therefore dominated by the properties

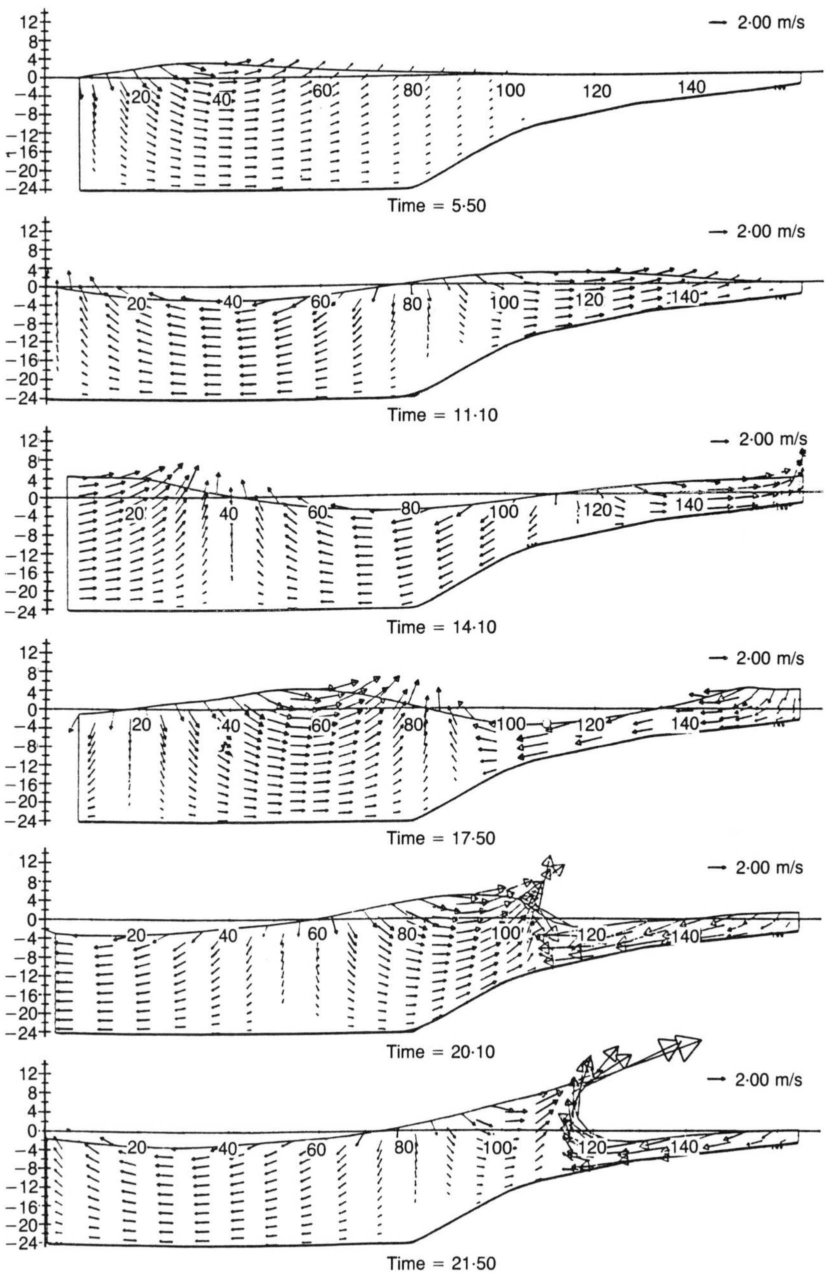

Fig 5 Wave action on the breakwater mound

of the sand. Although drained sands can give values of ϕ' up to 35°, undrained failure can occur at mobilised angles as low as $\phi' = 20°$. Even if the drained strength of the infilled rubble could be as high as $\phi' = 35°$, undrained failure would take place at values much lower than this. A reasonable value for ϕ' mobilised for undrained failure would be 27° to 29° for $\phi'_{max} = 35°$.

43. The boundary conditions on the seaward face of the mound were obtained from the numerical and physical model tests described in Chapter 6. Both considered the existing bathymetry, with the sea wall, and give water pressure distributions on the sea bed and mound surface. Under wave attack downrush velocities up to about 4m/s at the mound surface were calculated. These will produce surface shear stresses which peak at 3.2 kN/m^2. The pore pressure distribution within the mound was derived from the potential distribution calculated by the internal flow model.

44. For the sand infilled rockfill, the slope angle of 28° of the seaward slope of the mound coincides with the back-calculated value of ϕ'. The conclusion is that this slope is unstable during storm periods which generate waves of the magnitude input into the analyses. Values of ϕ' for undrained failure could be even lower than 28°, so waves with return periods of less than 1 in 50 years could also trigger failure of these seaward slopes in the existing structure.

45. To improve the geotechnical stability of the mound it will be necessary to decrease further the slope angle of the lower seaward face of the mound, and then prevent further generation of sand either by stabilising the foreshore material, or by decreasing the hydrodynamic loads on the foreshore.

Hydraulic modelling

Numerical models

47. Two numerical models (Ref 8) were used to quantify and analyse the performance of the existing breakwater mound: a two-dimensional boundary element model of wave flows outside the mound, known as BEMTOOL; and a two-dimensional finite difference model of flows inside the mound, known as BWATER.

48. BEMTOOL calculated wave velocities and pressures on the existing mound, which was assumed to be impermeable. These pressures were then used as input boundary conditions for BWATER, which calculated the pore pressures, velocities and hydraulic gradients within the mound. Results from some physical model tests were also used to provide comparisons between numerically predicted pressures and those recorded in the physical model tests. The calculated pore pressures within the breakwater mound were used to provide information in establishing the geotechnical stability of the mound.

49. BEMTOOL was first used to study the velocities and pressures at the mound surface for a range of regular waves, as shown for example conditions in Figure 5. The results suggested conditions which might cause instability of the mound. These then provided input conditions for BWATER to calculate conditions in the mound. The permeability and porosity were taken to be k=0.0002m/s and n_v=15% respectively, based on the permeability tests.

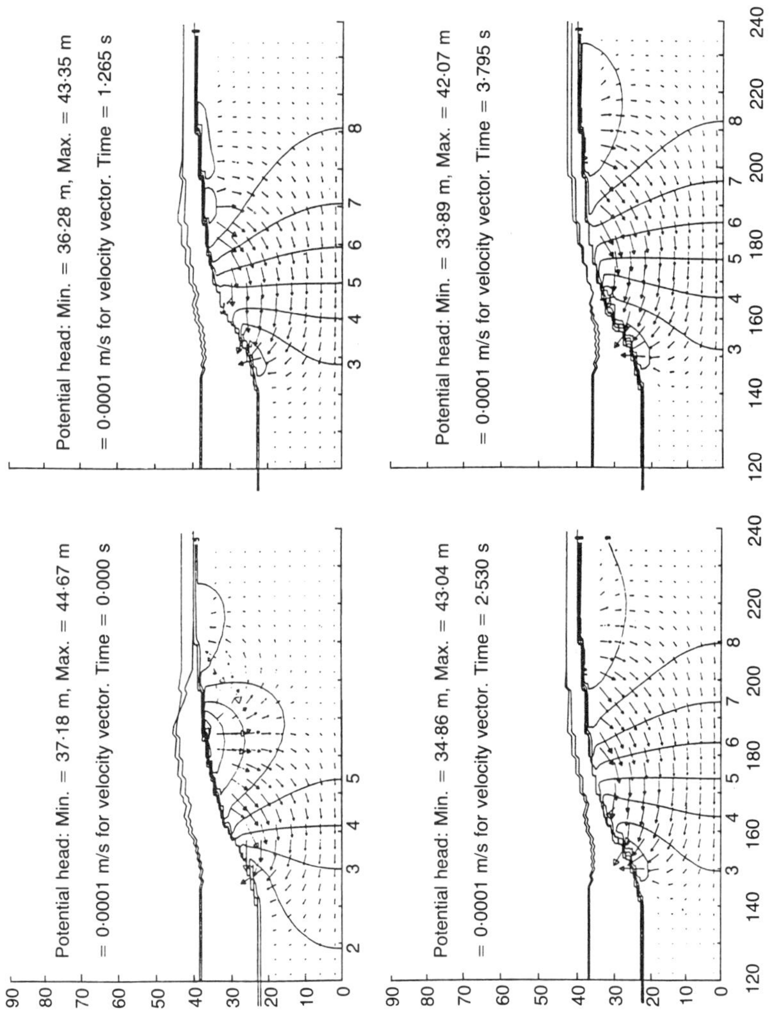

Fig 6 Flows within the mound

50. During the physical model tests, wave pressures were measured on the surface of the mound. These were then used as input conditions for BWATER. The finite difference grid for BWATER was much smaller than the separation between pressure transducers, so pressures were interpolated between transducers. Example pressure distributions calculated by the numerical model are illustrated in Figure 6. Hydraulic gradients calculated for the present mound, and for protection scheme 6, are given below:

	Hs (m)	Tm (s)	Water depth (m CD)	Hydraulic gradient
(existing)	8.4	12.7	0.9	0.20
(existing)	8.4	12.7	3.5	0.21
(existing)	8.4	12.7	6.1	0.19
(scheme 6)	8.4	12.7	3.5	0.18
(scheme 6)	8.4	12.7	6.1	0.18

Physical model

51. Physical model tests were used to quantify direct wave effects on the existing breakwater and to study the performance of possible protection schemes. Wave conditions and pressures on the mound and breakwater wall were measured. Protection schemes were tested for hydraulic performance and armour stability. The model tests were conducted in the deep random wave flume at HR Wallingford, at a scale of 1:40. Tests were run at three water levels: 0.9, 3.5 and 6.1m CD. Wave conditions corresponded to the 1, 5, 20 and 50 year return period and an 'extreme' event.

Existing condition

52. The existing mound was represented by an average cross-section for Mask 2. This represented the steepest bathymetry along the breakwater, where waves are most severely influenced by the sudden change in water depth. Wave probes gave the instantaneous water surface elevation at each probe. Six pressure transducers measured pressures on the surface of the breakwater mound and the seawall. The data was logged and processed on a computer connected to HR's network.

53. Incident wave spectra and statistics were measured at the toe of the breakwater mound, and at points over its surface. These tests were run without the breakwater wall, thus obviating corruption by reflections. Then the wall was replaced, and wave elevations and pressures re-measured.

54. Pressures measured in these tests were used to give input conditions for calculations using BWATER. Special software extracted data for the period when the pressure difference between two transducers was at a maximum, corresponding to maximum outflow from the mound. This provided input data for the numerical model tests.

Protection schemes

55. Six different proposed protection schemes were tested for hydraulic performance and armour stability. Schemes 1-3 explored the performance of mounds of large rock armour placed on the upper slope of the present mound. These reduced wave impacts on the wall, and inhibited the movement of material on the present mound. The toe of the protection was relatively unstable. Schemes 4-6 included a further

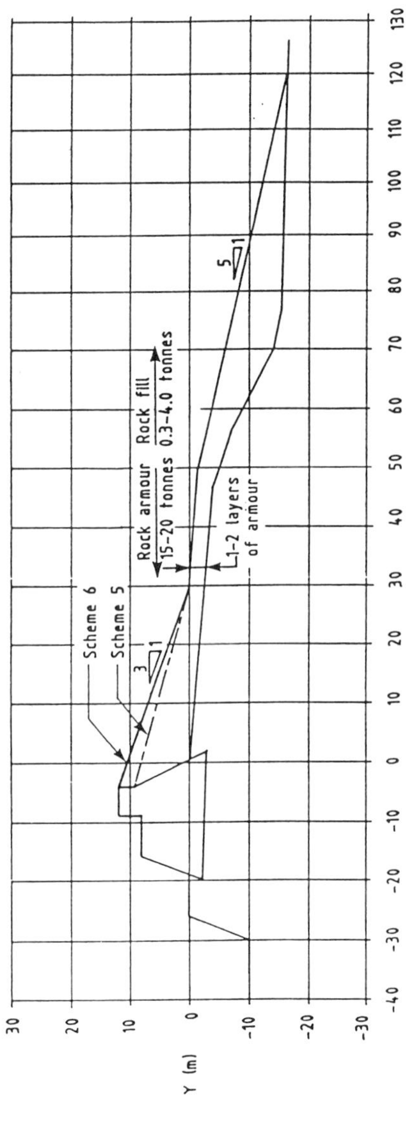

Fig 7 Protection schemes 5 & 6

mound of small rock, placed over the lower slope of the present mound. The scheme that gave the best performance, scheme 6, used 15-25 tonne rock at a slope of around 1:3. Schemes 5 and 6 are depicted in Figure 7. The upper surface of the mound was protected with 15-25 tonne armour rock. Beyond that point, wider graded rock, was used to form a toe mound with a seaward face at 1:5. This extension to the mound will give support to the toe of the present mound, as well as significantly reducing the severity of wave impacts on the wall and the mound.

Conclusions

56. The behaviour of the Alderney breakwater has been studied using laboratory experiments, and numerical and physical models. Historic rates of erosion from the mound have been quantified. The loss of material from the lower slopes of the mound has been ascribed to geotechnical failure of the sand-filled material. Hydraulic gradients within the mound have been shown to be large enough to cause internal instability. These studies have demonstrated that it will be necessary to stabilise the lower slope of the present mound, to ensure the stability of any protection on the upper slope. Studies of possible protection schemes, however, suggest that practical engineering solutions can be devised to resolve a problem that has persisted for more than 120 years.

Acknowledgements

57. The studies reported here were conducted for the States of Guernsey Board of Administration. The team for the studies described here was formed by: Coode Blizard; the Coastal Structures Section of Hydraulics Research, with assistance from other sections within HR; Dr Angus Skinner of Imperial College; Dr Alan Poole and John-Paul Latham of Queen Mary & Westfield College. The numerical modelling of waves at the breakwater, and the wave flume tests, were conducted principally by Dr Robert Shih, previously at HR. The numerical models were developed within a research project funded in part by the Department of the Environment under Research Contract PECD 7/6/108.

References

1. Allsop NWH, Shih RWK, & Skinner AE "Alderney Breakwater: Phase 3 hydraulic model studies" Report EX 2231, Hydraulics Research, Wallingford, December 1990

2. Bishop RW "Maintenance of some rubble breakwaters: Alderney breakwater" ICE Maritime & Waterways, Paper 16, Session 1950-51, pp 16-35 & 36-53, December 1950

3. Vernon-Harcourt LF "Account of the construction and maintenance of the harbour at Braye bay, Alderney" ICE paper 1347, 25 November 1873

4. Kenney TC & Lau D "Internal stability of granular filters" Canadian Geotech Jo, Vol 22, 1985

5. Kovacs G "Seepage hydraulics" Developments in water science Volume 10, Elsevier, Amsterdam 1981

6. den Adel H, Bakker KJ & Breteler MK "Internal stability of minestone" Modelling soil-water-structure interactions, SOWAS 88, Balkema, Rotterdam, 1988

7. Allsop NWH "Chesil beach interceptor drain: percolation study" Report EX 1248, Hydraulics Research, Wallingford, October 1984

8. Allsop NWH & Williams AJ "Hydro-geotechnical performance of rubble mound breakwaters" Report SR 183, Hydraulics Research, Wallingford, February 1991

19. Damage and repair work performed on some mound breakwaters in Spain

M. A. LOSADA, University of Cantabria, F. ORTIN, Government of Asturias, C. VIDAL, University of Cantabria, and J. A. DEL BUSTO, Government of Asturias

SYNOPSIS. During the last ten years the Harbour Authority of the Fishing Ports of Asturias (Spain) has carried out an extensive programme to improve the breakwaters of Cudillero and Luarca, fishing villages on the Coast of the Gulf of Biscay, primarily to reinforce their stability and to improve the harbour operation conditions. This paper resumes the work performed, particularly causes of damage, criteria for repair and design, and construction procedures.

INTRODUCTION

1. Cudillero and Luarca, Fig. 1, are two fishing villages on the Cantabrian Sea (Spanish border of the Gulf of Biscay) whose economy has been strongly dependent on fishing for over a thousand years. In this respect, the use of small rowing boats for whale fishing is well documented and now, of course, the activity extends over other species and into the North Atlantic Ocean.

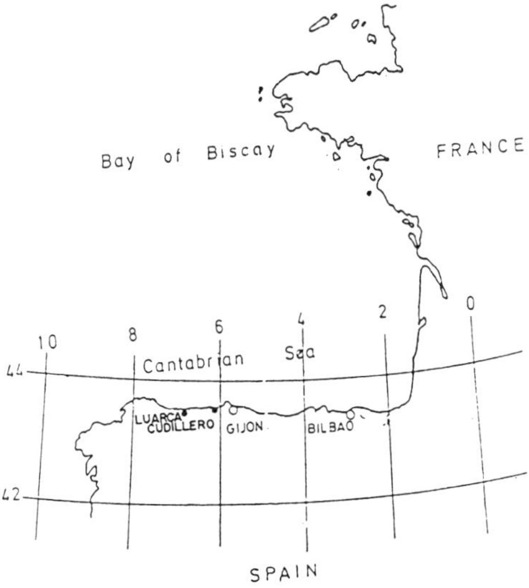

Fig. 1. Location of Cudillero and Luarca.

2. Both harbours as it will be explained later, were found to be lacking in terms of services and protection, and in heavy storms most of the boats had to be taken out of the water or the fishermen had to stay on board. In recent years, the Regional Government of Asturias has made a very serious effort to change this situation. However, the heritage of some old breakwaters (Luarca) or of an incorrect design and construction of a new breakwater (Cudillero) made the work even harder.

3. The following will deal with the repair of the Cudillero and Luarca breakwaters where there was an important level of damage and inefficiency. As well as the general criteria of design and rebuilding, some procedures are included in Appendix I and II.

WAVE CLIMATE IN THE CANTABRIAN SEA

4. The waves reaching the Cantabrian Sea Coast are generated in the North Atlantic by the action of extratropical storms. The polar mass oscillations through the year define the path of the storms through the Bay of Biscay. Under such conditions, the winds are from West to North West. Van Koten, 1977, estimated the mode of the dominant gradient wind speed (double exponential distribution) for the Cantabrian Sea Coast $u_N = 40$ m/sg.

There are maximum frequency and intensities of waves in the course N25W to N60W. After visual data, Hogben and Lumb, 1967, and instrumental data (waverider buoy) recorded through five years at depth = 35 m, the significant wave height and peak period distribution has been obtained, Fig. 2a, and the extremal significant wave height distribution, Fig. 2b. Both distributions are roughly representative of the wave conditions in deep water for the Cantabrian Sea. The spring semidiurnal tidal range is of the order of 4.5 meters.

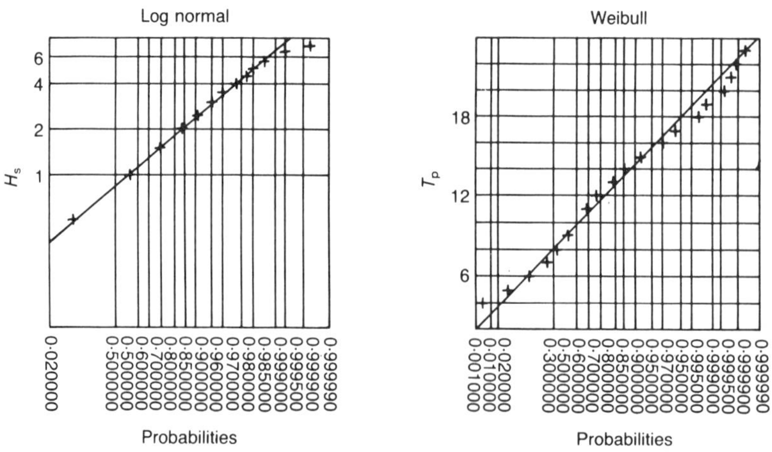

Fig. 2a. Significant wave height and peak period distribution.

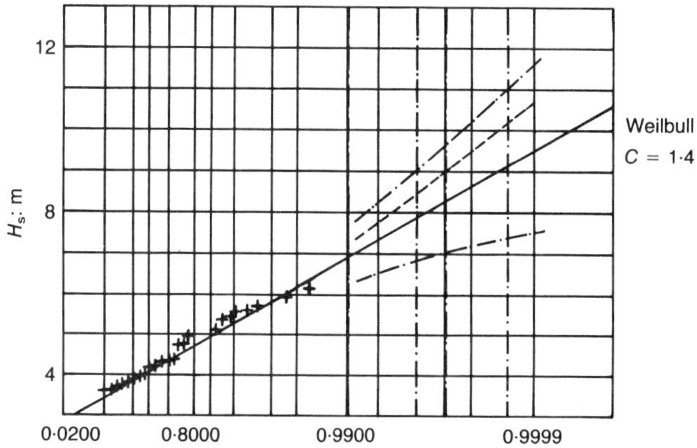

Fig. 2b. Extremal significant wave height.

CUDILLERO BREAKWATER

5. The old harbour of Cudillero is located in a short narrow bay, protected by a vertical breakwater. The town spreads around the harbour on the cliff like an amphitheatre. In 1990 the fishing activity in Cudillero brought in over $300,000.

In the 70's it was decided to build a new harbour, taking advantage of some rocky islands close to the coast. After several modifications of the original design, all of them looking for ways of decreasing the construction cost, and the bankruptcy of one of the construction companies, the harbour was finished in 1983. Fig. 3 shows the lay-out of the harbour and a section of the breakwater. It was built on maximum water depth of 10 m and its main feature was the absence of a core of fine quarry. The whole section was built with 60 Tn concrete blocks. The section was completed with an L-shaped superstructure founded at level 4.00 m and crowned at level 11.00 m.

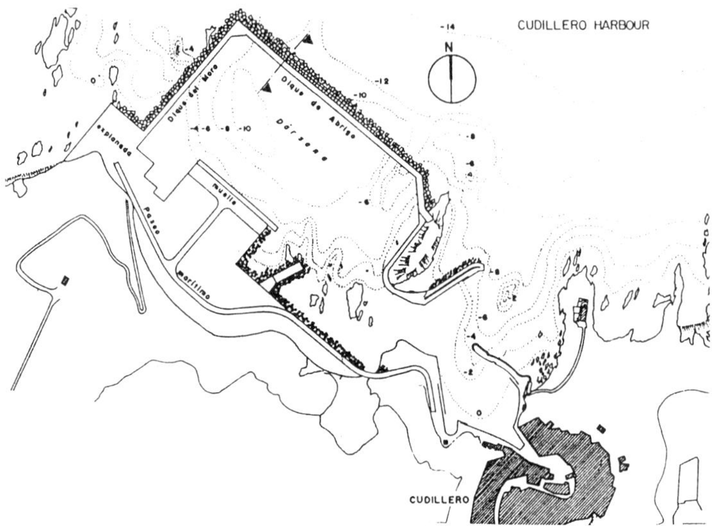

Fig. 3a. Cudillero harbour (plant).

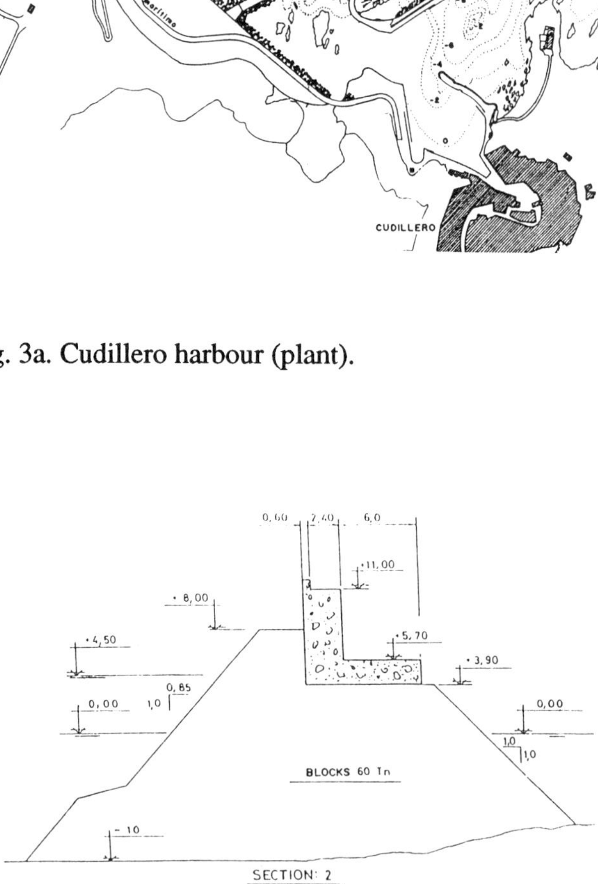

Fig. 3b. Cudillero harbour (breakwater section).

Inspection of damage

6. In summer 1985, an overall inspection of the breakwater was carried out. The main conclusions of the report were (Fig. 4):

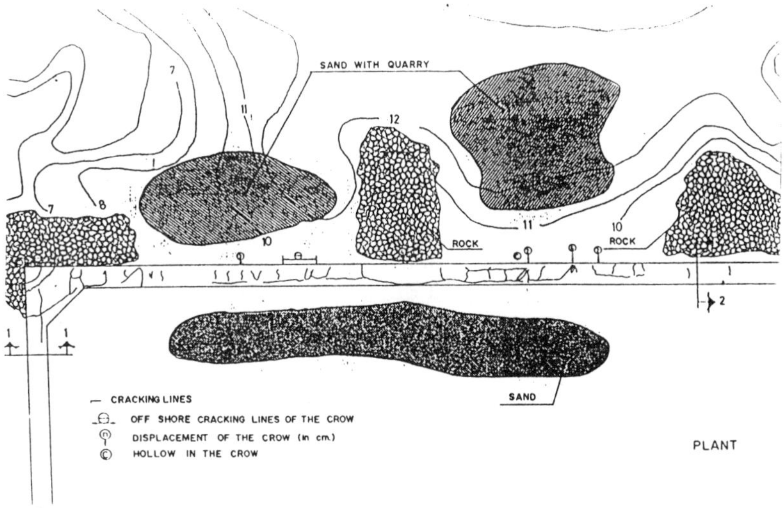

Fig. 4. Cudillero cracking lines.

I. Main Layer

(a) In many sections the original plane slope presented an S-profile with an almost horizontal part at level -5.00. Many armour blocks had rounded corners.

(b) Other sections had blocks without damage, but had slope angle of cotanα = 1.2-1.3.

II. Upper berm

(a) Some sections did not have an upper berm. Furthermore, between crown and first block a gap all along the breakwater of at least the size of a block was detected.

III. Crown

(a) The crown was cracked all over. The horizontal part of the L-crown had alineation parallel to the crown; the width of these lineas was in many places over 30 m. Also, transversal cracking lineas were detected. A longitudinal cracking linea of at least 100 m appeared all along the inner edge of the L (Fig. 4).

(b) Main failures on the crown were located along the bend and in two areas founded on sandy bottom.

(c) The foundation of the outer part of the crown had many grooves and cavities.

IV. Harbour

(a) Flow through the breakwater was apparently quite great. However, it seemed not to affect boat mooring.

(b) A big shoal of sand had appeared along the inner slope of the breakwater.

Analysis of the damage

7. In the report the damage to the crown was defined as total failure, considering that the structure was not able to resist the attack of the design storm. Indeed, it was found that the constructed crown would be not stable under waves higher than 7 m (Notice that the depth limited wave height is approximately 10 m).

8. In winter 1985 a medium storm of significant wave height of $H_s = 5$ m turned over the bended zone of the breakwater. The next stage was to analyze whether the steepness of the slope and the S-profile was produced by the occurrence of a big storm or the design storm. It was not the case. The biggest storm that occurred in the period 1983-85 was of $H_s \sim 6$ m. The calculations showed that for $H \sim 9$ m 60 Tn blocks on a slope of $ctg\alpha = 1.2$ were on initiation of damage conditions. However, under the same wave conditions 60 Tn blocks are stable on slopes of $ctg\alpha = 1.5$.

Reasons for damage

9. The report concluded that the breakwater failed for one or more of the following reasons:

Crown

(a) The crown had insufficient overall dimension. The concrete failure came from its working conditions as a beam. It was discontinuously founded, because of the sinking of inner armour blocks in the sandy bottom.

(b) Improper construction procedures, especially concrete quality (segregation of arids and moving casts).

Slope

(a) The blocks were not placed on a slope of $ctg\alpha = 1.5$, because of the limited bearing capacity of the crane. In some sections the blocks were just rolled over.

(b) Bottom erosion and sliding of the whole section.

Proposed solution

10. The main problems at the Cudillero breakwater can be resumed as follows:

(a) Toe and bottom erosion in the front of the breakwater.
(b) Sinking of the blocks in the "core".
(c) Cracking, sliding and overturning of the crown.
(d) Wave energy transmission through the breakwater.

11. Because of the time schedule, investment limitations and harbour functionality the following restrictions were imposed:

(a) A complete breech in the breakwater was not allowed.

(b) Removal of the crown in maximum lengths of 25 m.

(c) Access of materials and equipment through a new way on the cliff as transit through the village was not possible.

12. The final adopted solution for reparation was (Fig. 5):

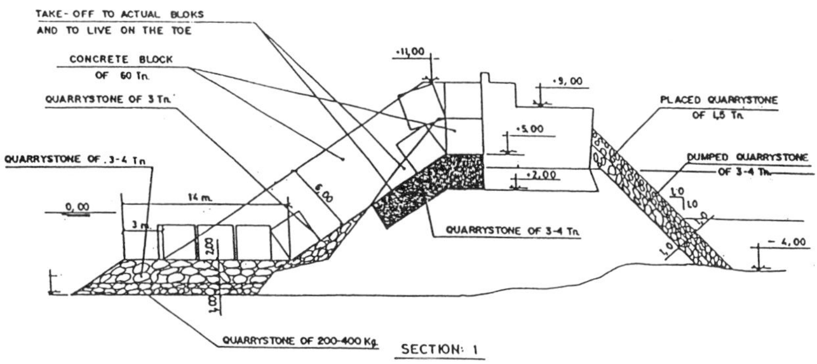

Fig. 5. Cudillero solutions of rebuilding.

(a) Crown

Enlargement of the crown to almost double the former size, and proper foundation.

(b) Toe berm

Building a berm of 50 m wide and 2 m high along breakwater's toe.

(c) Main layer

Building a main layer of 60 Tn cube armours placed on a slope of ctgα = 1.5, finishing in an upper berm of at least two armours wide.

(d) Inner slope

Building an inner slope in the harbour side to satisfy filtering conditions, that is the largest sizes covering the core armours.

Construction procedures

13. In order to avoid an excessive enlargement of the crown width just for supporting a crown of 1800 m.T, the constructor proposed to use a floating crane which worked from inside the harbour and placed the armours over the crown.

14. The horizontal part of the old crown and zone core's armours was blasted and broken in pieces in such a way it was used to refill the gap between armours. Next, the new crown was founded at level 2.00 using a proper filtering sequence and building a continuous, horizontal and stable, crown foundation.

Performance of the solution

15. The breakwater was rebuilt in almost all its length, except the last 75 m which was delayed because of investment capacity and

because of the minor consequences on harbour functionality if this part had failed. In 1990 during a winter storm of $H_S = 6.5$ m and $T_Z = 18$ s, this part failed, sliding and overturning the crown. The repaired section had no problems.

LUARCA BREAKWATER

16. The village of Luarca is located approximately 15 miles West of Cudillero (Fig. 1). It is also a fishing village where the fishing sales are of the same order as in Cudillero. The harbour is inside a medium size bay protected by a vertical breakwater built in the 30's. In the 60's, the breakwater was reinforced with 20 Tn blocks, Fig. 6. In 1988 a new mooring line was built and the mooring area was enlarged. To protect the new harbour the following sequential construction was undertaken: (1) A vertical porous screen at the basin entrance; (2) A sloping porous protection, and (3) A new breakwater on the old zone. Once the new breakwater is finished the functionality of the first two works will be minor.

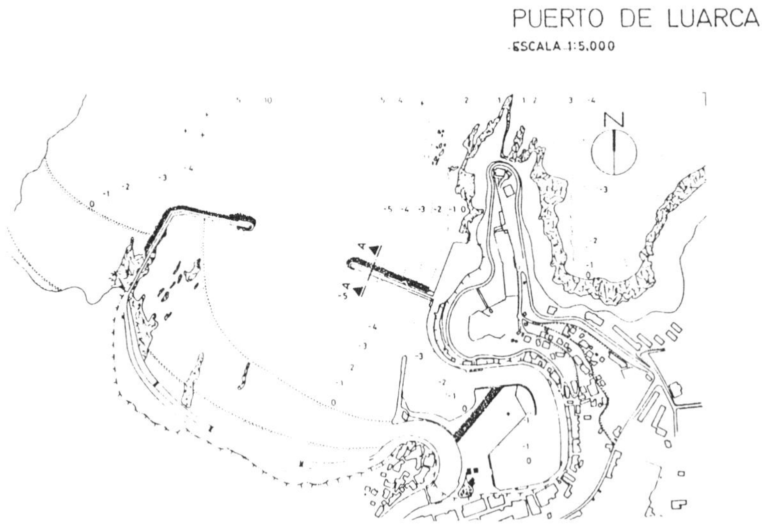

Fig. 6a. Luarca harbour (old).

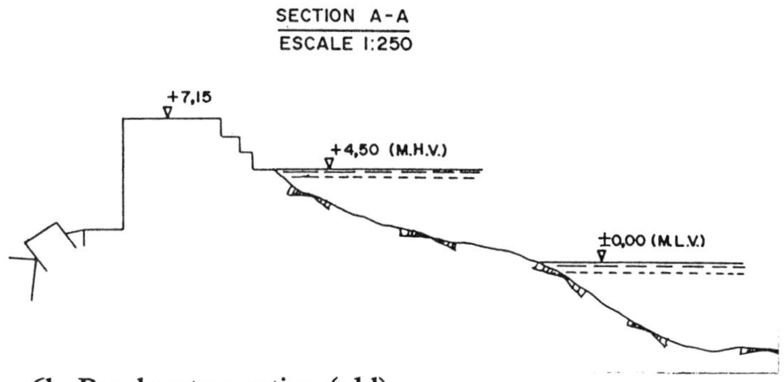

Fig. 6b. Breakwater section (old).

Performance of the old breakwater

17. The main problems of the old breakwater can be summarized as follows:

(a) Functionality

Minimum harbour protection, because of insufficient high (2.5 m over high tide level), length and width (see Fig. 6).

(b) Stability

Although the vertical wall is stable (because it is completely covered by big waves) it has grooves and cavities all along, making maintenance necessary every year.

(c) Slope stability

Protection blocks, built in the 60's, have almost disappeared, leaving a few units which are rounded or cracked.

Restrictions to the construction operation

18. The following restrictions were imposed on the design and construction of the new breakwater:

(a) To avoid the transfer of materials and equipments through the village. Therefore a new access was built on a cliff.

(b) To take into account, if possible, the beautiful landscape as a result of the combination of the old harbour, the graveyard, the village, the cliffs and the sea.

Proposed solution

19. In order to fullfil the previous requirements it was decided to design a breakwater which avoided overtopping mainly by width rather than by height. Furthermore, once the work is finished the new access will be transformed into a pedestrian way. Consequently, the retaining wall and neighbourhood was built with high quality materials, imitating the local rocky cliff. The old vertical wall was maintained, in such a way that it is still seen from the village. The transition of that wall and the new crown has been clearly indicated by using steel clamps. Finally, a high tower with suitable shape was built on the breakwater head.

20. The final breakwater section is shown in Fig. 7, where the most characteristic figures are also given. The breakwater was enlarged 25 m. Under design storm conditions, which correspond to a maximum wave height limited by bottom, H = 9 m it is expected that there will be an overall reduction coefficient of K = 0.1 at the entrance to the basin.

PUERTO DE LUARCA

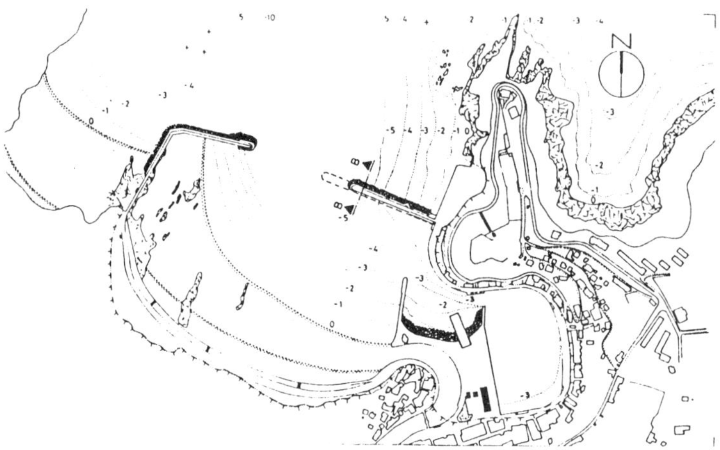

Fig. 7a. Luarca harbour (new).

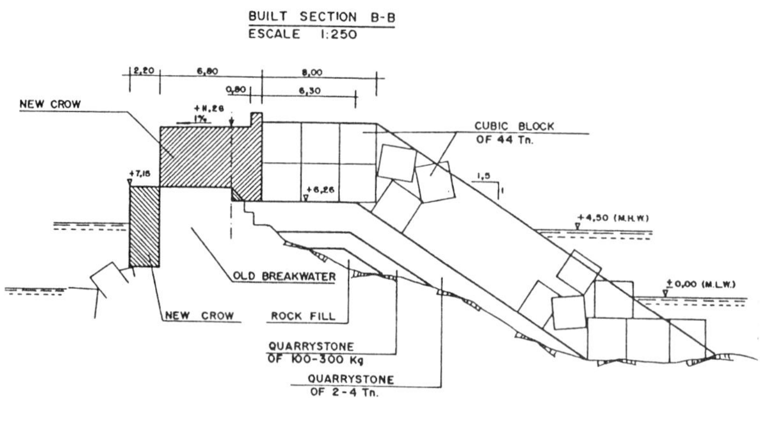

Fig. 7b. Breakwater section (new).

21. In order to avoid reflection of waves from the wall which protects the new access area, it is reinforced with 20 Tn blocks.

Construction procedures

22. To place the blocks on the slopes a crane type Lieber 1200 was used, mounted on a track which moved all along the crown. For this purpose the crown was first widened.

23. One of the most important problems during the construction was the storage of blocks, because of the limitations of transit through the village. Finally, it was solved by manufacturing the blocks two miles away from the harbour and transporting them by track using the new access way.

Performance of the work

24. The work is expected to be finished in April 1991. During the winter 1990/91, the improved sections performed correctly and agitation inside the basin was under all circunstances less than 0.2 m.

APPENDIX I

Stability function

Using dimensional analysis (Ref. 2) proposed the following relationship for evaluating W, the weight of an armor units as a function of the incident wave height, H and the specific weights of water and armor units, γ_w and γ_s respectively,

$$W = \gamma_w . Q . H^3 . Y$$

$$Q = S/(S-1)^3 \qquad (1)$$

$$S = \gamma_s/\gamma_w$$

where Y is a dimensionless function called Stability Function which for a trunk section depends on Iribarren's Number, slope angle, α, damage criterion, type of armor unit, unit placement and roughness and permeability (Ref. 3). For a head sector Y also depends on R/L (where L is the wave length and R the radius of the head cone), the head shape, and the section position on the head (Ref. 5). Notice that Iribarren's Number (also called Surf Similarity parameter) is defined by,

$$Ir = \tan\alpha/\sqrt{(H/L)} \qquad (2)$$

Figs. 8 and 9 show experimental results of the stability of head and trunk sections for Initiation of Damage and Iribarren's Damage.

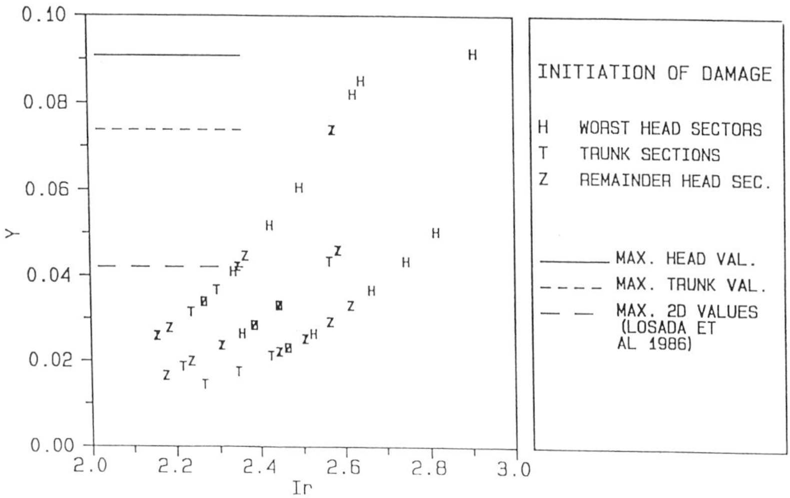

Fig. 8: Experimental results of the stability function: Initiation of Damage, cubes.

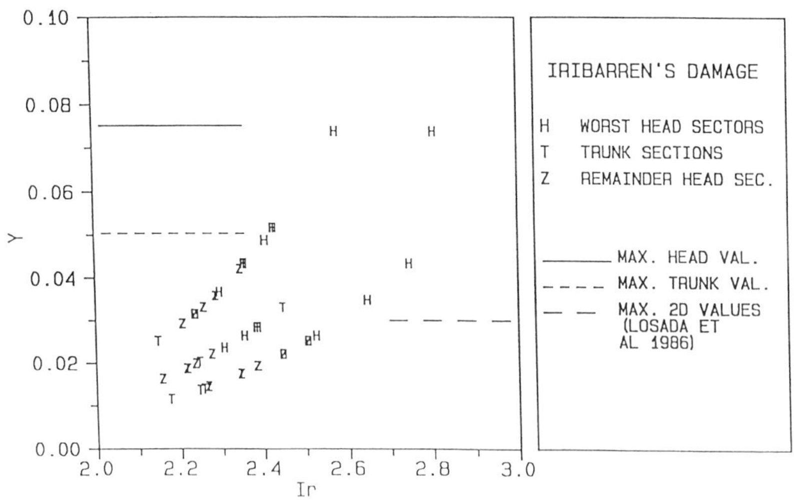

Fig. 9: Experimental results of the stability function: Iribarren's Damage, cubes.

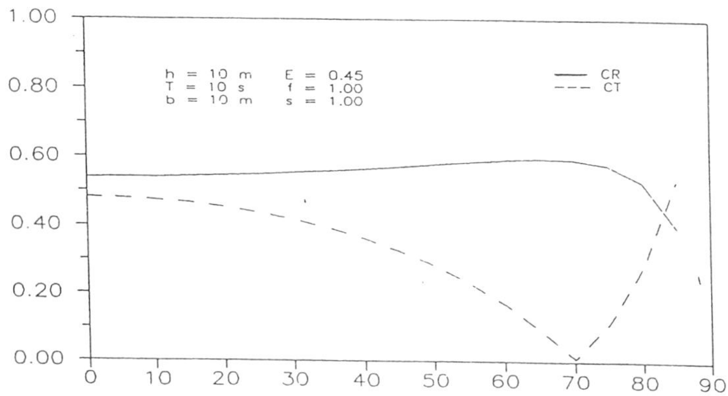

Fig. 10: Reflection and transmission coefficient versus wave period.

APPENDIX II

Transmission of waves through a porous structure

Dalrymple et al. (Ref. 1) examined the case of water waves impiging obliquely on a vertically-side porous structure by the method of eigenfunction expansion. One interesting result is that the reflection coefficient has a minimum for large angles of incidence (60-70°) which corresponds approximately to a maximum in the transmission coefficient.

Losada et al. (Ref. 4) extended previous work to the case of water waves impinging obliquely on crown breakwater. Fig. 9 shows the transmission and reflection coefficient obtained for a breakwater with the following characteristics: porosity 0.48; crown height 3 m; water depth 10 m, incidence angle $\theta = 15$ and 30° as a function of the wave period. This case simulates the flow through the old breakwater of Cudillero.

REFERENCES

1. DALRYMPLE R.A., LOSADA M.A. and MARTIN P.A. Reflection and transmission from porous structures under oblique wave attack. Approved to be published by the Journal of Fluid Mechanics, 1991.

2. LOSADA M.A. and GIMENEZ-CURTO L.A. The joint effects of wave height and period on the stability of rubble mound breakwaters using Iribarren's number. Coastal Engrg., vol. 3, 77-96, 1979.

3. LOSADA M.A., DESIRE J.M. and ALEJO L.M. Stability of blocks as breakwater armor units. Journal of Structural Engrg., vol. 112, no. 11, ASCE, Nov., 2392-2401, 1986.

4. LOSADA M.A., LOSADA I.J. and DALRYMPLE R.A. Water waves on crown breakwaters. Submitted to the Journal of Waterway, Port, Coastal and Ocean Engineering, ASCE, 1991.

5. VIDAL C., LOSADA M.A. and MEDINA R. On the stability of mound breakwaters head and trunk. Approved to be published by the Journal of Waterway, Port, Coastal and Ocean Engineering, ASCE, 1991.

20. The safety of breakwaters against wave overtopping

M. DE GERLONI, ENEL CRIS, L. FRANCO and G. PASSONI, Polytechnic of Milan

SYNOPSIS. The paper reports on experimental investigations to assess the safety of variable configurations of composite and vertical breakwaters against wave overtopping. Individual overtopping volumes were measured and their effect on model targets of persons and cars located behind the crown wall statistically evaluated.

INTRODUCTION.

1. Despite the increasing concern for safer design of harbour breakwaters little is reported on their performance against wave overtopping. Only approximate design criteria are in fact available to select the elevation and shape of the crown wall with correct safety margins, particularly when transit or even berthing is allowed on the rear of the superstructure (as usually occurs in Mediterranean marinas). Still admissible values of overtopping discharge are only based on old Japanese observations behind sea walls (ref. 1). Following similar experimental researches on the overtopping of rubble mound breakwaters (refs. 2-6), an extensive series of bidimensional model tests was carried out in the random wave flume of the Hydraulic and Structural Research Center (CRIS) of the Italian Electricity Board (ENEL) in Milano, Italy.

EXPERIMENTAL SET UP

2. The random wave flume used for the tests (Fig. 1) is 43.6 m long and 1.5 m deep; the width is 3.0 m for the first 34.0 m and 1.0 m in the "finger" end where the models are installed. The waves are generated by a wedge type wave generator whose reciprocating motions are produced by a linear servo actuator under closed loop position control. The command signal comes from a personal computer via a digital to analogue converter.

3. Figure 2 shows a typical test condition with a caisson installed in the glass walled end of the flume. The figure also shows the device used for measuring the overtopping volumes which essentially consists of a tray suspended trough a load cell to a supporting beam. The load cell signal chart recording allows to measure not only the total volume - hence

the mean overtopping discharge - but also the individual volumes and the number of overtopping waves.

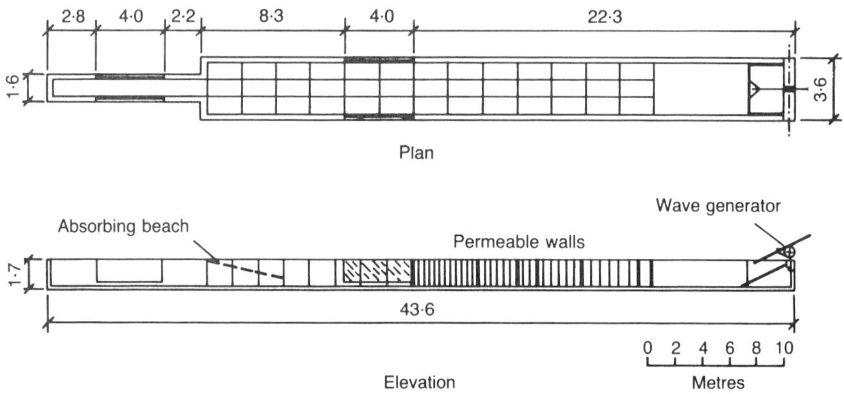

Fig. 1 The random wave flume used for the tests

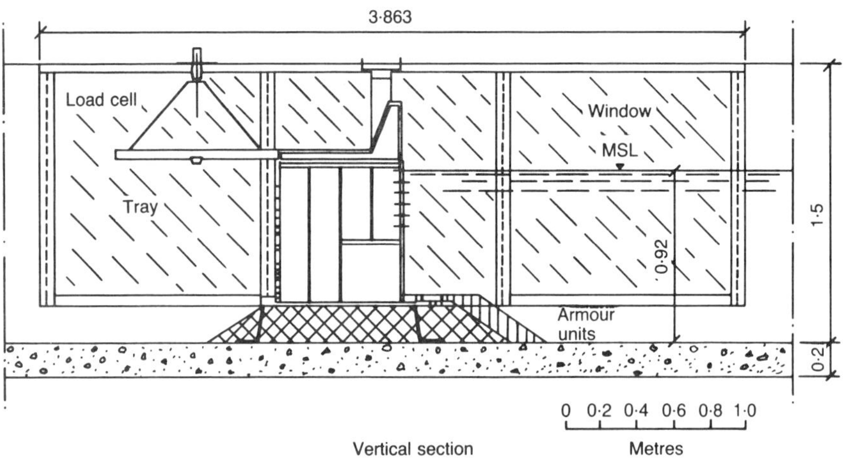

Fig. 2 Typical installation of the overtopping volume measuring device.

4. The instrumentation was completed by five twin wire resistive wave probes for statistical and spectral analysis of water surface elevations also allowing for the calculation of reflection coefficients.

5. A relatively large (1:20) scale was adopted: the choice was dictated by the availability of model targets to evaluate the effect of overtopping and by the sensibility and precision of the volume measuring system.

6. In the tests JONSWAP spectra with several peak periods were generated. Some tests were also run with Pearson-Moskowitz or bimodal wave spectra but no apparent difference was found with the results from JONSWAP spectra having the same Hs and Tp. The duration of each test was variable being

the generation length (pseudo-random) dependent by the period; a minimum duration of four hours (prototype) was adopted to have a reasonable probability of occurrence of extreme maximum wave heights and a reasonable statistical reliability of the results.

OVERTOPPING PERFORMANCES OF BREAKWATERS

7. Various breakwater configurations were investigated, particularly plain and perforated caissons with different crown walls and composite structures with rock armoured rubble mounds having different geometries. Some model sections are shown in fig 3. Though actually some of the structures were studied to optimize specific designs, useful indications for design purposes can be derived since the same detailed test procedure was applied.

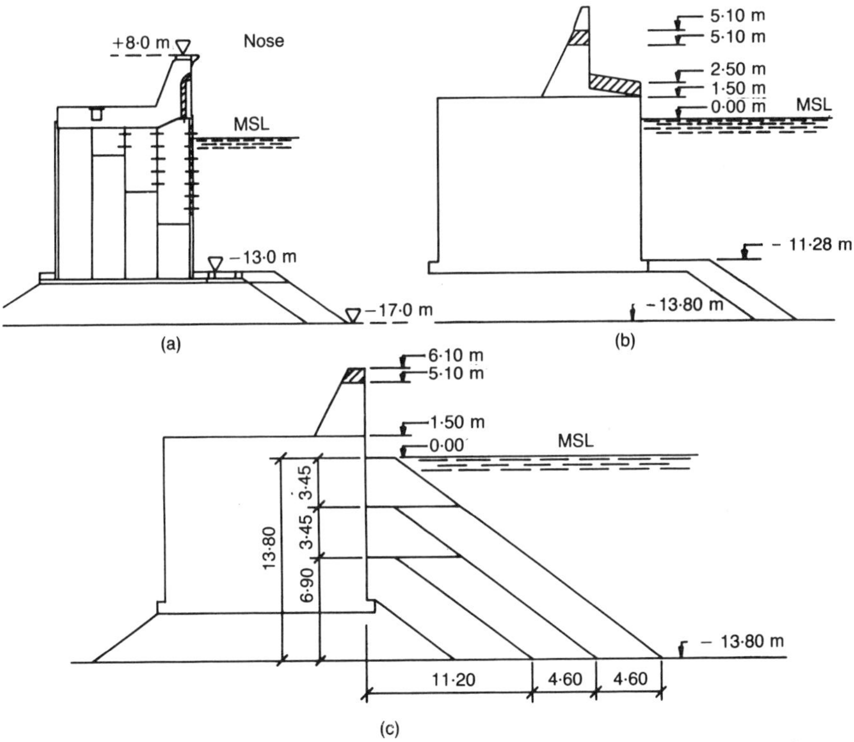

Fig. 3 Vertical sections of the breakwater tested in the experiments.

8. For each breakwater configuration graphs were produced plotting the mean overtopping discharge against wave height as a function of other parameters. As an example, for the perforated caisson shown in Fig 3a (ref. 7), 14 different geometries were tested, changing the number of absorbing chambers (1, 2, 3), the diameter of the holes (0.6, 0.8, 1.0 m), the shape of the wall "nose" and the number and diameter of the vent pipes. The overtopping performances of

some of the tested structures are shown in Fig. 4; as reference the worst response of a plain wall caisson with the same dimensions is plotted. The design wall crest elevation was in this case maintained fixed. The best solution turned out to be the one with three chambers with 1.0 m holes, two ø 0.5 m vents whit 3 m length and type 3 "nose", which reduced the mean discharge by two orders of magnitude with respect to the result obteined for the plain vertical wall. The reflection coefficient was also reduced to 0.5-0.6.

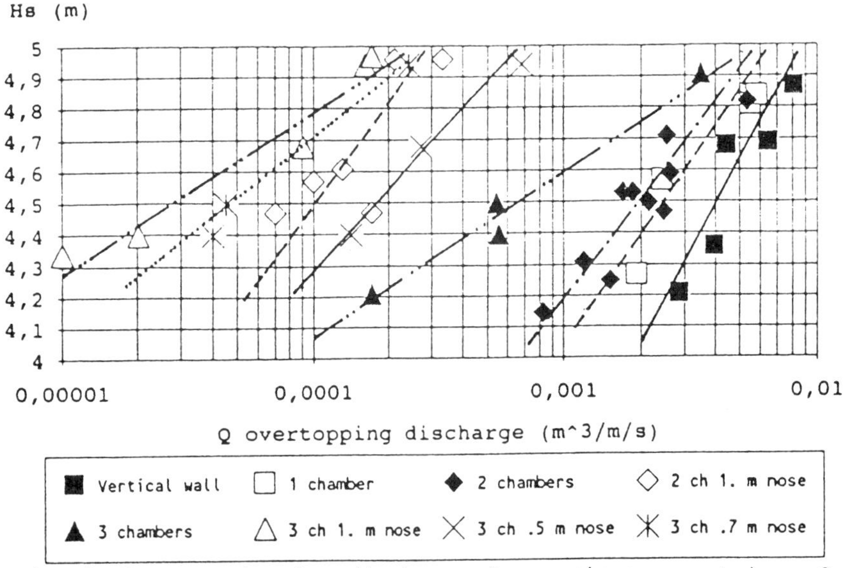

Fig. 4 Mean overtopping discharge for various geometries of the breakwater of Fig. 3a.

9. With reference to the caisson breakwater shown in fig. 3b it was found that the crown wall elevation has a much greater influence than the elevation of the upper corner of the seaward wall. The wave period plays an important role too: a minimum overtopping was consistently found for a period of approximately 11.5 s, while the worst conditions occur for the longest tested period of 13.0 s.

10. As far as the composite breakwaters are concerned (Fig. 3c) the most interesting finding is related to the effect of the elevation and width of the rubble mound protection. If the rock armouring is submerged up to M.S.L. the overtopping increases, whereas it gradually reduces for higher and wider berm crests. This is likely due to the counteracting effects of increasing wave energy dissipation and transformation of the vertical stationary motion along the wall to a progressive breaking wave pattern along the slope. The crown wall elevations always plays a major role, but the steeper the slopes the higher appears to be the overtopping.

11. The mean discharge data allows to compare these results with former experiments but it is most interesting to look at the correlation between the mean discharge and the maximum recorded individual volume for each run. It can be seen in Fig. 5 that, although the two parameters look well correlated for each configuration, the correlation does change for even slightly different geometries.

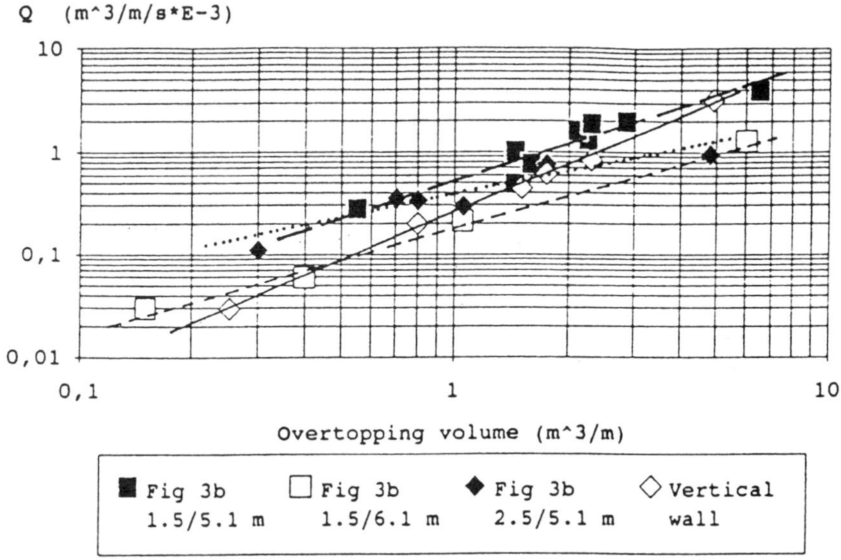

Fig. 5 Correlation between mean overtopping discharge and maximum recorded volume.

12. This means that there is no unique relationship between the two parameters; therefore the measurement of individual overtopping volumes seems necessary when the safety of the structures (rear breakwater stability analysis, safety of transit and berthing) is concerned. The mean overtopping discharge represents in fact a useful parameter for designing drainage systems behind sea walls but it is only a simple indicator for other effects. The safe response of each breakwater configuration can be better described by the statistical analysis of the damage caused by the maxima overtopping volumes; such new approach was adopted in this study and is presented in the following.

OVERTOPPING EFFECT

13. The second aspect of the research program concerned the evaluation of the effect of the overtopping waves. It was decided that cars and persons located on the rear of the superstructure could be considered as typical "targets".

14. For representing the cars available 1:20 scale toys were purchased which only required the adjustment of the weight to match exactly those of the real cars. Though low grade

rubber tyres were installed on the models a small friction scale effect is to be expected.

15. A more difficult problem appeared to be the modelling of the behaviour of human beings. A simple size and weight scaling could not be applied as a person reacts instinctively when moved from its equilibrium position. It was however thought that a model could be "calibrated" with a simplified full scale test procedure reproducing the same water jet impact on the person and on the model.

16. A series of large plastic pipes were connected to direct variable known amounts of water against a volunteer weighting 65 Kg; the water was released very rapidly from an especially arranged concrete bucket at an elevation of 4.5 m.

17. A life size plastic dummy equipped with an iron base was then tested under the same conditions to find the amount of ballast necessary to make it fall like the man. It was experimentally found that a total weight of 130 Kg was necessary to have the same "falling" effect. The life size dummy was then reduced to the model scale using especially cast plastic person toys.

18. The experimental procedure to evaluate the effect of overtopping waves consisted in placing a number of model cars and persons on the rear of the superstructure. An operator had to note, for each overtopping wave, the time of occurrence and the effects, repositioning the displaced targets immediately. The effects were divided in two categories: movements of less or more than a typical dimension which was set as the length for the car and the height for the persons.

19. By relating these effects with the recorded volumes graphs were produce like those of figures 6 and 7 showing respectively the statistical effect of overtopping waves on cars and humans. From the graphs, which refer to the tests on three different structures, it can be seen that a volume of

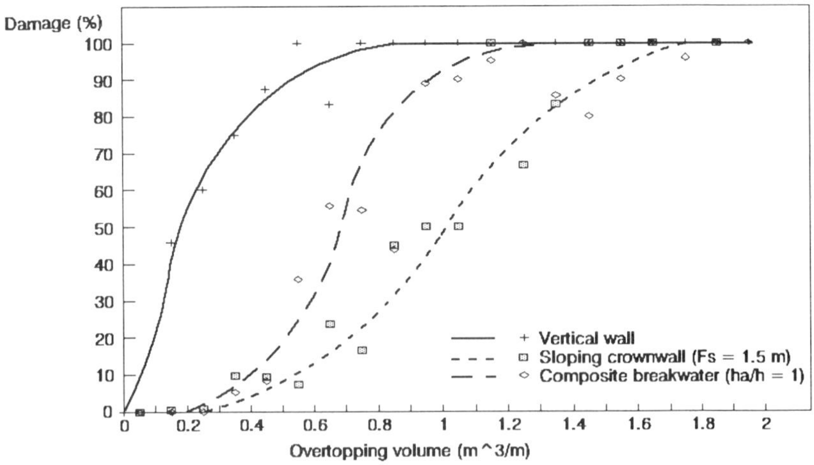

Fig. 6 Overtopping effect on model cars

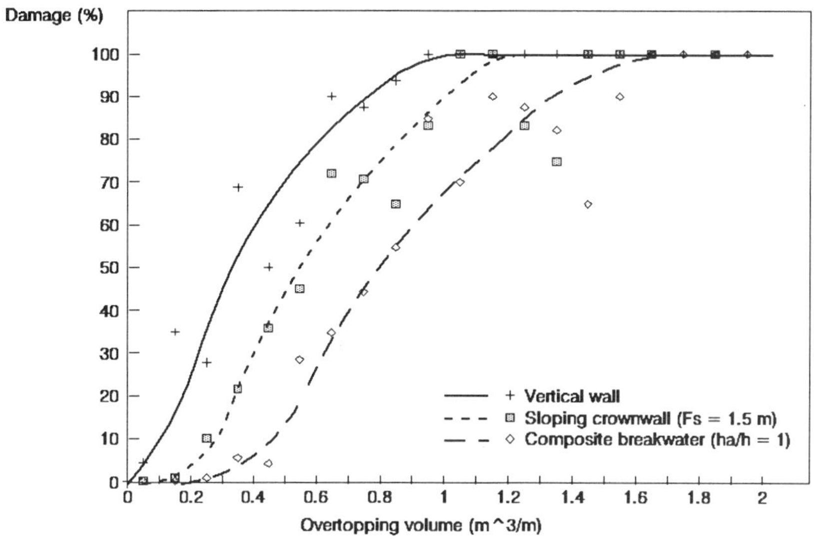

Fig. 7 Overtopping effect on model humans

0.5 m^3/m per meter of breakwater length has a 90% probability to move a car, while 0.7 m^3/m is enough to have 90% probability to make a person to fall, behind a pure vertical wall. The graphs also show that the caisson with sloping crownwall (3b) causes less overtopping damage than the composite breakwater (3c) or the vertical structures.

CONCLUSIONS AND FURTHER WORK

20. An extensive random wave flume test program was carried out in order to asses the behaviour of various modern caisson breakwaters against wave overtopping, with special attention to the definition of statistical safety criteria for the stability of cars and persons located on the rear side (model scaled after a full scale calibration).

21. The main innovation is represented by the measurement of each individual wave overtopping volume and by subsequent correlation of maxima volumes with observed "damage". This approach represents a step forward more reliable design guidelines for breakwater crownwall geometry which are still lacking.

22. The limited space allowed for this poster paper does not allow a detailed presentation of data results, which can be summarized as follows:

a) The overtopping performance is strongly influenced by any geometrical change of the tested structures. Large benefits are achieved by increasing the size and the number of absorbing holes/chambers, whereas more overtopping occurs when a rock protection is placed in front of the wall (up to m.s.l.).

b) The correlation between the mean overtopping discharge and the maximum overtopping volume is not consistent for different breakwater structures. Therefore individual overtopping volumes should be measured to asses stability of/on the rear side.

c) Overtopping rates are also sensitive to wave periods.

d) This complex non-linear phenomenon cannot be described by simple universal formulae.

e) Admissible values of overtopping volumes are found for different structures related to the "failure" probability of cars and persons placed on the rear of the crownwall.

23. Of course the reliability and applicability of the results can be improved by simulating the wind effect and by extending the range of wave conditions and structural configuration. It is intended to continue the research also on rubble mound breakwaters of various type and to perform a more detailed probabilistic analysis of the occurrence of single wave overtopping volumes and related "damage effects".

ACKNOWLEDGEMENTS The authors wish to thank Mr B. Belotti, Mr. S. Pastori and Mr. A. Triola who performed the tests and elaborated the overtopping data. The support of the National Electricity Board (ENEL) and of the Ministry of University and Scientific Research (MURST) are gratefully acknowledged.

REFERENCES

1. FUKUDA N., UNO T., IRIE I. Field observations of wave overtopping of wave absorbing revetment, Coastal Eng. in Japan, 17, 1974

2. AMINTI P. and FRANCO L. Wave overtopping on rubble mound breakwaters, proc. XXI I.C.C.E., ASCE, Malaga 1988.

3. JENSEN O.J., JUHL J. Wave overtopping on breakwaters and sea dikes. II Conf. Coastal and Port Eng, in Developing Countries, Bejing, 1987.

4. OWEN M.W. Design of sea-walls allowing for wave overtopping, Rep.EX924, Hydraulics Research, Wallingford, 1980.

5. BRADBURY A.P. and ALLSOP N.W.H. Hydraulic effects of breakwater crown walls, BREAKWATERS '88 Conference, ICE, Eastbourne, 1988.

6. ALLSOP N.W.H., FRANCO L., HAWKES P.J. Probability distribution and levels of wave run-up on armoured rubble slopes, Int. Conf. on Numerical and Hydraulic Modelling of Ports and Harbours, Birmingham, 1985.

7. DE GERLONI M., FRANCO L., NOLI A., ROSSI U. Porto industriale di Porto Torres: prove su modello dei cassoni della nuova diga di Ponente, II AIOM Congress, Napoli, 1989.

21. Probability of an armour block instability in uniformly sloping block layer

A. KIMURA, T. OHTA and Y. MATSUMI, Tottori University

1. Synopsis

In this study, experimental and analytical investigations on following properties of armor blocks on a slope are carried out, (1) spatial position and wave condition which bring about the shock wave force, (2) the dynamic behavior of a block under the shock wave force, and a method to estimate the reliability of blocks placed in 2 layers on a slope is developed.

2. Shock wave pressure

The 1/20 model of 2t tetrapods are placed in 2 layers on a 4/3 impermeable sea wall model with a bottom slope 1/10. Water depth at the foot of the sea wall is 23cm. Normal (F_n) and tangential (F_p) wave forces acting on blocks are measured separately with two block-spring wave force gauges. When ξ ($=\tan\theta/\sqrt{H/L_0}$) is around 3.0, waves break on the slope and shock wave forces appear. The breaking types of the waves are plunging or collapsing. The maximum wave force along the slope appears a little below the shore line. Figure 1 shows an example of $|F_n^2+F_p^2|/W$ when the shock wave force is measured

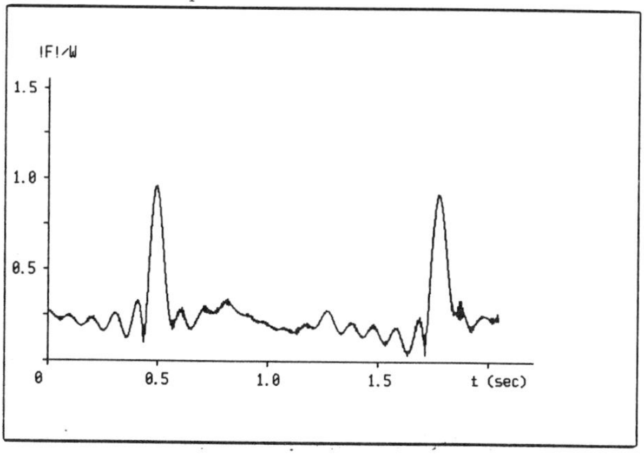

Fig.1 Shock wave force profile

below the shore line.The duration of the shock pressure is around 0.2-0.3 T where T is the wave period. Since these type of wave forces act to press blocks toward the slope, this situation has been considered to be safe for blocks. However, as explained later, if a block has a space to move around a placed position, it starts to rotate and this could become the first stage of the blocks being swept out from the block layer by the shock wave force.

3. Behavior of blocks

To observe the behavior of blocks on the slope due to shock wave forces, waves with a little larger wave height but the same breaking types are used in the experiments. Design wave height calculated by the Hudson equation for the present block model is about 10.5cm. The block behavior is observed carefully using a video system and typical behavior is clarified as shown in Fig.2.

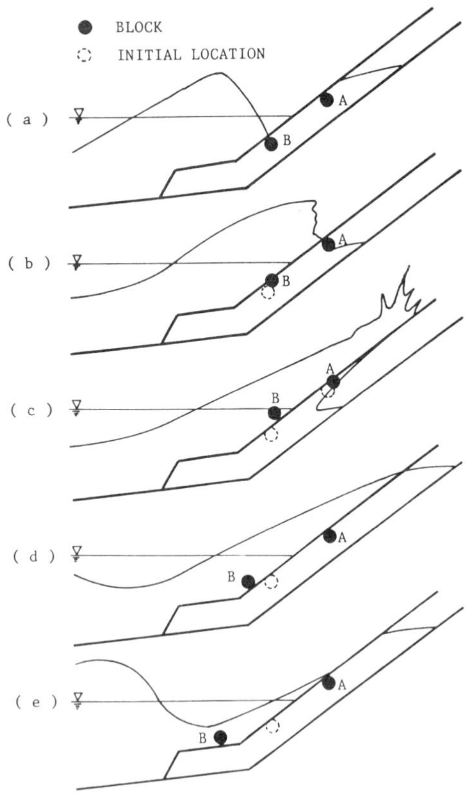

Fig.2 Behavior of blocks under shock wave force

Blocks A and B are originally placed a little above and a little below the shore line (Fig.a). Solid lines in individual figures show the wave profile and water level in the block layer. When the collapsing or plunging starts to act on a slope (Fig.b) blocks A and B begin to move along the slope upward and continue moving until the back wash of the wave starts (Fig.c). Since the displacement of block A is small, this block usually settles again in the same position. However block B tends to move significantly (Fig.b,c), it passes over the original position in the back wash of the wave (Fig.d) and settles around the foot of the block layers.

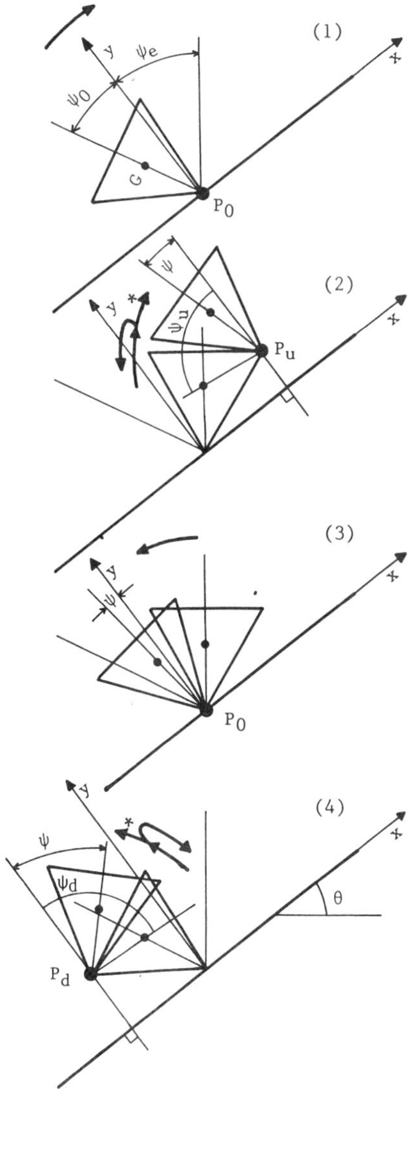

Fig.3 Rotation of the model block

Since no shock wave force appears in this water depth, this B block settles despite of any following wave actions. In the study these two unstable situations of block A and B are distinguished as the rotation and the fall respectively.
If a block has a weight calculated by the Hudson equation and it has no space to move in the layer, no clear movement from the placed position takes place despite the action of shock wave force induced by the design wave on it. However, the blocks, such as tetrapods in the above layer is unstable in rotation, since they are placed up-side down. Therefor even when the design wave force acts on the block, it very easily starts to rotate around the supporting point (Fig.3). Until the wave force starts to act it touches the downward block of the layer (point p_u) and the block in the lower layer (point p_0). Therefor if the movement of this block is not restricted by the upward block of the layer, it starts to rotate around p_0. If the block receives a sufficient momentum from the shock wave force within a very short time, it can start clock wise rotation and keep the rotation around the new supporting point after it touches the upward block.

4. Equation of motion

The tetrapod is modeled simply with a triangular cross-section column in the analysis (Fig.4). Movement in the z direction (vertical to the x-y plane) is restricted. The equation of motion of this body around p_0 is given as

$$I \frac{d^2\psi}{dt^2} = aW'\sin(\theta-\psi)+F_p\cos(\psi)-aF_n\sin(\psi) \qquad (1)$$

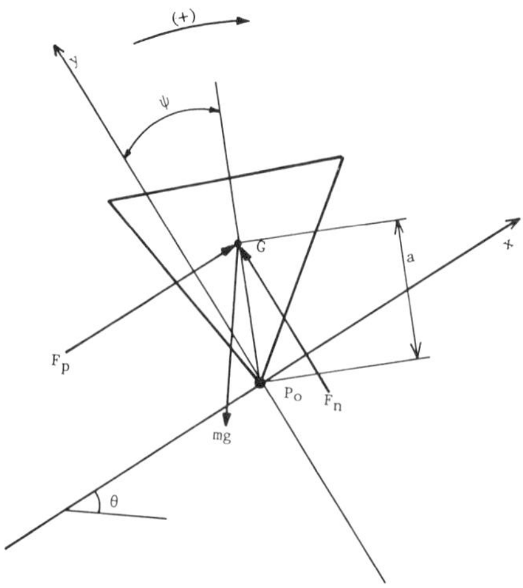

Fig.4 Coordinate system for the model block

in which, x axis is taken positive upward along the slope and y axis is taken positive normal to it, ψ is an angle between p_0-G and y axis (G is the gravitational center of the block), a is the distance between p_0 and G, W' is a weight of the model tetrapod in water, θ is an angle of the slope, ψ_0 is an initial angle between p_0-G and y axis, I is an inertial moment of the block. Measured shock wave forces F_n and F_p are used in the experiments.

The calculated process of the block rotation are shown in Fig.5. The first stage is the biginning of the wave force action. The block starts clockwise rotation around p_0. When an angle between p_0-G and y axis becomes ψ_e, the forward point of the triangle touch the upward block at p_u. If the momentum of the rotation is sufficiently large, the block starts to rotate around new supporting point p_u. The coordinates are shifted parallel to the new origin p_u from this instance. If the momentum is very large, the block can come out on the block layer in water and the further calculation is stopped but if it is insufficient, the block starts anticlockwise rotation (stage 2,3) , and it keeps anticlockwise rotation until it touches the downward block at point p_d (stage 4). If the block has a sufficient anticlockwise momentum of rotation at this stage, it starts to rotate around p_d. For further calculation the coordinates have to be switched (p_d becomes a new origin), but if the momentum is insufficient, it stops the rotation and moves back to the original position again. In the former case, it keeps the rotation around p_d and finally comes out on the block layer and no further calculation is possible. Instead of giving the space shape for the block, angles for ψ_0=- ψ_e (Fig.3) are used in this calculation. Figures 5 are the example when ψ_0=- ψ_e=10° . The thick lines show the angle ψ . Since ψ is measured clockwise from y-axis, the angle shifts suddenly when the coordinates shift takes place.

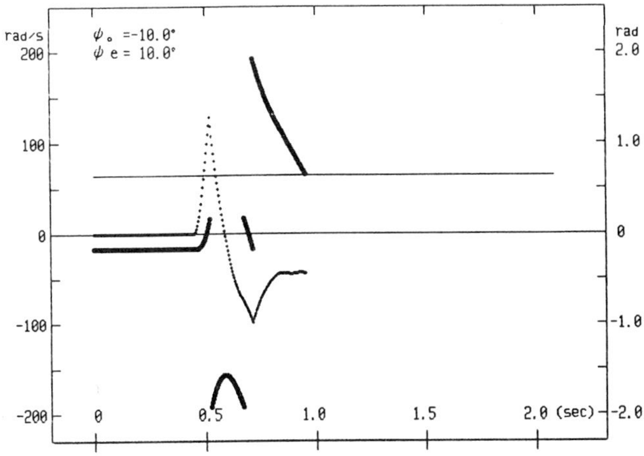

Fig.5 Calculated angle of the model block

Figure 5 shows typical behavior of the block which is brought away from the block layer during the wave down rush.

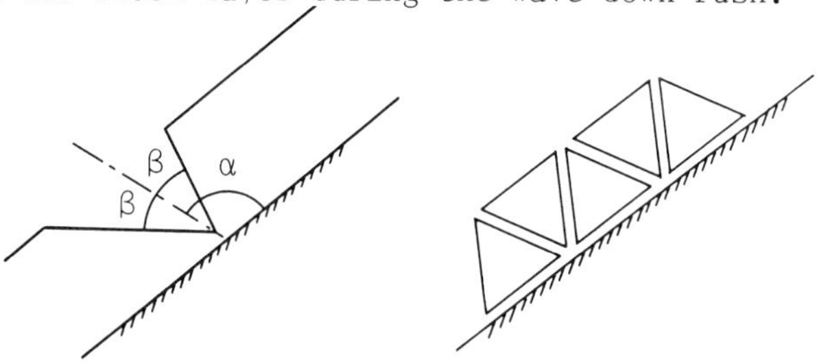

Fig.6 Model of the 2-layers placement

5. Probability of an block instability

The movement of a block in the 2 layers placement (Fig.6) is calculated. In this model the space has triangle shape. ψ_0 and ψ_e are calculated from the angles α and β in this figure. In Figure 7, calculated results are shown. Black circles and half black circles show the cases in which downward and upward displacements from the layer take place respectively. Those areas are approximately

$$\begin{array}{ll} \alpha \geqq -4\beta + 260^\circ & \text{downward displacement} \\ \alpha < 4\beta - 100^\circ & \text{upward displacement} \end{array} \qquad (2)$$

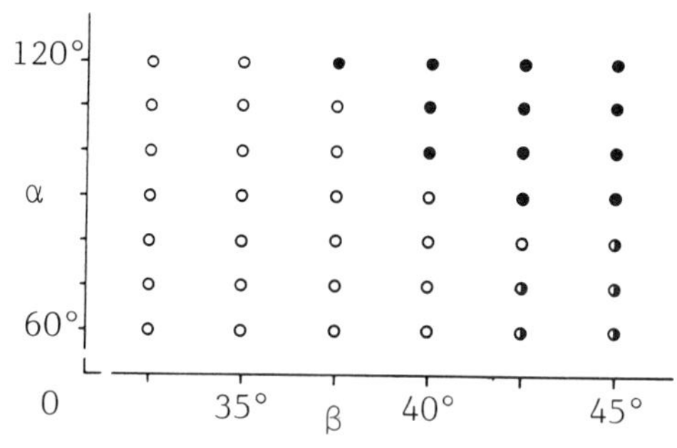

Fig.7 Stability of blocks

Therefore the probability of the block instability is given by the next equation.

$$K_D = \iint_S P(\alpha)\, P(\beta)\, d\alpha\, d\beta \qquad (3)$$

S is a region in which displacement of blocks take place. $P(\alpha)$ and $P(\beta)$ are the probability distributions for α and β. These distributions are experimentally measured using model tetrapods placed in 2 layers on a 4/3 slope, and can be approximated by the normal distributions with β_m=26-27° and σ_β=1-3° for $P(\beta)$ and α_m=87-89° and σ_α=4-11° for $P(\alpha)$. The calculated result for the several distributions with

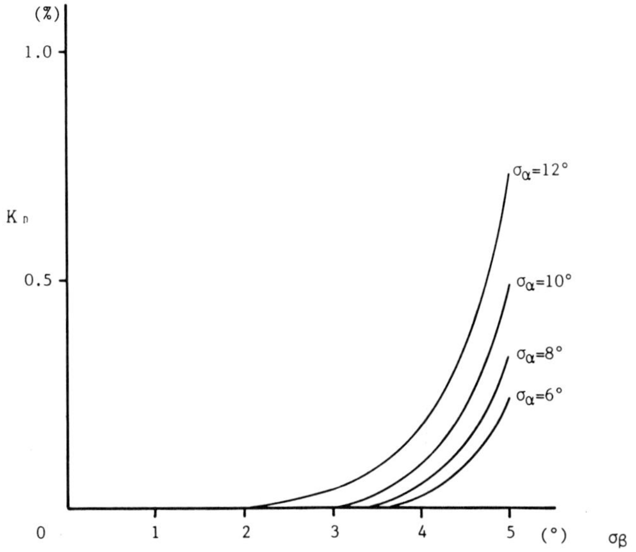

Fig.8 Probability of a block instability

several different σ_α and σ_β are shown in Fig.8.

22. Investigation into the collapse of the rubble mound breakwater over soft soil muds at Rhu Marina, Scotland

C. DINARDO, Dinardo Partnership

Objective:

Many similar rubble mound breakwaters have been completed successfully, however a significant few have not. Their failure has invariably been due to very soft seabeds which have been unable to sustain the applied loads or the combined disturbing effects of impact from free falling buoyant rocks.

The Author wishes to draw attention to the imperative importance of investigating seabed soil conditions and designing within proven methods of construction which allow successful completions.

Background:

The Rhu Marina project had been hindered by attempts at various methods of producing breakwaters including floating types which broke up and were found to be inappropriate.

A fixed rubble mound breakwater was considered more viable and construction started in February 1982 - with an intended construction duration of some six months.

The initial Contractor employed on the works went into financial difficulties and operations ceased during 1982. This Contract was terminated in March 1983.

The second Contractor was selected after some identification of the prevailing laid rock conditions. However the survey did not include the examination of the sub soils. The construction was envisaged to proceed along the lines of the previous Contract. Work restarted in May 1983.

During this accelerated Contractual program rock quantities could not be brought to site in sufficient time (by the flat bottom barges) and new supplies delivered by road, which involved double handling, were employed to speed up completion.

In July 1983 as the rubble mound level rose the first report of rock level failure was recorded. After further movements in August and September the matter was formally acknowledged as 'settlements'. The coincidental movement in the nearby off limits mooring piles which had deflected to a considerable angle was also indentified.

As delays in construction again began to occur, the fast-track program was reviewed with a revised completion before end of December 1983. Substantial completion was achieved but by mid December a major sudden overnight collapse of the crest of the breakwater took place which led to the termination of the works a few days later.

Research and Surveys:

The Author was appointed to investigate and report on the problems before any further operations were contemplated.

Various surveys were undertaken which included hydrographical (echo sounder) survey, a video of a physical inspection by divers, borehole penetration through the rock mound (and outside it) and through the seabed soft soils down to bedrock.

Progressive measuring, testing and monitoring were included. Interim 'Draft' Reports were produced in May and September 1984.

Data and Survey Results:

1 Seabed conditions indicated uniform soft silty clays some 20-25 metres thick overlying bedrock.

2 The intended profile of the Breakwater mound was established as having the base width ranging from 50-75 metres and a height varying from 8-14 metres progressing 250 metres long from the shallow end near shore to the deepest section at furthest offshore, in a direction towards mid channel.

3 The level contours of the seabed surface and mound is shown on Fig. 1.

4 The top crest level was planned to be some 2m above M.H.W.S. in an area with 3m tidal range.

5 Method of construction used flat bottomed barges with free rock drop through water depths of circa 3-12m.

6 The seaborne rock was allowed to drop from barges directly on top of the fabric lying on the seabed.

7 Core rock varied up to 1T ex-quarry blast and armouring rock was selected in the 1½T to 3T range (with the average over 2T).

8 The original design excluded the use of filter beds and membrane.

9 Before construction a membrane was introduced but due to difficulty of retaining in position it was substituted by the use of 2.4m x 4.8m sheets of standard mesh fabric reinforcement laid directly onto the seabed. The sheets were laid loose with each other.

10 Before the second Contract was awarded, rock outside the 'building line' was classified as 'wild' dumping.

11 Inside three months of the second Contract commencement rock mound failures began to be noticed and was classified as settlement.

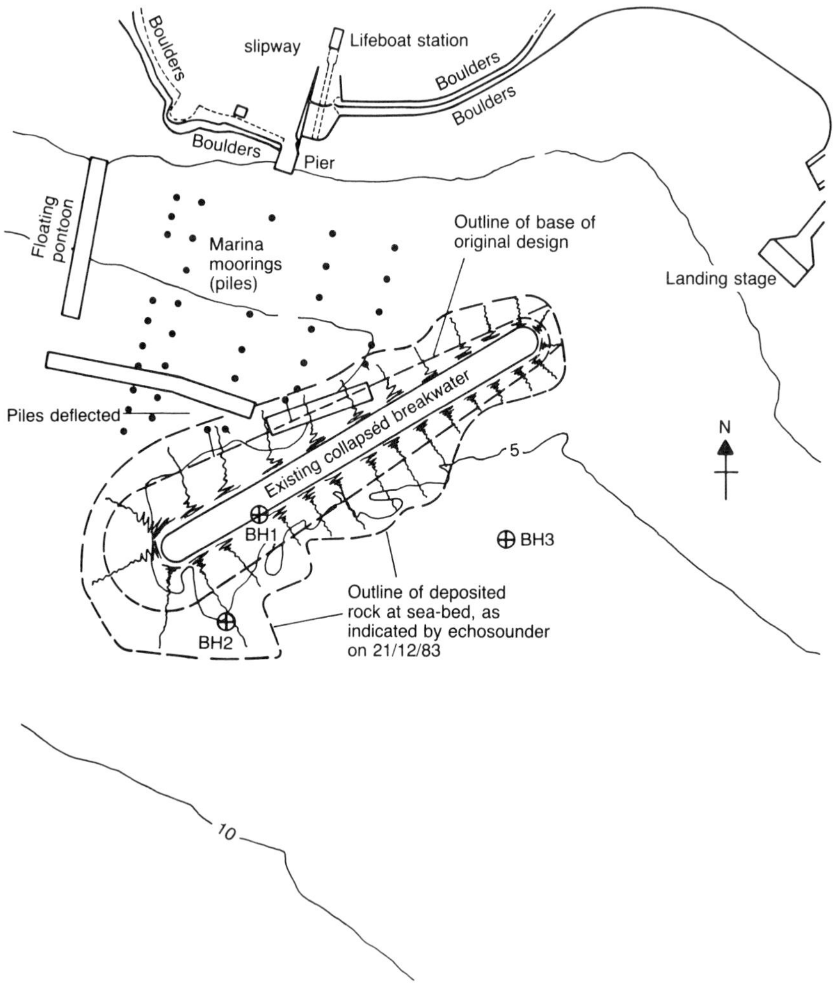
Boulders
slipway
Lifeboat station
Boulders
Boulders
Boulders
Pier
Floating pontoon
Marina moorings (piles)
Outline of base of original design
Landing stage
Piles deflected
Existing collapsed breakwater
5
N
BH1
BH3
BH2
Outline of deposited rock at sea-bed, as indicated by echosounder on 21/12/83
10

Figure 1

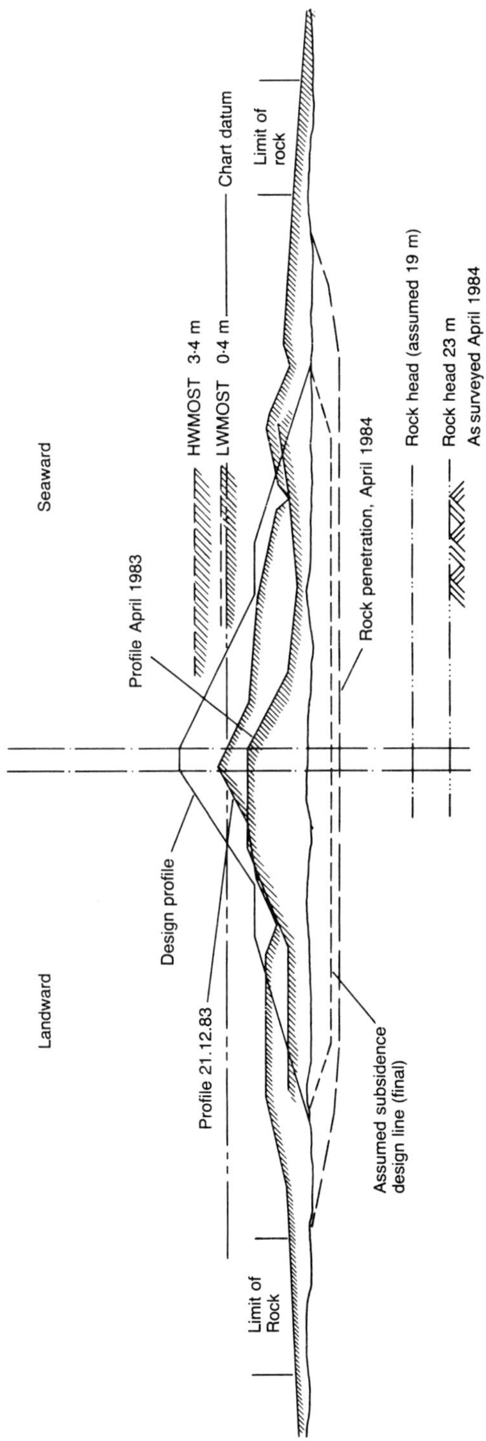

Fig. 2. Superimposed cross-sections and profiles, Rhu Marina report October 1985

12 The major collapse of the mound took place some 7 months into the second Contract by a visible drop in crest level of 3-4 metres. This occurred overnight but the rumbling sounds were apparently heard by local residents.

13 Only a short length of mound at the shallow end is now visible above high tide and only about two thirds of the mound crest is intermittently discernible at low tide.

14 The subsequent site investigation showed seabed soil to have shear strengths ranging from 2 to $9Kn/m^2$.

15 The theoretical Factor of Safety F.O.S. (without the effects of soil rupture during rock dumping) was found to be well below 'unity' especially towards the deep end.

16 The hydrographical level survey indicated irregular peaks and troughs over the surface of the collapsed rock and a spread base width stretching up to 150m - ie twice the original design.

17 The muds and mesh fabric in the interstices of the mound rock were encountered some 3m up from the general level of the seabed.

18 As expected, mound rock was also encountered 2½m into the seabed which confirms a high degree of initial penetration and/or replacement (for the displaced muds).

19 The coefficients of

a) Compressibility Mv (M^2/mn)
b) Consolidation Cv (M^2/year)
c) Permeability K (M/S)

were assessed which estimated a period of gradual dissipation of pore water and build up of shear strength in the soils to carry the maximum loaded area at some 26 years.

20 Monitoring has proceeded and levels indicate a continuing stable condition settling currently at the steady rate of circa 100mm per annum.

Assessment on Modes of Failure:

Extensive research was undertaken to compare worldwide experience of failures of breakwaters over soft seabeds and what measures were subsequently taken to overcome the problem.

The lack of a filter bed (with or without a membrane) below the mound prejudices the integrity of the soft soil bed especially if large heavy rock is allowed to drop directly onto the seabed surface.

The introduction of untied sheets of mesh fabric (in lieu of membranes) was detrimental to seabed surface integrity and assisted in the initial rupture of the surface crust.

The combination of increasing dynamic impact loadings and tearing effects of rock edges and mesh fabric into the crust will have disturbed and sheared the soil from the very outset of construction.

The build up of mound height took place by the load bridging over the wider spread base and mobilising a correspondingly larger area of seabed.

When, at each stage, the pressure exceeded the shear resistance of the soil, rupture took place and the mound became mobile and spread again. In soft seabed soils this has become known as the 'Phenomenon of Lateral Spread'.

Such rolling movements at the edges of the base is evidenced by the content of sea mud and the sheets of mesh fabric found in the interstices of rock up to 3m above the nearest natural seabed. (The bulbs of soil heave normally associated before classical slip circle failure at the toes or sides of settled structures were not encountered during the survey and any previously in existence must have been drawn in and mixed inside the failed toe portions of rock)

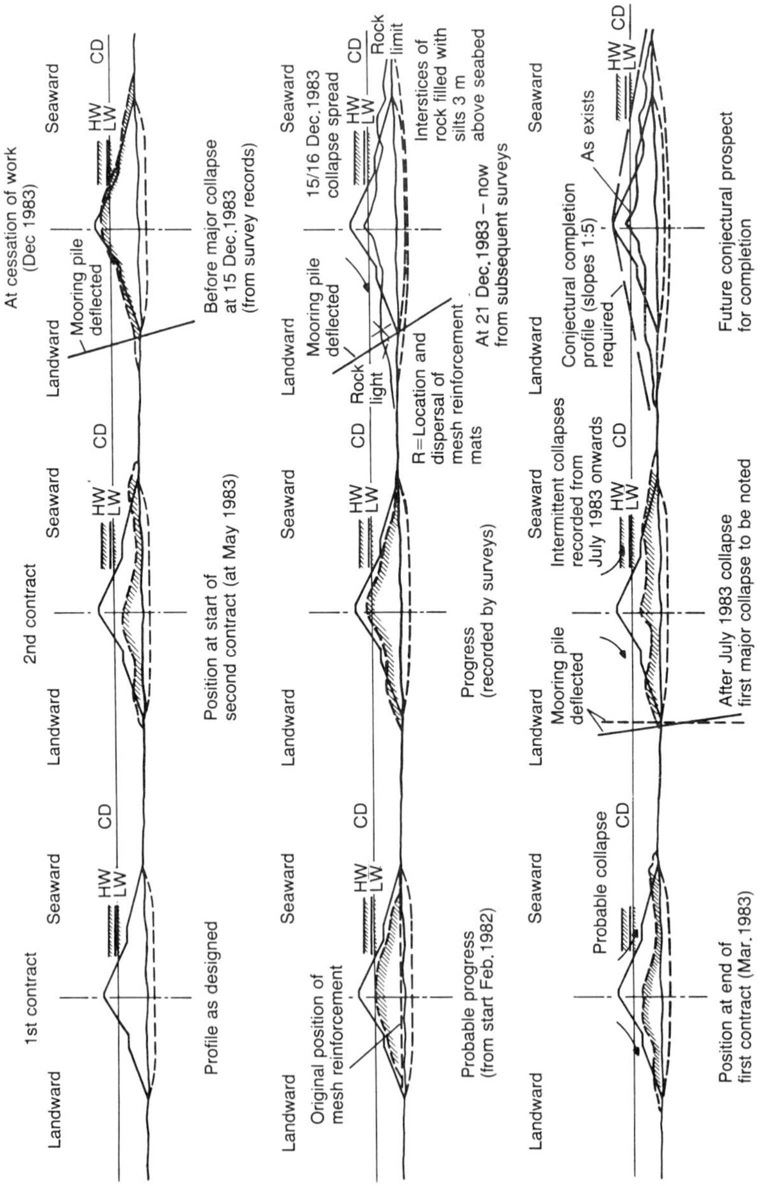

Fig. 3. RHU Marina: Interpreted sequential behaviour of breakwater

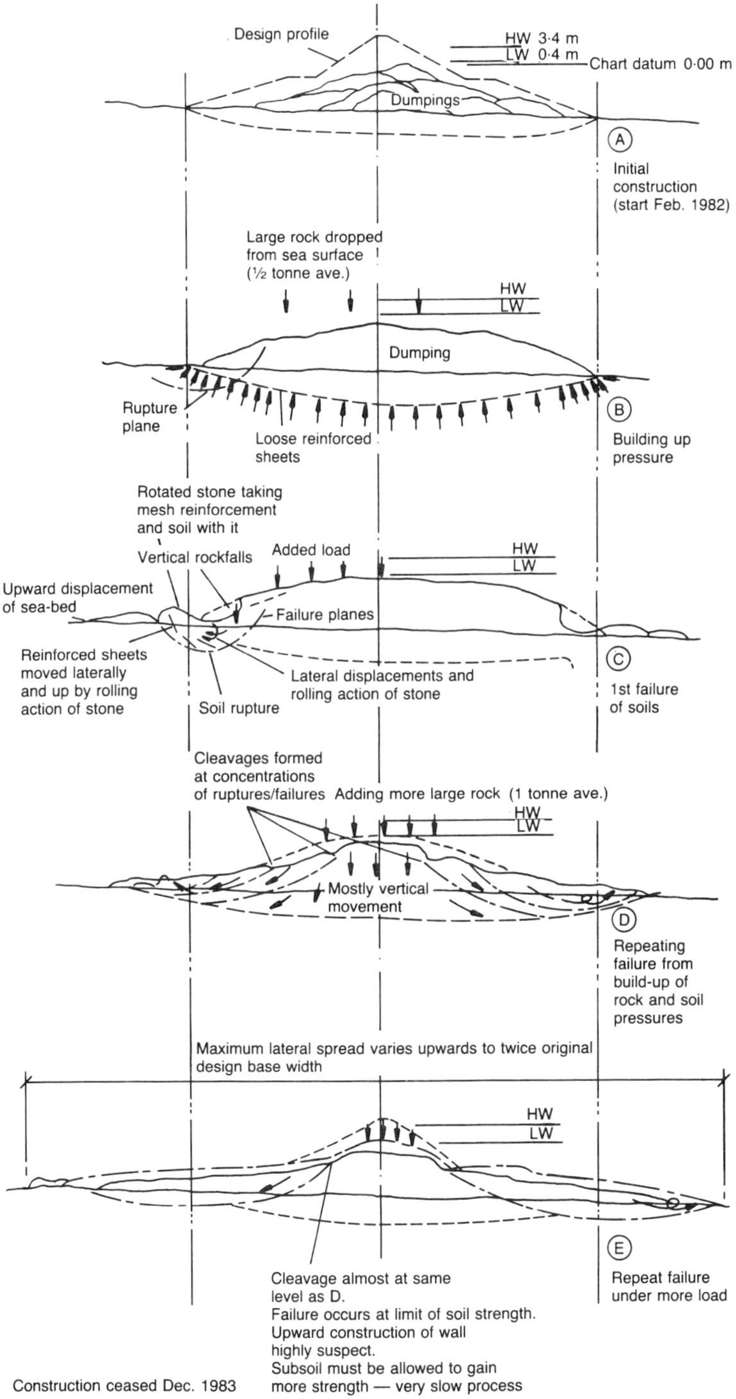

Fig.4. Sequential failure modes of rubble mound walls over soft mud as interpreted at Rhu Marina, Scotland 1983

Consideration in Future Completion of Work and Risk Analyses:

The prospect of completion over the existing mound was considered in 1983 but not within a viable time zone.

Many alternative options for completion using lightweight materials were considered which reduced the risk of re-energising the pressure on the soils but were all found to be expensive and of marginal lifespan.

It was also noted that some length of the existing mound, at the shallow end, could be completed with a higher degree of confidence. However, due to the small amount of marina achieved, combined with probable subsequent chain reaction failure of the outer deep section, this proposal was not considered viable.

The passage of time allowing the build up of soil strength was concluded as being the most essential factor in any future review of the existing collapsed breakwater.

The progressive natural gentle side gradient produced by the rock after 'lateral spread' of 1:5 slopes indicates that a Breakwater Mound could be built in one go, provided the quantity of additional rock required was specified and the risk of continuing failures absorbed (including need to return to site intermittently to top up the crest!). It is also to be noted that the natural slope at the deep end has a gradient of 1:6.

Amongst the many facets involved in creating a successful breakwater over soft muds, the Author strongly recommends that the utmost attention be given to

i) Full knowledge of the seabed soils including laboratory testing and modelling.

ii) The introduction of suitable filter bedding preferably combined with layer/s of membrane. The width of base and mantle thickness being very important

iii) Delicate methods of construction including sequential work and load phasing commensurate with the 'time' scale to mobilise the soil strength required

iv) Adopt monitoring systems to confirm behaviour and performance at all stages

v) Build in suitable factors of safety and contingent flexibility in all aspects of the work including design, construction, programme and costs.

Future Activities at Rhu:

The Owners have reconsidered tne prospects on a regular basis and have decided that a bona fide breakwater and marina should be produced within the confines of their overall site. Meanwhile the existing mound should be abandoned until a future date when some of it might be economically salvaged on a high "suck and see" risk basis and used on a spare capacity concept only.

The salvaging of the rock by moving it to another location is also considered highly unlikely in current/near future economic consideration.

The Author wishes to acknowledge contributions from

o Professor Hugh Sutherland - (Glasgow University)

o Mr G Duncan - (Baptie Shaw & Morton)

o Messrs N C Smyth/A I Will - (Williamson & Partners)

Some References

o M C Barberis - Examples of foundations of works resting on poor subsoil (quay wall at La Spezia, etc);

Fig. 5. Rhu Marina - collapsed breakwater at high tide: 17 December 1983 (tidal range = 3m, MHWS 3.4M HW-17.12.83-3.2M, MLWS 0.4m)

Fig. 6. Rhu Marina - collapsed breakwater at low tide: 17 December 1983 (tidal range = 3m, MHWS 3.4M LW-17.12.83-1.2M, MLWS 0.4M)

- A Casagrande - An unsolved problem in embankment stability on soft ground across Great Salt Lake;

- Terzaghi & Peck - Soil mechanics in Eng Practice - 2nd Edit. 1967;

- Stamatopoulos & Kotzias - Construction and performance of an embankment in the Sea on soft Clay - 6th Int. Conf. Soil Mechs (1965) Montreal;

- W R Thorpe - Foundation Problems - Breakwaters Design and Construction proceedings - ICE 1983.

23. Wave load measurement on a prototype wave power device

G. U. MULLER and T. J. WHITTAKER, Queen's University of Belfast

SYNOPSIS. The analysis of the design calculations of the Islay shoreline wave power device showed that design wave pressures for such a structure vary widely depending on the theory used. In order to clarify the situation, a wave load monitoring system for the Islay device was developed and installed. The pressures recorded so far were one order of magnitude lower than the design pressure. Using a statistical analysis of the results, 1/50 year pressures were determined. A comparison of these pressures with design pressures showed poor agreement between extrapolated and calculated values.

INTRODUCTION

1. The first wave power station on the British Isles is now operational on the Island of Islay, off the West Coast of Scotland. The power station consists of two main parts, a concrete structure which contains the oscillating water column and the turbine/generator unit which is situated on three portal frames. The concrete structure resembles an upturned box with a submerged inlet below low water level and an air duct above high water level. The working principle is illustrated in Fig. 1.

2. The front wall of the device is directly exposed to incoming breaking waves. For the design, the breaking wave loads proved to be the dominant load case. The most unfavourable wave condition was evaluated as an 18 second wave of 5.3 m height. Wave loads for this design wave were calculated using five different theories. The pressures evaluated for the different theories varied widely. For the design of the front wall of the structure, the loads of 435 KN/m^2, derived from the ventilated shock theory which takes account of the wall inclination of 1:1.5, was used. This high load led to a span to depth ratio of 5:1 for the front wall.

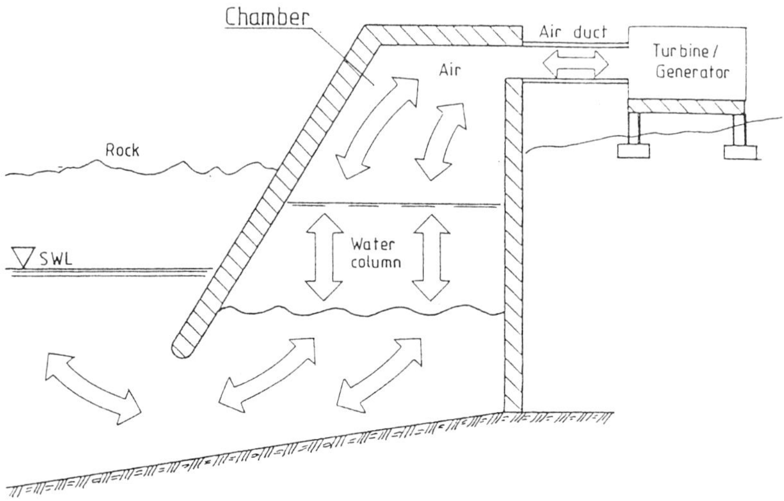

Fig. 1: Shoreline wave energy device

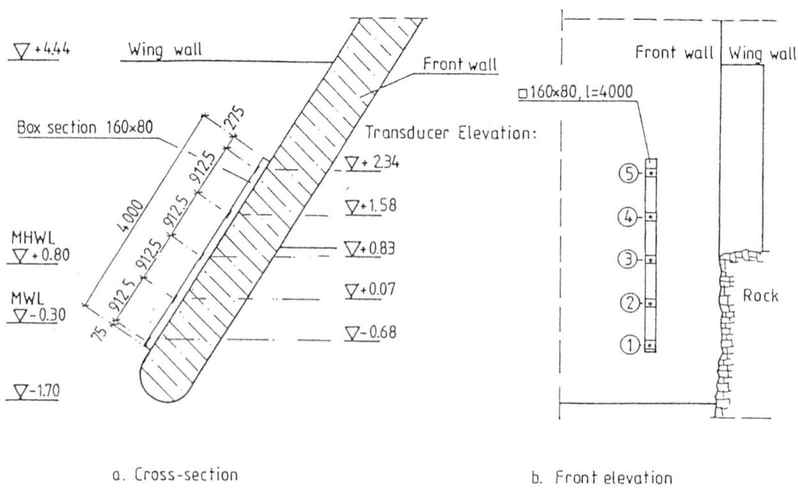

Fig. 2: Position of transducer module

3. The front wall of the Islay device was designed with an inclination of 1:1.5 in order to reduce the wave loadings. This resulted in a reduction of 7.5 % when compared with the BS 6349 (ref. 1) value for a vertical wall. From table 1 it can be seen that the breaking wave design loadings vary by a factor of 9, depending on the formula employed. Also, the effect of a wall inclination on the loads is unclear. It was thus decided to measure breaking wave pressures and to compare the results with the design pressures.

Table 1: Design wave pressures [KN/m^2]

Wall angle	vert.	1 : 1.5
BS 6349	470.0	-
SAINFLOU	51.0	-
C.E.R.C.	310.0	219.7
HIROI	75.0	-
Vent. shock	-	435.0

4. In the following, the development of the wave load monitoring system is described, first results are given and analyzed. The statistical analysis of the measured pressures and the design pressures calculated for different formulae are compared. The influence of the wall angle on the magnitude of wave loadings is discussed briefly.

BREAKING WAVE PRESSURES

5. As the first step within the project, a literature survey was conducted in order to establish the design criteria for the wave load monitoring system. It was found that the pressures exerted by breaking waves consist of a high peak pressure, with a rising time of 0.1 seconds or less, followed by a longer hydrodynamic pressure. In spite of the importance of breaking wave loadings for coastal structures, only thirteen monitoring programs have been reported in the literature since 1843 (ref. 2). The highest pressure recorded in a field measurement was 690 KN/m^2, measured at the Dieppe sea wall (ref. 3).It was reported that only a small percentage of 0,04 - 4.0 % of all waves created impact pressures. From the literature survey it was concluded that wave impact pressures may occur, these pressures may exceed 500 KN/m^2, these loads may last for a fraction of a second only and that the probability of occurrence may be very small.

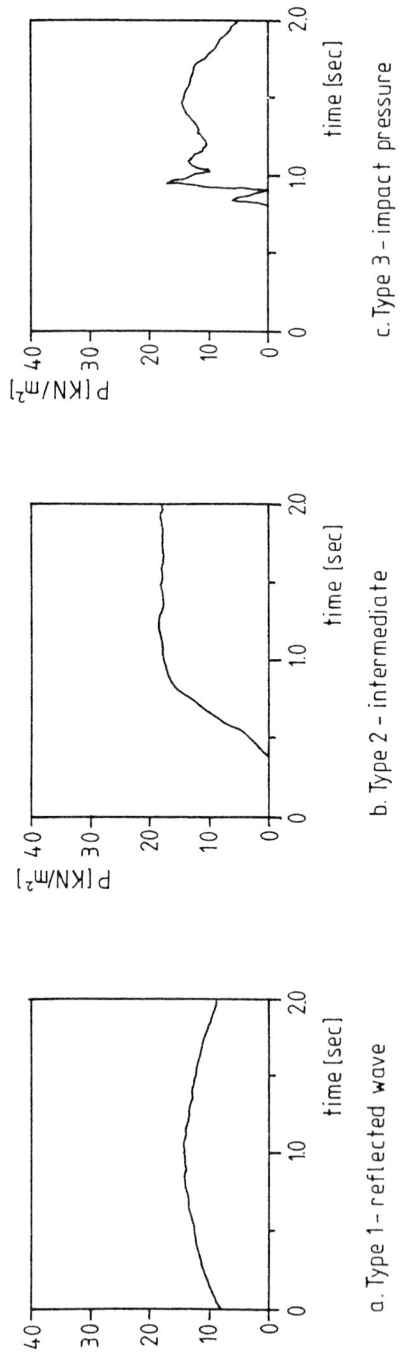

Fig. 3: Pressure /time traces

DESCRIPTION OF WAVE LOAD MONITORING SYSTEM

6. The wave pressure monitoring system comprises two sets of five pressure transducers. Each set is mounted in a 160 mm * 80 mm galvanized steel box section which is bolted to the front wall as shown in Fig. 2. The pressure transducers used are DRUCK AB transducers with a maximum capacity of 690 KN/m^2. The assembly was mounted on a very calm day in July 1990. A data acquisition system was already available on site to monitor the performance of the device. It consists of a Eurobeeb A/D converter and a data link cable to a Compaq 386-20 computer in the site office 350 m away. This computer is used for data preprocessing and storage.

7. The pressure transducers are continuously scanned by the on-site A/D converter with a scanning frequency of 100 Hz using a machine code program. The frequency was chosen initially and proved sufficient. Readings are stored in a 4 k block of memory which is continuously overwritten and corresponds to 2 seconds of wave data. Between the A/D board and the transducers there is an electronic circuit which constantly compares the analogue data with a preset value. Once this value is exceeded at any one transducer, a signal is given to the Eurobeeb to continue scanning for one second, then send the complete two seconds of data which are in memory up the data link, and presume scanning. The data block sent up to the master computer thus consists of the readings of each pressure transducer taken one second before the limit value was exceeded and one second after. This gives a complete picture of the wave impact event.

8. This system allows continuous scanning, stores only the events which are of interest and uses the time between two wave crests to transfer the data up to the main computer, where it is stored and processed. The scanning system thus extracts the important bit out of a massive amount of data created.

RESULTS

Pressure / time traces

9. Due to the unusually calm winter only two storms, from 17-19th January and on 26th March, were monitored with the system. For both storms, the trigger value was set at 13.8 KN/m^2. Transducer No 4 was damaged by lightning early in January 1991 and did not yield any results. The wave period was measured by hand at regular intervals and remained approximately constant at 12 seconds for both storms. The maximum wave height measured on the 18th January was 4.0 m.

10. From the pressure / time traces three different types of wave impact could be identified. Type 1 corresponds to a reflected wave condition where only hydrodynamic pressures are acting. The rise and fall of the wave crest can easily be seen in Fig. 3a. With rising times of 1 second and more this is a

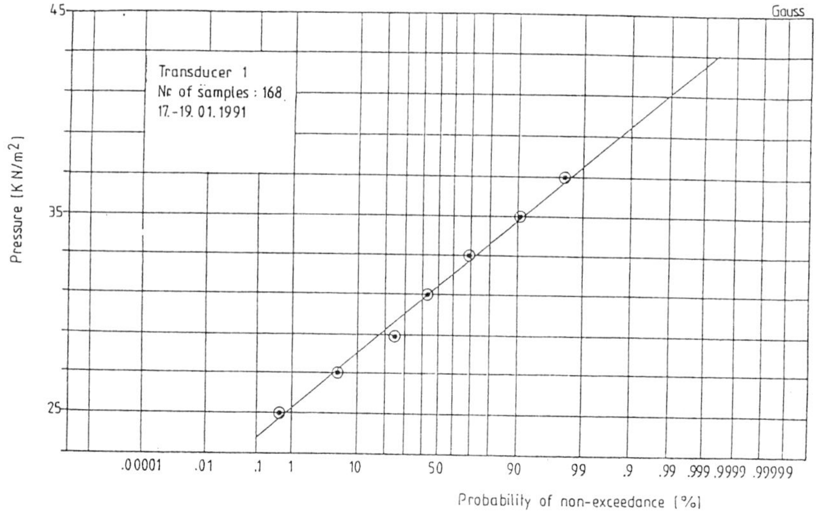

Fig. 4 : Probability plot 17.-19. 01.

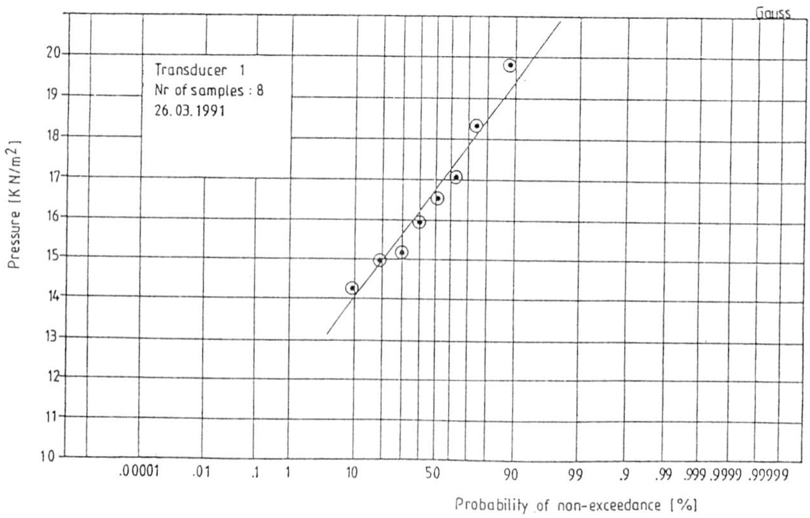

Fig. 5: Probability plot 26. 03.

nearly static pressure. Fig. 3b shows type 2, which is characterised by a short rising time of 0.25-0.35 seconds from zero to maximum and a near constant pressure for 1 second or more afterwards. The maximum pressures recorded so far showed these characteristics. Type 3 finally, as shown in Fig. 3c is a typical wave impact pressure with a rising time of 0.1 seconds, a peak pressure and a following slightly smaller hydrodynamic pressure. In magnitude, these impact pressures were smaller than the type 1 and 2 pressures. The occurrence probability for shock impact pressures was calculated as 1/2280 waves.

Statistical analysis

11. Design pressures by definition represent pressures of a very low probability of non-exceedance. The storms monitored so far cover only a tiny fraction of the expected lifetime of the structure. It was thus decided to analyze the data recorded so far by statistical means. A grouped representation of the results obtained so far is shown in Figs. 4-5.

12. Assuming storms to cover 4 out of 12 months and a return period of 1/50 years, a probability of non-exceedance of 99.9999906 % was determined. As only extreme values are of interest, a type 1 extreme value distribution was employed for the analysis (ref. 4). Table 2 shows the highest measured pressures and the extrapolated 1/50 year pressures. Using the wave conditions as observed during the storm on 17-19th January, the maximum design wave pressures were calculated for these conditions and five different formulae. Table 3 shows these maximum pressures and the ratio of design pressure to extrapolated highest 1/50 year pressure.

Table 2: Maximum and 1/50 year pressures [KN/m^2]

Transducer	P_{max}	$P_{1/50}$
1	37.4	65.6
2	33.4	56.7
3	27.8	46.7
4	-	-
5	23.5	42.0

Table 3: Design pressures for observed conditions

Formula	P [KN/m^2]	P/P_{islay}
C.E.R.C.	208.2	3.17
BS 6349	265.8	4.05
GODA	42.0	0.64
Vent.shock	365.0	5.56
Broken wave	59.7	0.91

13. It can be seen that none of the existing formulae give a satisfactory agreement for the given wave conditions. Only the values for the broken wave pressure formulae, which is given in C.E.R.C. (ref. 5), lie within the range of the extrapolated pressures. This formulae can however only be applied if broken waves alone can reach the structure, something that needs to be assured by model tests. Note that the two formulae that take the

wall inclination into account, ie the C.E.R.C. and the ventilated shock formula lead to a reduction of 23 % (C.E.R.C.) or an increase of 24 % (Ventilated shock compared with C.E.R.C.) in wave pressures. Model tests conducted at Queen's University indicated that for a wall inclination of 1:1.5 a reduction in wave loadings of 56 % can be expected. The disagreement between design formulae and 1/50 year pressures can be attributed to two different factors :

a. The seabed slope on Islay is rather shallow with an inclination of 1/30. This leads to a premature breaking of the waves and a consequent loss in energy.
b. Only one formula, the one given in BS 6349 (ref. 1), takes the effect of the local geometry, ie seabed slope, seabed roughness and the arising reduction in wave pressure into account. The empirical factor given in BS 6349 ranges from 0.3 for rocky coastlines to 0.5 for smooth seabeds. The Islay site however is characterised by a very rocky seabed, a shallow slope and surfacing rocks.

CONCLUSIONS

14. A wave load monitoring system for the Islay wave energy demonstration unit was developed and installed. The highest pressure measured so far was 37.4 KN/m^2. The measured pressures were analyzed and a 1/50 year pressure of 65.6 KN/m^2 determined. A comparison with different design pressures for the given situation led to ratios of design pressures to extrapolated pressures of 0.64 to 5.56. None of the recommended formulae showed satisfactory agreement with the extrapolated values.

REFERENCES

1. BS 6349, 1984, British standard code of practice for marine structures, Part 1, 4th am., British Standard Institution, London
2. BLACKMORE P.A., HEWSON P.J., 1984, Experiments on full scale wave impact pressures, Coastal Eng., Vol. 8, 331-346
3. ROUVILLE M. A. et al., 1938, Etudes internationales sur les efforts dus aux lames, Vol. 108, Annales des Ponts et Chaussees, 5-113
4. BURY K.V., 1975, Statistical models in applied science, J.Wiley & Sons, New York
5. C.E.R.C., 1984, Shore protection manual, 4th ed., US Army Waterways Exp. St., US Govt. Printing Office, Washington D.C.

Discussion

L. SADLER

My comments are based on the experience gathered in Israel, during the last ten years, with rock armoured breakwaters.

Rock density is defined, as recommended by Simm et al., as the saturated dry surface bulk density; in general, 2.65 t/m^3 with a maximum allowable water absorption of 2%. The required size distribution for the armour rock is such that at least 50% of the units but not more than 70% shall weigh the mean or more of the class. This requirement is checked by weighing or measuring the three main dimensions of each individual rock.

The recommended aspect ratio may not be practical to implement for big units (over 10 tonnes) especially taking into consideration the fact that measurements will be generally carried out on lorries, where rocks are placed close to each other.

Simm et al. also refer to some obscurity concerning the layer coefficient K_t. Based on mock-up determination carried out some 12 years ago and the checking of as-built sections, we use a value of 1.1 for prismatic to cubical rocks.

I do not agree with the authors' recommendation about the ball diameter for surveying rock layers. Since the ball has a general tendency to penetrate into the voids between the rocks, the value of 0.5 x D_{n50} will cause unrealistically low levels. We have successfully used a ball diameter of 0.8-0.9 x D_{n50}. measurements so made compared very well to those carried out by topographic levelling techniques, obviously along the above-water sections.

Finally, I would like to call the authors' attention to the fact that quality control and the method of measurement and payment are different items to be covered in properly prepared specifications.

P. BARBER, Shoreline Management Partnership

I wish to pay tribute to Simm et al. for their success in producing the new manual. I would, however, also wish to raise two matters of concern to me: the use of standard rock grading; and the presentation of a new set of tests for armourstone.

With regard to the second matter I wish to enquire whether it is proposed to research links between the new tests and those established for aggregate assessment. There will be considerable costs involved if all test laboratories have to invest in new equipment; this may be unnecessary if correlations can be found.

The standing of CIRIA/CUR is such that the new manual may become an effective standard referred to in specifications. I am therefore concerned about the use of standard grading envelopes for rock armour. Whereas other rock properties are dealt with by definition and matters to consider, this aspect has been taken much further. I have been continually broadening the gradings of armourstone used in shallow-water breakwaters, and I consider that this trend will continue. Standard rock grading may therefore be outdated quickly, and may also lead to approximation in the design approach. If quarries need guidance for advance stockpiling of armour then smaller bands would be preferable for blending to a specific design.

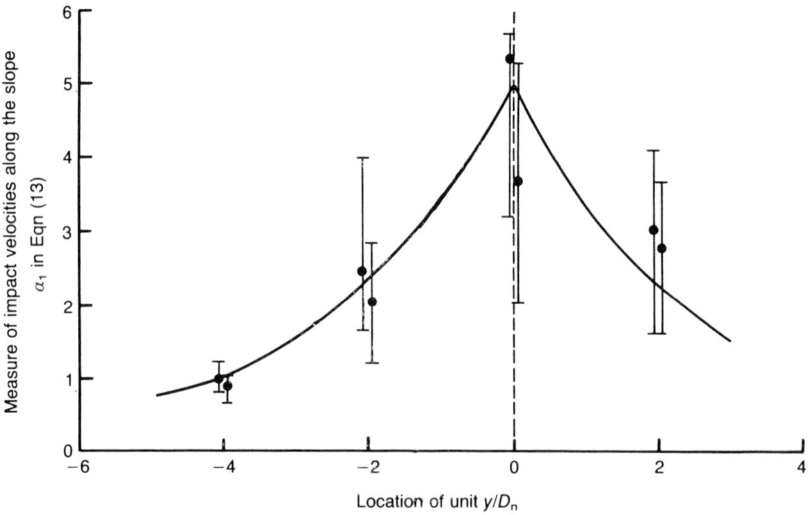

Fig. 1. Impact velocities in equation (13) as a function of the location on the slope: y = distance from SWL, D_n = nominal diameter

H. F. BURCHARTH, Aalborg University
I support the conclusion of Mr Herbert and Mr Waldron about the importance of fatigue with respect to cracking. **The area where most cracks are observed is** also the area where freeze-thaw is most serious. Consequently, this could also be a significant cause of cracking. Do the authors' fatigue curves represent new results, and how do they compare with impact fatigue curves presented by Burcharth over the last 10 years?

J. W. VAN DER MEER, Delft Hydraulics
Delft Hydraulics measured impact velocities of cubes and tetrapods on a slope 1 : 1.5. We found a similar distribution of impacts as shown in Fig. 1 of Paper 17. The highest impacts occur at still water level and it is more or less symmetrical around SWL.

Reference:
VAN DER MEER J. W. and HEYDRA G. J. Coastal Engng, 1991, 15, Nos 1, 2.

J. A. DUNSTER
Concerning Paper 17, Section 5, dramatic features of breakwaters when standard armour units break up apply specifically to random armour units and do not occur in the case of single armour units like the SHED.

I disagree with the statement in Section 7 that cracking of units occurs mainly in the tidal zone subject to maximum wave action. Inspection of the underwater units at Pygros shows failure below the tidal zone.

The main cause of cracking of 'SHED' units is the lower sections of the embankment, and there is staggered bonding of unit carrying point loads on the weakest sections..

The basic 'SHED' unit is inherently very strong structurally without reinforcement end. Cracking can be completely eliminated by flumes in vertical columns without brick-type bonding. This can also be achieved around round heads and curves by the use of wedge-shaped spacements glued to standard blocks. However, I do not consider cracking a problem with 'SHED' units as it can occur without affecting the structural integrity of the armour layer. Cracks mainly occur due to the bonding matter and can result in tripod units being forwarded which are still restrained within the armour layer.

Units rapidly lock together due to wave action and embankment settlement. This occurs to such an extent that even embedments are impossible to extract without demolition of the unit.

Polypropylene fibres have been used in 'SHEDS' to minimize impact resistance, but I do not consider that this justifies the increased cost, and I would recommend the use of very high strength concrete using fly ash or blast furnace cement replacement of Portland cement to produce very dense durable concrete resistance to sea water and abrasion.

C. T. BROWN, Seabee Developments
The research of Herbert and Waldron into the structural response of hollow cube armour units is of interest. It is to be regretted that the authors continue to describe their research association as the Single Layer Armour Club, as this is patently untrue - there is no discussion of Accropodes, Seabees, Tribars or Placed Rock. The 'Hollow-cube Research Club' sounds just as well and does not needlessly antagonize the providers of other single layer systems.

It would be useful if the authors would compare the limit of structural stability (resistance to cracking) with that of hydraulic uplift (reported by Stickland 1969).

Have the authors examined the case of a unit uplifted by one wave and slammed back into the slope by a subsequent wave? What increase in impact stresses would they expect?

Have the authors considered the case of a unit supported on only two rocks beneath and laterally supported by the adjacent units?

Mr Dunster has commented that individual cracks do not compromise the armour layer stability due to the redundant structure of the hollow cube. He added that a major source of cracking was when the unit were placed in a bonded pattern, when large shear stresses could be induced, endorsing the points made by Brown in discussing the paper by Dunster et al. at the Eastbourne Conference in 1988.

I would like to draw Mr Dunster's attention to the comment made by both Stickland and Webb in testing hollow cubes, that when placed in columns, if one unit is extracted the whole column above falls down and is prone to follow in out, leading to rapid failure.

This post-damage response has been addressed in the design of both Seabees and Accropodes, with quite extensive testing, at suitable scales to ensure failure.

Can the authors shed any light on the structural and whole-block failure modes and actual failure-mode testing of the units and their assembled structures?

A. BEZUIJEN, Delft Geotechnics
In Alderney the original rubble mound is filled with sand, resulting in a mound with a lower permeability and a lower stability. Could Allsop et al. say if this is very typical for Alderney or should this possibility be incorporated in the general design procedures for

breakwaters?

In the Netherlands we have found that for our placed-block revetments it is common that the filter layer underneath the blocks is filled with sand from the foreshore due to wave action.

J. W. VAN DER MEER, Delft Hydraulics
Could Allsop et al. give more details on the final solution for the breakwater slope angle stone size, armour type, etc.?

J. N. WOLFE-BARRY, Sir William Halcrow & Partners Ltd
In view of the exposure of the present harbour to the north-east, and in view of the apparent scale and cost of the solution worked out, could Allsop et al. say if the total economics of the problem have been looked into? It seems that it might be more beneficial overall to abandon the outer length of the breakwater and revert to a scheme approximating to that originally proposed by the Admiralty. This would provide a much smaller harbour, but it would be properly protected in all weathers.

If the failure mode described is taking place in the rubble mound, one would expect to find isolated large rocks lying on the sea bed in front of the breakwater. Is this in fact the case?

J. W. VAN DER MEER, Delft Hydraulics
In Fig. 2 of Paper 19 is given the distribution of H_s and T_p. Have Losada et al. looked at a relationship between H_s and T_p for extreme events (say higher than one event per year?).

I suggest that for extreme events (storms) the wave condition is described by a constant steepness ($s_{op} = 2\pi H_s/gT_p^2$). This means that H_s is proportional to T_p^2.

H. LIGTERINGEN, Frederic R. Harris BV
Could de Gerloni et al. explain in more detail how the effects of overtopping water on the human being were measured in the full-scale tests and translated to the model scale? Were velocities measured in the sheet of overtopping water, which would allow estimation of the forces on a body using some kind of drag coefficient?

J. W. VAN DER MEER, Delft Hydraulics
de Gerloni et al. gave a new and very good contribution to coastal engineering research. Also new, and working very well, was the way they measured the volume of overtopping waves individually. They gave a relationship between the mean overtopping discharge and the volume of the maximum wave. Could they also give results on the distribution of the volumes of individual waves?

Table 1. Extracts from recent specifications (post 1985)

		PRIME SPECIFICATION BY WEIGHT	BY SIZE	MIN	FURTHER SPECIFICATION	SIZE RATIO	COMMENTS
1	(a)	20.0 - 25.0t	(D_{50} = 2.11)	2.4	) Individually weighed and marked	$\frac{D_{100}}{D_O}$ = 1.08	Very tight grading on top weight range) These sizes are very difficult to procure - re-design on high density?
	(b)	10.0 - 20.0t	(D_{50} = 1.84)	2.4		$\frac{D_{100}}{D_O}$ = 1.26	
	(c)	5.0 - 10.0t	(D_{50} = 1.46)	2.4		$\frac{D_{100}}{D_O}$ = 1.26	
	(d)	3.0 - 5.0t	(D_{50} = 1.18)	2.4	) Sized against "target" rocks which themselves have to be weighed and selected to 10% accuracy	$\frac{D_{100}}{D_O}$ = 1.19	Range in D_n = 0.2m (15% of D_{100})
	(e)	1.0 - 3.0t	(D_{50} = 0.94)	2.4		$\frac{D_{100}}{D_O}$ = 1.44	Range in D_n = 0.33m (31% of D_{100})
	(f)	0.5 - 1.0t	(D_{50} = 0.68)	2.4		$\frac{D_{100}}{D_0}$ = 1.27	Range in D_n = 0.16m (21% of D_{100})
2	(a) (b) (c)	≥ 15.0t ≥ 10.0t 0.1 - 1.0t			) Specified geological types, with min. densities 2.25 and 2.70		) No specific rock quality requirements. This would have allowed use of a material from the 2.25 min. density series with an Na SO_4 soundness of 90, as opposed to 16 max. for the rest of the series.
3	(a)	5.0 - 10.0t	(D_{50} = 1.45)	2.6	W_{85} = 9.0t ± 0.5t W_{50} = 8.0t + 0.5, - 0.1t W_{15} = 6.5t ± 0.5t	$\frac{D_{85}}{D_{15}}$ = 1.11	
	(b)	Bedding	D_{85} = 0.35 D_{50} = 0.25 D_{15} = 0.20	2.6		$\frac{D_{85}}{D_{15}}$ = 1.75	$\frac{D_{50}\ \text{Armour}}{D_{50}\ \text{Bedding}}$ = 7.25 No max. or min. given for bed
4	(a)	8.0 - 10.0t	(D_{50} = 1.51)	2.6	W_{50} = 9.0t, 8-9t range ≤50%	$\frac{D_{85}}{D_{15}}$ = 1.08	50% by weight or number? Very tight weight range.
	(b)	Core 1.0 - 2.0t	(D_{50} = 0.83)	2.6	W_{50} = 1.5t, 1-1.5t Range ≤50%	$\frac{D_{85}}{D_{15}}$ = 1.25	50% by weight or number?

5	(a)	W_{85} = 9.8t W_{50} = 4.5t W_{15} = 1.8t	(D_{50} = 1.2)	2.6	Max 16.2t, min 1.0t, equivalent sieve sizes given for $W_{85, 50, 15}$ working out at 80% x corresponding D_n.100t sample to confirm ± 10%	$\frac{D_{85}}{D_{15}}$ = 1.76	W_{85} etc. written as M_{85} etc. Some sieve!
	(b)	Bedstone (W_{50} = 0.04t)	D_{85} = 0.35 D_{50} = 0.25 D_{15} = 0.20	2.6		$\frac{D_{85}}{D_{15}}$ = 1.75	$\frac{D_{50}\text{ Armour}}{D_{50}\text{ Bedstone}}$ = 4.8. No max. or min. sizes for bedstone
6		4.0 - 8.0t	(D_{50} = 1.32)	2.6	60% of rocks > 6t	$\frac{D_{100}}{D_O}$ = 1.26	60% by number?
7	(a)	Nom 5.5t	(D_{50} = 1.27)	2.69	Max. 6.9t, min. 4.2t, 70% by weight > 5.5t	$\frac{D_{100}}{D_O}$ = 1.16	Very precise requirements, no guidance as to how they are to be achieved.
	(b)	Nom 4.5t	(D_{50} = 1.19)	2.69	Max. 5.5t, Min. 3.25t, 70% by weight > 4.5t	$\frac{D_{100}}{D_O}$ = 1.19	Ditto
	(c)	Bedding Stone Nom 0.7t	(D_{50} = 0.64)	2.69	Max. 1.0t, Min. 0.5t, 70% by weight > 0.7t	$\frac{D_{100}}{D_O}$ = 1.26	$\frac{D_{50}\text{ Armour}}{D_{50}\text{ Bedding}}$ = 2.06 & 1.86
8	(a)	Nom 5.0t	(D_{50} = 1.22)	2.75	Max. 6.5t, min. 4.0t, 50% by weight ≥ 5.0t	$\frac{D_{100}}{D_O}$ = 1.18	Also a requirement for 6.5t rocks to be used in specific parts of the structure
	(b)	Underlayer 400 - 650 kg	(D_{50} = 0.57)	2.75	50% by weight ≥ 500kg	$\frac{D_{100}}{D_O}$ = 1.18	$\frac{D_{50}\text{ Armour}}{D_{50}\text{ Underlayer}}$ 2.14. No max. or min. requirements
9	(a)	W_{85} = 6.2t W_{50} = 3.6t W_{15} = 1.5t	D_{85} = 1.32 D_{50} = 1.05 D_{15} = 0.82	2.6	Min. 0.8t, 100t sample to conform ± 10%	$\frac{D_{85}}{D_{15}}$ = 1.61	D_{50} and W_{50} do not correspond - they should be either 3.6t and 1.11 or 3.0t and 1.05. No maximum size given.
	(b)	Bedstone (W_{50} = 0.04t)	D_{85} = 0.40 D_{50} = 0.25 D_{15} = 0.20	2.6	D_{85} etc. are "Seive Sizes"	$\frac{D_{85}}{D_{15}}$ = 2.0	No max. or min. given. This is the only size ratio which conforms with normal aggregate production practice.
10	(a)	(W_{85} = 3.7t) (W_{50} = 2.35t) (W_{15} = 1.0t)	D_{85} = 1.4 D_{50} = 1.2 D_{15} = 0.9	2.6	W_n - D_n relationship specified by the diameter of the sphere of equivalent volume to the individual rocks.	$\frac{D_{85}}{D_{15}}$ = 1.56	If the more usual interpretation of D_n were adopted, this would give W_{85} = 7.1t, W_{50} = 4.5t, W_{15} = 1.9t. No max. or min.
	(b)	Underlayer (W_{50} = 0.17t)	D_{85} = 0.47 D_{50} = 0.40 D_{15} = 0.17	2.6		$\frac{D_{85}}{D_{15}}$ = 2.76	$\frac{D_{50}\text{ Armour}}{D_{50}\text{ Underlayer}}$ = 3.0. Other notes as (a)

J. JUHL, Danish Hydraulic Institute
Some years ago model tests were carried out at the Danish Hydraulic Institute to measure the amount of water in individual wave overtoppings of a rubble mound breakwater. A distribution of the individual wave overtoppings showed that the largest wave overtopping accounted for a significant part of the total amount of wave overtopping.

Have or will the authors of Paper 20 analyse the measurements in order to represent the results as distributions of individual wave overtoppings? Another important aspect to be studied is the influence of wind on the results.

P. F. TATHAM, Binnie & Partners
The wave pressure formulae used by Whittaker and Muller may not be applicable to the situation of the wave power device in the rock gully at Islay. I note that the wave pressure transducers were very close to the rough rock wall. What was the planned installed power of the generator, the actual installed power, and the measured average net power output?

Y. GODA, Yokohama National University
I congratulate the successful field experiments for the development of wave power exploitation. However, it seems too early to make the conclusion regarding the applicability of wave pressure formulae, because of the short duration of measurement for the derivation of extreme values with a return period of 50 years.

J. D. SIMM, Paper 16
Paper 16 examines problems with current specification of quarried rock for use in coastal structures and proposes a unified approach to resolve these. Rock properties are viewed as splitting into three chronological categories, the properties in each depending on those in the previous category (Fig. 2).

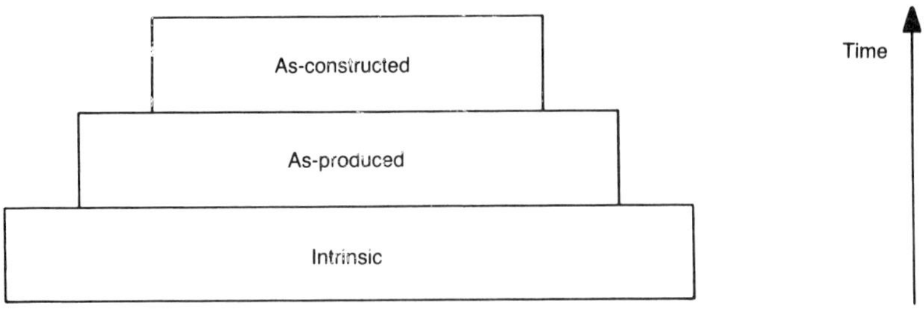

Fig. 2. Rock property category chronological building blocks

Placing pattern		W_{50}: t	Layer thickness: m	Number of stones: 100m^2
1	As specified	4·5 — specified	2·69	110
2	As specified	3·30	2·40 — specified	130
3	As specified	6·10	3·09	88 — specified
4	'To suit'	4·88	2·40 — specified	88 — specified
5	'To suit'	4·5 — specified	2·21	88 — specified
6	'To suit'	4·5 — specified	2·40 — specified	95

Table 2. Six feasible combinations of as-constructed properties of an armour layer

Table 1 presents the essence of 10 recent UK specifications for intrinsic and as-produced properties similar to the recent market reviews recently prepared by at least three major UK companies. The review notes that

(a) specified intrinsic properties appear not to be related to industry standards, and neglect the possibility of the use of high-density materials

(b) specified as-produced properties, particularly size, shape and grading width are highly variable, and the detailed wording belies much copying of old specifications and arbitrary and unexplained changes; excessively narrow grading widths ignore both quarry industry practice and the results of recent hydraulic model tests, and are often based on the 'special case' experience of the use of dedicated quarries for large breakwaters - the specifications reflect a lack of central (national/European/international) guidance.

The quoted specifications are workable but add costs due to: sorting of stockpiles (at least 1.00 per tonne), under-utilization of placing plant on site (typically 4000/day including supervision overheads and profit) due to slow material supply to site, additional haulage (by exclusion of otherwise suitable local quarries) and contractors' claims.

Claims also arise because of wide diversity of placing specifications and techniques of measurement for payment. The physics of packing density and layer thickness (Paper 16, para. 14) mean that it is not possible to constrain simultaneously more than a few properties of an armour layer. One recent 2M claim of a major UK contractor revolved around a specification in which four properties were simultaneously specified (without the benefit of a trial area of the works), only two of which could be physically satisfied (see Table 2).

The method set out in Paper 16 to resolve these problems is based on logical building blocks in the

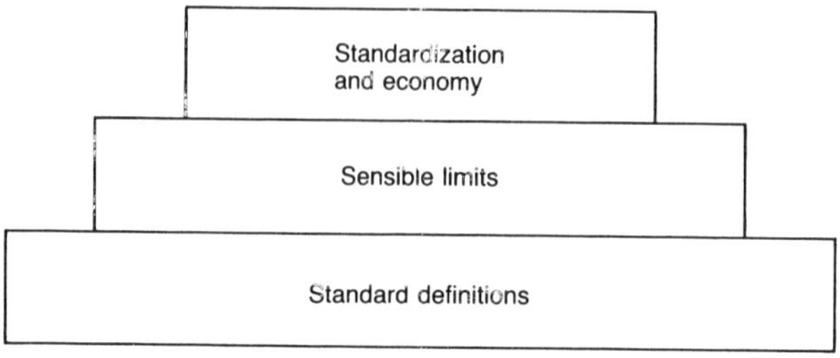

Fig. 3. Specification procedure building blocks

specification procedure (Fig. 3), and it is hoped this logical approach will be widely adopted. The proposed standard test methods and gradings are now feeding into the work of a Group of Experts set up early in 1991 by CEN technical committee TC154 (whose overall result is European standards for aggregates) specifically to develop a European standard for stone in hydraulic structures. The work of the group also covers channel bank, dam face and offshore rockfill protection structures, and by-product materials such as slags and demolition waste.

The Authors of Paper 16 can advise that at the time of writing the Group of Exports has prepared a draft standard with test methods to be considered in further detail by task groups. Many test methods currently in use are of doubtful relevance, and it has proved necessary to introduce some tests that require larger-sized pieces than those typical of aggregates. The addition of seven further European countries to Britain and the Netherlands in the search for a Eurostandard has widened the scope of possible tests and systems to be considered. However, the need for rapid compromise agreements using only well-established tests may constrain the approach and content of the final standards.

Mr Stadler discusses an alternative approach to the specification of rock grading, and presumably is looking for a further explanation of the grading system proposals, a full account of which can be found in the Manual (reference 1 in Paper 16).

It is possible to adopt Mr Stadler's approach of controlling the middle sizes of a grading by setting a requirement for the minimum W_{50} and maximum W_{30} (30% by weight lighter than) to be equal to the mean value of a given class range. The problem with this specification is that it is very severe on the selector, implying that very narrow or possibly unevenly graded material must be produced. The normal process of eye selection based on rejecting undersize and oversize material with feedback from weighings leads to a natural distribution of acceptable material unless the upper sizes are very poorly represented. The alternative approach suggested by Mr Stadler requires uneconomic selection procedures in order to bias the weight distribution away from its natural tendency to that he wishes to specify. Other requirements would also be necessary for controlling the upper and lower tails of the weight distribution.

The CIRIA/CUR system presented in Paper 16, by contrast, has been carefully designed to be optimal for functionality and producibility. The requirements are less severe on the producer and incorporate controls on the actual mean weight as a means of constraining the middle sizes. This allows the bulk weighing and counting method to be put into effective quality control. Note that because the standard gradings given in the Manual are designed to become wider as the weight classes become smaller, thus complying with producibility and function, there are no simple formulae to fix the middle range to the overall grading glass for all classes. A range of acceptable mean weights has therefore been worked out empirically from produced stockpiles.

Mr Stadler's suggestion of the measurement of block dimensions to obtain estimates of block weights could be extremely valuable for checking during quality control procedures. However, the volume correction factors incorporated must be checked by actual weighings, individually or in bulk.

We acknowledge Mr Stadler's point that measurements of the proposed l/d aspect ratio for big blocks (over 10 tonnes) is not easy; it is still possible, with assistance, to project the l and d dimensions with sufficient accuracy by carrying out these checks at the quarry (as the Manual recommends) and not on transport. However, the reduced ambiguity surrounding the definitions of l and d compared with X, Y and Z and the ability to use it for all rock types is the main consideration.

We note Mr Stadler's comments in regard to the layer coefficient, but the simple point to be made is that k_t depends on many factors including the survey method used to define the profile and the shape of the rock (for the definition of shape terms see para. 18 of Paper 16). A new, but clear, definition compatible with flume test survey methods upon which stability formulae are based, has now been proposed. The k_t values are therefore almost certainly in need of review.

In regard to the ball diameter for surveying rock layers, we do not agree with Mr Stadler's suggestion of using a diameter approaching D_{n50}. The agreement between his surveying techniques is unsurprising since both effectively pick up the highest points or surfaces of rocks in the outer layer. Our argument for a 0.5 x D_{n50} ball diameter is in order to give a more realistic mean outer profile based on a diameter which has been used successfully in practice and is compatible with procedures in hydraulic laboratories on which armour layer design is based. It may also be of interest to note that the CIRIA/CUR Manual is also recommending a maximum void size of D_{n35}, which will ensure in practice that a $0.5D_{n50}$ sphere will not penetrate the armour layer excessively.

We agree with Mr Stadler's final comment and would point out that the separate but related topics of quality control on site and method of measurement for payment are fully addressed in the CIRIA/CUR Manual.

In response to Dr Barber's query in regard to links between new tests and established aggregate tests, these are indeed likely to be the subject of future research, and some work has already been undertaken in Holland (Rijkswaterstaat Delft) and at Queen Mary and Westfield College. Further research proposals have been submitted. Little, if any, work relating intact rock properties to aggregate properties has been published. It is perhaps important to stress that the very scale of Armourstone is a factor which makes strength scale effects both important and also difficult to research, not to mention the problem of timescale in evaluating durability.

The move away from aggregate tests has not been taken lightly and it has been discussed again within the forum of European experts mentioned in our introduction. In fact, the requirements for Armourstone have perhaps more in common with those of dimension stone than those of aggregates. Because one has identified that mechanisms of breakage and deterioration of rock armour are extremely dependent on the weakening effect of cracks and floors and geological defects as well as the mineral fabric strength and constituents, the tests now being recommended will allow discrimination against rock with a preponderance of cracks and flaws at a scale often not noticed in typical aggregate fragments. In some cases of homogeneous rock, material that has been crushed due to aggregate sizes, whether by quarry plant or in a laboratory, will yield test results that will relate to performance meaningfully, but there will be many rock sources that do not.

Preliminary findings suggest that correlations between AIV type aggregate tests and intact strength tests such as uniaxial compressive strength or point load strength, typically carried out on 50 mm diameter specimens, is not very good. To determine the intact strength of the mineral fabric, which indicates the resistance of breakage along new fractures and thus the

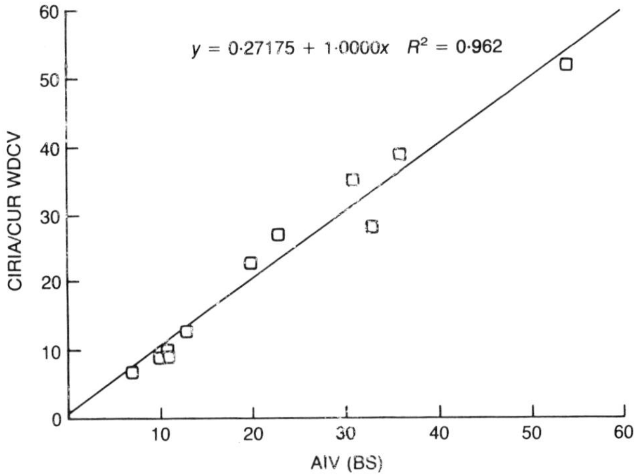

Fig. 4. Correlation of BS 812 AIV with CIRIA/CUR wet dynamic crushing value

removal of corners, edges and asperities, the most fundamental measure is that obtained by the fracture toughness test, and greater experience in the use of this test is being gained by certain appreciative groups. The point load strength index is a very simple measure and the failure mode is also highly relevant. Because of the popularity and simplicity of the AIV type test, the Manual has sought to retain this test, but in a revised form which includes two improvements: one is to use a non-flaky sample which lowers the values, and the other is to use wet material which increases the values because of the water weakening effect on certain rock types. Fig. 4 shows how the new WDCV in the Manual gives an overall correlation suggesting AIV and WDCV have interchangeable values. However, the WDCV is undoubtedly a better test and conveys results which relate more closely to performance. The methods whereby intact rock is reduced to aggregates can vary, and these can influence aggregate strengths in a manner which has no bearing on how the larger intact material behaves as Armourstone or even filterstone. In fact, a significant problem with the AIV type aggregate tests is that with coarse grain igneous rocks, the most excellent and proven Armourstone can yield AIV values in the high 20s and low 30s. Knowing the various test method deficiencies, a compromise selection of tests less familiar to the quarry industry have been proposed.

Perhaps the most difficult property to evaluate by testing is that of block integrity or 'wholeness', which indicates the resistance to breakage along partial or traversing flows; the drop test breakage index has been introduced. The reduction in mean weight after a number of block drops is a direct measure of integrity. The indirect approach is to use sonic velocity measurements. In the Manual, preference has been given to the drop test method, although the European Expert Group is seeking European funding to undertake further research into both of these two possible objective methods of block integrity measurement.

A new test, the Mill Abrasion Test developed at QMW, is one of the best ways of measuring wear resistance. Fig. 5 shows a graph of the BS 812 aggregate abrasion value correlated with the new test index K_S. It is not surprising that the correlation is rather poor. The new test method has certain similarities with the west Deval test (by Afnor); however, it is more discriminating of test apparatus and procedural improvements. Full correlation studies will be published as soon as time permits.

Weathering resistance (e.g. freeze-thaw, soundness), water absorption and density tests in the Manual have moved towards larger test pieces for the reasons already outlined, fundamentally to ensure that pieces are more representative without being prohibitively large.

Many of the new or modified tests can be carried out using existing laboratory equipment. However, in view

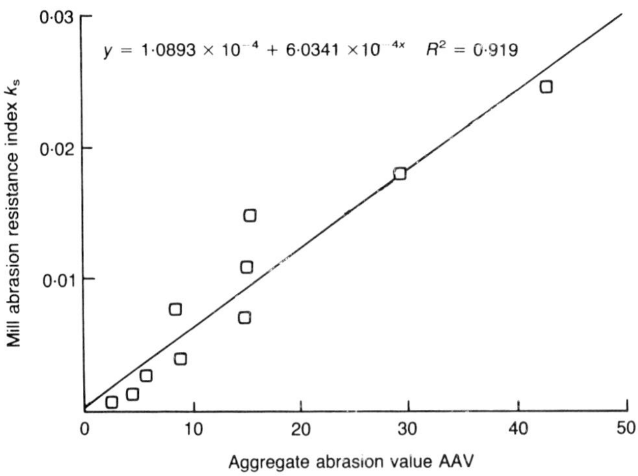

Fig. 5. Correlation of BS 812 AIV with CIRIA/CUR mill abrasion resitance index

of the importance of the tests being representative, we feel that where new test equipment is required, its cost (referred to by Dr Barber in his question) is a relatively small price to pay for improved confidence in material performance. As quarry producers come to understand the objectives of the new tests, they are increasingly sharing this view.

Dr Barber addresses the related questions of width of armourstone gradings, standardization and stockpiling. We agree with Dr Barber that armourstone gradings need to be widened in comparison with those still adopted by many engineers and refer to our discussion of this point in paras 11 and 25 of Paper 16. However, the standard gradings proposed in the Manual and listed in Table 3 of Paper 16 reflect this trend and, for example, the 1 - 3 tonne grading is now representative of armourstone material produced in many small North Wales quarries. However, because armourstone materials, defined as 'heavy' gradings in the Manual and our paper, are those which are handled piece by piece rather than in bulk ('light' gradings), there is considerably more flexibility for varying the grading in practice. For this reason the Manual proposed a method for calculating class limit values for non-standard gradings so that the standard specification can still be employed in these cases.

We note Dr Barber's concern in regard to approximation in the design approach, and this non-standard grading approach provides one way forward. However, the size of armourstone is not entirely governed by the minimum required by theoretical calculations of hydraulic stability. Often with coastal structures the total volume of rock beneath a desired outer surface profile crest and slide slope envelope is fixed. Thus the additional cost of moving to the nearest standard grading on the heavy side of that required by the theoretically calculated W_{50} or D_{n50} value is in practice small, since fewer of the larger rocks are required.

It is also worth pointing out that the points raised by Dr Barber in regard to grading standardization in face occupied a considerable amount of time during discussion in the Materials Sub-Group involved in preparation of the Manual. Our final decision was based o the consideration that, whereas the individual approach can work where quarries are local and produce limited ranges of rock sizes, the number of classifications required would get out of hand where large ranges of rock are produced. (The Armourstone Experts Group of CEN committee TC154 have only found it necessary to establish one further, heavier, grading in addition to those recommended in the Manual.) The Manual Sub-Group also looked at the current practices established for size ranges in Northern Europe generally, which had been working satisfactorily for a number of years. The overall aim was to establish a system incorporating best practice, simplicity of operation and minimum stockpiling requirements (important for many operators). The result is inevitably a compromise, which may need to be re-assessed at some future date, in which case it could be done through the EC standards.

D. M. HERBERT, Paper 17

In response to Professor Burcharth's point, as part of the investigation being carried out into the structural performance of the SHED and COB breakwater units, a study has been made of the behaviour of plain concrete in tension under loads of the type that may be experienced by slender breakwater units in the coastal environment.

Some of the results of recent fatigue tests carried out on cylindrical specimens in direct tension at three different load rates are shown in Fig. 1. These previously unpublished data compare well with some of the data reviewed by Professor Burcharth in his interesting paper on fatigue in Dolosse units.[1] When correction is made for the important load rate effect, agreement with fatigue tests on splitting 150 mm cubes by Tepfers et al.[2,3] and on a 25 kg model Dolosse by

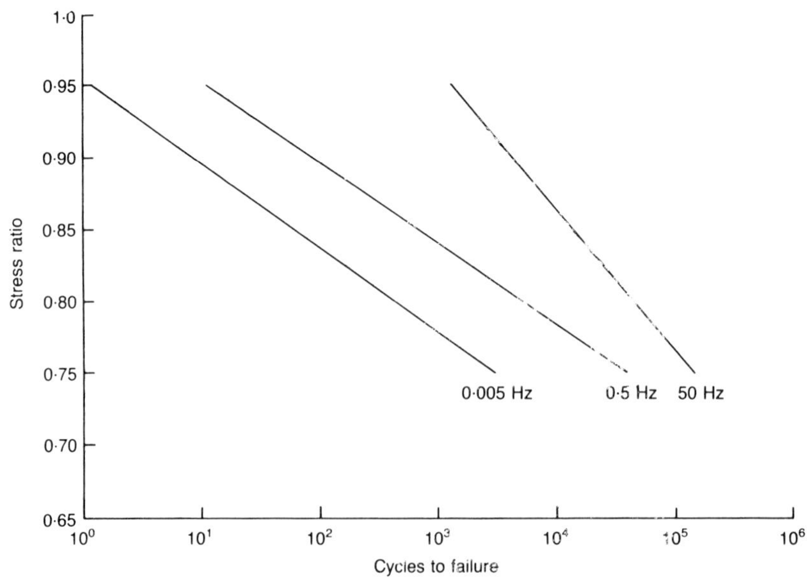

Fig. 6. S - N curves for fatigue of concrete in direct tension at three load rates

Tait and Mills[4] is excellent. The data do not, however, correlate with the repeated impact results from uniaxial tests by Fagerlung and Larsson[5] and Zielinski et al.,[6] or the pendulum tests on 200 kg Dolosse carried out by Professor Burcharth and reported in Paper 4.

There was considerable variation in the magnitude of the impact pressures but generally the maximum impact events occurred at or slightly below still water level. The distribution of maximum impact events appeared to be symmetrical around still water level. It is encouraging that Dr van der Meer has reported similar results to our own.

The statement in para. 7 of Paper 17 that no cracking has been detected in the permanently submerged lower units at Pyrgos is based on the results of a survey[7] carried out in February 1991 in which a diver examined units which were submerged at low tide. Although the whole breakwater was not inspected, no cracking was found in SHEDs below the tidal zones in the areas examined. It should of course be noted that a diver would have difficulty in discovering fine cracks below water.

We agree with Mr Dunster that maximum stresses in the SHED unit will be higher when point loads are applied at the relatively weak mid-limb sections. However, it should be pointed out that cracking will only occur if the load applied in this way is unusually high.

Mr Dunster is correct in saying that cracking can occur in SHED and COB units without affecting the integrity of an armour layer. It has been found in practice that even the most severely damaged units retain their shape and remain in place long after cracking first occurs. Repair and replacement of damaged units has, however, been carried out periodically on a number of existing structures, though this has been done more to give the client confidence than for a demonstrated engineering requirement.

We would agree that the use of polypropylene fibres to increase the impact resistance of the breakwater units is unnecessary. The use of non-corrosive reinforcement bars, could, however, be cost-effective if it convinced the client that the units would remain intact after cracking.

Regarding the name of the Single Layer Armour Unit Research Club, the Authors refer Mr Brown to the reply given to him at the last breakwaters conference.[8] It is relatively meaningless to compare directly the hydraulic stability with the structural stability of an armour unit under wave attack since, as stated in the conclusion of Paper 16, it is only the combination of wave loading with other loads[9] which would cause damage to the units. It is, however, the ultimate aim of the

Single Layer Armour Unit Research Club to develop a rational design method which will allow structural stability under combinations of load cases, as well as hydraulic stability of single layer armour units, to be assessed.

As a result of the close placement patterns of the SHED and COB units, and because out of slope forces are relatively small even under severe wave attack, displacements under wave loading will be small, if they occur at all. However, the potential load case is worthy of some consideration since, although no rigorous analysis of the problem has been carried out by the Authors, they consider that movement of the units could result in relatively high impact stresses. While these stresses have not been quantified by the Authors themselves, they refer Mr Brown to work carried out at Politecnico di Milano.[10]

One of the problems in assessing the structural performance of armour units on a rubble mound is the range of support conditions and load types that are possible in practice (although pattern placed units do allow more control than randomly placed units). For this reason the effect on peak stresses in SHED and COB units of changing support conditions and points of load application was examined in a thorough study carried out using the ANSYS FE programme. The case suggested by Mr Brown was considered and found to be relatively onerous, though not necessarily damaging. A further interesting conclusion from the study was that points of contact between adjacent units are more critical than points contact with the underlayer.

As pointed out by Mr Dunster, there is no evidence that the cracking of an individual unit will result in the collapse or loss of the unit and possible subsequent damage to the armour layer as a whole. It has proved almost impossible to fail an armour layer during hydraulic model testing; failure only occurs under very extreme wave conditions. Even if a unit is removed from the slope it is not always the case that the column above will fall and lead to failure.

References

1. BURCHARTH H. F. Fatigue in breakwater concrete armour units. Proc. 19th International Conference on coastal engineering. Houston, 1984, 2592-2609.

2. TEPFERS R. and KUTTI T. Fatigue strength of plain ordinary and lightweight concrete. ACI J., 1979, May, 635-652.

3. TEPFERS R. Tensile fatigue strength of plain concrete. ACI J., 1979, Aug. 919-933.

4. TAIT R. B. and MILLS R. D. W. B. An investigation into the material limitations of breakwater Dolosse. ECOR Newsletter, 1980, No. 12.

5. FAGERLUND G. and LARSSON B. Betongs slaghallfasthed. Swedish Cement and Concrete Research Institute at the Institute of Technology, Stockholm, 1979.
6. ZIELINSKI A. J. et al. Experiments on concrete under repeated uniaxial impact tensile loading. RILEM Materiaux et Constructions, 1981, Vol. 14, No. 81, 163-169.
7. WILKINSON A. R. Analysis of breakages in COBs and SHEDs in breakwaters. Private communication, April 1991.
8. DUNSTER J. A. et al. Single layer armour units. Proc. conf. breakwaters '88. Eastbourne, 1988.
9. TONER W. L. et al. Structural behaviour of single layer concrete breakwater armour units. Symposium on developments in coastal engineering. University of Bristol, 1991, 13-23.
10. FRANCO L. et al. Impact response of fibre-reinforced concrete elements in the marine environment. Private communication, October 1991.

N. W. ALLSOP, Paper 18

In response to Mr Bezuijen, the Alderney breakwater mound seems to be particularly unusual among breakwaters in the way in which the main geotechnical or structural skeleton was formed. Originally as-quarried material was placed to form the mound, chiefly using barges. Some efforts were made to blind off the upper part with fine material to reduce the local permeability and to give a more regular surface on which to found the blockwork. This procedure was abandoned during the construction, as strong local tidal currents and wave action were found to wash the finer material away. We can then presume that the mound material was essentially quarry all-in, less a proportion of the finer material. An estimate of the grading of the material as quarried, B1, is shown in Fig. 4 of Paper 16.

From 1864 to 1964, less a few interregna, further quarry rock was dumped onto the front face of the mound from the breakwater wall. Under the frequent severe local wave conditions, much of this material was quickly ground down to sizes such as the sand sampled from the mound, grading D3 in Fig. 4. The critical difference between Alderney and other rubble mounds is that this fine material was then included into the structural skeleton of the mound formed as dumped material was carried from the toe of the wall by wave action, thereby reducing the structural strength of the mound. This should be distinguished from the situation with many other rubble structures, where fine material is later washed into the voids, reducing overall permeability, but without changing the structural

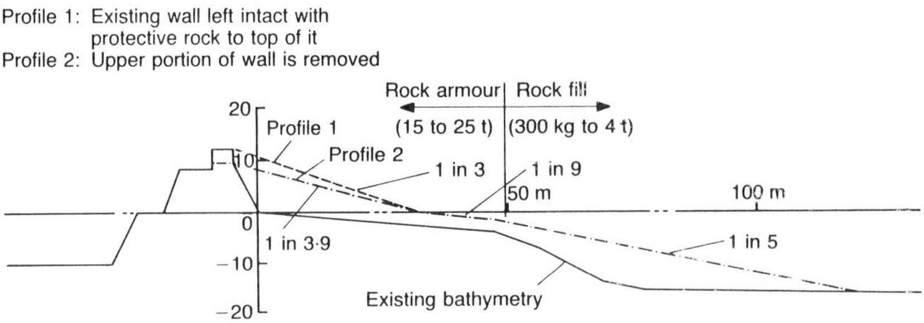

Fig. 7. Alderney Breakwater: typical cross-section of protective works as model-tested

strength of the mound.

We would generally expect fine material washed into the underlayer of placed block revetments to wash out again relatively easily under wave action, unless the underlayer material was itself mobile, and allowed the fine material to move into the structural matrix.

In reply to Dr Van der Meer, Paper 18 describes studies to identify reasons for the repeated damage to Alderney Breakwater, to outline possible engineering measures. These are not presented as final solutions, and much work is still needed to develop the final solution. The main principles of the preferred measures are illustrated in Fig. 7, which shows two possible profiles and the sizes of rock used in the model tests.

With regard to Mr Wolfe-Barry's point, a number of alternative schemes have been considered, and indeed are still under consideration, including schemes similar to that suggested. All schemes at Braye will require work on the Admiralty Breakwater, although not necessarily for the whole length. Various proposals for reducing the breakwater length have been studied at Wallingford in the past. Data from previous physical model tests will be used to calibrate any future numerical model studies of the harbour.

The question 'where does the failed slope material go?' was one that gave the study team considerable trouble, particularly as there is no sign of large material on the sea bed close to the mound. Experience of undrained slope failures on land sites may provide the clue for an answer. Tailings dams, mining waste, and waste from the recovery of materials like kaolin all produce large mounds of sand and sand filled rubble. Undrained failure in these materials may lead to dramatic flow slides, similar to that in 'colliery waste' at Aberfan. Flow slides in china clay waste tips and tailings dams all produce debris sheets which travel many hundreds of metres from the failure site. The travel distance of the debris sheet depends strongly on the volume of material in the failure.

Our analysis of the Alderney mound has shown that slab slides are a likely mechanism for local **geotechnical failures of the mound under storm** conditions. At Alderney these slides may contain (relatively) small quantities of material, but because they occur underwater the slide material will be well-mixed with water. This process will, however, be very variable, depending as it does on rapid, hence undrained, failure. Where the failure is less rapid such that pore pressures are reduced, the slide may 'stick'. This may produce a step on the seaward face as the slide only moves partially down the slope. Some of the survey profiles seem to show evidence of this type of failure.

M. DE GERLONI, Paper 20

To Mr Harris, it has to be stressed that the purpose of the full-scale tests was not the evaluation of the effects of overtopping water on the human being. The goal was rather that of 'calibrating' a manikin in such a way that it would behave like a person when hit by a water jet which was not intended to simulate a real overtopping wave. Our thought can be put in this way: take a person and find its equilibrium limit when hit by a water jet; take a plastic dummy and modify its weight and base until it capsizes with the same water jet; scale down the plastic dummy and look at the effects of the overtopping waves in the model. However, for the scale effects it can be assumed that the behaviour of the model is the same of that of a person.

In response to Dr Van der Meer and Mr Juhl, the individual volume data are available for all the tests but they have not been elaborated statistically. The Authors have, however, some doubt about the reliability of a possible statistical evaluation since the wave generation was a pseudo-random (repetitive) sequence rather than a truly random generation. Pure random tests with very long duration are planned for the future. The value of the ratio between the largest overtopping wave and the total volume depends upon the breakwater type and upon the amount of the total volume itself. It was found that such a ratio is greater for vertical walls (up to 0.50) than for composite breakwaters and it decreases as the amount of total overtopped volumes increases. Wind can indeed have a big influence; the Authors hope to be able to study this parameter too in the future.

T. J. T. WHITTAKER, Paper 23

The wave pressure formulae as discussed in Paper 23 were developed for specific situations like rubble mound breakwater caissons (Goda's formula) or seawalls with a rather steep and smooth seabed (CERC formula). The pressures calculated with these formulae did not show good agreement with the measured and extrapolated breaking wave pressures at the front wall of the wave power station. This was thought to be caused by the site conditions which are characterized by a shallow seabed slope, a very rough seabed and surfacing rocks close to the shore. As none of the formulae takes these factors into account, the structure had to be designed conservatively for the highest calculated pressure. Owing to the site conditions, the measured pressures are one order of magnitude smaller than the design pressures. Thus the use of standard formulae for this led to an overconservative design.

The position of the transducer module about 1.25 m away from the side of the gully is not considered to influence the measurements, as the rock face is very smooth and the waves come in parallel to the side of the gully.

During the feasibility study, the peak electrical output was estimated at around 180 kW, which would have been accommodated by a 120 kW generator. For the actual design, however, a flywheel with 2 MJ inertia energy storage was added to the turbine to accommodate short-term peak pressures. Consequently, the generator rating was reduced to 75 kW. The generator has been operational since July 1991; however, the unit has been operated intermittently only in order to test the components of the system, such as turbine, generator, generator management and safety appliances under various conditions. The maximum electrical power output measured so far has been 50 kW. An average value for power output would be meaningless at this time, as the system was not operated continuously but only within the system test programme. Based on the results from the commissioning period, several modifications are now to be made, and it is expected that the average power output will be 20 - 39 kW.

In reply to Professor Goda, the wave load measurements presented in Paper 23 were taken over a period of four days. For the statistical analysis, the rather generous assumption was made that stormy conditions as encountered on those days prevail for four months of the year. This analysis was made using the limited amount of data available in order to estimate the magnitude of the design pressures, and not to evaluate accurate values for 1/50 year pressures. With the measurements taken during the winter of 1991/92 and a detailed weather record, the Authors hope to increase the amount of available data and thus to improve the statistical prediction of extreme pressures for this particular site.

24. Performance assessment and maintenance of coastal defences

L. DE QUELERIJ, Fugro McClelland, E. VAN HIJUM and K. PILARCZYK, Delft Hydraulics

SYNOPSYS. Coastal defences are constructed and maintained to protect man and economic values by preventing inundation and controlling shoreline retreat. Proper management has to guarantee that the defence works will remain fit for their purpose for a stated period of time under specified conditions of use. This paper gives a review of the tools needed by the coastal engineer to execute his management job in a proper way. One of these tools, that is the maintenance and monitoring system, is described in more detail and will be illustrated by means of a case study for a sea dike. The developed system can also be of use for seawalls, breakwaters and other coastal structures.

MANAGEMENT OF MAINTENANCE

Goal. Sea defence works have to be designed, constructed and maintained in such a way, that for a stated period of time (reference period) they will perform all agreed functions with satisfactory reliability and without excessive maintenance. The reference period may vary between 1 and 50 or 100 years and has to be regarded as a help parameter for design and assay decisions. The true lifetime of the structure may be larger than the reference period.

During the lifetime of the structure especially the maintenance is a continuously returning task. The responsible maintenance engineer must devise a maintenance programme in which the necessary use and employ of personel, materials and equipment (including inspection techniques) are scheduled and accounted.

Actual situation. The management of coastal defences is one of the few areas of civil engineering in which there are no standardized solutions and methods. Approaches still strongly rely on empirical methods, experience and local habits and customs. This was no problem in a time with less interest for the coastal environment and steady state budgets.

However, nowadays various disciplines become aware of the variety of (potential) functions of the coastal environment. Important functions may derive from environmental and ecological purposes, economic activities (human settlement, land use, sea-borne transport and mining) and recreational potentials. On the other hand the struggle for budgets has become more intensive: more functions, constraints, standards and criteria and less money to parcel out, after weighing of the various interests.

So the managing coastal engineer has to look for a rational approach for this problem and has to wrap it up in such a way that it will be equivalent and successful in the resource competition with the demands from other types of public service.

MANAGEMENT SCHEME

Objectives and elements. The management scheme has to be suitable for the next two items:

- satisfaction of the regional council (or other responsible public authorities) that maximum value for money is achieved (social output)
- critical reflection on the own performances (technical output).

It appears to be almost impossible to identify and valuate the social output of coastal defence works in an explicit way. Therefore it is proposed to do this in an implicit way in a REC-assessment model. REC stands for "risk, economic and conservation". Together with a data-base (or register) and a management policy plan this model provides the basic information for the maintenance plan, the annual budget estimates, the annual account and justification reports.
All these elements together form the management scheme.

Data-base/register. This register contains all physical qualities of the administered objects and related aspects inside the influence zone:

- the boundaries of the influence zone. This zone includes those areas of adjoining sea bed and land where processes are linked in some way to the behaviour of the defence works.
- a description of the as-built situation and the actual situation, including longitudinal and cross-sections, geotechnical profiles etc.
- a list of issued licenses.
- an ownership and farming-out register.
- a map with all cables and pipes owned by public utilities, oil companies etc.
- a damage record.
- a record of executed maintenance.
- boundary conditions (hydraulic, geotechnical, traffic, etc.).

Management policy plan. Control of the coastline and coastal activities may lay with different authorities and regulatory bodies. In this case it may be difficult to fix responsibility, control and accountibility throughout the considered coastal area, or to decide among proposed uses when questions of permitting, engineering, maintenance and financial responsibilities arise (see [2] and [3]).

However it may be, a management policy plan is needed to link the objectives of the political arena and users of the coastal zone with those of the defence works management.
The plan should contain the following elements:

- A statement on the scope of the plan with the links to all relevant laws and regulations.

- The procedure to be followed at the adoption and amending of the plan by the responsible administration.
- A statement on the management policy outlines. Policy options are: maintain existing coastline, set back defence line, retreat or advance. Management and engineering options are: create, modify, reinstate or do nothing.In some situations it may indeed be desirable to do nothing: future decades may see an increasing use of this "zero option".
- A review of all relevant functions and uses, dealing with environmental aspects, flood defence, industry, recreation and land use (transport , agriculuture etc).
  The policital weighing of the different competing functions have to be described in the management policy plan. The weighing reveals criteria and boundary conditions for the management (input and assay) functional criteria, hydraulic and geotechnical boundary conditions, defence standards and acceptable risk levels.
  Note that original functions may have changed and still may change. In addition some af the boundary conditions may change. For instance a changing climate will influence the threat to coastal areas, both directly in the hydrodynamic sphere and indirectly via changes in sediment transport and biotic coastal dynamics.
- The management and maintenance strategy and tactics to be followed for the short and long term, including scenarios in case of calamities have to be included.

<u>REC-assessment model</u>. Fig. 1. shows the flow scheme for a model on risk-, economic- and conservation assessment, that is suited to visualize the change in social output value when certain measures are taken: for example the increase of flood risk when the maintenance budget is cut down.

Essentially, the method contains the following steps:
- Definition of objectives and specifications of the structures.
- Definition of criteria for performance, structural integrity, boundary conditions, accepted risk level, finances and conservation.
- Assessment of input data: structural (resistance) parameters and environmental (surcharge) parameters.
- Models to describe the behaviour of the structures.
- Model to evaluate the results and to define remedial options.

The objectives are based upon the functional and structural requirements and the corresponding performance and design specifications. All this information can be found in the data-base and the management policy plan. The structural design and/or assay specifications may be expressed in terms of a probability of failure or as a damage level.
The model used for the evaluation of the results can be a (multi) criteria analysis, supplemented by a more sophisticated model to optimize the effect of changes in structural or environmental parameters.

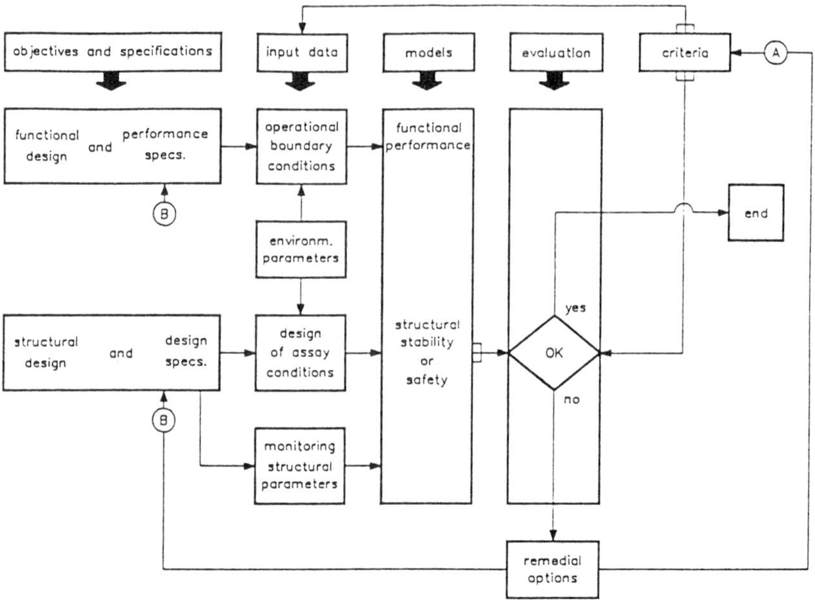

Fig. 1. Flow scheme for a REC-assessment model

Remedial options, in case the evaluation reveals no o.k. status, are (see Fig. 1.):
A: adjust criteria, what usually means: accept an higher risk level, accept a lower leisure quality, accept a lower level of naturalness etc., or: increase management budgets.
B: modify the functional and/or structural design.
Decisions on these options have to be made by the responsible authorities; resulting new policy has to be incorporated into the management policy plan.

Annual budget estimate. The total annual costs not only consist of the expected maintenance costs (Ei), following from the maintenance plan. The annual costs should also incorporate the capital costs (Ec), that are the annual payments to cover interest and writing-off. In addition the total annual cost should include an allowance for potential costs consequential upon failure or unserviceability (Ep): these costs may be treated as equivalent annual insurance premium payable throughout the reference period p and just sufficient to compensate for the loss, valued at present day costs and equal to Ef*Pf/p, where Ef is the estimated consequential cost and Pf is the total risk of failure in the reference period.

It is possible to determine for each structure an optimum in the relation between the annual cost and the annual probability of failure. The short-term character of the budget estimate does not match the long-term character of maintenance works. Therefore the annual estimate should always be completed with a detailed review of expected adaptions for the coming five years and a less detailed estimate for even longer periods.

Maintenance plan. The maintenance plan will generally comprise:

- type of maintenance strategy:
  - corrective : maintenance that is performed after disfunctioning of (parts) of the structure;
  - preventive : maintenance that is carried out before disfunctioning of the structure;
  - used-based : preventive maintenance that is carried out after a prescribed period of time or intensity of use;
  - condition-based : maintenance that is performed when a condition parameter reaches a prescibed limit.
- monitoring and inspection programme, aiming at a periodical assessment of the actual condition of the structure,
- functional requirements and standards that prescribes the norm values (criteria or limits) of which the actual conditions have to meet,
- prediction models to estimate the alteration of the condition of the structure in the period till the next inspection,
- planning of repair measures consisting of reconditioning or restoration of the structure by means of remedial measures,
- cost and time schedule of personel, materials and equipment.

Based on additional information during service life of the structure it may be necessary to change the originally (during the design stage) developed maintenance plan. It is emphasized that the asssessment of such an alterated plan is restricted by a set of actual boundary conditions at which the managing coastal engineer has much less flexibility than during the design stage of the structure.

MAINTENANCE AND MONITORING SYSTEM FOR DIKES

General conceptions. According to the REC model the maintanance plan forms a major instrument for the management of the coastal structure. The elements of the maintenance plan with emphasis on the monitoring system will be described briefly. Although the set-up of the plan will be worked out for a dike system focussed on the main function of flood protection the set-up is also applicable to other coastal structures and other functions.

A dike system forms part of a total water retaining system, consisting of different types of structures. If one part of the system fails flooding of the inland area, the so-called dike-ring may occur. In order to guarantee an acceptable safety level under extreme loading conditions the contribution of the total lenght of the dike sections to the overall probability of failure of the potential flooding area may not exceed a specified value. Each of the dike sections is composed of a number of structure components, termed as elements. These elements can be related to the detailed water retaining functions as derived from the main function. Examples are:

dike element		function
slope revetment	:	hydraulic erosion control
dike crest	:	control overflow and overtopping
sub soil layer	:	ensure geotechnical stability

The elements can be described by a number of characteristic condition parameters, mainly geometrical and material parameters. For example the dike crest can be described just by the geometrical parameters height and width of the crest, whereas the description of a slope revetment necessitates specifications of the type, geometry, weight and strength of the successive layers of the revetment.

During the service life of the dike a number of deterioration mechanisms may occur that will affect the strength of the structural elements. In other words these deterioration processes may change the values of the condition parameters.
Examples of these processes are gradual settlement of the dike crest due to consolidation of the underlaying soft soil layers, washing out of finer particles of a stone revetment during daily tidal conditions, weathering of asphaltic revetments due to ice, wind and rain effects and damage effects of slopes due to biological activities such as vegetation and animal burrowing.

These alterations caused by deterioration mechanisms, which will be recognized as characteristic damage patterns, can be of such significance that the dike safety will no longer be sufficient. This means that the actual quality of the condition parameters may decrease to a level at which the extreme loading conditions cannot be sustained.

Management strategy. By means of an adequate maintenance and control system of the dike the managing authority aims to secure that the actual quality of the relevant condition parameters do not decrease to a level lower than the acceptable failure limit. For this purpose the so-called preventive condition-based maintenance strategy has been adopted. This strategy distinguishes three limits of quality levels of the condition parameters (see Fig. 2).

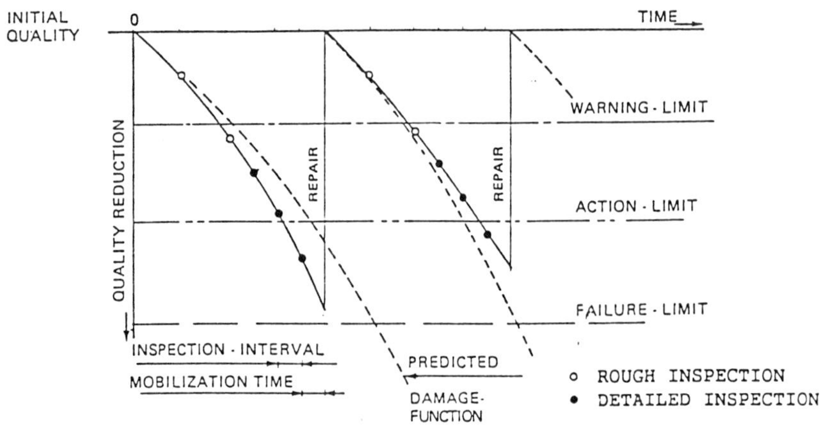

Fig. 2. Preventive condition-based maintenance

These limits are:
- warning limit: quality level at which a more intensive control of the condition parameter is needed (higher inspection frequency);
- action limit : quality level at which repair measures should be prepared and carried out before the failure limit has been reached;
- failure limit: quality level that is just acceptable from the safety requirement. If the condition decreases below this level the dike system will not provide sufficient safety.

The margin between the action limit and the failure limit will depend on the inspection frequency and the mobilization time for the execution of the repair measures. An optimum maintenance and control strategy will be obtained by considering the minimum costs of repair and inspection, on the condition that the probability of exceeding the failure limit is sufficiently low.

Although most of the before mentioned aspects are familiar in structural engineering (steel and concrete structures) the applications of a systematic condition-based maintenance approach for hydraulic and geotechnical engineering are very limited. This may be caused particularly by the considerable length of the dike sections, the heterogeneous composition of the dike elements including the variability of the soil layers and as a consequence the variability of the failure mechanisms and associated damage patterns.

To obtain a practical approach for the periodical management control of the Dutch dikes along the coastline, the main rivers and the estuaries a research programme, commissioned by Rijkswaterstaat, has been carried out by the Dutch Civil Centre for Research and Codes (CUR) [6].
For this purpose a systematic control method has been developed whereby the observable damage patterns, related to both the deterioriation and ultimate failure mechanisms play a major role. Much attention has been paid to a stepped approach of data collecting. In an early stage only rough information is needed for the total dike length, while in a later stage more detailed data have to be collected out only for a limited part of the total dike length. In this way the total amount of data collecting and in situ monitoring can be reduced significantly.

Set up of management control system. The introduction of the management control system for dikes can be divided into three stages:
a. The first set-up of the control system for the total length of dikes to be maintained. This includes the collection of the basic information on the dike, the assessment of relevant damage patterns and the inpection strategy.(activity to be performed once)
b. The periodical inspection and monitoring of the dike and comparison of the actual behaviour with the predicted behaviour.

(activity to be performed continuously, several times a year)
c. Up date of the control system to incorporate new developments with respect to functional design requirements and new knowlegde of loading conditions, modelling of physical behaviour of the dikes or monitoring techniques. (activity to be performed every five or ten years, depending on the technological developments)

The first set-up of the management control method for dikes consists of the five following successive steps:
1. Rough description of dike elements and boundary conditions
2. Selection of decisive dike sections and dike clusters on the basis of the relative strenght with respect to ultimate failure mechanisms on one hand and deterioration mechanisms on the other hand
3. Determination of relevant damage patterns and damage failure limits
4. Assessment of inspection and monitoring strategy
5. Determination of scheme for repair measures

The first two steps are dealing with rough information and apply to all concerning dike sections. As a result of step two decisive dike sections will be selected. The remaining steps, whereby more detailed data is needed, are applicable to a limited number of representative decisive dike sections.
The practical meaning of the five steps will be demonstrated by means of the case study "Hondsbossche sea dike". Before that step 4 will be discussed briefly in general terms.

Inspection and monitoring strategy. The inspection stategy should provide information on what (which damage pattern characteristics), where (locations and depth), when (how often), how (which measuring method) and by whom (which organization) data should be collected and evaluated.
Monitoring is assumed to form a part of the total inspection and will not be discussed seperately.
It is not possible to give a detailed description of the inspection strategy because it depends very much on the actual situation. Alternatively some general remarks will be given as a guideline for the set up of an inspection system.

The principle of the inspection stategy should be based on a phased inspection with respect to the amount of detail. Starting from a rough inspection a more detailed inspection may be required depending on the results of the preceeding inspection.
A distinguish should be made between the periodical inspection of the damage patterns (deterioration mechanisms) according to the maintenance plan on one hand and the incidental inspection after the occurrence of special loading conditions (extreme mechanisms) on the other hand.
The last type of inspection should be performed after each severe loading condition (such as an extreme storm surge) and will generally start with a rough visual inspection of the total dike length.

The answer to the question which data on which location have to be periodically inspected will primary follow from the type of damage patterns to be controlled. Each damage pattern has its own observable characteristics. For example the quality of a grass on clay revetment can be characterized by the presence of uncovered spots (that is the area where the grass is absent). The quality with respect to consolidation of the soft soil layers can be characterized by the settlement of the dike crest height. Some damage patterns have to be controlled only at a specific location that is representative for the total dike cluster, while other damage types should be controlled along the total dike length.

To decide upon the inspection frequency and method the predictability of the increase of damage increase and the actual condition, relative to the warning-, action- and failure-limits are inportant aspects.
As an illustration two extreme combinations of damage patterns and associated inspection strategies are outlined in Table 1.

Table 1 Inspection stragegy by different damage patterns

Aspects of Inspection strategy	Aspect of damage increase	
	progressive great uncertainty	gradual increase small uncertainty
Inspection method	detailed	rough
Inspection frequency	high	low
Action-failure limit	great margin	small margin

The decision how the data should be collected, that is to say which measurement techniques should be applied, again depends strongly on the type of damage patterns to be controlled. In this context the availability and terrain accessibility by equipment for geotechnical site investigation may play a major role.
For each situation a cost effective technique has to be selected. The most applicable measurement techniques related to the damage patterns clustered per extreme mechanism are listed below.

Dike geometry
- conventional geodetical land surveys (for a.o. crest height)
- advanced techniques with high resolution (for local settlements)

Slope instability, seepage and piping
- inspection of seepage flow of inner slope during high water levels by visual observation or by (infra-red) air photogrammetry
- installation of observation wells and piezometers for the measurements of pore water pressures and the freatic level in the dike during both normal and high water levels
- visual inspection of inland sand-boiling wells

Erosion of grass revetment
- visual inspection of uncovered area
- shallow borings and sampling of top layers

Erosion of stone pithing or block revetment
- infiltration test to determine permeability of top layers
- in-situ tensile test to determine local up-lift resistance of the individual stone or block
- geophysical measurement to detect local cavities underneath the top layer (by ground radar or acoustic methods)
- visual observation of the gaps between the stones and blocks or the detection of complete loss of these elements

Erosion of asphaltic revetment
- assessment of layer thickness from sampling or geophysical measurements
- laboratory testing on samples to determine durability effects
- in situ deflection measurements to determine the stiffness
- nuclear or electrical density measurements in bore holes to determine the quality of the foundation layer underneath the asphaltic top layer
- global visual inspection of superficial cracks

As shown in this list, visual observation plays a major role in the inspection. However, special attention should be given to those damage patterns, which are not easy detectable from the outside geometry of the dike. An example for this type of damage is the presence of erosion channels underneath a top layer of a revetment which may substantially reduce the strength of the slope revetment, without showing any superficial damage under daily conditions. There is a need to develop special measuring techniques to control this type of "hidden" damage patterns. Geophysical methods, such as the ground radar, may contribute to the solution of this type of detection problems.

CASE STUDY HONDSBOSSCHE SEADIKE

Description dike elements and boundary conditions (step 1). To demonstrate the application of the management control system in practice the case Hondsbossche Seadike will be worked out briefly. The Hondsbossche Seadike along the North Western coastline of the Netherlands forms part of the primary protection against flooding from storm surges in the North Sea. The dike has a total length of 4.6 km. The latest improvement (enlargement) of the dike took place from 1977 till 1980. The construction of the dike consists of the following elements as shown in the cross-section in Fig 3.:

1. groynes in front of the dike (hand pitched basalt stones, stake rows and bottom protection matresses)
2. toe protection (stake row and cover stones)
3. outer slope - lower part 1 : 8 (hand pitched basalt stone revetment and asphaltic concrete revetment)
4. storm surge berm 1 : 20 (asphaltic concrete revetment)
5. outer slope - higher part 1 : 3 (asphaltic concrete revetment and grass revetment on clay)
6. crest, inner slope 1 :3 and inner berm (grass revetment on clay and maintenance road)

7.dike core (sand)
8. subsoil (clay/peat toplayer overlaying a sandy layer, partly intermitted by a thin peat layer and clay layers of varying thickness)

Along a part of the seadike a road is situated just at the land side of the dike. In addition some road connections are present.

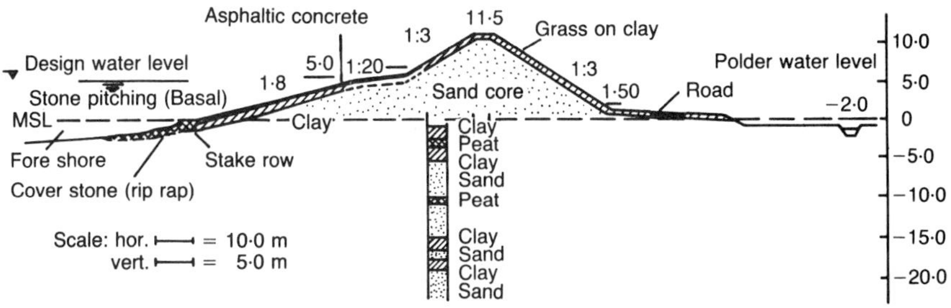

Fig. 3. Typical cross-section of Hondsbossche Seadike

The loading conditions can be divided in hydraulic loadings and other type of loadings. The hydraulic loadings are represented by variation of normal sea waterlevels due to tidal conditions (MHW 0.60 m + MSL and MLW 0.90 - MSL) and extreme storm conditions: waterlevels from 2.70 m + MSL for a return period of 10 years till 5.05 m + MSL for a return period of 10,000 years (= design waterlevel) and associated wave run up of respectively 4.80 m and 5.95 m. The waterlevels at the land site are determined by the polder waterlevel in the nearby polder ditches.

The other types of loadings are caused by:
- durability effects (rainfall, temperature, mechanical erosion);
- animal activities by cattle (sheeps);
- human activities (recreational damage or vandalism).

<u>Selection decisive dike sections and clusters (step 2)</u>. The selection of the decisive dike sections is based on differences in the strength with respect to the ultimate failure mechanisms, the sensitivity for deterioration mechanisms and the presence of particular elements.

The relevant ultimate failure mechanisms for the Hondsbossche Seadike are:
- geotechnical instabilities;
- hydraulic erosion of crest and inner slope;
- erosion outer slope.

Taking into account the soil conditions and the hydraulic loadings the potential failure due to piping may be neglected.

Differences in strenght will occur due to variation of:
- crest height of the dike (group a: 11.5 - 11.8 m and b: 11.8 - 12.1 m + MSL);
- foreshore geometry (two sections with relatively steep slopes)
- thickness of underlying soft soil layers (from 4 m till 10 m);
- polder water levels (varying from 0.60 m - till 2.60 m - MSL);

-inner berm geometry (with and without a main road).
The remaining condition parameters of the dike elements related to the extreme mechanisms are considered to be constant over the full lenght of the seadike.

The relevant deterioration mechanisms of the Hondsbossche Seadike are identified as follows:
- mechanical durability of both the asphaltic, the pitched basalt stone and the grass on clay revetments;
- climatological durability of the asphalt, the pitched basalt stone and the clay layer;
- consolidation and creep of the soft soil layers resulting in settlements of the dike crest;
- erosion during daily conditions of the toe protection, the foreshore area (groynes) and the revetments;
- biological effects caused by sheep and human effects caused by vandalism.

The sensitivity for the consolidation mechanism is mainly determined by the variation in thickness of the soft soil layers. Since both the condition parameters and the loading conditions with respect to the remaining deterioration mechanisms are uniform along the total dike lenght no additional differences to select dike sections can be obtained.

In conclusion, based on the variation in strength with respect to the ultimate failure mechanisms, the total Hondsbossche Seadike has been divided into 9 dike sections. Since these differences are not very significant and the hydraulic loadings at sea side are equal along the length of the dike, all the 9 sections can be clustered into one single group.

<u>Selection damage patterns and damage limits (step 3)</u>. From a further study of the relationship between the ultimate failure mechanisms and the deterioration mechanisms the following relevant condition parameters for the periodical management control have been selected:
- gradient foreshore
- crest height
- quality toe protection
- quality stake row
- strenght stone revetment
- permeabilty stone revetment
- thickness asphaltic layer
- quality grass revetment

For each condition parameter failure limits have been assessed based on the analysis of the ultimate failure mechanisms. If a mathematical model of a limit state is available, these failure limits have been back-calculated straight on. If these limit state expressions are not available practical failure limits have been chosen based on experience and engineering judgement. The selected damage patterns and failure limits of the Hondsbossche Seadike are summarized in the four left hand columns of Table 2.

Table 2. Inspection and maintenance scheme of the Hondsbossche Seadike

condition pararameter	damage pattern	failure limits	ultimate failure mechanism	inspection	repair measure
slope of fore shore	change of 2.5 m - MSL line	min. distance of 2.5 m line and toe pretection	geotechnical instability	bathymetrical survey fore shore one month before and and after storm season	sand supplement/ bottom protection of fore shore
crest height	change of crest height	min. height 11.5 m + MSL after plan period	erosion of crest and inner slope	geodetical survey once per 1 to 5 year depending on previous inspection results	heightening of the dike
quality toe protection	height of cover stones	max. reduction of height 0.4 m	erosion	visual observation of height reduction	new supply cover stones
quality of stake rows	visual deterioration	no rotting	erosion of outer slope	visual observation of stake rows, if needed pull a test pile	replacement of rotten stake rows
strength of stone revetment	interlocking stones	no stone lifted close connections	erosion of outer slope	visual observation one month preceeding the storm seasen	replacement of lifted basaltic stones
permeabilty of stone revetment	pore pressures below filter layer	freactic level max uplift water pressure 0.4 m	erosion/geotechnical instability	measurement of hydraulic head below the clay layer during high water levels	inprove filter of the toe protection
thickness of asphaltic revetment	layer thickness cracks	min layer thickness ha	erosion of outer slope	coring of the aphaltic layer	supply new coverlayer of asphalt
quality of grass revetment	visual lacking of grass	covering degree "reasonable"	erosion	visual observation of grass revetment during growing season	adapt grazing oppose the moles seed bold areas

Scheme for inspection and repair measures (steps 4/5). The inspection scheme of the dike follows from the periodical control of the specified damage patterns. For some damage pattern only the decisive dike sections have to be inspected. If no decisive dike sections are identified the total dike cluster will be inspected.

The repair measures to be taken before the failure limits are exceeded have been assessed for each damage pattern. The resulting scheme for inspection and maintenance of the Hondsbossche Sea dike is presented in the two right hand colums of Table 2. Apart from the before listed damage patterns also unpredicted damage patterns may occur. To detect these last types of damage in time a global visual inspection of the total dike lenght has to be performed preferably two months before the storm season. In addition every 2 years after completion of the dike the geometry of the cross section of the dike has to be surveyed to detect unexpected deformations of the construction.

EVALUATION

The assessment of a good safety control and maintenance system of the dike requires adequate data on the design loadings and the actual condition of the construction. Particularly the composition of the dike core, the revetment and the geotechnical profile of the subsoil are of major importance. If these basic data are not available additional hydraulic and geotechnical in situ investigations have to be carried out.

It is recommended that a suitable management organization is set up by the responsible authorities in order to guarantee a proper performance of the safety and maintenance system during the life time of the structure. Computerized data bases, that can continuously be up-dated from inspection results, may be very useful during the total life time.

REFERENCES

[1] Barrett, M.G. et al., Coastal Management. Proceedings Conference on 9-11 May 1989, Bournemouth, U.K.
[2] Fischer, D.W., Governing the US Coastal Zone: an unresolved issue, Ocean and Shoreline Management 13 (1990) 21-34, USA
[3] PIANC, PTC I-WG 3. Consideration of risk in determining bank protection, Brussels, 1986, Belgium.
[4] Government of the Netherlands, Law on Flood Control (in Dutch), Draft 1989, SDU, The Hague, The Netherlands.
[5] CUR, Periodical judgement on the strength of dikes (in Dutch),part 1 "methodology" and parts 2 to 5 "case studies", 1990, Gouda, The Netherlands.
[6] Quelerij, L. de and Hijum, E. van, Maintenance and monitoring of water retaining structures, Proc. Coastal Protection pp.369 - 401, 1990, Balkema, Rotterdam, The Netherlands.

25. Recent developments on shore protection in Japan

T. ASAKAWA, M. HASEGAWA, H. SATO and
N. HAMAGUCHI, Nippon Tetrapod Co. Ltd

SYNOPSIS. In Japan, ideas on coastal protection have undergone a major change in recent years. Protection of the shoreline with mainly vertical seawalls has given way to new considerations for the coastal zone using various structures based on the concept of zonal protection. In this paper, the authors introduce the characteristics and planning behind the new structures.

PROFILE : SHORE PROTECTION IN JAPAN

1. Japan consists of approximately 4,000 small islands and four large land masses. Steep mountain ranges are to be found in the centers and planes spread towards the coasts. Most of population and industry is distributed along or near the approximately 34,000km total length coastline. These areas often feel the effects of typhoons, storms, tsunamis etc. The need to protect is a matter taken seriously.

2. Phase I : In the 1950's the planes suffered a great deal of damage. After the Ise-Bay typhoon, earnest attempts at shore protection were made. At first, priority was placed on economical efficiency to pursue minimum cost under limited budgets. Vertical type concrete structures, (hereinafter referred to as conventional type) occasionally using Tetrapods to attenuate wave energy as shown in Fig. 1 became the accepted method.

3. Phase II : Erosion problems brought on by the decrease in volume of sand flowing out of rivers, the sheltering effects of coastal structures, the scouring of toes of seawalls etc., became a cause for concern. Countermeasures included the application of jetties and detached breakwaters. The beaches however suffered erosion and attitudes toward the structures began to change i.e.:

(a) The structure did not always blend in with the surrounding scenery.
(b) Access from the crest elevation of a seawall to the fore shore was restricted by the steep slope or armour blocks placed in front.
(c) The beach itself was not wide enough to allow bathing, recreation etc.

4. PhaseⅢ : A move toward creating a better environment for coastal utilization was born.

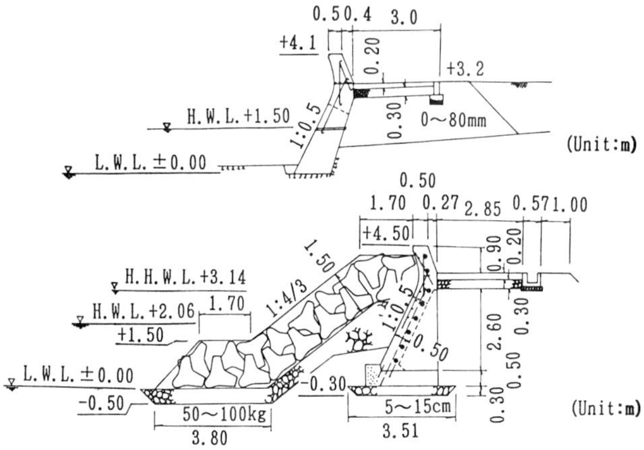

Fig. 1. Examples of vertical type seawall

New demands and Zonal Protection

5. Shore protection using conventional vertical seawalls did concern itself with utilization except in certain areas e.g. bathing beaches. In recent years, the Japanese people have become better off in terms of economics and free time. More and more people want to enjoy the coastal facilities. As a result, as well as shore protection, coastal utilization such as accessibility to the beach and the protection for the natural environment have become of prime importance.

6. Modern coastal structures are placed not only on the shoreline but in the nearshore and offshore areas. This is in sharp contrast to the convention ideas such as vertical seawalls, jetties and detached breakwaters etc. Application calls for a series of wave reductions at stages from offshore to backshore to compensate for individual structure demerits and realize good beach conditions. Coastal creations responding to the new public demands are being tried using not only conventional structures but also littoral nourishment and recently developed structures with mild slope revetments, artificial reefs and head-land type jetties, etc.

RECENT DEVELOPMENT ON SHORE PROTECTION STRUCTURES

Mild slope revetments

7. Mild slope revetments having characteristics differing from the conventional type have been developed recently and are becoming more popular. The slopes of the revetments normally range from 1:3 to 1:5 when permeable concrete armour blocks are used. This concept was initially proposed by Toyoshima, introduced in the Coastal Engineering Conference, 1984(ref. 1) and 1988(ref. 2), and has been installed in many

sites. A typical cross section of the mild slope revetments is shown in Fig. 2 and Fig. 3.

8. The aims of the mild slope are to limit environmental impact while reducing wave runup, achieving low reflection coefficient of waves, while realizing accessibility to the beach : In other words, a structure which resembles the natural beach itself unlike a conventional seawall. Structures stemming from this concept are now being chosen over the conventional types and significant changes in the planning of shore protection have resulted. Further modification of the concept to meet considerations for safety and comfort has prompted the development of novel blocks.

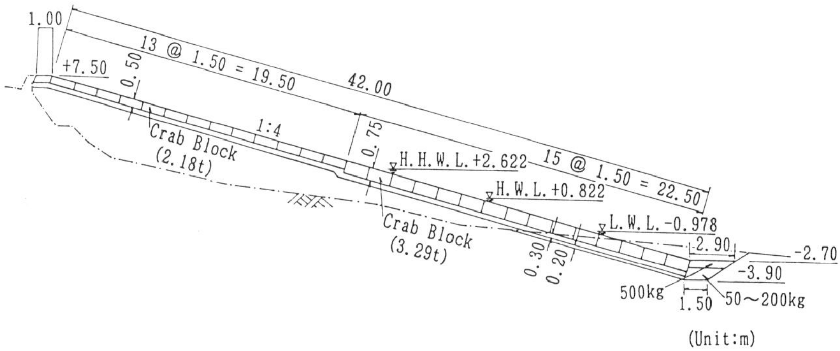

Fig. 2. A typical section of mild slope revetments of Crab Blocks

Fig. 3. View of mild slope revetments of Crab Blocks

9. New type blocks. Fig. 4 shows a typical cross section utilizing stepped type concrete armour blocks developed by Hydraulic Laboratory, Nippon Tetrapod Co. Ltd. The new stepped type named Terrace Block was developed with the desire to achieve :

(a) easy access to the beach.
(b) a low crest elevation to blend with the landscape thanks to reduced wave runup.

Fig. 5 shows a view of a Terrace Block. The weight per unit is 2.5 tons and the block has the following properties.

(a) Holes absorb uprushing water into a permeable layer beneath.
(b) Stable armour layer utilizing wedge shape for interlocking.

A revetment of Terrace Blocks has the following functions.

(a) Reduction of wave runup.
(b) Aesthetically pleasing view from of adopting curved fronts as shown in Fig. 6.
(c) Accessibility, like a wide stairway.
(d) Safety is ensured as drainage gaps are tucked away under the eaves of the blocks.

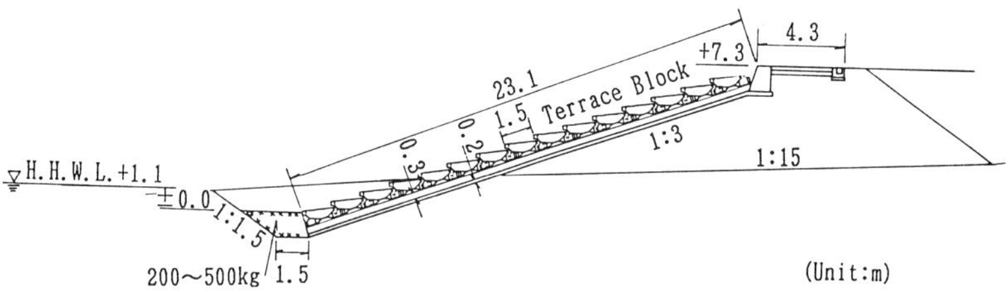

Fig. 4. A typical cross section of mild slope revetments utilizing stepped type concrete armour blocks

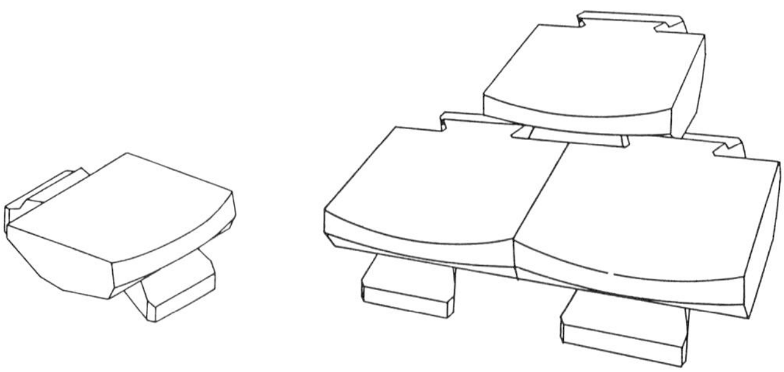

Fig. 5. Sketch of Terrace Block

Fig. 6. View of Terrace Block revetment

10. In certain cases Terrace Blocks C-type is available as shown in Fig. 7. The C-type is particularly suitable for application to curved alignments as shown in Fig. 8 and Fig. 9. A 2.5 ton block or a taller 4.0 ton block to improve stability if necessary, can be used.

11. Runup reduction effect. Our laboratory experiments were conducted to obtain wave runup data on the newly developed mild slope revetments. At first, the effect of the slope angle was checked using impermeable slopes inclined at 1:3 and 1:5 as shown in Fig. 10(a) when h/Lo equals 0.025, where R is the wave runup height, Ho' the equivalent deep water wave height, Lo the wave length of offshore and h the toe depth of the revetment. The result revealed the wave runup height for the 1:5 to be considerably smaller than that for the 1:3 slope. This result has been well documented by Saville(ref.3), etc., and supports the view that a mild slope has a significant effect on the wave runup.

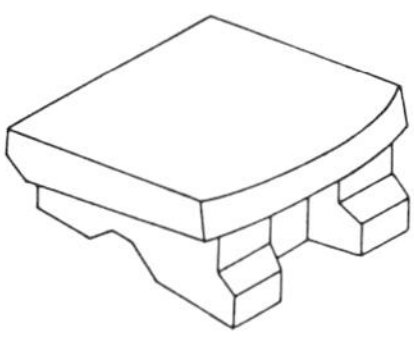

Fig. 7. Sketch of Terrace Block C-type

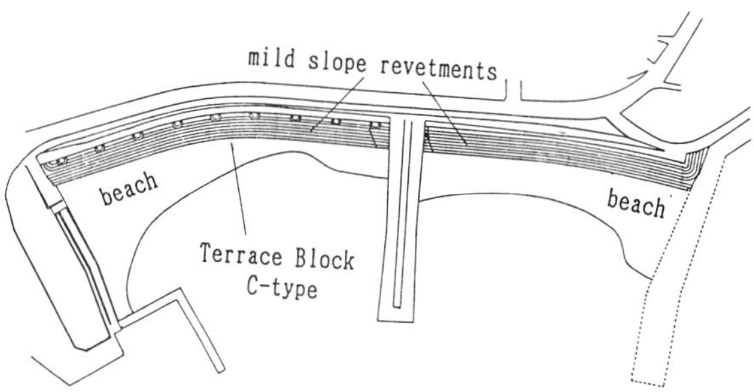

Fig. 8. Example of curved alignment

Fig. 9. View of application of curved alignment

12. Secondly, the effects of surface shape and permeability were checked. Fig. 10(b) illustrates the findings for both impermeable slope of smooth and stepped and revetment covered with Terrace Blocks with the cross section the same as a stepped slope when h/Lo is from 0.011 to 0.015. All of them were given same angle of slopes, 1:5. Although the factor Ho'/Lo has been given a different runup in each slope condition, the surface of the slope with steps and of permeable condition, shows superior runup reduction in comparison with the smooth and impermeable slopes.

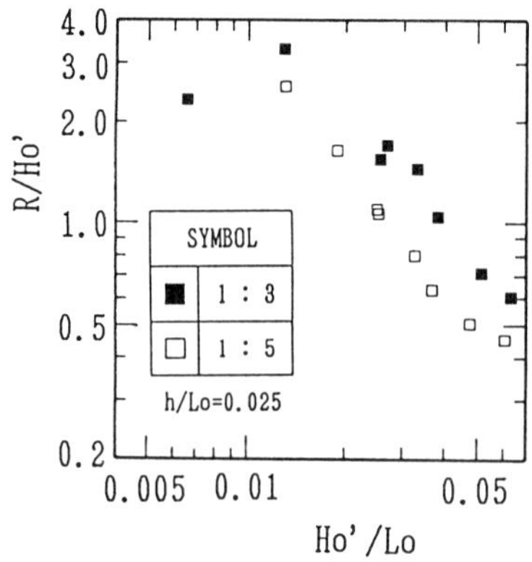

Fig.10(a) Effect of slope angle on smooth, impermeable slope

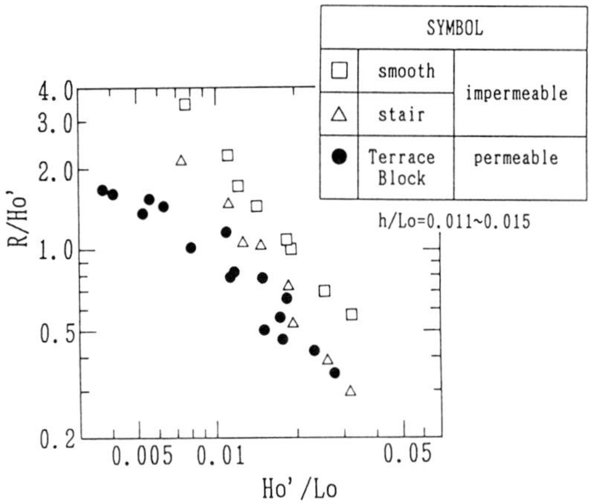

Fig.10(b) Effect of surface shape and permeability, 1:5 slope

Fig. 10. Runup reduction effects of mild slope revetments covered with Terrace Blocks

13. Runup height on revetments covered with Terrace Blocks on slope of 1:3 and 1:5. These are shown as in Fig. 11 in a comparison of changes of toe depth. The results reveal that as the wave runup changes depend on the toe depth, the crest elevation should be determined taking into consideration not only existing toe depth but also anticipated the scouring depth.

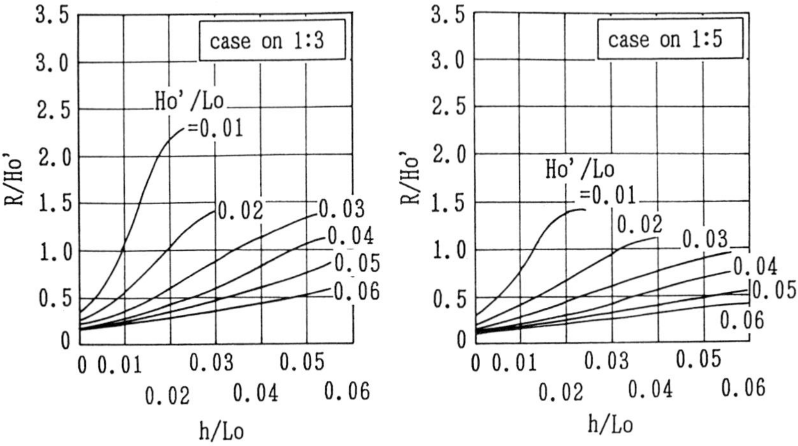

Fig. 11 Runup height of mild slope revetments covered Terrace-blocks

14. The stability of Terrace Blocks on a mild slope revetment. The stability of the rubble mound, expressed by the following surf-similarity parameter ξ depends not only on the

slope angle but also the wave period(ref. 4).

$$\xi = \tan\theta / \sqrt{(H/Lo)} \tag{1}$$

where θ is the angle of slope, H, the wave height at the toe of the slope, and Lo, the wave length of the offshore. Using the parameter and the test results regarding stability, our results were plotted as shown in Fig. 12. In the figure, the black and white circles show the unstable and stable situations respectively together with a line indicating the boundary. As a design wave height is normally governed by the breaking wave height in areas shallower than the surf zone, changes of water depth in front of structures should be taken into account.

15. For reference, the stability of Crab Blocks is illustrated in Fig. 13, where $1/K_D$ is described with the following Hudson formula(ref. 5).

$$1/K_D = \frac{W(Sr-1)^3}{\gamma r \ Ho^3} \tag{2}$$

The more ξ changes, the more $1/K_D$ differs. The stability of Crab Blocks depends on ξ, in the same way as the stability of Terrace Blocks does.

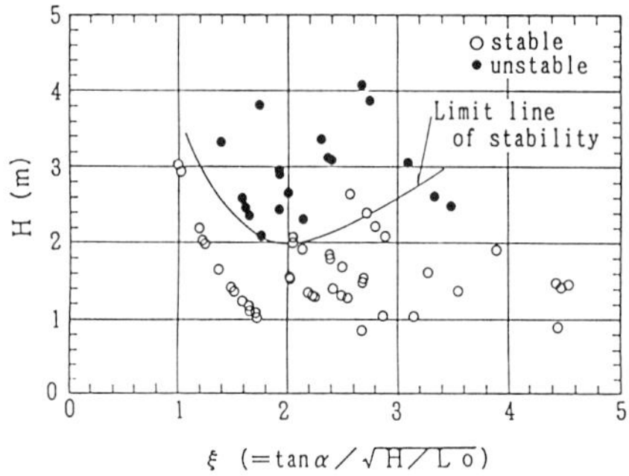

Fig. 12. Effect of ξ for stability of Terrace Blocks

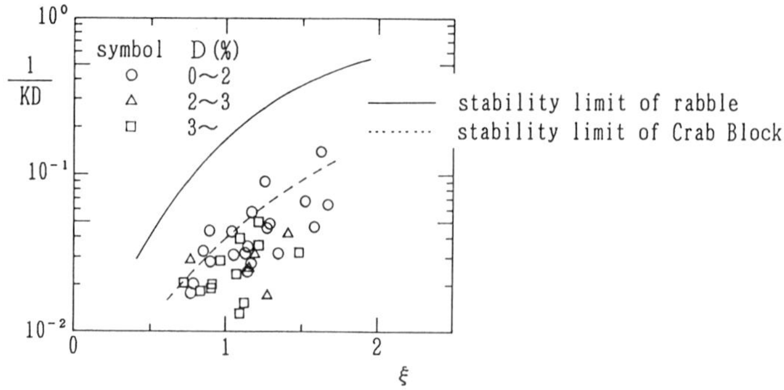

Fig. 13. Stability limits of Crab Blocks

16. Design considerations for permeable underlayer. The permeable underlayer plays an important role in supporting the armour layer and protecting the foundation. The permeable underlayer has to be strong enough to support the upper layer while accommodating filtering and preventing core materials from being washed away by the uprush and backrush waves. Another important role is to act together with the upper layer to weaken the backrush waves in order to alleviate damage from toe scouring. From this view point, an armour layer utilizing Terrace Blocks is rather suited for the purpose because it creates less wave runup compared with impermeable stepped revetments.

17. Fig. 14 illustrates an example of the type of failure that can occur if design of the underlayer is unsuitable. This kind of damage has actually been observed in several sites where the underlayer was too thin or the stone size unsuitable, as shown in Fig.15. Terrace Block revetments have been installed in more than 100 places in Japan. A few have received damage. An example of a damaged section is shown in Fig.16. In this case, some of the blocks were turned over and the underlayer stones as well as the toe apron were washed out. As will be explained later, this incident provided a good experience for us to base examinations into the stability of Terrace Block revetments and the reconstruction of sections.

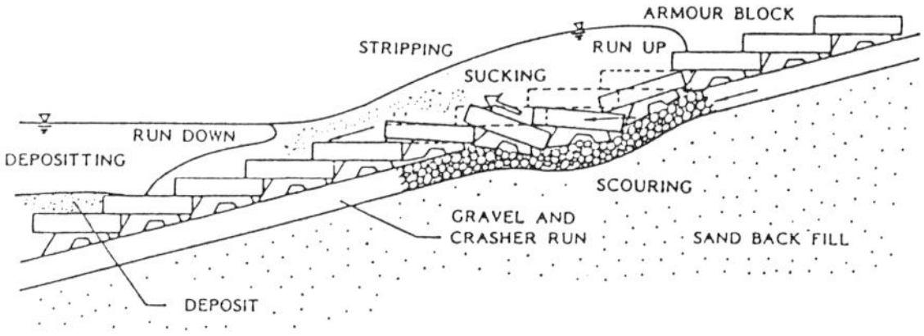

Fig. 14. Schematic drawing on instability of a gentle sloped seawall (ref.6)

Fig. 15. Damage from underlayer washout

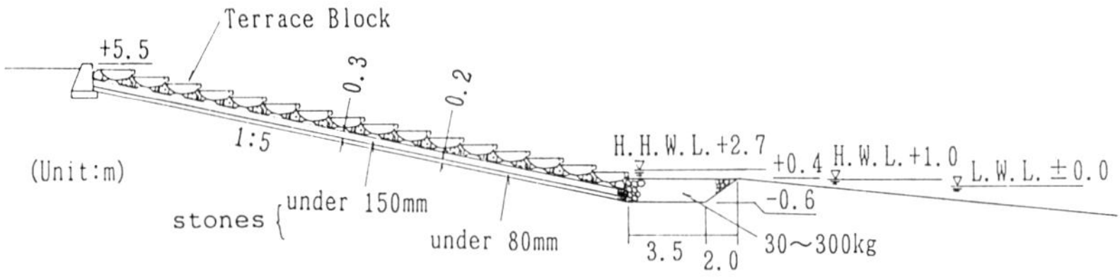

Fig. 16. An example of damaged cross section of Terrace Block revetment

18. Toe apron protection. The toe apron of a revetment is that part which is most exposed to wave attack and is therefore important in the prevention of any damage from scouring, and washing out of the underlayer as the total stability of the structure is affected.

19. Regarding the damaged Terrace Block revetment shown in Fig.16, improvement plans were examined using hydraulic model tests with irregular waves. The sea bed slope was approximately 1/20 and the design wave height in the offshore was 5.41m. In the model tests, the stability of the Terrace Block armour layer, under layer stones and the toe apron were examined and an improvement plan decided according to the test results. A cross section of the reconstruction is shown in Fig.17(a). 4.0t type Terrace Blocks, a thick underlayer and an improved toe apron were adopted.

20. In this test, the stability of the improved toe apron was examined by means of movable bed tests. An example of the results is shown in Fig.17(b). The improved toe apron received relatively large deformation. However no damage was observed in the Terrace Block armour layer and underlayer stones.

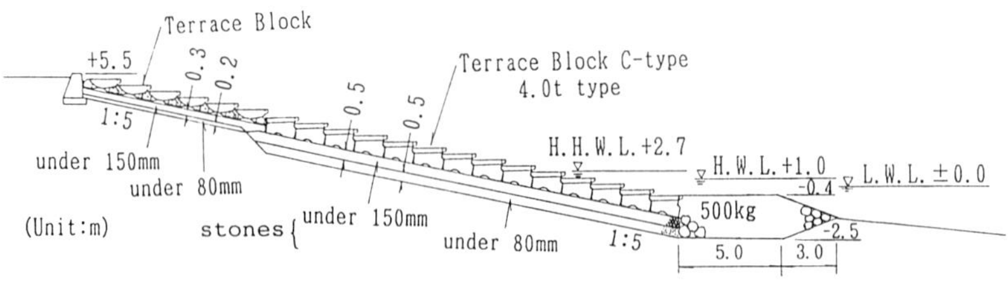

Fig.17(a) Reconstructed cross section after damage

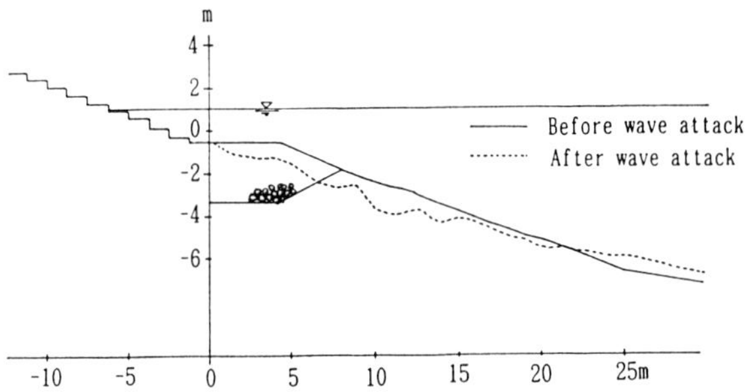

Fig.17(b) Toe apron deformation after wave attack in the movable bed model tests

21. <u>Scouring at the toe</u>. Scouring at the toe is always a problem to be solved in designing shore protection structures. An example of hydraulic model test results regarding toe scouring problems is shown in Fig.18 in comparison with an impermeable stepped revetment. A Terrace Block revetment can reduce the scouring depth because of less backrush waves and low reflection coefficients.

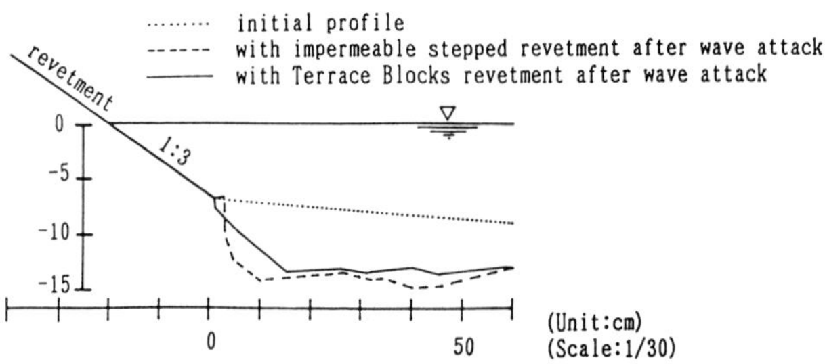

Fig. 18. Example of test results regarding toe scouring

22. Mild slope revetments should be designed after careful stability considerations regarding armour block weight, underlayer conditions and toe protection. As well as stability considerations for individual parts of the structure, another source of damage is deepening of the seabed at the toe and in the nearshore area. As the design wave height and suitable location of toe apron are governed by the water depth at the toe and nearshore sea slope, changes in the seabed profile should be taken into account to secure the stability of the structure.

<u>Trends in shore protection</u>

23. In recent years, the artificial reef as shown in Fig. 19 has started to replace detached breakwaters and present a more beautiful view of the sea. At present, the effectiveness of artificial reefs has not been confirmed in the site so experimental constructions are under way in some areas.

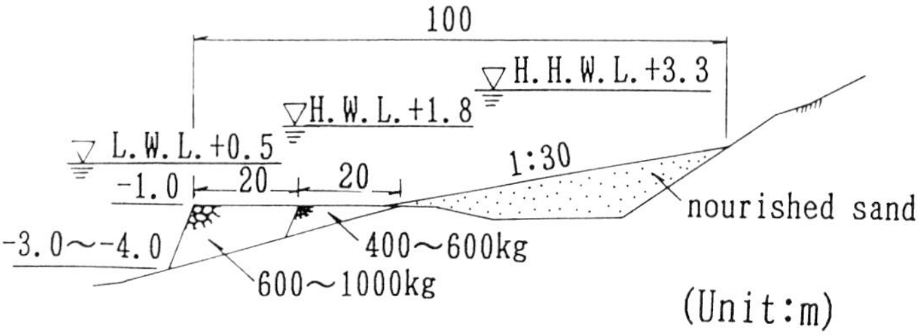

Fig. 19. Typical section of artificial reef

24. Inplace of artificial reefs covered with rubble, a new structure using reinforced concrete blocks is now being investigated. This structure, named Aqua Reef, was designed aiming at large wave energy dissipation within the structure and increased stability induced by the blocks leaning backward on to each other as shown in Fig. 20. An additional bonus with the Aqua Reef is its potential as a fisheries reef thanks to the large spaces within the structure.

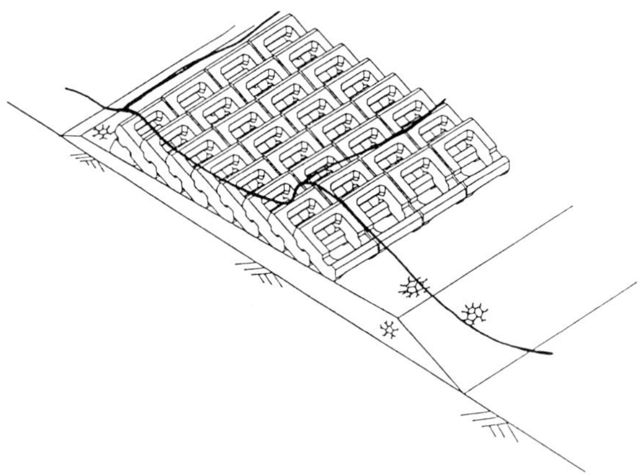

Fig. 20. Sketch of Aqua Reef

25. Results of model testing with the Aqua Reef, in terms of wave reduction as viewed in artificial reefs are described. An example of the transmitted wave height ratio at the crown end of the nearshore side is shown in Fig. 21 after adjusting the result for the case Ho'/Lo from 0.02 to 0.03 where Ht, is the transmitted wave height. The more R/Ho' decreases or B/Lo increases, the more the wave reduction is effective.

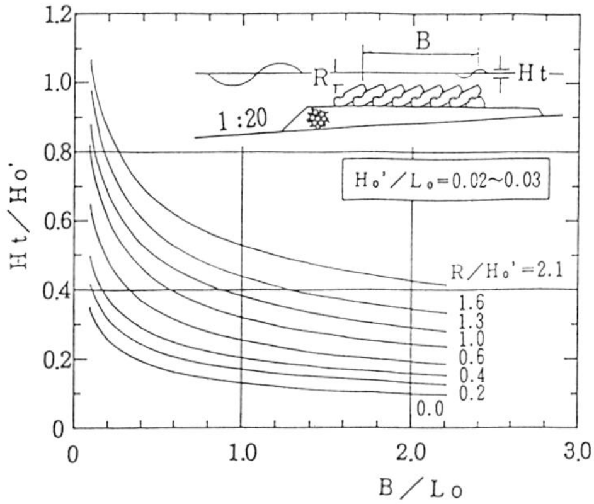

Fig. 21. Figures for planning for transmitted wave heights

26. In adopting mild slope revetments, protection with either a detached breakwater or artificial reef is recommended in those cases where the structure is exposed to severe storm waves as these structures generally show less stability than the conventional type. This kind of combined protection is very desirable to realize the accessibility to the beach and relieve beach erosion problems.

REFERENCES

1. Toyoshima O. New Type Blocks for Seawall Slope Protection. Proc. 19th. Costal Eng. Conf., 1984, 2536-2545.
2. Toyoshima O. Gentle Slope Seawalls Covered with Armour Units. Proc. 21th. Costal Eng. Conf., 1988, 1983-1996.
3. Saville T. Jr. Wave Runup on Composite Slopes. Proc. 6th. Conf. on Coastal Eng., ASCE, 1958, 691-699.
4. Bruun P. and Gunbak A. R. Stability of Sloping Structures in Relation to $\xi = \tan\alpha / \sqrt{(H/Lo)}$ Risk Criteria in Design. Proc. Conf. on Coastal Eng., 1977, 287-322.
5. Hudson R. Y. Laboratory Investigation of Rubble Mound Breakwaters. Jour. of Waterways and Harbor Div., ASCE, 1959, Vol.85, No. WW3, 93-121.
6. Maeno Y., Yamamoto M. and Nemoto K. Stabilities of Gentle Slope Seawall with Armour Unit. Proc. of 7th Symposium on Coastal and Ocean Management, Coastal Zone '91, Coastlines of JAPAN, 1991, 108-122.

26. Simplified design method for block revetments

M. KLEIN BRETELER, Delft Hydraulics, and A. BEZUYEN, Delft Geotechnics

SYNOPSIS. A summary of the theoretical and empirical knowledge of the stability of block-revetment structures under wave attack is given. This has been accomplished by selecting the very essence of the design formulas from [2] and completing it afterwards with a data base of results of large-scale model studies from all over the world.

1 INTRODUCTION

1. Extensive studies have been performed during the last decade on the stability of placed block-revetments, mainly in the Netherlands, but also abroad. The studies in the Netherlands have been carried out by DELFT HYDRAULICS and DELFT GEOTECHNICS in cooperation with and commissioned by the Netherlands Ministry of Transport and Public Works (Rijkswaterstaat). It has led to the establishment of a large data base with results of model studies. In addition, it also resulted in various design formulas (see for example ref. [1] and [2]).

2. Examples of the structures considered in this paper are shown in Figure 1.

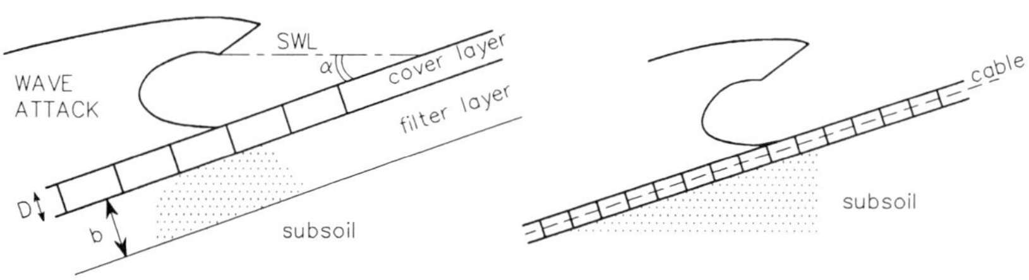

Figure 1, Examples of block-revetment structures (cross sections)

3. Such structures are found on banks of rivers and lakes and on sea dikes. Damage to these structures is mainly caused by wave attack. Damage by current is only found when applied in spillways, steep canals, etc.. These are not considered here.

4. A theory concerning the uplift of individual blocks will be presented briefly in the next chapter, resulting in a for-

mula including all significant parameters that influence the stability. This theory will be simplified for practical application and then fitted to the results of large-scale model studies in chapter 4.

2 THEORY FOR REVETMENT ON GRANULAR FILTER

5. It is concluded from several large-scale model studies with revetment structures based on a granular filter layer (for example [3]) that the uplift of individual blocks occurs just before, or during the wave impact. This is shown schematically in Figure 2.

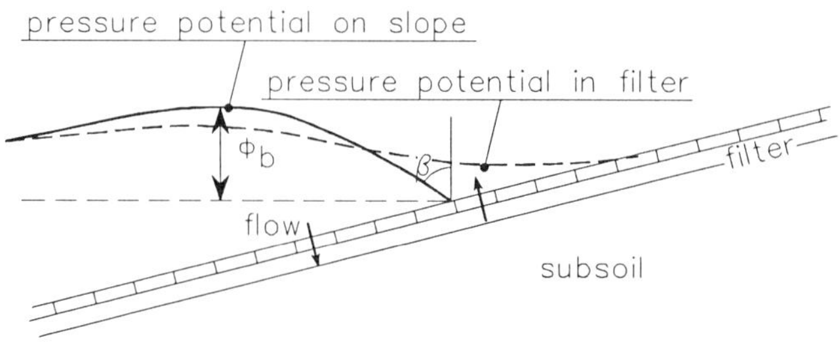

Figure 2, Schematized damage-mechanism caused by pressure potential (piezometric head) over the cover layer just before wave impact.

6. During the wave rundown there is a large piezometric head gradient on top of the revetment (see Figure 2), caused by the simultaneous occurrence of rundown of the preceding wave and the arrival of the present wave. The piezometric head underneath the top layer is a damped representation of the potential on top of the revetment, causing an uplift pressure at the location of maximum wave rundown. The extend of the damping is influenced by the permeability ratio of the top-layer and the filter layer and also by the compressibility of the air/water-mixture in the filter. The latter is important for very fine granular filters (such as fine sand) and will not be considered here (see for example [4]).

7. The piezometric head over the cover layer during wave rundown can be quantified by considering the mass balance of the water in the filter and the Darcy flow equation (Figure 3).

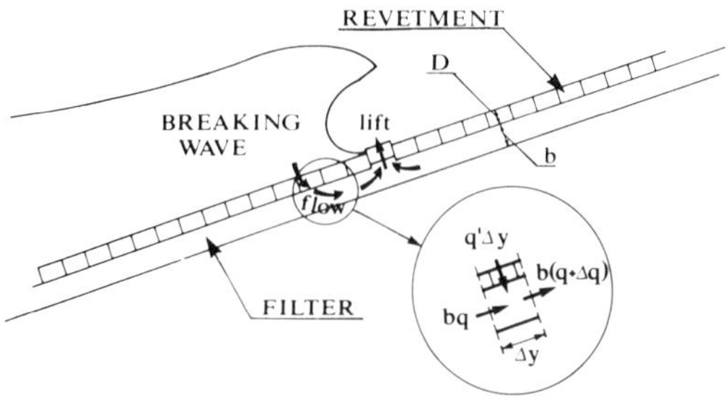

$$\Rightarrow \quad \frac{\delta^2 \phi}{\delta y^2} = \frac{1}{\Lambda^2} (\phi - \phi')$$

Figure 3, Mass balance in filter

8. This has been described in detail in [2]. Here only the resulting formula for the maximum piezometric head over the cover layer is given:

$$[\phi - \phi']_{max} = \{\frac{\Lambda}{2\tan\beta}\,[1 - \exp(-\phi_b \tan\beta/\Lambda)] + \frac{1}{2}\,\Lambda\sin\alpha\}\cdot$$
$$\cdot\{1 - \exp[-2\phi_b/(\Lambda\sin\alpha)]\} \qquad (1)$$

with:

ϕ	= piezometric head in the filter layer	(m)
ϕ'	= piezometric head on the slope	(m)
$[\phi - \phi']_{max}$	= maximum uplift piezometric head	(m)
q	= specific discharge through the filter layer	(m/s)
q'	= specific discharge through the cover layer	(m/s)
Λ	= leakage factor = $\sqrt{(kbD/k')}$	(m)
k	= linearized permeability of the filter	(m/s)
k'	= linearized permeability of the cover layer	(m/s)
b	= thickness of the filter layer	(m)
D	= thickness of the cover layer	(m)
y	= coordinate along the slope	(m)
ϕ_b	= maximum piezometric head on the slope	(m)
$\cot\beta$	= max. piezometric head gradient on the slope	(-)

9. The displacement of a block occurs if the uplift pressure exceeds the own weight of the block added by the additional forces, such as friction and inertia. The limit state is:

$$[\phi - \phi']_{max} = \Gamma\Delta D\cos\alpha \qquad (2)$$

with:

Δ	= relative density of the cover layer = $(\rho_c - \rho)/\rho$	(-)
ρ	= density of water	(kg/m³)
ρ_c	= density of cover layer (concrete)	(kg/m³)
Γ	= coefficient representing friction, inertia etc. [2]	

10. The maximum piezometric head distribution on the slope is characterised by the maximum piezometric head ϕ_b and maximum gradient β, see Figure 2. Empirical formulas for ϕ_b and β are given in [2]. A re-analysis of the original test data has shown that the following simple formulas, which can easily be applied in combination with eq. (1) and (2), describe the tendencies in the measurements as good as the rather complicated formulas in [2]:

$$\phi_b = 0.36\cdot\sqrt{(HL_o\tan\alpha)} \qquad (3)$$

$$\tan\beta = 0.17/\sqrt{(H/L_o)} \qquad (4)$$

with:

H	= incoming wave height (regular waves)	(m)
L_o	= wave length at deep water (= $gT^2/(2\pi)$)	(m)
T	= wave period	(s)
g	= acceleration of gravity	(m/s²)

11. The substitution of eq. (2)....(4) in (1) leads to a complex and difficult-to-handle stability formula, in spite of the simplification of the formulas for the maximum piezometric

head on top of the slope. The complexity is not only hindering the practical use of the result, but also the comparison with results of model studies.
In the above given formulas some dimensionless parameters can be distinguished, which are used to simplify the stability formula:

1) $\Gamma\Delta D\cos\alpha/\Lambda$
2) H/Λ
3) $\tan\alpha$
4) H/L_o

12. The following simple power function has been derived by way of trial and error with the above given parameters:

$$\Gamma\Delta D\cos\alpha/\Lambda = 0.43\cdot(H/\Lambda)^{0.6}\cdot(H/L_o)^{-0.2}\cdot(\tan\alpha)^{0.5} \qquad (5)$$

13. This formula gives a satisfying agreement with the original stability formula (combination of eq. (1)....(4)) as is shown in Figure 4.

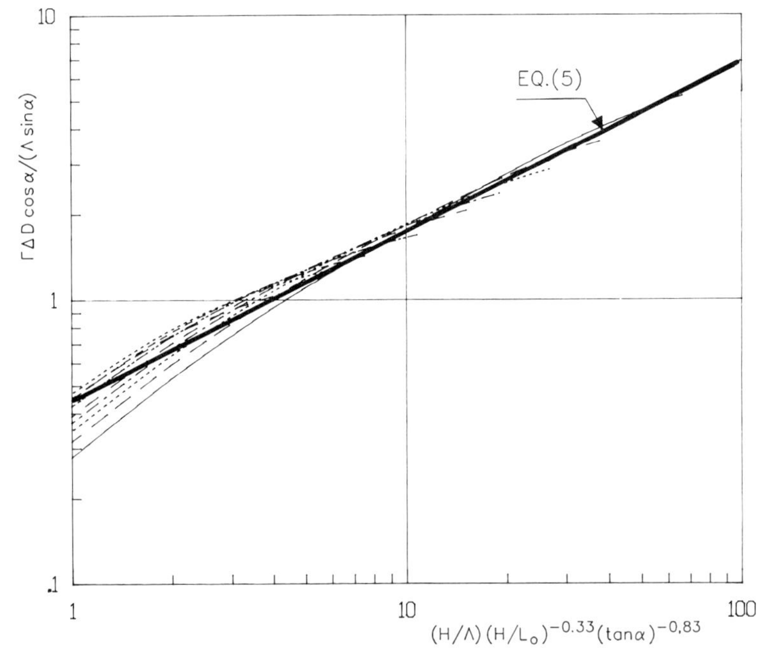

Figure 4, Comparison of original stability eq. and eq. (5) (lines are given only within the range of applicability of the original eq.)

14. This stability formula that has been derived holds for regular wave attack. Unfortunately, little is known about the load ratio of regular and irregular waves. It is believed that especially the large waves cause instability and that the amount of waves play a minor role. This leads to the assumption that the regular wave height H, can be compared with the

irregular wave height with a rather small exceedance percentage, for example $H_{2\%}$ or $H_{0.5\%}$. The results of model studies do not contradict this idea, although there is only a small number of relevant tests, showing a large scatter. These tests will be dealt with in chapter 4. On comparing the piezometric head on the slope under regular and irregular wave attack the following is concluded for the wave height at threshold of damage:

$$H/H_s = r \approx 1.4 \tag{6}$$

with:
H = wave height at threshold of damage with regular waves (m)
H_s = significant wave height at threshold of damage with irregular waves (m)
r = ratio between H and H_s (-)

15. Introducing the coefficient for irregular wave attack, r, and the breaker parameter, ξ_{op}, into eq. (5) leads to:

$$\Gamma\Delta D\cos\alpha/\Lambda = 0.43\cdot(H_s/\Lambda)^{0.6}\cdot\xi_{op}^{0.4}\cdot(\tan\alpha)^{0.1}\cdot r^{0.4} \tag{7}$$

with:
ξ_{op} = surf similarity parameter (-) = $\tan\alpha/\sqrt{(H_s/L_{op})}$

This yields, with the substitution of $\Lambda = \sqrt{(kDb/k')}$ and after some rearrangements:

$$\frac{H_s}{\Delta D} = f\cdot(k'/k)^{0.33}\cdot(D/b)^{0.33}\cdot\xi_{op}^{-0.67} \tag{8}$$

with:

$$f = 3.9\cdot(\tan\alpha)^{-0.17}\cdot\Gamma^{1.67}\cdot\Delta^{0.67}\cdot r^{-0.67}\cdot\cos\alpha \tag{9}$$

16. This stability formula contains all our theoretical understanding of the physical phenomena involved. However, the practical applicability is limited:
- it is only useful for block revetments with a granular filter layer underneath the blocks.
- the permeability of the cover layer can be calculated with the formulas in [20], but they are very complicated.
- a lower boundary for Γ has been given in [2], but a good quantification is still not possible.
- the theoretical and empirical bases for the assumption $H/H_s = 1.4$ is poor.

17. Most of these problems can be avoided by the application of only the most essential parts of the stability formula and to complete it with empirical data from large-scale model studies. The very essence of eq. (8) is believed to be:

1) $$\frac{H_s}{\Delta D} = F\cdot\xi_{op}^{-0.67} \tag{10}$$

2) F depends on the type of structure, characterised by the ratio's of k'/k and D/b. With the permeability formulas from [20] it is concluded that the parameter (k'/k)·(D/b) ranges between 0.01 and 10, leading to a subdivision into

3 ranges of one decade each. Therefore the following types are defined:

a) Low stability: $(k'/k)\cdot(D/b) < 0.05....0.1$

b) Normal stability: 0.5 or $1 > (k'/k)\cdot(D/b) > 0.05....0.1$

c) High stability: $(k'/k)\cdot(D/b) > 0.5....1$

The conditions for high stability are very difficult to meet and it turns out that there are no model studies performed with this type of structure. Therefore, it has been left out in the following, leaving only two types of structures. With the formulas from [20] they can be defined as follows (D_{f15} = grain size of filter exceeded by 85% of weight, (m)):

- low stability, if:
  - thick filter layer, namely: $b/D \gtrapprox 0.5$ <u>and</u>
  - course filter material, namely: $D_{f15} \gtrapprox 4$ mm <u>and</u>
  - closed cover layer, namely:
    - solid blocks with small joints: open area $\Omega < 2\%$
    - blocks with holes with a spacing less than 0.3 m: $\Omega<5\%$
    - blocks with holes with a spacing wider than 0.3 m: $\Omega<10\%$
- normal stability (structures other than defined as "low stability")

3 EXTENSION OF THEORY TO REVETMENTS WITHOUT GRANULAR FILTER

18. Up to now we have dealt with block revetments on a granular filter layer only. However, there are also structures with a cover layer directly placed on clay, or with a geotextile on sand. For these structures there is no such a theory as for the blocks on a granular filter. Therefore, we can merely assume that eq. (10) also is valid for these structures. In the next chapter this assumption proves not to be contradictory to the available test results.

19. Another subdivision that we make is:

a) cover layers with individual, not interconnected, blocks.

b) cover layers with linked blocks or interlocking blocks.

In the first type an individual block can be lifted out of the revetment with a force exceeding own weight and friction, which is not possible with the second type. Examples of the second type are: block mattresses, ship-lap blocks and cable mats.

20. We can conclude that the theory has led to a simple stability formula (eq. (10)) and a subdivision into 8 types of structures:

a) cover layer with loose blocks (without linkage or interlocking):
   - a1) cover layer on granular filter, low stability
   - a2) cover layer on granular filter, normal stability
   - a3) cover layer on geotextile on sand
   - a4) cover layer on clay

b) cover layer with linked blocks (for example with cables or interlocking, such as block-mattresses):
   - b1) cover layer on granular filter, low stability
   - b2) cover layer on granular filter, normal stability
   - b3) cover layer on geotextile on sand
   - b4) cover layer on clay

The coefficient, F, will be quantified for each structure type by way of fitting eq. (10) to the results of model studies in the next chapter.

4 MODEL STUDIES

21. The theory presented in the previous chapter is fitted to the results of a large collection of results of model studies from all over the world.
Only large-scale studies are used because both the waves and the wave induced flow in the filter should be well represented in the model. This requires:

a) Waves:
Froude number: $Fr = u/\sqrt{(gh)} = 1$

b) Flow in the filter:
Reynolds number: $Re = qD_{f15}/\nu = 1$ or both in model and prototype $Re > 600$ (turbulent flow) or both in model and prototype $Re < 4$ (laminar flow).
with:

u	=	water velocity or wave celerity	(m/s)
h	=	water depth	(m)
D_{f15}	=	characteristic grain size of filter	(m)
ν	=	viscosity of water	(m^2/s)

Usually it is not possible to meet these requirements. Consequently, it is concluded that in most cases the results are influenced by the model scale.
Therefore only large-scale model studies, with $D > 0.1$ m or $H > 0.5$ m, can be used to find the coefficient F. This results in a length scale of less than 5 for the most important applications, namely $1 < H_s < 2$ m.

22. The majority of tests have unfortunately been carried out with regular waves. The load of these tests has been transformed to an equivalent random wave load with eq. (6). Table 1 gives some results of stability tests which are performed with both random and regular waves. The type of the structure is according to the subdivision given at the end of chapter 3. Since some comparable tests were not carried out with the same ξ_{op}-value, it was necessary to use eq. (10).

Discription of structure	Ref.	Type	Average H/H_s
Basaltonn on geotextile on sand	[12]	a3	1.46
Open blocks on granular filter	[6]	a2	1.50
Blocks on filter	[5]	a1	1.31
Blocks on clay	[8]	a4	1.87
Blocks on filter	[8]	a1	0.86

Table 1, Stability ratio for random and regular waves.

23. The accuracy of these results is highly influenced by the fact that the stability of a revetment depends on coincidental friction between the blocks. Consequently, the results of different tests do not reproduce very well. This poor re-

productivity especially decreases the accuracy of the measured ratio of H/H_s.
This was less present during the tests in [5], because the model was constructed in such a way that friction was minimized.
The average of all available data, being $H/H_s = 1.4$, lies close to the result from [5] and is therefore used for all tests.

24. In the following Figures and Tables all available tests are summarised and for each type of structure a lower and upper boundary for the value of F is given. The lower boundary gives with eq. (10) a stability curve under which stability is guaranteed. Between the upper and lower boundary the stability is uncertain. It depends on various unpredictable influences if the structure will be stable or not. The upper boundary gives a curve above which instability is (almost) certain.

25. All values listed in Table 2 are on model scale. In some reports the exact information about cover layer or filter was lacking. In these cases the open area percentage, Ω, and/or the characteristic grain size were estimated from photographs, drawings or descriptions (indicated in the table with "≈"). Then a range for the value of $(k'/k)\cdot(D/b)$ is given in the Table.

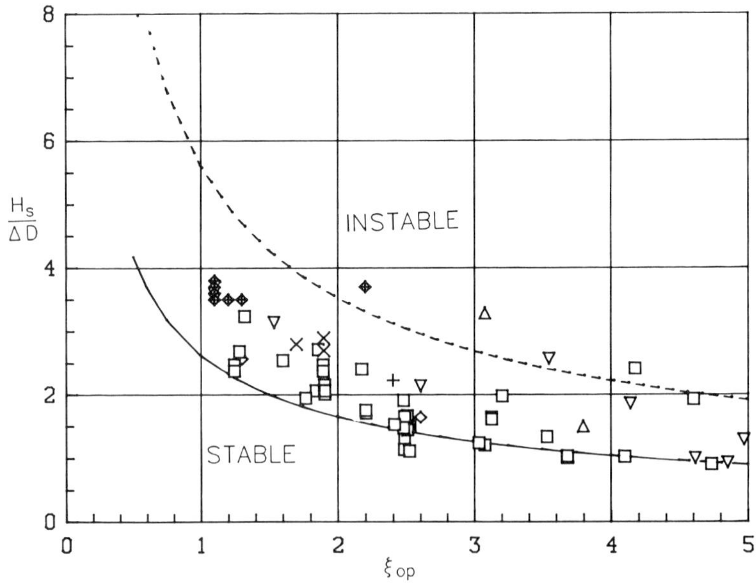

Fig. 5, Test results on type a1 (loose blocks on granular filter, low stability)

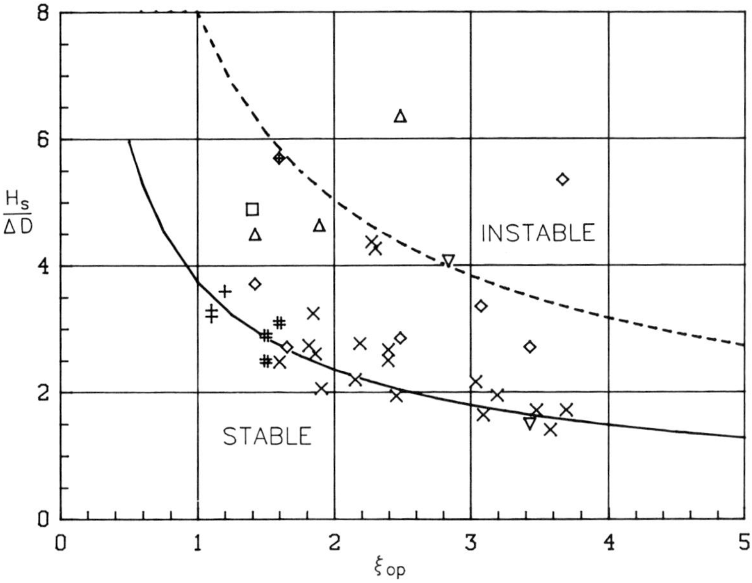

Fig. 6, Test results type a2 (loose blocks on granular filter normal stability)

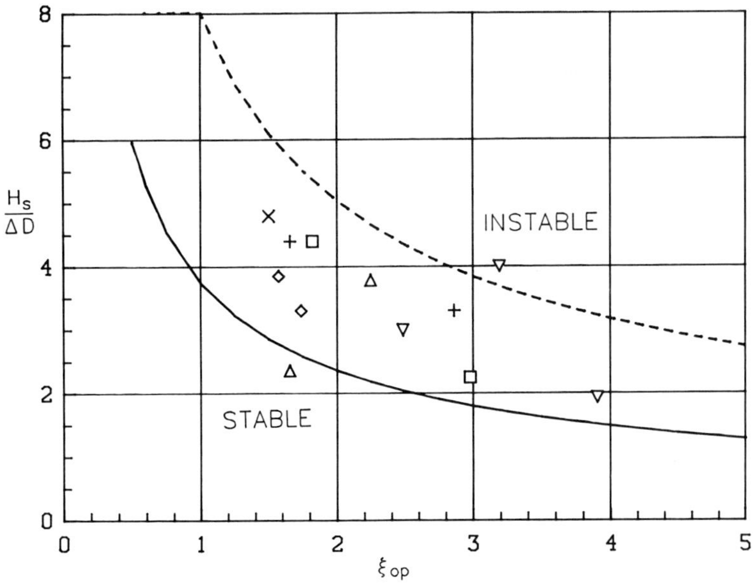

Fig. 7, Test results type a3 (loose blocks on geotextile on sand)

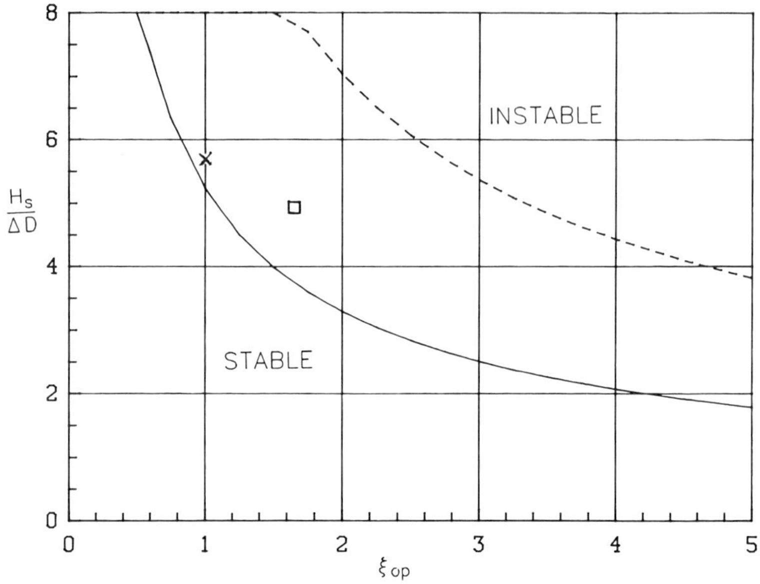

Fig. 8, Test results type a4 (loose blocks on clay)

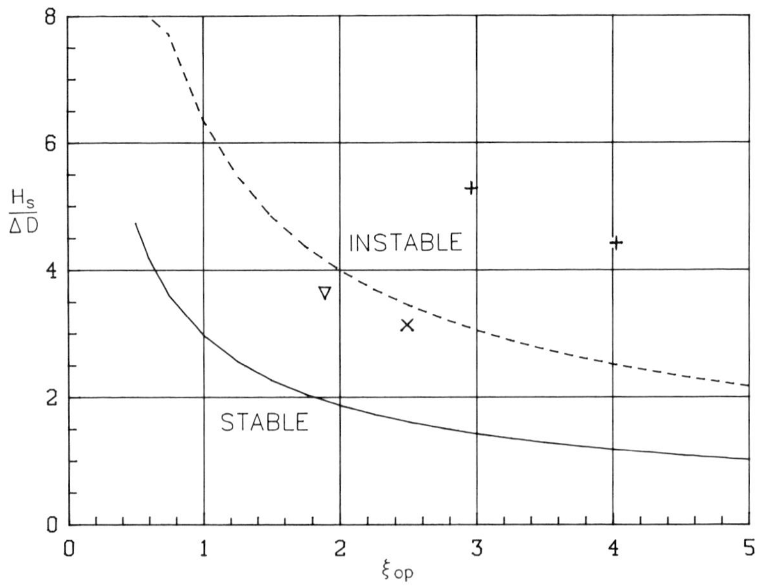

Fig. 9, Test results type bl (linked blocks on granular filter, low stability)

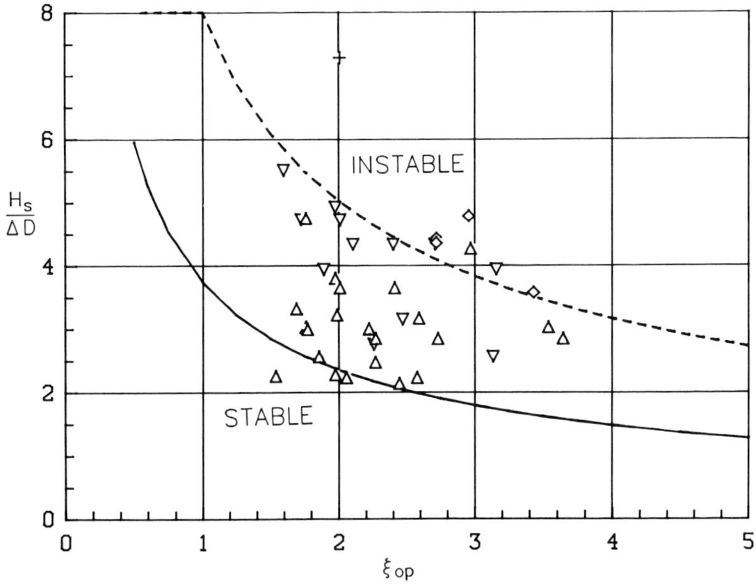

Fig. 10, Test results type b2 (linked blocks on granulair filter, normal stability)

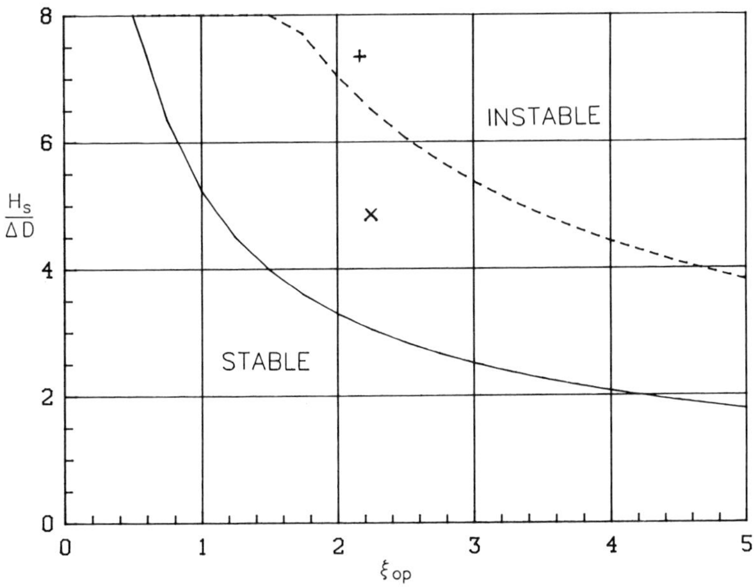

Fig. 11, Test results on structure type b3 (linked blocks on geotextile on sand)

Type	Ref.	tanα	D	Δ	Ω	b	D_{f15}	(k'/k)·(D/b) value or range	Figure symbol
-	-	-	m	-	%	m	mm		
a1	5	0.33	0.15	1.35	≈1.5	0.35	≈25	0.01 - 0.02	▽△
a1	8	0.25	0.10	1.37	≈0.5	0.15	14	0.01	◇⊕
a1	8	0.25	0.15	1.37	≈0.5	0.15	14	0.02	◇
a1	16	0.20	0.20	1.30	≈0.8	0.20	≈12	0.05 - 0.20	+
a1	15	0.33	0.15	1.38	2.0	0.50	3.6	0.06	□
a1	15	0.33	0.15	1.35	2.0	0.50	9.2	0.03	□
a1	15	0.33	0.15	1.20	2.0	0.50	9.2	0.04	□
a1	15	0.33	0.15	1.38	0.8	0.50	3.6	0.02 - 0.03	□
a1	15	0.33	0.15	1.39	5.3	0.50	8.5	0.02 - 0.03	□
a1	15	0.33	0.15	1.36	9.1	0.50	8.5	0.04	□
a1	15	0.33	0.30	1.38	2.0	0.35	17	0.07	□
a2	5	0.33	0.15	1.35	≈16	0.35	≈25	0.04 - 0.51	▽
a2	8	0.25	0.10	1.40	≈ 7	0.15	14	0.03 - 0.06	+
a2	7	0.33	0.18	1.40	≈12	0.15	≈22	0.21 - 0.44	□
a2	6	0.33	0.12	1.40	21	0.25	25	0.15 - 0.30	⊕
a2	9	0.29	0.10	1.30	≈30	"∞"	≈ 1	-	◇
a2	10	0.29	0.22	1.53	≈20	0.05	≈15	0.6 - 2.5	#
a2	10	0.29	0.25	2.00	≈15	0.05	≈15	0.5 - 2.0	#
a2	10	0.29	0.20	1.38	≈ 0.5	0.04	≈ 9	0.05 - 0.15	#
a2	15	0.33	0.15	1.35	2.0	0.25	3.6	0.12	×
a2	15	0.33	0.15	1.35	5.3	0.50	3.6	0.08	×
a2	15	0.33	0.15	1.36	9.7	0.50	8.5	0.06	×
a2	15	0.33	0.15	1.39	16	0.50	8.5	0.09	×
a2	15	0.33	0.15	1.39	16	0.50	8.5	0.06	×
a3	11	0.33	0.11	1.41	≈ 0.5	-	-	-	▽
a3	12	0.33	0.20	1.30	15	-	-	-	△×
a3	16	0.25	0.20	1.20	≈ 0.7	-	-	-	◇
a3	16	0.25	0.10	1.20	≈ 1.5	-	-	-	+
a3	16	0.25	0.20	1.20	≈ 2	-	-	-	□
a4	8	0.25	0.10	1.37	≈ 0.5	-	-	-	×□
b1	9	0.50	0.14	1.30	≈ 0.5	0.15	≈20	0.02 - 0.06	×
b1	9	0.50	0.15	1.30	≈ 0.8	0.15	≈20	0.01 - 0.02	+
b1	15	0.33	0.12	1.35	≈ 1.7	0.54	≈ 8.5	0.02 - 0.04	▽
b2	18	0.33	0.08	1.45	14	0.15	≈75	0.05 - 0.10	▽
b2	19	0.33	0.06	1.28	19	0.10	≈10	0.03 - 0.07	△
b2	19	0.33	0.10	1.28	19	0.10	≈10	0.08 - 0.16	△
b2	13	0.33	0.10	1.13	≈15	"∞"	≈ 3	-	◇
b2	9	0.50	0.14	1.30	≈ 1.5	0.15	≈20	0.06 - 0.13	+
b3	14	0.33	0.10	1.30	≈10	-	-	-	×
b3	17	0.33	0.05	1.43	14	-	-	-	+

Table 2, Some details of the stability studies

26. Most of the tests have been performed without a berm in the slope, except the studies reported in [8] and [18] and most of [19].

27. The studies concerning structure type b2 and b3 were very scarce. Therefore, also the tests with a rather small scale are used.

28. There are no test data available on structures with linked blocks on clay (b4).

29. The results for structure type a3 and b3 (blocks on geotextile on sand) may only be applied if $H_s < 1....1.5$ m or to structures with course sand ($D_{b50} > 0.3$ mm) and gentile slope ($\tan\alpha \leq 0.25$), because the uplift of blocks is assumed to be the dominant damage mechanism (instead of soil mechanical failure).

30. The results for structure type a4 can be applied on the condition that clay of high quality is used. If there is no such clay present, then a geotextile is recommended to prevent erosion during (lengthy) wave loading. The stability is than equal to that of structure type a3.

31. The Figures lead to conclusion given in Table 3.

Type	Description	F min	F max
a1	loose blocks on granular filter, low stability	2.6	5.6
a2	loose blocks on granular filter, normal stability	3.7	8.0
a3	loose blocks on geotextile on sand	3.7	8.0
a4	loose blocks on clay	5.1	11.0
b1	linked blocks on granular filter, low stability	3.0	6.5
b2	linked blocks on granular filter, normal stability	3.7	8.0
b3	linked blocks on geotextile on sand	5.1	11.0
b4	linked blocks on clay	?	?

Table 3, Lower and upper value for F.

32. The choice of the F-value, which is rather small compared to the results of the tests, is based on the assumption that the stability must be less than type b2 and the fact that the number of tests is too small to exclude coincidence.

5 DISCUSSION

33. The combination of the stability formula, that was derived from theory, together with the results of many large-scale model studies from all over the world has produced a reliable design tool for the preliminary design of placed block-revetments.

34. Its reliability is only influenced by the fact that most tests have been performed with regular waves. An inaccuracy is introduced by the transformation of the regular wave load to an equivalent irregular wave load. Further studies are undertaken to improve the transformation method.

REFERENCES [x]

1 Bezuijen A., M. Klein Breteler and K.J. Bakker; Design criteria for placed block-revetments and granular filters; Proc. of 2nd Int. Conf. on Coastal and Port Eng. in Developing Countries, Beijing 1987.

2 Burger, Klein Breteler, Banach, Bezuyen and Pilarczyk; Analytical design formulas for relatively closed block-revetments; Journal of Waterway, Port, Coastal and Ocean Eng. ASCE, vol. 116 no. 5 Sept/Oct. 1990.

3 Boer, K. den, C.J. Kenter and K.W. Pilarczyk; Large-scale model tests on placed blocks revetment; Conf. on Coastal Structures, Washington 1983; DELFT HYDRAULICS, publ. 288.

4 Bezuijen A., M. Klein Breteler and K.W. Pilarczyk; Large scale tests on a block-revetment placed on sand with a geotextile separation layer; Third Int. Conf. on Geotextiles, Vienna 1986.

5 Burger, A.M.; Taludbekleding van gezette steen, fase 2; Grootschalig gidsonderzoek; Verslag M1795/M1881, deel IX, dec. 1983; DELFT HYDRAULICS, DELFT GEOTECHNICS.

6 Lindenberg, J.; Stabiliteit van Armorflex-steenzettingen onder golfaanval; DELFT HYDRAULICS and DELFT GEOTECHNICS, M1910, jan. 1983.

7 Weide, J. v.d. and Visser, P.; Basalton, stabiliteit onder golfaanval; DELFT HYDRAULICS and DELFT GEOTECHNICS; M1900, febr. 1983.

8 Grootschalig onderzoek tbv de Oesterdam; M1795/M1881, deel VI, 1982; DELFT HYDRAULICS, DELFT GEOTECHNICS (see also [3]).

9 Boer, K. den; Taludbekleding van gezette steen; Hydraulische aspecten; M1795, deel II, juli 1982; DELFT HYDRAULICS, DELFT GEOTECHNICS (see also [3]).

10 Burger, A.M.; Sterkte Oosterscheldedijken onder Geconcentreerde Golfaanval; DELFT HYDRAULICS, M2036, 1985.

11 Meer, J.W. van der, en A. Bezuijen; Taludbekleding van gezette steen; Grootschalig modelonderzoek naar een steenzetting op zand; M1795/M1881, deel XII, dec. 1983; DELFT HYDRAULICS, DELFT GEOTECHNICS; (see also [4]).

12 Lindenberg, J.; Verweking van zand onder een steenzetting talud 1:3 onder invloed van golfbelasting; M1795/M1881 deel XXII C, 1988; DELFT HYDRAULICS, DELFT GEOTECHNICS.

13 Large scale model study of Armorflex slope protection; Tekmarine inc., Sierra Madre, California, May 1984.

14 Large scale modelling of Armorflex slope protection system; Tekmarine inc., Sierra Madre, California, Oct. 1983.

15 Burger, A.M.; Taludbekleding van gezette steen; Eindverificatie-onderzoek Deltagoot; Eerste concept meetverslag; DELFT HYDRAULICS, H195.25; 1988.

16 Führböter, A. and U. Sparboom; Full-scale investigations on the stability of concrete block-revetments; Proc. of Int. Symposium on Modelling Soil-Water-Structure interactions, SOWAS, Delft 1988.

17 Large-scale model studies of arctic island slope protection; Tekmarine inc., Project TCN-015, April 1982; Sierra Madre, California, USA.

18 Large-scale model investigation of compound slope profiles; Tekmarine inc., Project TCN-024, May/June 1983; Sierra Madre, California, USA.

19 Two-dimensional model study of slope protection systems for the Sohio Endicott Project; Tekmarine inc., January 1985; Sierra Madre, California, USA.

20 Klein Breteler, M. and A. Bezuijen; The permeability of closely placed blocks on gravel; Proc. of Int. Symposium on Modelling Soil-Water-Structure interactions, SOWAS, Delft 1988.

27. The use of bituminous materials with open structures in coastal engineering

J. J. L. M. KEMPS and P. C. BARBER, Hesslberg Hydro 1991 Ltd and Shoreline Management Partnership

COASTAL STRUCTURES AND BREAKWATERS

PAPER:
The use of bituminous materials with open structures in coastal engineering:

Synopsis:
For decades, Engineers and Technicians have been aware that care must be taken when applying asphaltic surfacing to roads in wet conditions.
They are also aware of the flexibility and resistance to erosion of bituminous bank protection and the possibility of achieving imperviousness by using bituminous linings and cores for dams and reservoirs.
However all those applications are known in connection with the requirement to place such materials in the dry.
This paper describes the development and application of bituminous products like Open Stone Asphalt and Lean Sand Asphalt that can be placed in wet conditions, even under water and provide a porous structure, well suited to hydraulic applications.

I: General view:
Much of the Netherlands is below sea level. The country is therefore very much dependent on good sea defences. Without dikes, the west of the country would be flooded which means that the Netherlands as it is now would cease to exist. Driven by the necessity to withstand the water for centuries the Dutch engineers built up their knowledge on coastal works.
Certainly during the last fifty years, due to the shortage of quarry stone, the abundance of sand and the wide availability of a well equipped dredging industry, a particular method of dam and dike building evolved in the Netherlands.

Open asphaltic mixtures like Lean Sand Asphalt (LSA) and Open Stone Asphalt (OSA) have been developed to meet the technical, economical and environmental requirements.
These products are very suitable for coastal structures and breakwaters because they have the properties that enable them to withstand large hydraulic loads.

II: Open Stone Asphalt (O.S.A.)

Par. 1: The most essential properties of O.S.A. are:

a: A porous structure similar to the subsoil. This virtually eliminates the uplift pressures that can cause damage to sea defences.
b: A flexible construction that allows irregular settlements to take place without causing cracking or other defects.
c: A smooth surface that unlike concrete blocks or rock armour cannot be disturbed by wave energy.
d: The capability of acting like a large elastic plate in order to absorb static and dynamic loads (wave energy).
e: Environmentally friendly with a safe access for any kind of recreation due to the smooth surface and gentle gradient.
f: Easy and cheap to maintain.

Par. 2: Porosity:

Uplift pressures usually occur in circumstances where water is not able to leave the body of an embankment easily.
Many types of artificial defence constructions can result in uplift pressures (slabs, blocks, stones etc.)
Compensation can be found in designing with overweight.
Reduction in unit weight of protective surfacing can be achieved by the use of permeable constructions.
However such reduction is limited due to the permeability differential between the permeable revetment and the core underneath, (soil and filter material).
Apart from natural rigid defence elements like rock, chalk etc. naturally formed banks consist of porous material (sand, clay, gravel, shingle etc.) However such natural material is completely controlled by mother nature due to deformation by waves and tides, erosion by water and air and mechanical damage by ice and other floating elements.
With this O.S.A. material we have managed to imitate the characteristics of natural porous material but made it resistant to erosion and mechanical loads.

Deformations caused by settlements will be followed because of its flexibility.

Par. 3: Flexibility

The flexibility of the Open Stone Asphalt revetment is determined by the characteristics of the binder bitumen (see Par. 12).

Par. 4: Surface structure

Open Stone Asphalt provides a relatively smooth surface that is Hydraulically more rough then concrete slabs but smoother than rock armour. In addition the O.S.A. surface structure is permeable and so coastal works using O.S.A. show improved run-up and reflection performance compared with smooth impermeable works.

In order to compare with rock armour slopes the O.S.A. needs to be laid to a flatter gradient and cost comparisons for similar run-up performance will vary from site to site.

Par. 5: Elasticity

The elasticity of the O.S.A. revetment is determined by the characteristics of the mineral aggregate, that makes up the major portion of the mix, and by the binder.

Par. 6: Safety and Environment:

a: Safety:

Recreation related to sea and coast is still increasing and it is often important to try to create methods of defence, that will enable safe access to beach and water (see Fig. 1) and an O.S.A. revetment can provide this facility.

FIG. 1

b: <u>Environment:</u>

- Bituminous materials are chemically inert in the presence of water, to which they impart neither taste nor odour. In order to ensure complete acceptability, health officials have made many tests of the hygienic qualities of drinking water stored in asphalt-lined structures. In all cases, asphaltic materials have received unqualified approval.
- Due to the weathering effect of water mixed up with silt or other material, the mastic skin on top of the surface stones of an O.S.A. - revetment will disappear with time, and the colour of the lime stone aggregate and the sand will dominate (see Fig. 2).
  This means that the original black appearance of the surfacing is only temporary.

Where site exposure is significant the black initial appearance of O.S.A. slopes will vanish over a few months. In quieter locations it may be appropriate to take special measures; to produce an immediate sand colour as for the breakwaters at Zeebrugge.

Trials are presently under way to investigate the feasibility of spreading sand over the "sill-hot" O.S.A. immediately after laying.
Initial results are promising and show that the sand adheres to the surface mastic and a sand coloration of the O.S.A. surface is achieved that is not wasted away under reasonable actions of wave and tide.

Across the upper sections of coastal works where salt water exposure is minimal the O.S.A. surface will permit the development of vegetation without any rupture of the surfacing. This arises from the flexibility of the O.S.A. material. Also where utilised as a protection to reservoirs and other inland situations, vegetation can be encouraged in O.S.A. if necessary by lightly spreading topsoil or by hydro-seeding.

FIG. 2

Par. 7: Maintenance

Coastal works using O.S.A. have been in service now for over fifteen years and there is no requirement for maintenance to date.
However it has been noted on some locations where shingle occurs within the intertidal zone that the O.S.A. can be abraded. If such abrasion removed sufficient thickness to warrant reinstatement then the nature of the material allows simple "grafting-on" of new material once the existing surface has been cleaned and primed.

III: LEAN SAND ASPHALT (L.S.A.)

Par. 8: The most essential properties of L.S.A. are:

a: A porous structure similar to the subsoil, with the advantages as mentioned before (Par.1,a).
b: A flexible construction (Par. 1,b).
c: Excellent filter material.
d: L.S.A. can be placed in thin layers where a filter is to be constructed in the dray and in bunds, dikes etc. where construction is in the dry or under water.
e: L.S.A. is easy to handle and doesn't require special equipment for placing.
f: L.S.A. can be produced from locally available sand (e.g. dredged from the sea).
g: L.S.A. is durable and has good geotechnical properties.

Par. 9:

L.S.A. is a mixture of sand and 3 to 5 percent by weight of bitumen. Its resistance towards erosion and its sand tightness renders L.S.A. a popular substitute for clay, mine waste or quarry-run in dam construction.
The extension of dams is started by laying a bottom protection over the existing subsoil or over a dumped sand-gravel foundation. A core of hydraulically transported sand is deposited subsequently on the bottom protection between surrounding dikes which are made in advance of the sand core to protect it against eroding currents and wave attack.
The protecting dikes will form part of the final dam-structure and must for that reason satisfy minimum requirements as to sand tightness and stability, especially in the zone of waves and tides.

The mechanical stability of the L.S.A. mix is determined by the quantity and the "hardness" of the bitumen (Par. 12). The permeability of L.S.A. is of the order of that of the constituent sand.
Its permeability and pore size distributions are so similar to the properties of sandy core material that there is no suffosion. As a protecting filter, L.S.A. surpasses gravel materials. The resistance to erosion enables the rapid construction of the sand core in the tidal zone since the protective layers can be constructed afterwards.

Lean Sand Asphalt can be manufactured in large quantities at a rapid rate so that construction of temporary bunds can be completed quickly. L.S.A. will slowly erode back to its constituent sand so that it is ideal for temporary protection. The construction of temporary havens for small fishing fleets can use L.S.A. cost-effectively in parts of the world where banks and channels can change dramatically and sand is locally available. In such applications L.S.A. provides a material with which the designer can accommodate the changing environment and also provide a construction which is environmentally sympathetic.
In less developed parts of the world the use of L.S.A. in temporary breakwaters avoids massive disruption to the local community from large construction plants and precast concrete manufacturing areas.

IV: Design with open bituminous mixtures:

Par. 10: Open Stone Ashpalt (O.S.A.)

O.S.A. as the bituminous protective layer has to meet the following requirements:

a: The dimensions of the voids have to be sufficient with regard to the permeability to water, but not so that in the long term, turbulence could damage the filter layer underneath.

b: It must be sufficiently strong, to meet all requirements of the design criteria.

The permeability of O.S.A. has been investigated by the Royal Dutch Shell Laboratory of Amsterdam and found to be in the range of course gravel. Rijkswaterstaat, the Dutch Public Water Authority, has tested Open Stone Asphalt in a high discharge flume.

During 24 hours it was tested up to velocities of 7 m/sec without damage or loss of material.
It is more difficult to determine the resistance to wave attack. Due to the complicated visco-elastic behaviour of asphalt-mixtures, it cannot be scaled down. The assessment of wave attack can only be carried out at full-scale.
Therefore in 1983, Delft Hydraulics Laboratory in collaboration with the Delft Soil Mechanics Laboratory investigated the stability of a dike cover layer of Open Stone Asphalt on a slope with a gradient of 1:3 in the Delta Wave Flume at 1:1-scale.
The maximum wave height achieved was 2.65 m'. No damage occurred in any of the tests executed.
This test also confirmed an empirical design approach which has been compiled from a number of executed projects:-

$t = C.H_s$ in which (1)

- t = Thickness of O.S.A. layer
- H_s = Significant waveheight
- C = coefficient, related to the stability of the sub-base

For O.S.A. on filtercloth : $C = 1/6$
For O.S.A. on Sand Asphalt: $C = 1/10$

This empirical approach may also be considered in parallel with the theory of a plate lying upon an elastic foundation as represented in Fig. 3.

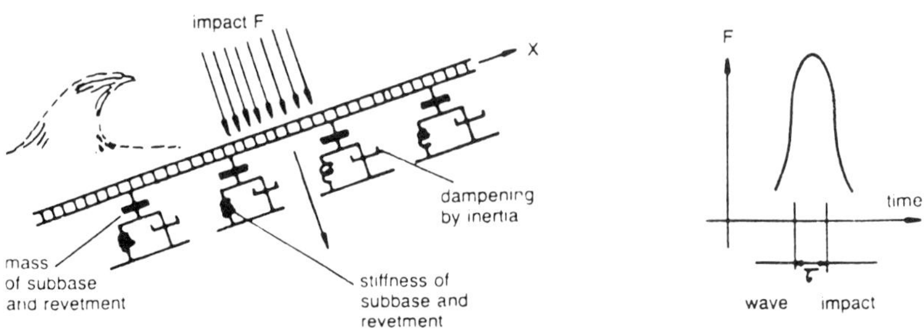

$$\text{D.E.:}\quad \underbrace{K.\frac{\delta^4 w}{\delta x^4}}_{\text{stifness of revetment}} + \underbrace{M.\frac{\delta^2 w}{\delta t^2}}_{\text{inertia}} + \underbrace{D.\frac{\delta w}{\delta t}}_{\text{dampening}} + \underbrace{C.w.}_{\text{elasticity of foundation}} = F \qquad (2)$$

Fig. 3: "Plate on elastic foundation"

The formula set out in Fig. 3 includes parameters with the following definitions:

K = $\frac{E.h3}{12(1-v2)}$ is the bending stiffness of the plate
E = the modulus elasticity (N/m2)
h = plate thickness (m)
v = poisson's ratio for asphalt
M = mass of the plate + added mass of the subsoil (Kg/m2)
D = subsoil damping coefficient (Ns/m3)
C = modulus of subgrade reaction (N/m3)
w = deflection of the plate (m)
t = time (s)
x = horizontal axis (m)

The following wave-impact formula has been developed:-

$$h = 0.75 \sqrt[5]{\frac{27}{16} \frac{1}{(1-V^2)} \cdot \left(\frac{P}{Tb}\right)^4 \cdot \left(\frac{S}{C}\right)} \qquad (3)$$

in which: h = thickness of revetment (m)
Tb = asphalt stress at failure (N/m2)
P = wave impact (N/m')
S = stiffness modulus of the asphalt (N/m2)
V = poisson's ratio for asphalt
C = modulus of subgrade reaction (N/m3)
0.75 = reduction factor, as a wave impact is not linear but distributed

When this wave impact formula is applied to situations with varying waves and water levels, the fatigue factor needs to be included as expressed by the parameter

$$f = \left[\Sigma \frac{ni}{ns} \left(\frac{Pi}{Ps}\right)^5 + 1 \right]^{4/25} \qquad (4)$$

so that:

$$h = 0.75 \times f \sqrt[5]{\frac{Ps^4}{\gamma b^4} \cdot \frac{27S}{16\,(1-V^2)\,.C}} \tag{5}$$

In which:
Ps: wave impact related to the significant wave height Hs from the characterised wave spectrum of the design storm.
ns: number of loading cycles in which Ps occurs
Pi: wave impact which occurs ni times in the lifetime of the revetment and is
therefore determined from the wave spectrum.

General values of layer thicknesses can also be given in a graphic form. These values are based on assumptions for application in Dutch conditions. (Fig. 4) (1).

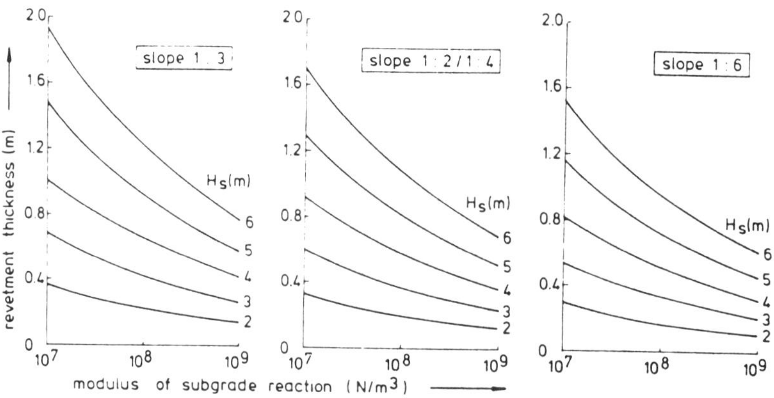

Fig. 4:

Par. 11: <u>Sand Asphalt L.S.A.</u>

The most essential properties need to be:
- stability of the filter against the egress of sub base material
- resistance to current and wave loads in the temporary and permanent condition

Laboratory tests have shown that water permeability is strongly dependent on the degree of compaction.
In practice compaction depends on the method of placing, for example use of bottom-door dumping vessels, conveyor belts, vertical pouring pipes or cranes.
This compaction is in fact simply determined by the materials own weight and with that low degree compaction, the void content is appr. 40%. This particular value for the compaction also applies to filter layers with a limited thickness.
Standard water permeability test equipment can be used to determine the water permeability of L.S.A.

Because of its cohesive character L.S.A. can resist large steady and cyclic hydraulic gradients without loss of function as a filter material.
Because of this the material has a high filter stability. Investigations into filter stability have been carried out by Bundesanstalt fur Wasserbau, Karlsruhe (Germany).(8). Their conclusion is that L.S.A. is an ideal filter on which large gradients can be allowed without migration of the underlaying materials.

Both L.S.A. sublayers and cover layers, as might be used to form a breakwater or dike, contribute to structural stability. The visco-elastic properties of the material are complicated and extensive tests have been carried our by Rijkswaterstaat and Delft-hydraulics in Holland. The results of this research indicate that the material cannot be specified from the rheological point of view, simply by soil mechanical characteristics. Tests have been performed with triaxial equipment as well as with cell apparatus. Mainly stress-controlled tests have been performed.
Cell apparatus is shown schematically in Fig. 5.

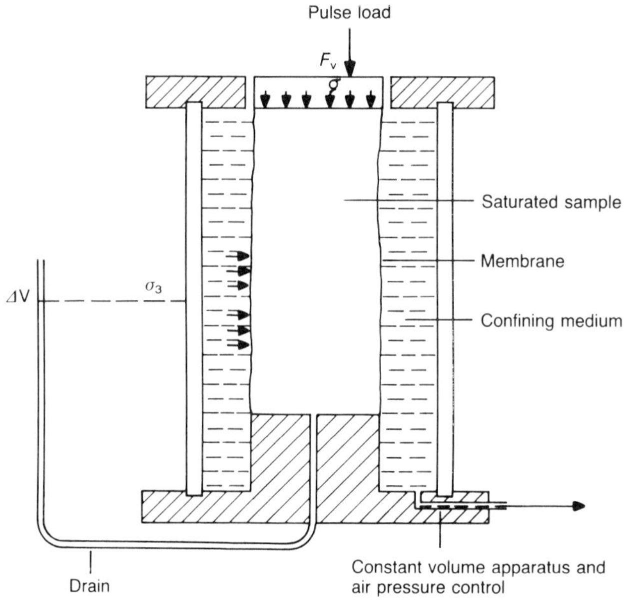

Fig. 5: Cell apparatus for testing L.S.A.

A typical result is given in Fig. 6.
The figure shows a time-deformation diagram and a stress strain diagram for sand-asphalt.

The results indicate the characteristic performance of L.S.A. in which the shear stress mobilised by external loading is carried initially by the bitumen. Relaxation of the binder however introduces rearrangement of stress and leads finally to an elastic grain-on-grain performance of the L.S.A.
Extensive use is made of the Bishop calculation method to determine the stability of slopes. (7)
L.S.A. is obviously a complex material; a safe assumption however is to choose an angle of internal friction which produces the same order of deformation under the applied loadings, as in sand.

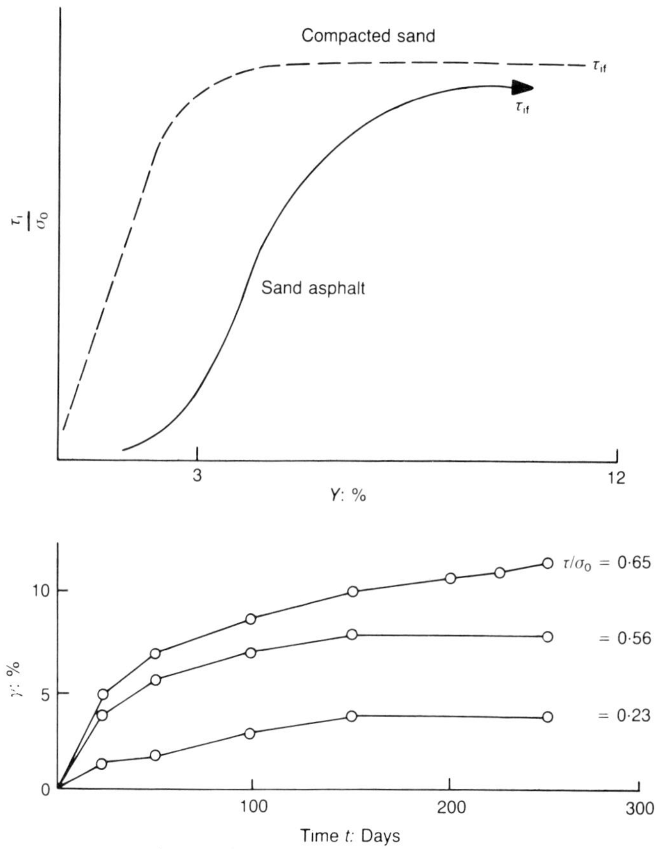
Compacted sand
τ_{if}
τ_{if}
$\frac{\tau_i}{\sigma_0}$
Sand asphalt
3
12
Y: %
τ/σ_0 = 0·65
= 0·56
= 0·23
10
5
0
γ: %
100
200
300
Time t: Days

Fig. 6

The following characteristics apply in general:

Loading duration	Stability is determining factor	Deformation is determining factor
Short (hours)	= 32o	= 27o
Long (weeks)	= 27o	= 22o

In the selection of safety factor it is important to remember that sand-asphalt gives no sudden loss of stability (for example, it is safe in case of an earthquake or other dynamic loadings).
An accurate assessment of the stability can be obtained using a modified Bishop calculation, such as developed by Delft soil mechanics. In this calculation the stress distribution is determined in a number of concentric slip circles, deformation rates are also determined and finally integrated. (Fig. 5) (7).

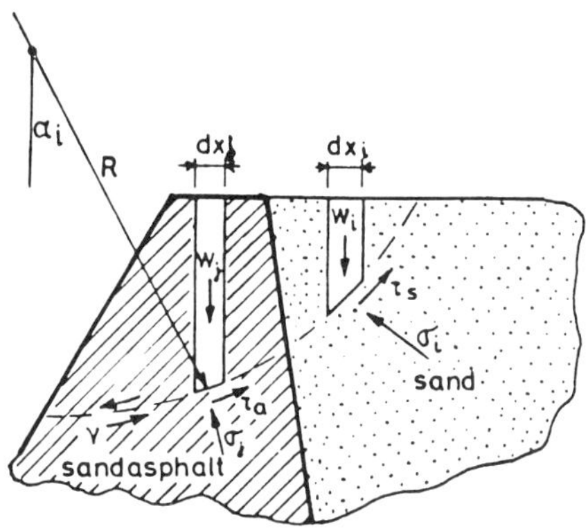

Fig. 7: Input parameters for sandasphalt and sand into Bishop's stability analysis based on 'the equilibrium of slices'.

The resistance to currents has been assessed from a series of large scale tests carried out by the Dutch Rijkswaterstaat. Extensive measurements were made on a L.S.A. embankment built at the head of a sluice. Various tests were also carried out in a current flume adjacent to a weir.

The flume dimensions were: width 2m, water depth 3m', difference in water levels across the weir about 4m'.
Tests were also made on L.S.A. sections built into the northern gap of the Crevelingen closure dam. The test sections were constructed using the method employed for placing concrete blocks in the closure gap. This involved dumping sand-asphalt, contained in large netting bags, from a cable way.
It appears from these tests that L.S.A. is resistant to currents up to 3 to 3.5 m/sec, with no loss of materials. In higher, steady currents erosion develops which increases gradually as the current velocity increases.
Non-steady currents occur during wave attack. In addition to the high current velocities, waves also create forces perpendicular to the slope. It is possible to calculate the stresses for L.S.A. layers or bunds (1) using calculation models normally accepted in mechanical engineering such as:

- plate on an elastic foundation (see Open Stone Asphalt)
- Boussinesq

Tests on L.S.A. indicate that it is more resistant to the alternating currents and accelerations experienced under wave attach than to prolonged abrasion by steady, though lower, current velocities.
Observations indicate that, in practice, unprotected L.S.A. - -slopes can resist wave heights of 3 to 4 m' without significant damage.

V: TECHNOLOGY:

Par. 12: Open Stone Asphalt. (O.S.A.)
O.S.A. is an underfilled mix of limestone and mastic. Limestone is preferred mainly because of the good affinity to bitumen. However granite or other stone material could be chosen based on the results of proper material investigations. Manufacture is in two phases: first the mastic is prepared and then it is mixed with the preheated stone element. By using this method of mixing, the mastic can be spread evenly over the surface of the crushed stone and all the contact surfaces between the stones can be fixed with a minimum quantity of the mastic.
The production of O.S.A. requires accurate control of the mix-composition. In addition it is important that the stone does not contain any undersized particles.

In general the composition is:

- mastic	20% by mass
- limestone (20/40mm)	80% by mass

Mastic is a mix of 60 to 70% by mass sand, 15 to 25% by mass filler and about 20% by mass bitumen. It is an overfilled mix which, in itself, has the properties of a highly viscous fluid. The mastic mix-design is therefore not based on stability but on viscosity.
In practice, the mastic/stone ratio is determined by a sag test in which mastic is added to the mix and the quantity noted at which the mix is still open.
In practice, the mastic percentage lies between 17 and 21% by mass.
The stability of the mix is determined by the stone skeleton and the adhesion between the mastic and the stone.
Mix-properties such as breaking strain, stiffness modulus etc., need to be calculated or obtained from reference to design graphs (1).
The voids ratio of O.S.A. is generally appr. 25% The material is thus very permeable. The water permeability is of the same order of magnitude as that of loosely dumped stone of the same nature.
Investigations indicate that, under normal conditions, the flow in O.S.A. is between laminar and turbulent. If O.S.A. is used with a filter cloth, the permeability of the structure will be largely determined by the permeability of the cloth.

Par. 13: Lean Sand-asphalt (L.S.A.)

Generally, L.S.A. is a mix of sand and bitumen. The content is normally appr. 4%. The bitumen functions as a binder and tends to concentrate on the contact surfaces between the particles.
Mix-design, aimed particularly on durability and water permeability, is based mainly on experience. Factors which can influence the properties of L.S.A. are:

1: The bitumen content
2: Degree of compaction

A high degree of compaction increases the wear resistance and the deformation and strength properties but lowers the permeability.

Therefore no compaction is specifically needed with the material functioning as a filter. When placing in bunds or thick layers it will automatically be compacted during the placing procedure.
In addition to investigations into basic materials, mix-design tests can comprise a comparison with known L.S.A.

These investigations can include:

- permeability
- erosion test (by rolling or water-jet)
- retained stability
- marshall tests
- triaxial tests

The stability, in relation to forces applied slowly is less than for natural sand. With forces, which are applied more quickly, the stability is greater.
The stiffness modulus of the mix is dependent on the duration of loading and the temperature and is related to the mineral content and the stiffness modulus of the bitumen.
Because of the high voids ratio of L.S.A. the bitumen hardening process is faster than for most dense mixes.
Creep-tests carried out on samples taken from a 1:1 scale test which comprised 96% fine sand and 4% bitumen at temperature of 15o and 20oC, indicated values as shown in Fig. 8 and Fig. 9.

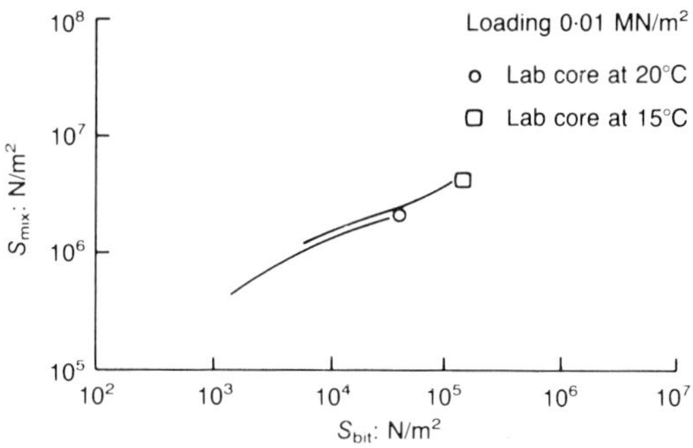

Fig. 8: Stiffness moduli of lean sand asphalt

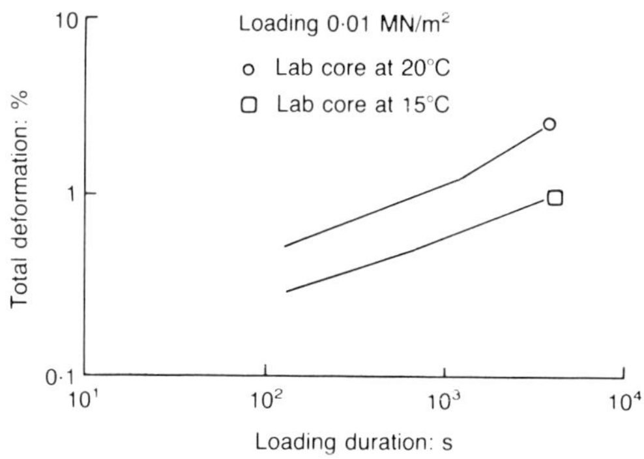

Fig. 9: Deformation as a function of the loading duration

Because of the limited quantity of bitumen in the mix the sand grains are only covered by a thin film of bitumen a few microns thick.
With the low degree of compaction which is found in practice the permeability tends to be similar to the sand from which the L.S.A. is made.

VI: DURABILITY

par.14:

Investigation of the durability of bituminous materials depends largely upon field experience.
Initial estimates of durability can be made from laboratory scale similarity tests.
Full scale measurements however give more reliable results.
Observations on existing structures show a lifetime exceeding 30 years.
During this period, hardly any maintenance has been needed. O.S.A. hasn't shown sensitivity to waves, running up and down a slope.
A test of the Dutch Rijswaterstaat, simulating an attack of 250,000 waves with a height of 2.00m' showed no damage.
As a matter of comparison: 10,000 waves with a height of appr. 1.87 m' simulate a period of 1.5 years near Noorland (Storm-Surge-Barrier Easter Scheldt - Holland!)
Essential for durability for such products is the accurate mix-design, mix procedure, quality control and placing method.
With regard to the durability of O.S.A. a few other tests have been carried out by "Bundesanstalt fur Wasserbau" at Karlsruhe. (8)

Those test were related to temperature resistance:

1: One sample, placed on a slope of 1:3 was subjected to the equivalent of 3,000 hrs sunshine at a surface temperature of 60oC. No flow phenomena were noted.
2: One sample, placed on a slope of 1:2, was subjected to ultra violet rays, equivalent of 4,000 hrs of sunshine and, again, no flow phenomena were noted.
3: Freeze/thaw tests on O.S.A. - cylinders did not show any damage to the material.

For L.S.A. the durability is mainly determined by the right bitumen content, translated into the thickness of the film on the grains.

Adhesion between sand and bitumen is of course important and normally good.
However the risk of stripping should be avoided. Stripping could occur when water is forced into the mix, such as when L.S.A. be placed under water at a temperature above 100oC (Boiling effect).

VII: EXAMPLES OF APPLICATIONS:

Par. 15: General execution aspects:

The products as described can be manufactured with conventional asphalt plants (Batch-mixer). L.S.A. can also be produced with drum-mixers.
Transport of the materials is by conventional equipment.
For O.S.A. care has to be taken with the temperature-loss during transport and to avoid segregation.
L.S.A. however can easily be stored in heaps near the plant, certainly when the temperature needs to be reduced before placing anyway (below 100oC). This also makes L.S.A. ideal for transport with floating equipment.
Placing of O.S.A. in the dry is normally done with a backhoe or with a standard asphaltpaver under certain circumstances.
When O.S.A. is placed under water, it will first be placed in prefabricated mattresses near to the location of placing.
When prefabrication is finished, the mats are placed with a lifting crane using a purpose built frame. For L.S.A, when placed in the dry, standard earthmoving plant is used.
When placed under water, bottom-door, dumping vessels can be used. The L.S.A. will then normally be packed into large netting bags to prevent excessive spreading during dumping.
Placing with a vertical pouring pipe is a continuous operation, which can be employed when continuous bunds are being placed under water.
Sand-asphalt can also be placed in the form of prefabricated "Sausage mattresses" - a mattress of geotextile-tubes filled with L.S.A.
The advantage is that such a mat will not loose its fill if the filter cloth
should be damaged; this is due to the cohesive property of the L.S.A.

Par. 16: Examples of executed projects:

Zeebrugge - Harbour extensions

About 450,000 tons of L.S.A. were placed in bunds for the construction of the Zeebrugge work harbour. Two bunds were constructed on an improved foundation and a bottom protection. Sand was placed between the bunds which extended beyond the sandfill to limit the loss of sand. Because of the large range two working sections were needed, a lower section up to mean water level and an upper section to above high water level. For construction above high water the sand core was first profiled and then covered with a L.S.A. blanket. The L.S.A. was carried from the depot in conventional trucks and placed with a backhoe on the crest of the embankment. The slopes were profiled in a later phase also using a backhoe. The production, which involved a drum-mixer, was in the order of 20,000 tons per week and in 8 months a work harbour was built in the sea with material from the beach at the nearby seaside resort of Heist. The beach was restored to its natural form.
O.S.A. was used as the main revetment.
The toe of the construction, the part under water was protected with rock armour. (See Fig. 10).

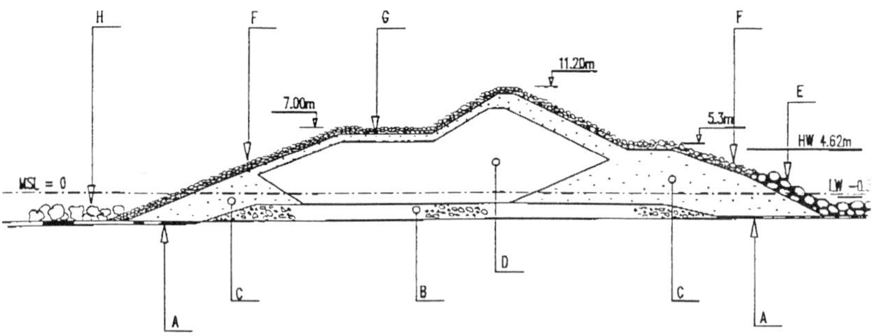

A : FASCINE MATTRESS
B : GRAVEL
C : LEAN SAND ASPHALT
D : SAND BODY
E : MASTIC GROUTED STONE
F : OPEN STONE ASPHALT
G : LEAN SAND ASPHALT
H : RIP RAP

Fig. 10: Cross Section Zeebrugge Breakwater

- Separating jetty, Hook of Holland

Another large-scale application is the construction of the separating jetty at the Hook of Holland. This jetty divides the New Waterway from the Caland Canal, which lead to different harbours in the Rotterdam Port area. The jetty has a massive core of L.S.A.; in the sub-layers the L.S.A. is mixed with gravel (40% gravel/60% sand-asphalt) in order to obtain a good joint with the rubble foundation. In this example also, the material was brought from the depot in lorries and bulldozed into place from the crest. Batch mixers were used.
As the major part of the works had to be constructed under water, it was decided to use mastic grouted stone as the permanent revetment.
(See Fig. 11).

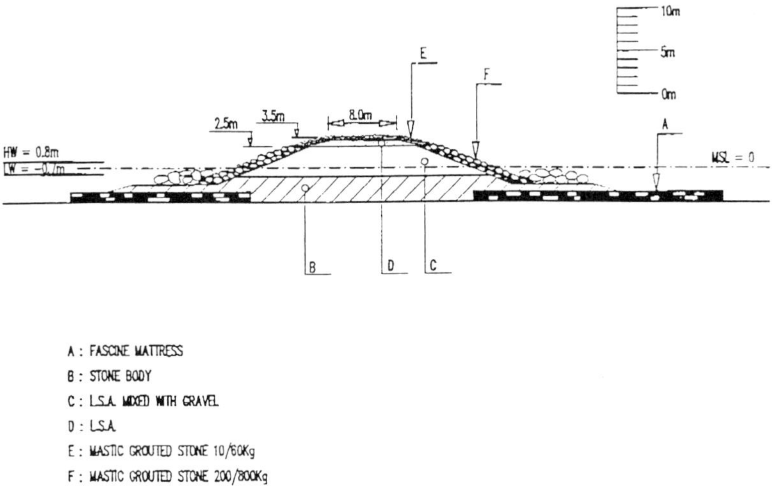

Fig. 11

Cross section of separating jetty - Hook of Holland

- Western breakwater at Zeebrugge:
A more modest use of gravel/L.S.A. is as a filter between the core of a rubble mound breakwater and an adjacent (industrial) area formed of reclaimed sand. In such a situation the filter layer has more than one function, the most important being the filter in the completed structure, which prevents sand being washed through the stone core of the breakwater. A secondary function is to protect the core during the construction phase. An example of this type of application is to be found at Zeebrugge where part of the western harbour breakwater has a filter of this type.
(See Fig. 12).

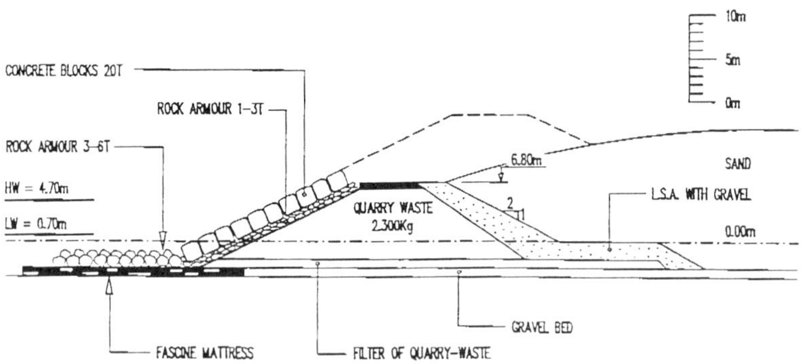

Filter Construction with L.S.A. against rip rap

Fig. 12: Cross Section. Western Breakwater - Zeebrugge

- Breakwater at Esbjerg:

L.S.A. has been used in the construction of a breakwater at Esbjerg in Denmark. At this site the foundations were formed of dumped sand with a toe protection. Because of the sea conditions, construction in the wave zone had to be completed as quickly as possible. In this case small bunds of L.S.A. were used, between which the sand core of the breakwater was placed. The sand core was raised above high water and covered with a L.S.A. filter layer. When completed, the breakwater was covered with O.S.A. (See Fig. 13).

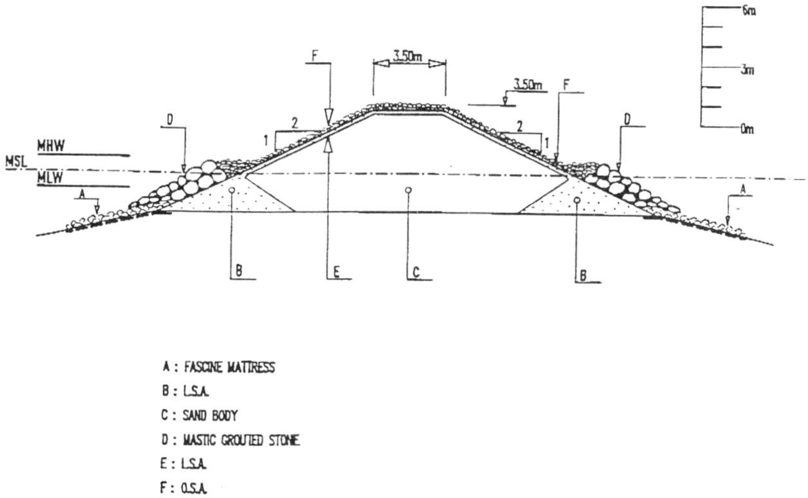

Fig. 13: Cross Section of Harbour Breakwater Esbjerg

- High Island project - Hong Kong

For the High Island project in Hong Kong it was necessary to use cofferdams to form construction pits for the high main dams. These cofferdams were built in water depths of 10m' and were designed as rubble mound breakwaters with a L.S.A. core to reduce permeability.

The toe section and the parts under water were formed of packed L.S.A. placed by bottom-door dumping vessels. The parts above water level were constructed from the crest. In the final phase the L.S.A. was covered by a suitable ballast of rubble.
(See Fig. 14).

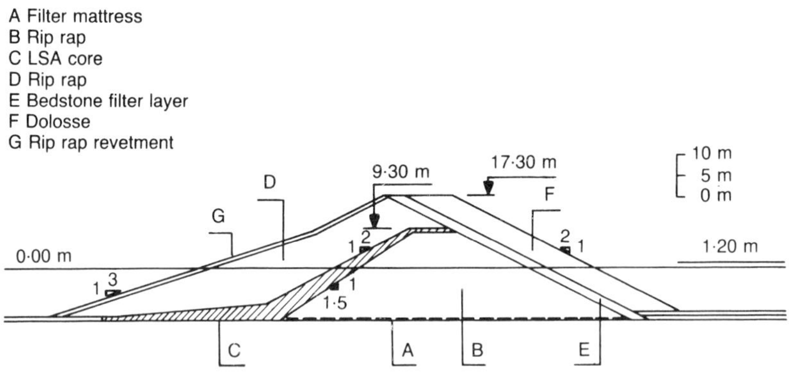

Fig. 14: Cross Section of High Island Cofferdam

-use of pouring pipe and/or conveyor belt

Temporary bunds were needed during the closure works in the delta of the Maas, Scheldt and Rhine to guide the currents. After the foundations had been placed the bunds were built up from a vessel with L.S.A. using either a vertical pouring-pipe or a conveyor belt. (See Fig. 15)

Fig. 15

The sand required was brought alongside the placing vessel by barge. The placing vessel was anchored in place and able to reposition itself using winches. The placing vessel is equipped with a pouring pipe. Because the bunds were only temporary, they were constructed with the minimum amount of bitumen (2½%)
(See Fig. 16)

Fig. 16

Furthermore there is now a widespread application of O.S.A. in combination with L.S.A. as the filter layer for coastal works.

Over the last 25 years, coastal projects with the L.S.A. - O.S.A. combination or O.S.A. on filter cloth have been executed in Holland, Belgium, Denmark, Germany and U.K. where the first application was at Granton (Scotland) in 1976.
After that, projects at:

Caister on sea	- Phase I, II	1978 - 1987
Prestatyn	- Phase I, II, III, IV	1985 - 1991
Kings Parade		1984 - 1985
Southport		1984
Tollesbury)	Blackwater	1989 - 1990
Goldhanger)		
Immingham	Humber	1989 - 1990
Dovercourt Bypass	North Sea	1991

VII: COASTAL ENGINEERING APPLICATIONS:

The application of O.S.A. and L.S.A. in coastal engineering is to date
quite limited in the U.K. In order to assess the potential of these materials for greater application it is important to consider their specific advantages and disadvantages. Both O.S.A. and L.S.A. are advantageous materials for maritime application since they can be laid in dry or wet conditions thereby requiring no time "to set" as required for insitu concrete. In addition the material viscosity allows it to be shaped by machine (backhoe) without the need for shuttering provided slopes are less than 1 in 1.
These construction advantages for coastal works in the intertidal zone are balanced by the perceived disadvantages of the "black appearance" and rougher surface finish than provided by traditional concrete surfaces.
Sufficient tests have now been carried out on these materials to show their technical suitability for wave exposures of at least 2.50 m breaking wave height and flow exposure of up to 7.0 m/s.
Their permeability and filter properties resolve internal drainage problems effectively and the ability to easily "graft on" repair or extensions make structures adaptable for changing exposure conditions (sea level rise; climatic change) and rapidly repairable after severe storms.

O.S.A. surfaces provide easy access (subject to gradient) for the general public but they are not ideally suited to access by heavy plant. It is necessary to make separate arrangements for such access across O.S.A./L.S.A. construction.

Unlike rigid construction O.S.A./L.S.A. construction will deform into any voids formed within the core to a structure or deform around any buried hard-spot (abandoned outfall or groyne). It is, therefore, important to carry out thorough site investigations to record any buried hard-spot so that they may be removed or specifically accommodated within the scheme design.

Owing to historical development of coastal engineering in the U.K. the manufacture of these types of bituminous material is not commonplace.

However they can be produced from any standard asphalt batching plant and this is an important consideration for ongoing maintenance and repair work.

Finally, the use of O.S.A./L.S.A. construction for coastal works must depend largely upon its equivalent unit costs compared with other alternatives. However, these materials have advanced considerably in recent years and have been technically proven over a twenty five year period to offer a firm technical alternative for coastal works in the U.K. and overseas.

IX: CONCLUSIONS

Par. 17:

The use of Open Stone Asphalt and Lean Sand Asphalt in maritime works has become extensive over the last ten years. The initial concerns over visual appearance have been largely allayed and there is a growing awareness of the advantages to be gained from the use of these materials in coastal zones. The following conclusions are therefore drawn from the paper set out above.

1: In countries, where rock and stone material are either not available or relatively very expensive, open bituminous mixtures have proved to be successful products to build coastal works.

2: The high level of stability against the range of hydraulic loads, the thermoplastic flexibility, the elasticity and durability give these bituminous materials a unique technical standing.

3: Compared to many other methods of construction bituminous materials require minimal maintenance.

It is to be hoped that the use of asphaltic materials in coastal engineering can become still more extensive in the years ahead. It is likely that opportunities exist for hybrid constructions using these materials together with concrete in ways whereby the benefits of each material type are emphasised and the weaknesses diminished.
The greater the number of technical operations available to the coastal engineer the more cost effective will be the solutions to the varied problems around the shorelines of the world and thereby benefit the costal communities.
This paper is submitted as a contribution towards that objective.

IX: LITERATURE

1: Rijkswaterstaat Communications nr. 37/1985;
The use of asphalt in hydraulic engineering.

2: Kerhoven, R.E.
Hydraulic Applications in the Netherlands, Report nr. 100 F,
Koninklinjke Shell laboratorium, Amsterdam. Novem,ber 1985.

3: Brouns P.;
Application of Asphalt Mastic, Open Stone Asphalt, Lean Sand Asphalt and Bituminous watertight revetments.
Technological Institure - K.V.I.V. Gent, 25 Jan. 1981.

4: Asbeck, W.F., Baron van
Bitumen in Hydraulic Engineering, Shell Petroleum Co. Ltd. London 1955.

5: Deutsche Gesellschaft fur Erd- und Grundbau E.V.
Empfehlungen fur die Ausfurhrung von Asphaltarbeiten im Wasserbau
EAAW 1983.

6: Bosche, D.J. van den,
Application of Lean Sand Asphalt and Open Stone Asphalt in the Zeebrugge work harbour and the extrapolation of the findings to slope revetments in waterways.
Technological Institute, K.V.I.V. Gent 27-28 Jan. 1981.

7: Calle, E.O.F.,
Modified Bishop slip - circle analysis for Lean Sand Asphalt reclamation bunds.
Report CO-406645/18 Delft Soil mechanics Laboratory, October 1980.

8: Bundesanstalt fur Wasserbau.
Versuchsbericht uber die Prufung von Mastixschotter auf Frost- U.V.
-und Wasserbestandigkeit, Karlsruhe, 27 November 1973.

9: Hoek, K.A. van den
Lean Sand Asphalt durability investigations; Notitie ONW-M-81001
Deltadienst Rijkswaterstaat; 5 January 1981

10: Hogervorst, D en R.C. Reintjes
Development of an open bituminous revetment construction of mastic gravel or mastic crushed stone.
Weg- en Waterbouw no. 2, 1972.

11: Angremond, K.D.' ., H.J.Th. Span, J. van der Weide and A.J. Woestenenk
Use of asphalt in breakwater construction,
Coastal Engineering Conference, Washington, D.C. 1970

12: Mulders, G., R. Termaat, P. Ruijgrok and P.O. Petschl,
Application of Lean Sand Asphalt in the construyction of the land heads of the S.V.K.O.
Voorlopige nota 12 DALA-IPROBU N 844, Deltadienst Rijkswaterstaat,
August 1980.

13: Delft Hydraulics Laboratory, Delft Soil Mechanics Laboratory:
Stability under wave attack
Report: M 1942 WL/CO 2266520 LGM,/ June 1983.
14: Breakwater with sand Core Proc. Coastal Engineering Conf.,
pp 2194-2212, 1980
Loewy E, Witthaus KG, Maddrell RJ, Wood JP.

28. The design and construction of the outer layby sea defences at Shoreham

R. MADDRELL, Sir William Halcrow & Partners, and
A. J. VAUGHAN, Shoreham Port Authority

SYNOPSIS. The design of the new sea defences in the known area of downdrift erosion following harbour construction in 1957, had to be carefully assessed. The established design conditions were examined using mathematical models and applied to alternative revetment types, in front of which was a new gravel beach. Flume tests showed the chosen design to be satisfactory and that no savings could be made. A slower than expected construction rate exposed the partially completed revetment to two major storms, virtually destroying it and delaying completion. Since that time the revetment has performed as expected but an additional groyne was introduced to limit beach movement.

INTRODUCTION

1. The easterly movement of shingle along the Sussex coast between Worthing and Brighton and the creation of a bank of shingle and river mud, resulted in the mouth of the river Adur being moved several kilometres eastwards between the 10th and 18th centuries. In 1760 an harbour entrance was constructed across the bar in its present location.

2. In the 1950's new entrance breakwaters and a deeper dredged entrance channel, designed by Sir William Halcrow and Partners Ltd (Halcrow), were constructed along with other major works (layby berths, lock), providing facilities for the coal vessels serving the newly constructed Brighton 'B' Power Station (ref 1).

3. One effect of the new breakwaters was to substantially prevent the movement of shingle across the harbour entrance. The expected accretion of shingle on the west side of the entrance and severe erosion on the downdrift side were very evident. A timber and concrete groyne system had been constructed on the downdrift beaches and a sheet piled revetment armoured with 7 tonne Stabit units was provided at the root of the east breakwater. Notwithstanding these works, beach levels dropped significantly and the steel sheet piles suffered overtopping as well as shingle abrasion.

4. Following the closure of Brighton 'B' Power Station in 1987, Shoreham Port Authority (SPA) acquired the Central

Electricity Generating Board estate in the port. One affect of this acquisition was to increase the length of SPA's coastal frontage, for which they are responsible, to 3km.

5. As part of a port redevelopment programme, SPA decided to address both the erosion problem downdrift of the harbour entrance and the lack of cargo storage space behind the two outer layby berths. An outline scheme was prepared to move the seawall further seaward thereby reclaiming approximately three hectares (see Fig. 1). The feasibility of the project was evaluated on the basis of its financial rate of return calculated from income derived from the additional storage area and savings in the maintenance of the existing sea defences.

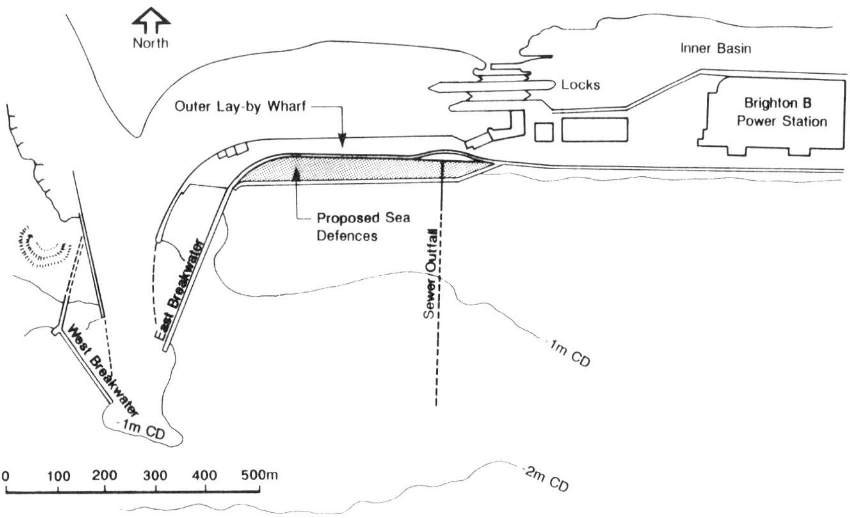

Fig. 1. Original layout and proposed sea defences

6. In 1988 SPA commissioned Halcrow to undertake a study and detailed design of new sea defences to protect the area to be reclaimed. Throughout the study, design and construction stages, close liaison with Adur District Council was maintained with regard to the provision of a new amenity beach and public access to the east breakwater.

DESIGN STUDY

7. The first study phase required a proper assessment of five key elements, namely

(a) refine, adapt, incorporate and statistically analyse existing wind/wave, surge and bathymetric data for use in numerical models.

(b) assess the deep and shallow water wave climates using numerical refraction models.

(c) establish the design criteria for the whole revetment using a wave propagation model.
(d) determine the beach response using a beach plan shape model.
(e) prepare alternative designs and cost estimates.

8. Examination of the continuous mean sea levels records for the Port from 1962 indicates an annual increase due to isostatic and eustatic changes of 2mm, or 100mm in a 50 year design life of the revetment. Unfortunately, it was not possible to analyse the Port records statistically for surges and more general analyses for the south coast were used (refs. 2, 3 and 4). Annual maxima for Newhaven (the nearest Port) were analysed indicating an annual trend of 5mm (see Fig. 2(i)), but they had to be adjusted as tidal levels (0.4m) and range (6.1m) were greater than at Shoreham (range 5.5m). They were also adjusted to reflect both the long term and recent rises in sea level (see Fig. 2(ii)). Table 1 shows the predicted levels for Shoreham in relation to Ordnance Datum (OD), the highest level recorded being 3.98m AOD in 1985.

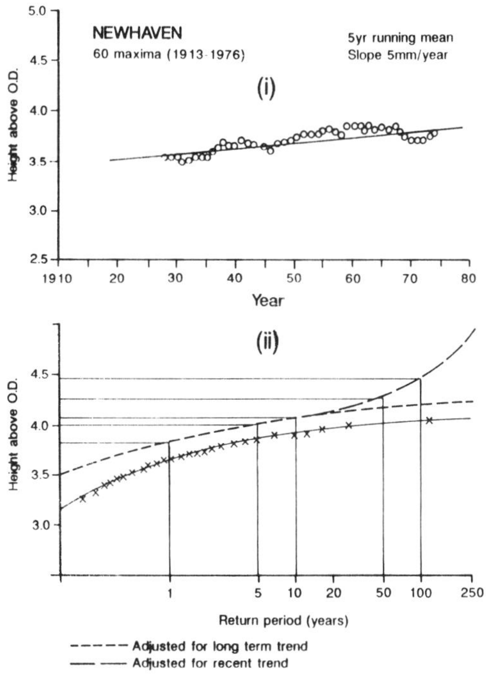

Fig. 2. Trend of annual maxima water levels and predictions of surge levels for Newhaven

Table 1 Predicted Surge Levels for Shoreham

Return Period Years	Level above OD	
	Normal GEV Analysis	Adjusted for annual trend
1	3.54	3.65
5	3.74	3.84
10	3.82	3.90
50	4.00	4.05
100	4.08	4.10

9. The wave refraction models used a total of 8 grid sizes, varying from 375mx375m to 50mx50m, reflecting both bathymetric changes and design requirements. As no wave data were available, hourly wind data were hindcast (ref. 5) based on earlier wind records from the Shoreham Coastguard station since 1980 (ref. 6). These were checked against more recent data.

10. Percentage wave power distributions for JONSWAP spectra were established for 30° sectors for the available angles of approach. As might be expected waves in the 210° to 240° sector were the most frequent (47%) and provided 66% of the energy. Table 2 shows the significant wave heights (H_s) and return periods, both inshore and offshore, at a water level of 2.93mAOD ie Mean High Water Springs (MHWS), for the mean zero crossing wave period (Tz), for an offshore direction of 225°. Clearly, because of the relatively flat seabed with slopes in excess of 1:500, significant shoaling and energy losses take place.

Table 2 Inshore and offshore wave heights and return periods

Return Period (years)	Offshore			Inshore		
	Deg	H_s m	T_z Secs	Deg	H_s m	T_z Secs
1	225	4.29	6.80	191	2.20	6.60
5	225	4.87	7.35	191	2.40	7.00
10	225	5.1	7.56	190	2.46	7.15
50	225	5.64	8.09	188	2.63	7.95
100	225	5.86	8.30	189	2.70	8.11

11. Bathymetric data were collected in areas of expected change ie the beaches and immediate areas offshore and topographic data for the tender documents. Bed samples taken, indicated the seabed was predominately fine sand with some gravel lag deposits. Current metering indicated maximum

ebb (westerly) and flood (easterly) velocities of 0.5m/sec and 0.33m/sec respectively, parallel to the shore.

12. Three conceptual layouts were examined using a unidirectional monochromatic mathematical wave model with a sheet piled revetment, namely:

(a) with breakwater spur
(b) spur and artificial breakwater
(c) headland and shortened spur

13. Since the model is sensitive to local bathymetric changes, a 5m model mesh covering an area of 730mx480m was used. Fig. 3 shows the layout described in (c) above and its associated wave pattern. The offshore wave direction was 225° with the inshore angle of 190° being almost parallel with the shore and the face of the revetment.

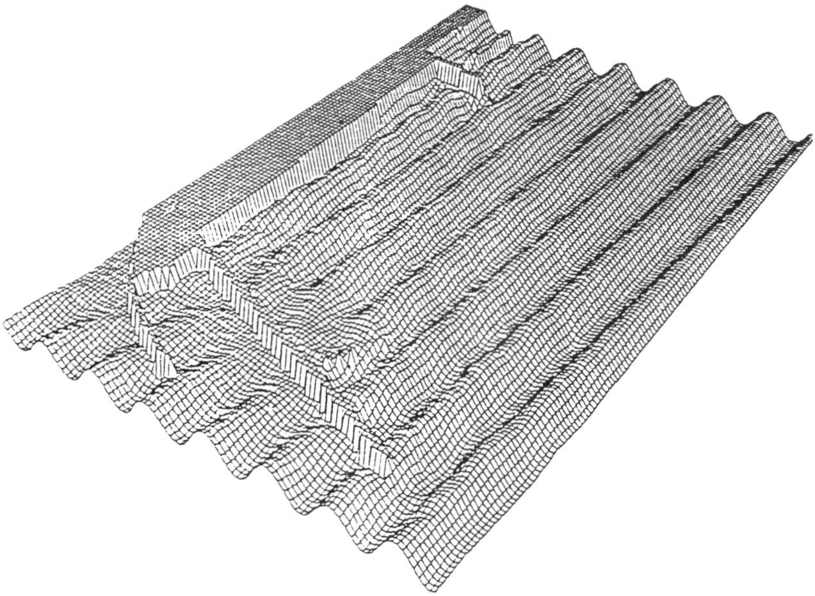

Fig. 3. Wave pattern associated with sheet piled revetment, spur and headland

14. An assessment of the variations in alongshore drift caused by changes in wave climate was carried out using a two dimensional beach plan shape mathematical model containing both closed and open boundaries ie the harbour arm and downdrift beaches, as well as existing long and short groynes. The average easterly drift was thought to be about 10,000m^3 annually, but this can increase in stormy years eg it was 15000m^3 in 1983. The model prediction for 1983 was 14,000m^3. The model also showed that the presence of a sheet piled wall increased the drift rate once it became exposed. The model was also run for the new revetment in deeper water for a period of 5 years with no recharge and,

by varying the recharge, showed that the new annual recharge should be about 15,000m^3.

15. The design life of the revetment was taken as 20 years which, with a 1:50 year storm, gives an accepted encounter probability (ref. 7) of 33%. Overtopping was to be limited as the area behind was to be used for port cargo storage and the discharge had to be assessed (ref. 8). Seven alternative designs were costed and an environmental impact assessment prepared for each ie (a) to (c) above, a Stabit revetment, Seabee revetment with spur and headland, offshore breakwater with spur and seawall and a headland with spur and beach. A spur was required for most of the options to eliminate the Mach-Stem wave generated along the eastern side of the east breakwater entrance arm. Likely maintenance costs were also assessed for design lives of 10, 15 and 20 years and an option containing a Seabee revetment and headland and a Stabit spur which utilised the existing units, costing about £3.2 million in 1988, was recommended.

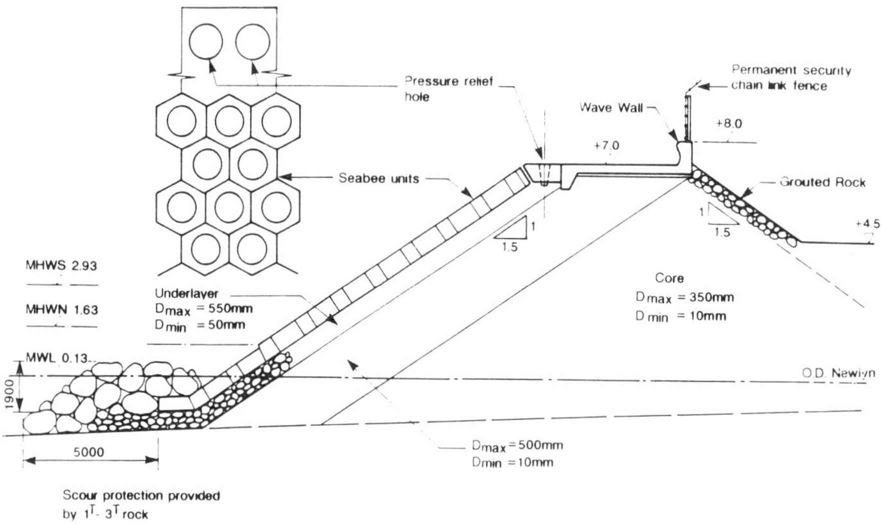

Fig. 4. Typical cross-section of the revetment

DETAILED DESIGN

16. After the acceptance by the Port Authority of the design recommendation in the first study phase, a detailed design was carried out to refine the overall concept. This included varying the position and orientation of the spur and fishtails on the spur and headland. However, the latter while being very effective in terms of retaining a beach, had to be eliminated on cost grounds. It was also necessary to establish the effect of the scheme on the existing Southern Water Authority outfall (see Fig. 1) some 350m to seaward in 1.5m of water below Lowest Astronomical Tide (LAT). A full Environmental statement was prepared and showed the impact of the new scheme to be negligible.

17. The Seabee armour units selected are concrete hexagonal blocks with a large central hole, which when placed with others gives a tightly packed 'honeycomb' pattern (see Fig. 4). Unlike conventional armour units, they are designed using the 'Blanket Theory' equation, which is based on Hudson. The equation gives the depth of unit required and the other dimensions are derived using the common geometrical relationship for its hexagonal shape. For the design, a loss of concrete of 1mm/year due to attrition from gravel was assumed.

18. Wave heights are depth limited near the toe and thus smaller units were used. However, as with all such units which rely heavily on interblock friction, the toe is critical and was protected by armour rock. The crest, which is also a vulnerable area, is protected by a cast in-situ unit, with a pressure relief hole, at the top of the Seabee column (see Fig. 4).

19. The spur is subjected to Mach-Stem waves and was designed to re-use the 7t Stabits protecting the root of the eastern breakwater. There were an estimated 850 Stabits available allowing some 80m of spur to be built (see Fig. 6). Stabits were on either side of the crest which was at MHWS and thus the spur was designed to overtop (see Fig. 5). A filter membrane was placed beneath the rock filter blanket because the sand beneath was fine and because of the exposed nature of the structure.

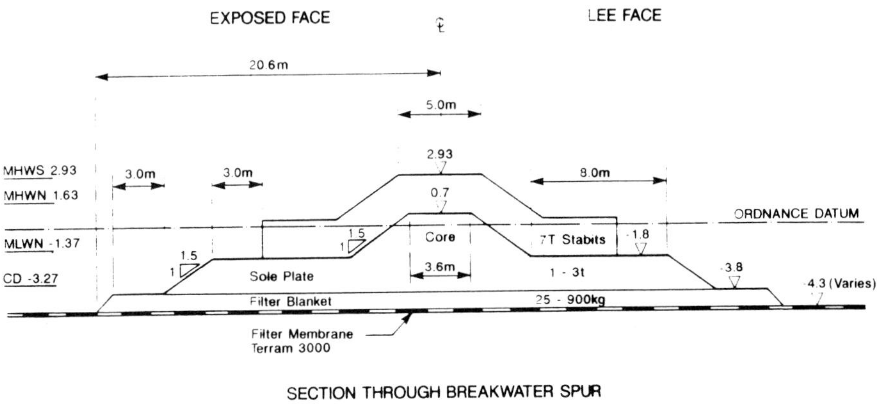

Fig. 5. Section Through the Breakwater Spur

20. The revetment design was the subject of 1:35 flume model studies carried out at Hydraulics Research (ref. 9),

the purpose of which were to confirm the design and overtopping rates; if possible refine the design; determine the long term characteristics; investigate shingle drawdown; and establish the wave reflection characteristics. The tests showed the design to be acceptable and that wave reflection at 0.65 was higher than for other similar units. In an effort to reduce the potential construction costs, the crest and beach elevations were reduced and the model re-run. However, large quantities of water and gravel (coal) passed over the rear wave wall and the discharge was not reduced significantly even when the wave wall was raised. The only damage seen were for test conditions far in excess of the design criteria, where several units were eventually extracted after 3000 wave cycles at a constant surge level. Wave action on a partly constructed revetment was also examined.

21. The final design chosen (ref. 10) had a buried toe at about 1.5m below OD protected by 1 to 3t rock, with 0.5m deep Seabees to 1m AOD and 0.7m deep units above with a 1.24m wide cast in-situ cope unit at the top at a level of 7m AOD (see Fig. 4). There was then a 4.4m wide slab, required for access to the main breakwater, with a 1m wave wall with a back slope protected by sand/bitumen grouted rock. The headland was also to be armoured with Seabees and there was to be 70,000m^3 of shingle placed as a beach infront. There are two rows of steps down the revetment for public access.

22. The chosen revetment, headland and spur scheme can be seen on Fig. 6. A gravel beach was to be provided with its toe extending from the spur to the headland and a crest level at 3m AOD. The volume of beach gravel was drastically reduced during construction as was the design of the headland was changed.

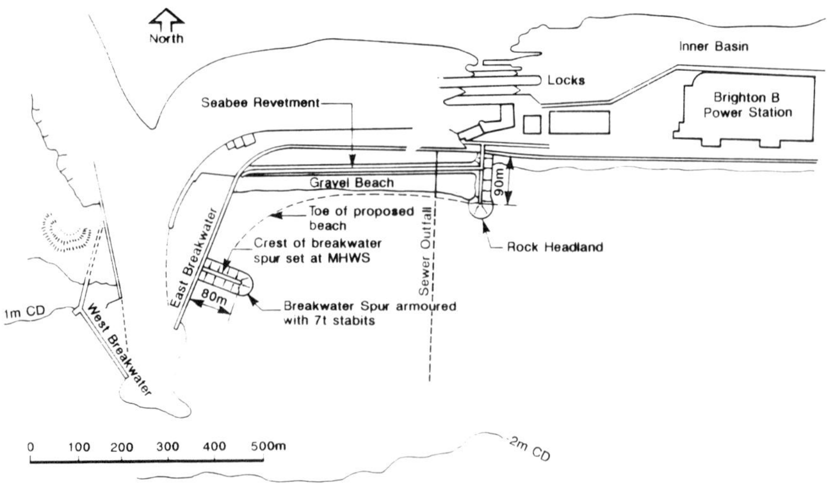

Fig. 6. Details of the final scheme

CONSTRUCTION

23. Tenders were invited in February 1989 and the contract for the sea defences, reclamation and paving works awarded to Balfour Beatty Civil Engineering Ltd in April 1989. Work commenced in May 1989 with completion due in July 1990. Priority was given by the Contractor to the completion of the breakwater spur in order to provide some protection to the western end of the revetment during construction. While Halcrow were responsible for the design, supervision of construction was done entirely by SPA.

Breakwater Spur

24. The breakwater spur was built using land based plant by constructing a haul road along the east side of the east breakwater. Following the seabed excavation for the spur, the terram 3000 membrane was placed by divers and covered with filter rock and then 1 to 3t rock of the sole plate and core, proceeding eastwards from the existing breakwater (see Fig. 5). Once complete, the core was trimmed and placing of the existing 7 tonne Stabits commenced back from the spur roundhead westwards. To the authors' knowledge this was the first fully overtopping breakwater constructed with Stabits on both the up rush and down rush sides. Stabits were brought up simultaneously from the sole plate on both sides of the structure with the crest units placed last to lock in the units at the top of the two slopes. Some difficulty in achieving full interlock of all the crest units was experienced.

Headland

25. The headland was originally designed as a Seabee armoured structure with a rock core. The Contractor proposed an alternative concept utilising rock armouring but the Engineer was made responsible for its design. This comprised a level crest headland incorporating 6-10 tonne rock armour, a 0.6-1.0 tonne underlayer and a core of hardcore, from the Brighton 'B' Power Station demolition project, placed on a terram 3000 filter membrane.

26. Rock was imported from Norway in two 5000 tonne vessels and discharged at the port directly to trucks, which transported to the headland. Rock was placed on the headland using a Liebherr 982 tracked excavator.

Revetment

27. The Seabee units were cast on site, a production rate of 850 units/week eventually being achieved using the following numbers of forms:

toe units	2 No
500mm Seabees	4 No
700mm Seabees	10 No

Each form accommodated 5 units.

The total number of units incorporated in the revetment were:

toe units	645
500mm Seabees	3662
700mm Seabees	8969

28. Construction of the revetment initially commenced in an easterly direction from a point some 100 metres east of the existing breakwater, as space was left for the haul road to the breakwater spur. Progress on the revetment face had reached approximately 30% completion by late October 1989 when the works were severely damaged by storms. In previous storms the top and sides of the Seabee face had been protected by 3 tonne armour rock and by pinning Seabee units together. This was not successful however for the event on 28/29 October 1989, when wave action, generated by 45 to 50 knot mean wind speeds, removed approximately 50% of the completed revetment face. A decision was then taken by the Contractor to suspend the works until March 1990.

29. The work duely restarted on programme with this time revetment construction commencing at the eastern end and proceeding westerly. Strongpoints, comprising rock filled gabions were introduced at approximately 75 metre centres to protect the core and underlayer material. The Seabee units were anchored back to the strongpoints using threaded bars and steel walings. The 75 metre spacing was selected as being the estimated progress of each revetment construction operation in one week and a total of five strongpoints were constructed. Weather conditions throughout the summer of 1990 were such that the capacity of the strongpoints to withstand wave attack was never tested.

30. Work proceeded extremely well and the 500 metre long revetment was finally completed ahead of the revised programme in July 1990. A new but reduced beach was then created using dredged shingle from the Owers Bank and pumped from a dredger berthed on the Outer Layby berth.

PERFORMANCE OF THE SEA DEFENCE WORKS

Breakwater Spur

31. This structure has clearly been successful in reducing the height of waves at the western end of the revetment. Wave diffraction around the spur has resulted in the build up of sea bed levels along the east side of the breakwater and seaward between the revetment and the spur. By January 1991 levels have increased by approximately 2m and they are still building. The effect of this accretion is expected to be reflected in a reduction in the incident wave heights at the revetment in this area.

32. Some damage to the crest of the spur has occurred with five of the 30 year old Stabits being either displaced or broken largely, it is considered, because the units were not adequately interlocked during construction. It is intended to replace these units and to raise the height of the spur adjacent to the breakwater during the summer of 1991, utilising some spare Stabit units stored on a nearby beach.

Headland

33. During its 6 months life the headland has acted satisfactorily, retaining beach gravels whilst at the sametime permitting some shingle to bypass at approximately the design rate ie. approximately 8000m^3 of shingle bypassed

the structure in 6 months compared with an estimated annual drift rate of 15000m^3. Wave diffraction around the headland which, because it is a level crested structure, occurs at all tidal levels, has encouraged the formation of a healthy beach immediately downdrift.

Revetment

34. Flume testing showed that, while the revetment was stable under design wave conditions, an adequate beach was essential to limit overtopping. The original scheme proposed by the study provided for a large beach extending from the headland around to the breakwater spur as a curved bay (see Fig. 6). In an effort to reduce the costs of the overall scheme a smaller beach was eventually constructed. The intention was to monitor it and take action as necessary.

35. The reduced profile beach was re-shaped within four months and demonstrated that without significant additional shingle volumes an adequate beach could not be held along all the revetment. The need for an adequate beach level was demonstrated in the December 1990 and January 1991 storms when, in an area where the beach crest had been drawn down by 4m, significant overtopping occurred. At the same time abrasion of the Seabees was noticeable over a height of approximately 2m (1.5m AOD to 3.5m AOD). Where the beach crest against the revetment was maintained at not lower than 2.5m OAD, abrasion was negligible. The extent of abrasion varied along the revetment face generally removing the smooth face of the units to expose aggregate. In areas where Seabees stood proud of their neighbours abrasion was significantly greater.

36. Beach feeding was undertaken in January 1991 to raise the beach crest level against the revetment. Following a review of the beach performance, it is intended to maintain high beach levels along the western half of the revetment by constructing a rock groyne approximately mid way along it. This solution was adopted in preference to the alternative of building the larger beach originally intended, for reasons of cost and the higher level of confidence associated with the provision of a groyne.

PROJECT COSTS

37. The costs of the sea defence works were as follows:

Breakwater Spur	£ 325 000
Revetment and Beach	£2065 000
Artificial Headland	£ 345 000

REFERENCES

1. RIDEHALGH H. Shoreham Harbour development. Proc Instn Civil Engs, 1958, Vol II, 285-298.

2. SUTHONS C.T. Frequency of occurrence of abnormally high sea levels on the east and south coast of England. Proc Instn Civil Engrs, 1963, 25 August, 451-483.

3. GRAFF J. and BLACKMAN D.L. Analysis of maximum sea levels in southern England. Proc 16th Coastal Eng Conf Hambere, 1978, vol 1, 931-947.

4. GRAFF J. An investigation of the frequency distributions of annual sea level maxima at ports around Great Britain. Estuarine, Coastal and Shelf Science, 1981, 12, 389-449.

5. HYDRAULICS RESEARCH. Wind wave hindcasting for Shoreham 1986.

6. POSFORD DUVIVIER. Dredging and materials movement study report, 1984, Appendix D.

7. BORGMAN L.E. Risk criteria. J Waterways and Harbours Div ASCE, 1963, Vol 89, No WW3, 1-35.

8. HYDRAULICS RESEARCH. Design of sea walls allowing wave overtopping. 1980, Report No EX 924.

9. HYDRAULICS RESEARCH. Flume studies of the Shoreham revetment. 1989, report No EX 1875.

10. SIR WILLIAM HALCROW AND PARTNERS LTD. Outer layby sea defences, Phase 2 - detailed design. March 1990.

Discussion

K. W. PILARACZYK, Rijkswaterstaat

Concerning Paper 24, for a proper maintenance of coastal structures it is very important to know the variation of input parameters, as used in the design, with time (during the lifetime of structure). Special attention should be paid to the components inside the structure which are difficult to detect by visual inspection only, i.e. erosion processes in filter and base layer.

New non-destructive techniques should be developed for this kind of monitoring. In the Netherlands some developments in this direction are underway concerning the application of radar techniques and other electro-magnetic methods. Because it is a common problem for all countries, co-operation in this field should be stimulated. The new methodology of maintenance needs acceptance by the users (local authorities). It has been proven in the Netherlands that creating a working group around the research programmes, composed of various representatives (researchers, consultants, contractors and local managers) strongly increases the acceptance of new methodology/techniques in practice. Such an approach to the problem will probably also work in other countries.

I. R. WHITTLE, National Rivers Authority

The NRA has recently established a database of sea defences in England and Wales. The database contains details of nearly 1200 defences covering a length of over 1200 km of low-lying coastline. Of this, just over 800 km of defences are owned by the National Rivers Authority, nearly 250 km by local authorities, and 250 km by private owners, companies and agencies. British Rail are surveying their own sea defences, and information will be added to the database.

The data for each defence is augmented by recording separately data on each principal element. The

Sea defence survey — general / levels of service

Anglian region

Map page no.	Defence code	Location/ name	Grid references		Length of defence km	Asset type of defence	Level of protection: m	Effective level of protection: m	Degree of exposure	Effective LOS	Land at risk (area)	Principle type	Property at risk		Reference port	HAT: m (GRAFF: m)
			Start	Finish									Domestic	Commercial		
14	053/1012	Sutton on Sea	TF524817	TF523822	0·48	Sea wall, groynes	7·10	7·10	Hi	S	A	B	B	B	Immingham	4·10 (4·90)

Table 1. Record of sea defence (053/1012)

Sea defence survey — condition / maintenance

Anglian region

Defence code	Sub-code	Defence element		Grid reference		Length of defence	Position	Primary/ secondary (p/s)	Defence condition				Foreshore				Date Const-ructed	Resi-dual life	Total or shared maintenance	Refurbishment		Sup. report no.
		Structure	Material	Start	Finish				Over-all	Land-ward	Crest	Sea-ward	Type	Cond-ition	Level	Depend-ency				Priority	Urgency	
053/1012	105	Wall	Cn	TF524817	TF523820	0·29	Backshore	p	Fr	Fr	Fr	Fr	Sa	S	Hi	Lw		A	NRA	2	2	1105
053/1012	106	Wall	Cn	TF523822	TF523820	0·19	Backshore	p	Pr	Pr	Pr	Pr	Sa	E	Hi	Hi	1953	B	NRA	2	1	1106
053/1012	107	Brwork	Cn	TF522822	TF523820	0·19	Backshore	p	Fr	Fr	Fr	Fr	Sa	S	Hi	Lw	1977	A	NRA	2	3	1107
053/1012	115	Groynes	Tm	TF524817	TF523822	0·48	Foreshore	s	Fr				Sa	S	Hi	Lw	1955		NRA			

Table 2. Record of elements of sea defence (053/1012/106)

Defence elements in a poor or bad overall condition

Anglian Region — Total element length = 696·79 km

Defence code	Sub-code	Location	Structure	Material	Length of defence: km	Foreshore		Principle land use type	Residual life	Reference no. if included in the medium term plan	Maintenance works in progress (Y/N)	Comments
						Type	Condition					
053/1012	106	Sutton on Sea	wall	Cn	0·19	Sa	E	B	B	39685		Port 19, Mablethorpe – Skegness

Table 3. Planning of repairs to inadequate element of sea defence (053/1012/106)

NRA medium-term plans

Submission date: Sep. 1991 (all estimated expenditure at 1991 base prices)

Scheme ref. number	LDW number	Scheme name	Expenditure cat.		Statistical B/C ratio	Total cost ×10^{3}	Grant eligible expenditure ×10^{3}, year ending 31 March							
			Start	End			Pre-199	1991-1992	1992-1993	1993-1994	1994-1995	1995-1996	1996-1997	Post-1997
RO53/9139885	36520	MAB/SKEG PT 19 GRDEN CAFE-SUTTON PLVER	03/1991	03/1992	U S A A	785	130	655	0	0	0	0	0	0

Table 4. Expenditure profile for refurbishment of sea defence (053/1012)

attributes for each element are coded using alpha-numeric indices.

A nest of programmes has been designed to display and analyse the extensive range of data collected during the survey. The output from the computer can be in tabular, graphical, statistical matrix, or map-based GIS format.

Tables 1-4 show the range of data collected and the use made of one of a series of monitoring programmes. A defence containing an element requiring major works is traced from survey through to works programme.

Other analyses have related defence lengths to either land area protected or land use of that area, by each type of owner.

This database therefore complements and can be enhanced by the range of structural and probabilistic analyses of this type undertaken by Rijkswaterstaat.

K. W. PILARCZYK, Rijkswaterstaat

Knowledge of physical interactions between water, soil and structure is essential for a proper modelling and derivation of stability criteria. In this respect it would be useful for Asakawa to provide additional information concerning the stability criteria for Japanese revetment units, regarding the scale of model testing and reproduction of sublayers. Representation of stability for block revetments by the K_D factor in Hudson's formula is less proper than by using a linear representation, i.e. $H_s/\Delta D$, where H_s is the wave height, Δ is the relative density and D is the thickness of the block. It would be useful to have detailed information on revetment (including sublayers) and wave specifications as tested. It would allow one to make a comparison of these systems with others (Europe, USA), and to apply the theory and practice as presented by Klein Breteler (Paper 26).

In general, the exchange of basic research data between the researchers (or institutes) should be further stimulated. In this way we shall obtain better progress in solving our common problems.

L. FRANCO, Politechnico di Milano

Could Asakawa et al. give some additional information on the design and application of submerged barriers (e.g. Aqua Reef), such as typical water depth, submergence, berm width, block weight? Regarding the submergence below mean water level, have they considered the danger for bathers (swimmers) and/or navigation? Do they already have prototype applications of these detached underwater barriers as sills for containing artificial beaches, and what is the experience of their performance regarding sediment transport effects?

C. P. ORBELL-DURRANT, May Gurney Contractors
Regarding terrace blocks, how do Asakawa et al. secure the top line of clocks? Do the authors have a problem with air getting through points in the blocks when it is expelled violently from the filter layer? If so how do they secure public safety?

N. W. ALLSOP, Hydraulics Research, Wallingford
The stability of revetment facings has been the subject of a research project at Hydraulics Research Wallingford supported by the DoE Reservoir Safety Committee. This has involved the collection of site-specific data from reservoir embankment dams in the UK by Rofe, Kennard, Lapworth acting through CIRIA as the subcontractor to HR Wallingford. The study team has analysed a large number of incidents of damage to blockwork and slabbing. These will be compared with estimated wave conditions where possible, thus perhaps supplementing the data presented in Figs 5-7 of Paper 26, and will be presented in a report by HR and CIRA in early 1992.

W. LEEUWESTEIN, CUR
Will the empirical relationships for the wave front characteristics (ϕ_b, $\tan\beta$) change for irregular waves?

How is (in)stability defined? Presumably by using an uplift distance criterion.

How can one explain the difference in the H/D versus ξ curves shown in Papers 25 and 26, in particular referring to the rising part for $\xi > 2$ shown by Asakawa et al. in their Fig. 12.

The factor between wave heights in the regular and irregular case will be determined by the distribution of the irregular waves and by the transfer function for the loading (wave height - uplift pressure) only for Rayleigh-distributed waves and linear transfer. r may be approximately 1.4 (Paper 26, Table 1).

If one has to deal with other failure modes, in particular with uplift of a row of blocks (say 2 or 5) one will easily find F values above say F=10 (Table 3).

R. J. MADDRELL, Sir William Halcrow & Partners
Sand bitumen (with a bitumen content of 4%) was used in the core of a rubble mound cooling water basin on the west coast of South Africa to prevent recirculation and ingress of some of the estimated 1×10^6 m^3 of littoral drift into the basin. The impermeable core also had the benefit of allowing the backface of the breakwater to be grouted, thus acting as a spillway.

The breakwater is also unique in that at the time of its design (1976) a K_D of 12 was adopted for the dolosse (24 was being recommended) partly because the 6 m design wave was an annual event and because, on the

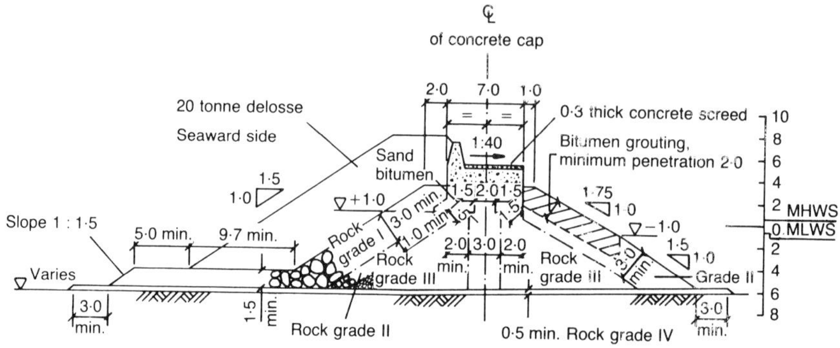

Fig. 1. Idealized cross-section through breakwater

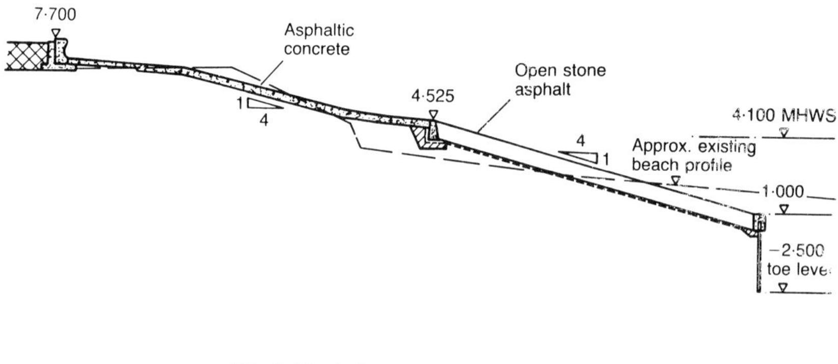

Fig. 2. Typical cross-section of new sea wall

most critical area of the roundhead, additional dolosse was placed, utilizing their interlock and increasing the slope steepness to about 1 in 1.

The breakwater has been in service for about 15 years without any significant damage being reported. It has also recently been subjected to a 1 in 50 year storm.

A. BEZUIJEN, Delft Geotechincs

We were involved in the repair works of Caister Seawall. When analysing the damage on that open stone asphalt revetment, it was found that abrasion of the open stone asphalt was caused by cobbles that are present on the beach. During wave attack the cobbles are transported over the revetment and hit the stones in the OSA, causing removal of the stones out of the OSA. This causes an abrasion much higher than could be expected from the result of model tests with clean water.

In the presentation of Kemps and Barber the disadvantages of OSA were mentioned. This possible large abrasion should be added to this list.

Because of the repair works that have been performed on Caister Seawall, it amazes me to read in your paper that 'coastal works using OSA have been in service now for over 15 years and there is no requirement for maintenance to date'. Therefore my question is, what is your definition of maintenance?

K. J. RIDDELL, Babtie Group

I fully support Kemps and Barber in their implicit recommendation that bituminous materials should be considered more often. They do offer significant benefits in quite a wide range of applications.

The works at Prestatyn, North Wales were mentioned in Paper 27 and I would like to present a few details regarding this project. Prestatyn is probably the most significant application of open stone asphalt in the UK and is also one of the most exposed sites. Wave heights of up to 5 m have been recorded offshore, and storms of this type have caused considerable damage to the existing concrete and masonry revetments.

The new wall consists of over 60 000 tonnes of open stone asphalt. It is formed in a complex curved profile with the lower part (in the inter-tidal zone) formed of 700 mm thickness of open stone asphalt and the upper part in asphaltic concrete. This design was model-tested at Hydraulics Research Ltd and yielded one of the lowest reflecting coefficients they had observed for a sea wall. Of course, this wall forms just part of an overall management scheme which includes dune stabilization, beach replenishment, and the construction of large rock groynes in order to maintain healthy beach levels in front wall of the wall.

Part of the philosophy of this design was to maximize energy dissipation within the structure during periods of mild wave attack, particularly long-period swell waves. These periods would typically be ones of beach build-up, and the permeability of this material is helping in this respect and is already showing signs of success. I would like to make two comments about the design and specification of open stone asphalt.

Each mix design should be optimized by site trials as its characteristics change when even the source of filler or sand changes, despite having apparently the same grading. Drainage of the mastic, due to incorrect viscosity, can be a real problem, causing the completed works to be impermeable.

Design by conventional methods, apart from some overall guideline limitations, would imply that, given sufficient thickness, any wave height can be accommodated. This would not seem reasonable as one would expect this material to suffer significant bursting stresses under wave impact. A monitoring exercise is being undertaken at Prestatyn by Hydraulics Research Ltd and the Polytechnic of Wales. This utilizes pressure transducers, buried at various depths in the structure, and hopes to yield improved understanding and hence better design methods for the future.

H. OUMERACI, University of Hannover

How did Kemps and Barber solve the differential equation (2)? Why did they not put the first term on the left-hand side of equation (2), together with the third term?

The assumption of the 'elastic foundation' for the revetment plate is not the most critical assumption, since due to cyclic mobility of the saturated sand induced by the cyclic wave impact load near still water level, the soil beneath the revetment can move downwards so that a large gap can develop under the revetment. That is the most critical situation.

D. J. VANDENBOSSCHE, Haecon NV

The Zeebrugge North Service Port Breakwater has been undergoing construction since 1979 on a design of Haecon. After more than 10 years of service no maintenance works on the OSA revetments were necessary. The only maintenance cost incurred is of course the yearly inspection of the structure.

Indeed both materials will suffer abrasion caused by pebbles, boulders and similar bombardment of the revetment under wave attack. In a proper design this can be avoided. At Zeebrugge the rubble toe protection has been fixed by either hydro-concrete or asphalt mortar.

D. M. HERBERT, Hydraulics Research, Wallingford
Could Kemps and Barber please give an indication of the permeability of OSA for a hydraulic gradient of 1?

H. L. WAKELING
When the Shoreham breakwaters were built over 30 years ago it was expected that they would intercept littoral drift from west to east. Accordingly, the Port Authority was required by Act of Parliament to transport shingle round the port entrance if there was erosion of the downdrift beaches which might result in damage at Hove and Brighton promenades.

In fact, there was erosion on the updrift side due to wave reflection off the main breakwater, which at times resulted in a reverse of the littoral drift for a distance of about 3 groynes to the west and very little accretion at the root of the breakwaters.

Apart from erosion at the root of the east breakwater which is the subject of Paper 28, there has been no shortage of beach material further east. Could Maddrell and Vaughan comment on how littoral drift has been passing the harbour entrance, as there is no sign of a shingle bar across the entrance.

J. METTAM, Scott Bertlin
30 years ago I was responsible for hydraulic model studies of the recently completed Shoreham Harbour extension. Our problem studied was the concentration of wave attack at the root of the east breakwater. (This problem would have been largely avoided if the breakwater had been aligned at right angles to the coast, or slightly east until it reached deep water). After trying to solve the problem with timber groynes and wave screens we eventually developed the solution with a concrete groyne, mentioned by Maddrell and Vaughan, angled to the east with a spending beach to protect the troubled area. It proved cheaper to use stabits, with a shorter groyne, rather than shingle. I was sorry to see this groyne, still in good condition after nearly 30 years, demolished.

Two years ago after collapse I was asked to advise on how the Outer Layby Sea Defences could be completed without further risk. Our first suggestion was to wait until the following summer and then start by building the new groyne at the east end and work towards the breakwaters.

We were not able to suggest a way of building the groyne using Seabees as then designed, without risk of major loss. We therefore proposed that the design be revised. My first preference for concrete construction was not accepted but a design with rock armour was adopted.

To reduce the extent of damage to a partly completed revetment, we first suggested constructing some intermediate groynes, which we thought would also help, in the permanent design, to maintain the beach. When this was rejected we designed gabion strong points which were built in the rubble embankment. With an improved grading of rock in the embankment we thought that there would be little loss if a storm occurred during construction.

In the event work started at the earliest date suggested by our analysis of past wave records and was completed without a storm.

E. P. VAN HIJUM, Paper 24

Some additional points, made during the presentation of Paper 24, are detailed below.

The actual products of coastal research seem to be insufficiently tailored for an effective performance assessment.

In this statement the word 'effective' stands for quality and acceptance: Effectivity = Quality x Acceptance. The technical input of the management scheme is insufficient, both in quality (models) and in acceptance (standards and criteria), to create sharp rules for maintenance, improvement or renewal of the defence works. The acceptance is affected by the 'third generation problem', a psychological problem: coastal defence works are initiated by catastrophes, and a law of nature is that decisions made by the suffering generation are generally respected by the second generation but tend to be neglected by the third generation, until the next catastrophe takes place.

Remedial actions are: improvement of the technical input; legislation.

Legislation is necessary for defining responsibilities, standards and procedures: who does which job, who pays, who exercises supervision, which standards, which procedures, which auditing system. In relation to the third generation problem, the Dutch society is now in the second and in the third generation, depending on the region. Legislation (the Flood Defence Bill 1989), still in discussions in Parliament) will guarantee the flood protection standard as achieved by the Deltaplan and additional decisions in future; this bill covers all the items mentioned above. The auditing system charges the responsible manager with a periodic assessment every five years and to report it to the supervising provincial government; the provincial government reports to the central government. In this auditing system the low-lying parts of the country are divided into dike-rings.

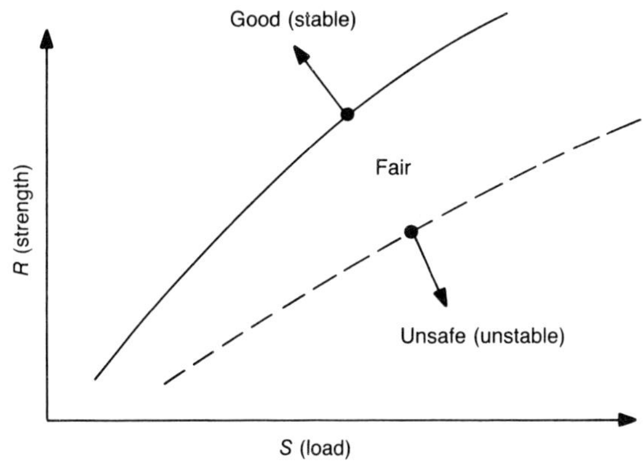

Fig. 3

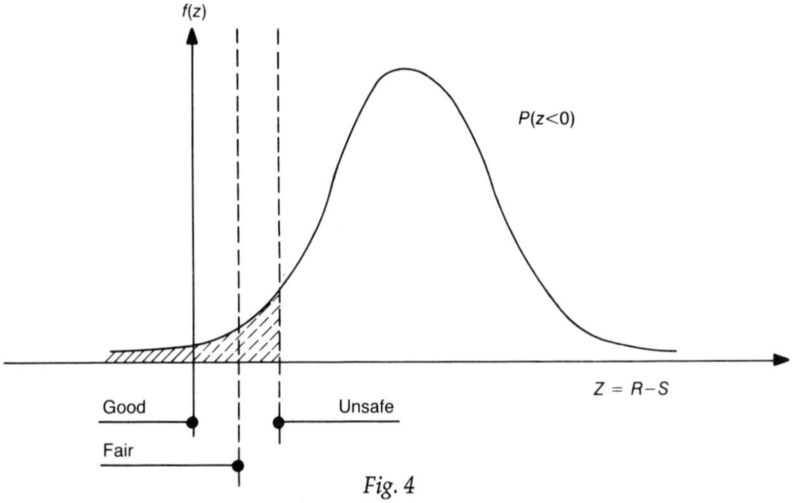

Fig. 4

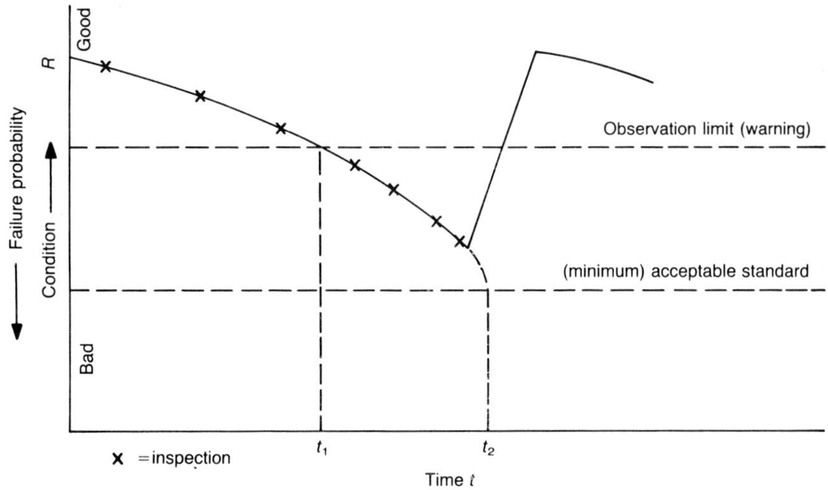

Fig. 5

For genuine assessment one needs sharp standards, but the products of coastal research are not complete (too many blank spots), insufficiently reliable (the probability of failure can be estimated once the uncertainty on each parameter has been defined, and this appears to be quite difficult), hardly accepted by the management (the distance between theory and practice is far too great), and insufficiently simple (the models are too complex).

As an illustration, see Paper 24, Fig. 3. The real inaccuracy in the calculated wave run-up height is 1 or 2 m, so at which level do we put the criterion for an unsafe crest height? The stability of the stone pitching depends on (among other things) permeability and friction forces (see Paper 26). What will the permeability and the friction forces be if the revetment is 5, 10 or 20 years old? Up to now measuring methods have not been very successful; this means that it is at this moment almost impossible to perform a reliable assessment on block revetments. Similar questions arise for almost all failure mechanisms.

This leads to the following list of requirements.

(a) Make a realistic and reliable description of wind and water behaviour at low probabilities

(b) Show for every strength parameter a stable/unstable line, or present a limit state function, with well-defined narrow distribution functions of the relevant parameters. Present these items also for the conditions after x years. Present the monitoring system.

(c) Wrap it nicely and digestibly for the management level

(d) Do this also for relevant products from the past.

T. ASAKAWA, Paper 25

In reply to Mr Pilarczyk, additional information on the model tests is shown below.

Terrace block model tests ware conducted under the following conditions using regular waves: sea bottom slope i = 1/30; water depth at the toe = 0, 6.7, 10.0, 13.3 cm; slope of the revetment = 1:3, 1:5; block weight = 97.9 g; wave period = 1.28, 1.83, 2.37 s; wave heights were increased up to the level where clear damage results.

The revetment section in the test is shown in Fig. 6.

Model tests for Crab blocks were conducted in Tokushima University by Nakano et al. jointly with Nippon Tetrapod using the block model shown in Fig. 7. A scale model was constructed in the 30 m long wave flume, as shown in Fig. 8.

The revetment model was constructed on sand with a 1.5 cm thick stone underlayer (5-10 mm diameter

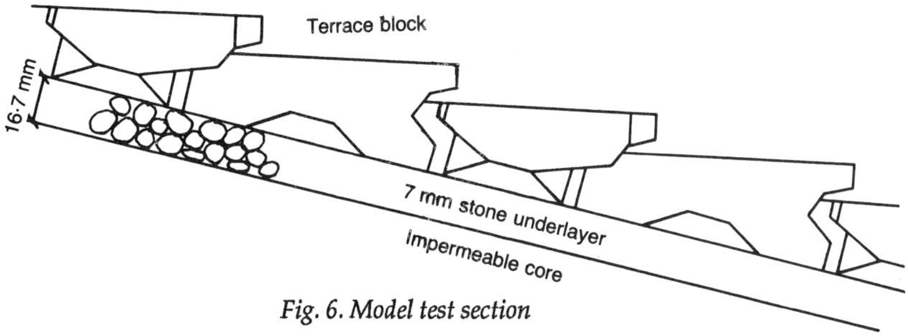

Fig. 6. Model test section

stones). Fig. 13 of Paper 25 shows the test results using regular waves. Wave periods were 0.8-2.1 s in the test. Irregular wave tests were carried out later and the results were partially reported by Nakano et al. in the annual conference of civil engineers, JSCE in September 1991.

We introduced a stability factor for the Crab blocks using K_D in Fig. 13. The K_D value is currently used in Japan and most engineers engaged in design and construction are familiar with it. As was indicated, H_S/D could be considered to assist in more general use when the block heights are changed, when the stabilities of the blocks are linearly proportioned to changes in block height and the effects of revetment slopes are properly explained. Until now one has not had enough data to confirm this. We are trying further tests.

In reply to Professor Franco, we introduced 70 examples of submerged barriers already in place at various sites in Japan, as far as we can obtain information. The dimensions of water depth, submergence, berm width and block weight are illustrated in Fig. 9. The water depth is measured from low water level.

The majority of the water depth distribution is from 3 to 5 m, but there are a few cases deeper than 10 m. Submergences are mostly less than 2 m. Cases deeper than 3 m submergence probably resulted from consideration of small fishery boat navigation.

A broad distribution of berm width from 3 to 100 m can be recognized with a predominant width of 40 - 50 m. In the block weight distribution only concrete armour units are shown. In addition to these, 29 cases use armour stones weighing mostly less than 2 tons.

Regarding Aqua reefs, no example has been constructed as yet, though there are some plans to do so.

As mentioned above, artificial reef type structures have been constructed recently at many sites in Japan.

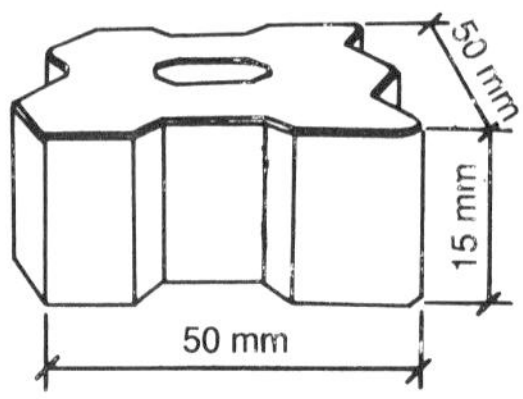

Fig. 7. Block model used in test

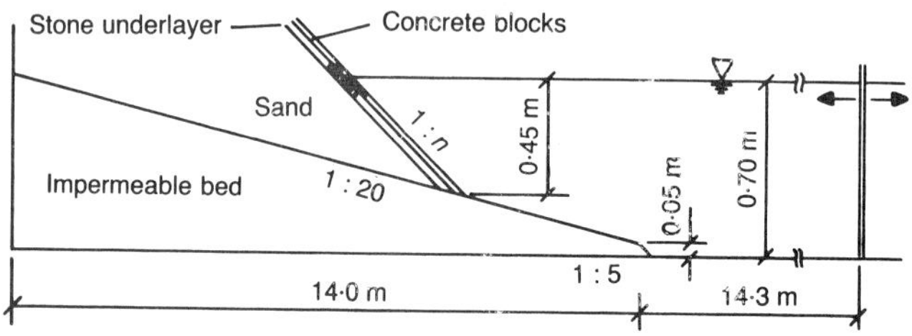

Fig. 8. Model construction

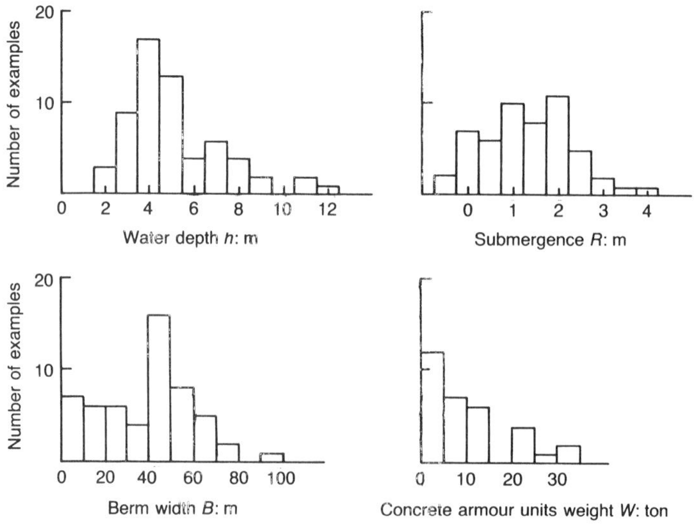

Fig. 9. Dimensions of submerged barriers

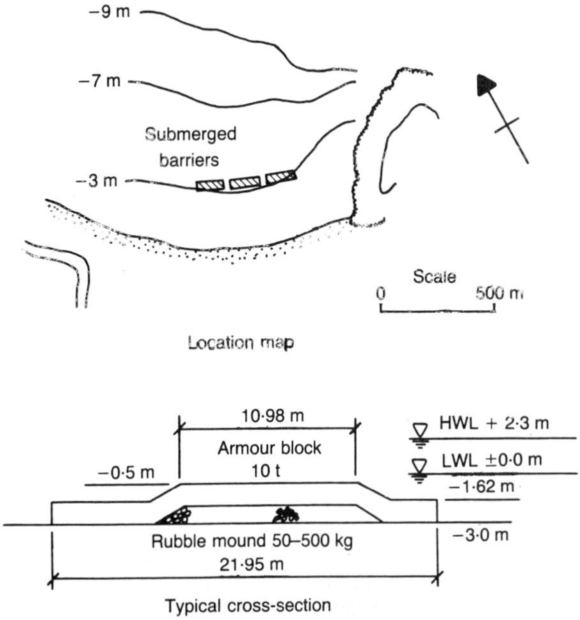

Fig. 10. Submerged barrier constructed at Aoshima Beach

Public safety was taken into account in every design, with respect to indication using buoy systems or adopting deeper submergence to allow small craft to navigate. However, if a project site were congested with bathers or ships, careful consideration of whether an artificial reef was appropriate in the interests of public safety would have to be taken.

At this moment we have no data to show the sediment transport effects of underwater barriers. In the case of Aoshima beach, effects on the shore erosion problem were presented by Kohno et al. at the 38th conference of coastal engineering in Japan, 1991. The results there were seen to be very good in preventing erosion. The location and typical section of the submerged barrier is shown in Fig. 10.

With regard to the point of Mr Orbell-Durrant, construction using terrace blocks is not so difficult in terms of securing the top line because wedge-shaped legs interlock. Terrace block revetments have been adopted at more than 100 sites in Japan, and we have not heard of any problems with expelled air. However, regarding public safety, we are sometimes requested to take steps to avoid slippery surfaces caused by algae adhering to the upper surface.

To Mr Leeuwestein, the limit line of terrace block stability was drawn as an average based on the test results. We recognize that the data to judge the effects of wave periods were limited, and additional tests will be needed. In actual practical design, we recommend terrace block for use in regions of less than 2 m wave height for every ξ.

M. KLEIN BRETELER, Paper 26

The reply to Mr Leeuwestein's first question is 'yes'.

The reply to his second is as follows. It depends on how detailed were the measurements reported. I have chosen a minimal damage (a small block motion, damage or total collapse). I have chosen for a block motion of approximately 10 - 20 mm if block displacements were measured.

Figures 5 and 6 of Paper 26 show a monotonous decrease of H/D with increasing . The points in Fig. 12 of Paper 25 show an unclear trend if $2 < \xi < 4$. The line is probably drawn with the trend for rubble mound breakwaters in mind.

The distribution of the wave heights is indeed important. Therefore we shall include the influence of wave breaking on a shallow foreshore in our future research with random waves (see Chapter 5 Paper 26).

Table 3 of Paper 26 shows that F = 10 is possible (structure types a4 and b3), but much higher values than 10 cannot be advised. Besides this, other damage mechanisms will be dominant if the stability of the cover layer allows for such high F-values.

J. J. L. M. KEMPS, Paper 27

Maintenance work needs to be separated from work required as a consequence of design errors or bad workmanship. The project at Caister that Mr Bezuijen refers to, has mainly suffered from the consequences of the chosen design for the toe-connection and arrangements for the material supply. The toe of the section, as built in 1979, consisted of a mastic slab able to follow beach level fluctuations. In this particular situation, however, there was debris (old brickwork, timbers, etc.) buried in the beach which became exposed when the beach levels dropped. As a result, the mastic slab could not accommodate the hardspot deformations, and eventually ruptured. Sand from the embankment core was leeched out and the OSA revetment suffered significant settlement.

The fact that a new section of coastal protection at Caister, for 1992, will also be built in OSA shows that both client (Great Yarmouth BC) and consultant (Posford Duvivier) are satisfied with the OSA-performance experienced with sections, built in 1979 and 1987 and repaired in 1988.

OSA is sensitive to abrasion by, for example, shingle material moving up and down the slope during wave activity (see para. 7 of Paper 27). Such a situation will require increased maintenance, and alternative materials may be more appropriate where abrasion forces are significant. This aspect however has to be seen in combination with two essential facts.

(a) An OSA revetment can be substantially cheaper than other types of suitable construction, for a coastal site.

(b) Maintenance of an OSA revetment will usually involve placing a fresh layer of this material on top of the existing worn OSA protection (see para. 7 of Paper 27). this means that maintenance is also relatively cheap, compared with the requirements of other types of construction.

In reply to the comment of Dr Oumeraci, for the answer to his question about differential equations refer to Appendix I of the 'Rijkswaterstaat Communications': the use of asphalt in hydraulic engineering. In this appendix the differential equation is resolved and also the derivation of the calculation formula to determine the layer thickness is described.

Research carried out by Delft Soilmechanics (P. Rulgrok) and by TAW (Mr Ebbens) led to the conclusion that there is no danger of fluidized sand from cyclic wave impact when the sand sub-base has been compacted properly. Only in situations where the sub-base is badly compacted is fluidization and recompaction of the sand possible. However, the bituminous revetment is a relatively stiff plate, and within the very short available time (frequency of waves) there is virtually no possibility for sand to be transported and relocated.

In response to Dr Herbert: regarding the relation between the filter velocity V^n and gradient i for a porous structure like OSA, with a void content between 25 and 30%, tests show a parabolic relationship (laminar to turbulent). The exponent n lies in the range 1.5 - 1.7. Delft soil mechanics carried out permeability tests on OSA at a test section along a canal in Holland. These tests show a variety of permeability coefficients (k-values) from which we may conclude that with the gradient i = 1, the k value lies in the range between 2×10^{-3} and 5×10^{-3} m/s.

R. J. MADDRELL, Paper 28

In reply to Mr Wakeling: the Act of Parliament enabling the then Central Electricity Generating Board (CEGB) to build Shoreham Harbour in the 1950s did not require gravel to be transported from the updrift to the downdrift side of the new breakwaters. It did, however, have a provision for the downdrift communities to make representations to the CEGB and Shoreham Port Authority (SPA) if they believed the works were adversely affecting their beaches.

Although updrift of the harbour entrance there has been a reversal of littoral drift, naturally due to both the less frequent east and south-east storms and wave reflection off the west breakwater by the more frequent south and south-west storms, their effect has been local. Thus, shingle has accreted updrift of the harbour, and between 1973 and 1987 this amounted to over 200 000 m^3 or say, some 14 000 m^3 annually.

On the contrary, there has been significant erosion of the eastern downdrift beaches. The SPA is, following its 1991 Beach Management Study, enacting enabling measures to hold and improve these beaches. As far as we are aware from dredging records and bed samples, no shingle is bypassing the harbour entrance, only sand, over 2 m of which has accreted in the lee of the new Stabit spur. In order to maintain the downdrift beach levels, the SPA is transporting 15 000 m^3 of shingle annually by truck from the updrift beach. Most of the shingle is placed in front of the new revetment from where it is transported further downdrift by nature.

We agree with the comment of Mr Mettam that the amount of wave energy concentration could have been reduced had the lee breakwater been at 90° to the shoreline, but this would have meant a longer and more costly breakwater. Even if this had been done, there would still have been a Mach Stem wave, but its occurrence would have been less frequent. The concrete groyne referred to has not been demolished, only buried and awaiting rediscovery!

The initially proposed alternative concrete groyne was not accepted for two reasons: firstly its sloping longitudinal profile would not have retained an adequate width required for the updrift beach; and secondly the steep side slope to the groyne was expected to encourage wave reflection and scouring on both its sides. The level crested rock armour headland overcomes these problems by holding a wide beach at the eastern end of he bay and by encouraging the formation of a healthy beach adjacent to the downdrift side of the headland, as described in Paper 28. The longitudinal profile of the rock armour structure was identical to the original design for a seabee armoured headland.

The gabion strongpoints that were designed for the Contractor by Mr Mettam's firm are described in Paper 28.

29. Construction of an artificial island for Kansai International Airport

Y. ARAI, K. OIKAWA, S. SUZUKI and K. HAYASHI, Kansai International Airport

SYNOPSIS. Kansai International Airport, the first fully developed offshore airport, is being constructed in the south-eastern part of Osaka Bay, some 300 miles southeast of Tokyo. The superstructure of Kansai International Airport will be built on an artificial island approximately 5 km off the coast of the Senshu District. The island area is 5.11 million m^2. The depth of water at the construction site is approximately 18 m. This paper provides an overview of Kansai International Airport, the island and seawall construction work, and the reclamation operations.

1. KANSAI INTERNATIONAL AIRPORT PLAN

Kansai International Airport was planned with a view to:

a.serving as a hub for domestic and international air transport;

b.being open to traffic 24 hours a day;

c.affording passengers and cargo ready access to the mainland for maximum convenience; and

d.paying due and full regard to pollution control and environmental protection in Osaka Bay and its surrounding areas.

Based on these guidelines above, Phase One of the construction plan was decided as follows:

1) Location: 5 km offshore from the Senshu area in the south-eastern part of Osaka Bay (Fig.1) to avoid any undesirable impact due to noise of air traffic.

2) Scale: One main runway 3,500 m long in a total area of 5.11 million m^2

3) Capacity: Approximately 160,000 takeoffs and landings per year

4) Access bridge: 3.8 km long providing both road and railway transportation

5) Target date of airport opening: Summer, 1994 (Fig.2)

6) Total construction cost: Approximately 1,430 billion yen

Fig.3 shows the overall layout of the airport planned for phase one, while Fig.4 shows the final phase of Kansai International Airport. In the final phase, we have three runways and about 12 million m^2. The capacity will be 260,000 takeoffs and landings per year. However, this final phase has not been authorized by the Government.

The airport will be accessible by bridge and by boat. An access bridge will connect the north side of the island with the mainland. A sea access terminal will also be built on the northern part of the island.

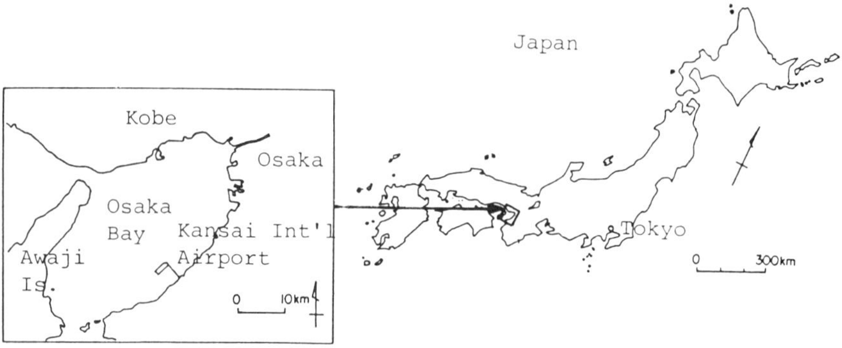

Fig.1 Location of Kansai International Airport

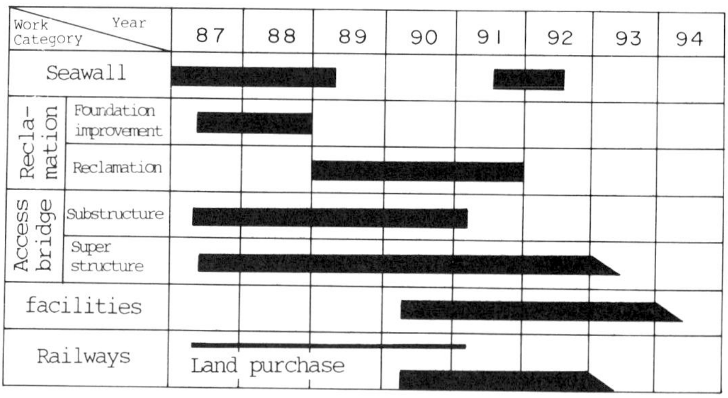

Fig.2 Construction schedule

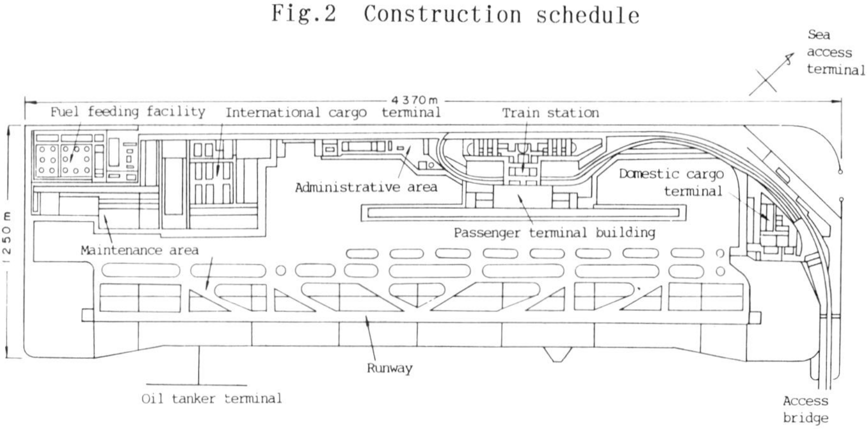

Fig.3 Overall layout of Kansai International Airport

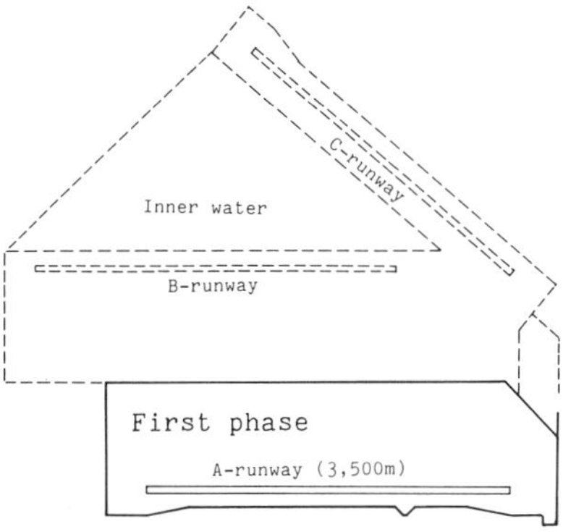

Fig.4 Final phase plan

2. NATURAL CONDITION

2.1 Seabed soil

The water depth at the construction site varies from 16.5 m to 19 m. The seabed slopes with a gradient of about 1/750. Seabed soil at the airport site was evaluated by boring 65 holes of 100 - 400 m in depth. The system of wire line boring developed by the Port and Harbor Research Institute of Ministry of Transport was used to ensure high quality and efficiency during the work. In this soil exploration, a series of geological studies was carried out on all of the soil samples obtained. The geological studies included measurements of micro fossils of foraminifera, diatom, pollen, calcareous nannoplankton and others. Existence of the nannoplankton implies the sedimentation under marine condition, and its pattern of emergence with depth reflects the sedimentation environments. As the result of these geological studies, the sedimentation environment and age of the soil layers were traced.

Fig.5 show the soil profiles at the construction site. The thickness of soft alluvial clay stratum (Ac) is about 16 - 20 m. Substrata found below the alluvial clay stratum are alternations of diluvial clay (Dc) and thin gravel strata.

2.2 Other Condition

The airport island is located in the sea, therefore it is necessary to consider the forces caused by the surrounding sea, in addition to the force by earthquakes. Because of its location inside a Bay, the island will be subject to waves only 50 cm or less in height for about 80 % of the year. In designing the seawalls, however, the highest significant waves that may occur in a 50-year period were predicted to be 3.5 m in height ($H_{1/3}$) with a period of 6.7 sec ($T_{1/3}$) in a WSW direction.

Tidal levels around the island are as follows.

Highest high water level observed so far(HHWL)	:C.D.L +3.2 m
High water level (HWL)	:C.D.L +1.6 m
Mean water level (MWL)	:C.D.L +0.9 m
Low water level (LWL)	:C.D.L +0.1 m

The surface elevation of the island is specified to be not lower than C.D.L +4 m after 50 years from the opening.

The seismic force acting on structures of port and harbor facility is calculated by the seismic coefficient method in Japan. The design seismic coefficient is 0.15, determined with the classification of region where the structure is located, that of the subsoil condition and the degree of importance of the structure.

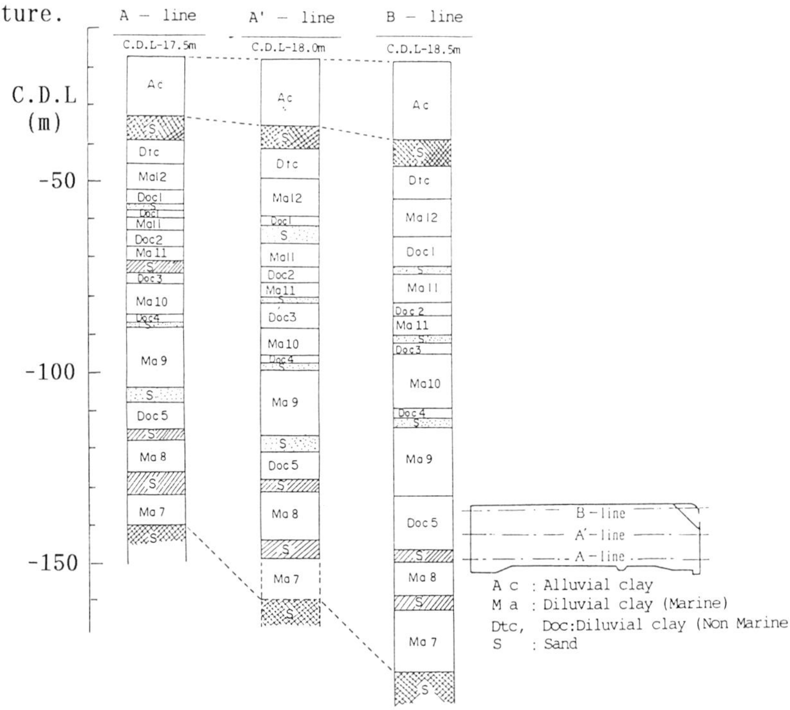

Fig.5 Soil profiles at the construction site

3. CONSTRUCTION OF THE AIRPORT ISLAND

3.1 Improvement of alluvial clay layer

The existing seabed surface stratum at the airport island is covered with a thick, weak alluvial clay layer, which has to be improved so as to ensure the stability of seawalls and embankments while they are being constructed and to minimize any residual or uneven settlement that may occur after the airport in put into operation.

The scope of foundation improvement works covers, horizontally, the surrounding seawalls and the reclaimed land inside, and when viewed vertically, involves the alluvial clay layer. In this construction the sand drain (SD) method and the sand compaction pile (SCP) method have been applied over most of the area, with the deep mixing (DM) method at two corners of the seawall.

1) Sand Drain Method

The sand drain method has been adopted for the seawalls area (rubble mound type) and the reclamation area. The sand drain piles were 40 cm in diameter and the spacing of piles was 2.5 m X 1.6 m (equivalent conversion : 2 m square grid) in the seawall area and a 2.5 m square grid in the reclamation area. As water contained in the clay layer is discharged to the sand pile by applied surcharge, the clay layer decreases in volume and increases in strength. While the method is more economical than others, it requires a longer construction period and careful management.

2) Sand Compaction Pile Method

The sand compaction pile method has been adopted for the remained seawalls (steel cellular type and wave dissipating caisson type). The compacted sand piles were 2 m in diameter and spacing of piles was as 2.1 m square grid. The method has the advantage of attaining sufficient strength for mounting the structure immediately. On the other hand, the method experiences a problem when the replaced soft clay is forced upward by sand. Conventionally this soft clay is removed by dredging. In this project, however, it was decided that it formed a part of ground to save the cost of removal and avoid adverse environmental effects of dredging.

To deal with the extensive foundation improvement in a short period, more than 10 large, high-performance work vessels were used. Each vessel is equipped with computer-controlled electronic devices that control execution. In the sea surrounding the island site, 22 surveying scaffolds with reflectors have been placed 500 m apart. A vessel determines its position by sending a light beam from its light wave range finder to a given reflector and analyzing the reflection received.

Photo 1 Foundation improvement work vessels

3.2 Seawall

The seawall surrounding the 511 ha airport island is approximately 11 km long. As shown in Fig.6, it consists of four sections, each of which has different functions in accordance with orientation, land use, and construction schedule. Functional requirements for each seawall section are summarized in Table 1 in terms of (a) need for reducing disturbance due to reflective waves in the nearby sea area, (b) need for the quay to function both as sea access wharves and landing places for construction equipment and materials, and (c) need for short-term construction in accordance with the construction plan. Structures of seawalls satisfying functional requirements were compared and selected in terms of economy, ease of construction and stability.

Seawalls of types A and B, about 80 percents of the seawall, are rubble mound embankment. The rubble mound type seawall is cheapest and can adapt itself to the settlement of seabed. However, armour blocks for wave dissipation are provided on outer face of the type A seawall, since this type of seawall is subjected to the most severe wave action.

The type C seawall is a steel plate cellular bulkhead, which is 23 m in diameter and 23 m in height. The type D seawall consists of concrete caissons, the front face of which is provided with a series of slits in order to dissipate wave energy.

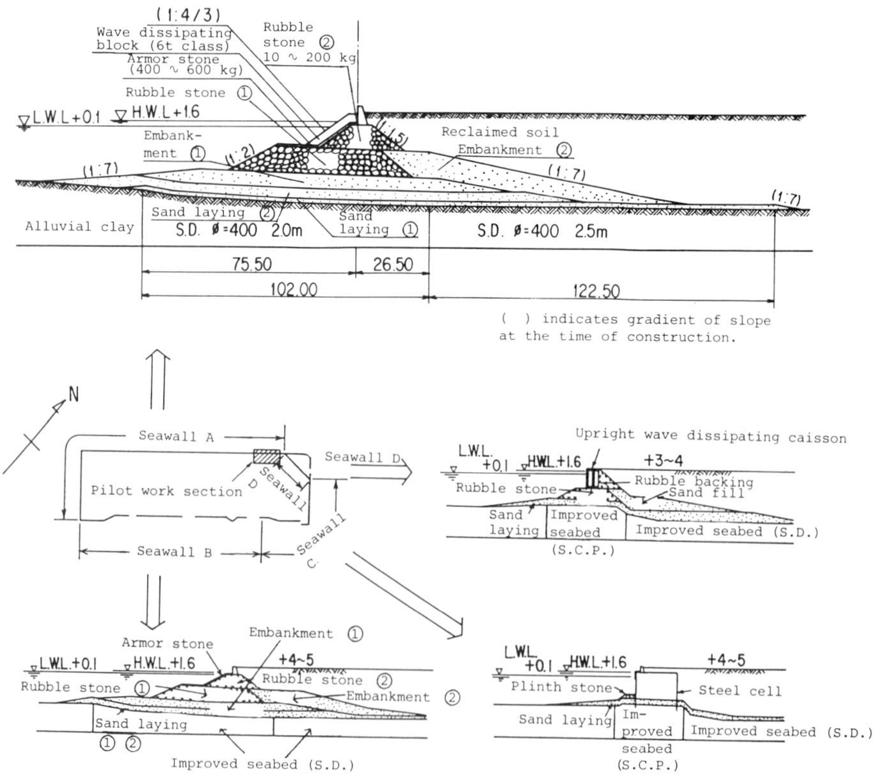

Fig.6 Seawall structure and location

Table 1 Functional Requirements for Seawall by section

Section / Type / Requirement	Seawall A,B	Seawall C	Seawall D
	Rubble Mound	Steel Cellular	Wave Dissipating Caisson
Reduction of reflective wave	Required	——	Required
Quay function	Required in selected parts	Required	Required
Need for speedy Construction	——	Particularly Required	——

3.3 Reclamation

After all the sand drain piles were driven into the alluvial clay layer, reclamation work was started. In principle, reclamation work should proceed evenly over the entire site to prevent uneven settlement. But unidirectional reclamation method was adopted because of the strict time constraints on the construction of the terminal building and other facilities. This method involved reclaiming the site from seawall A (western) side to seawall B (eastern) side. Reclamation sequence is as follows :

1) Sand laying, first stage: Sea sand is spread evenly at the site to form a layer 1.5 m thick in the same manner as in the seawall construction.

2) Sand drain: Sand drain piles, each 40 cm in diameter are driven into the alluvial clay layer at a spacing of a 2.5 m square grid.

3) Sand laying, second stage: Bottom dump hopper barges are used to spread pit sand carefully up to a surface level of C.D.L -15 m. The purpose of this is to protect the permeable sand layer placed in the first stage of the sand laying operations.

4) Direct dumping, first and second stage: After consolidation time of six months from second sand laying, bottom dump hopper barges are used to dump soil directly onto the site to depths of between 10 m and 6 m. The surface level limit of C.D.L -6 m has been set to allow for the draft of reclaimer barges and box barges.

5) Direct dumping, third stage: Six months after stage 4) direct dumping, bottom hopper barges of relatively small size will be used to reclaim the site up to a surface level of C.D.L -3 m.

6) Above water, first stage: Box barges are used to carry pit sand to the site. Reclaimer barges fitted with boom mounted conveyors will place the pit sand to C.D.L +7 m - +9.5 m. The maximum dumping height of reclaimer barge is about C.D.L +9.5 m, because the height limited by embankment stability is C.D.L +9.5m.

7) Above water, second stage: In the area that the planned height is higher than C.D.L +9.5 m, the maximum dumping height of reclaimer barge, shovel-dump truck method is employed for loading and unloading. The fill in use is unloaded by reclaimer barge over seawall.

Fig.7 is conceptual view of the reclamation works. Following the sand laying works, full-fledged reclamation work began in December 1988 when the entire seawall emerged above the water surface. The earth and sand used for reclamation is being transported by barges from excavation site in the hilly area in Hannancho, Osaka Prefecture, the Kada area of Wakayama Prefecture, and Awaji Island in Hyogo Prefecture, shown in Fig.8. These excavation sites are 10 km to 30 km from the airport island.

The total volume of pit sand used for reclamation is required approximately 183 million m^3. It is expected that the volume of pit sand handled per month during the peak period will reach approximately 5.5 million m^3. Such large-scale reclamation works require daily deployment of approximately 80 soil barges with an average capacity of 3,000 m^3 each, and about 10 reclaimer barges with 1200 to 2100 t.p.h capacity.

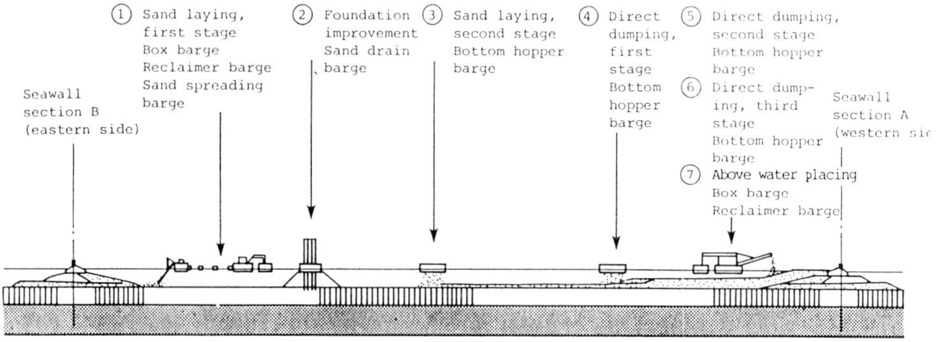

Fig.7 Conceptual view of reclamation works

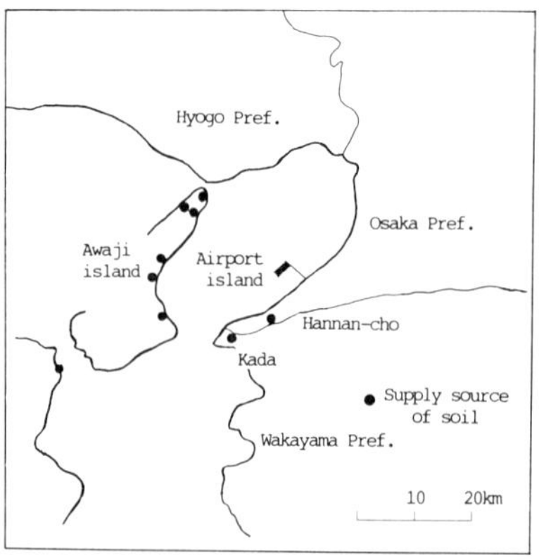

Fig.8 Supply sources of reclamation soil

4. DEVELOPMENT OF NEW TECHNIQUES IN CONSTRUCTION

4.1 Feature of technique

The construction of a large-scale man-made island is required within the relatively short period of six years under the adverse conditions of a deep water and a soft foundation. Therefore, there are many problems, construction technique, operation control, environment control, etc,. Table 2 outlines the new techniques developed in the construction.

Table 2 New techniques developed in Construction

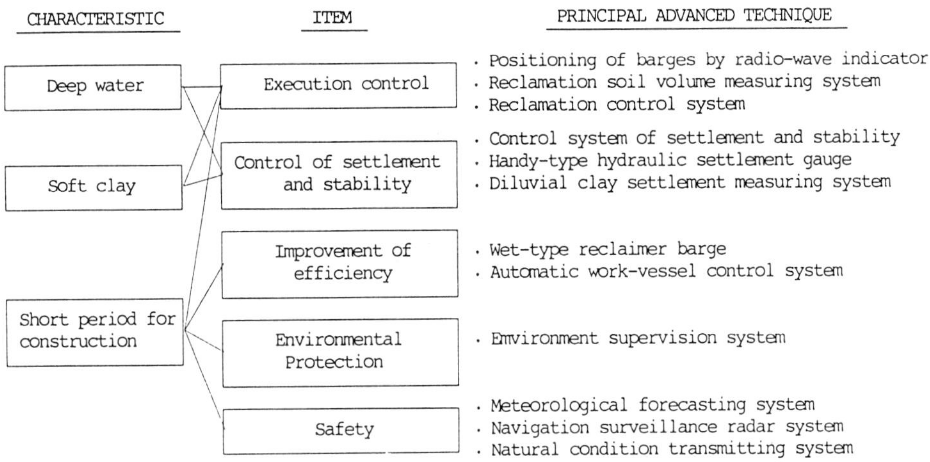

4.2 Control system of settlement and stability

1) Outline of settlement stability management system

Most of the foundation improvement at the island site was carried out using the sand drain method. With this method, it is extremely important to continuously check if the strength of the foundation is increasing as expected, and if the settlement is progressing as calculated. For that purpose a settlement stability management system has been newly developed.

Preceding the foundation improvement works on the entire seawall, a pilot site, approximately 500 m long, was selected apart of the seawall to observe the behavior of seabed soil improved by the sand drain method. The progress of the settlement by consolidation was measured and the results were applied to subsequent works. Settlement measuring plates were positioned 300 to 600 m apart along the seawalls to periodically measure the settlement. Every time the next load is to be placed, bore hole tests are carried out beforehand to see if the foundation has been strengthened as expected. The results are evaluated to see if works can proceed safely and as designed. If the results are negative, the work process may be reviewed for correction or the related structural design may be checked again.

In a similar procedure as in the seawall construction, preceding the entire reclamation, a reclamation pilot site was designated inside the seawall pilot site. The site area is approximately 6 ha. The progress of the settlement by consolidation was continuously measured. By use of the values measured at the pilot site, settlement and strength of the alluvial clay layer can be estimated accurately in subsequent reclamation works.

2) Settlement analysis

a. Designed settlement analysis

Accuracy in predicting future settlement is the governing factor in geotechnical problems involved in Kansai International Airport project. Owing to its vast area, an error in settlement prediction will lead to a change of a huge amount of the fill material for maintaining the specified surface elevation of the airport island.

In our design, the settlement prediction in a alluvial clay layer was carried out by the m_v method, generally applied in the settlement analysis of port facilities in Japan. The settlement curve with time is calculated by Barron's theory.

The result of calculation, shown in Fig.12, indicates that settlement of approximately 6.5 m would occur in the alluvial layer and almost stop after 6 months after fill completion.

Overburden stress due to the load of reclaimed land is as much as 40 tf/m^2 (392 kPa) and exceeds consolidation yield stress to the depth of approximately 150 m, so diluvial layers are also expected to compress. In settlement analysis of diluvial layer, the specification of drainage layers is a little complicated, because of an existence of a great many number of alternating layers of cohesive soils and sandy or gravely soils. In design, a soil layer with more than 70 % of sand fraction and more than 3 m of thickness was specified as the drainage layer.

The settlement prediction has been carried out mainly by the finite element method, based on Biot's consolidation theory. Calculation indicated that the settlement of diluvial layers was approximately 0.5 m at airport opening and 1.5 m after 50 years from the airport opening.(Refer to Fig.10)

b. Modified settlement analysis

In order to carry out such a large-scale reclamation project in such a short period on a soft seabed, it must be assessed at each step of the construction process whether the seabed or structure is behaving within the construction process whether the seabed or structure is behaving within the allowable range assumed in design or not. As part of the reclamation process for Kansai International Airport island, a pilot work section was designated for detailed study of the behavior of seabed soil.

The items measured in the pilot work section are not only settlement but also horizontal displacement, pore water pressure and increase in clay strength.

The observed settlement of the alluvial layer is shown in Fig.9. The comparison between the observed settlement and the calculated one shows that the observed one turned out to be less in amount and/or slower in speed than estimated. Hence, the soil constants set in the design analysis were reviewed, and the settlement calculation was corrected so that the calculated settlement could coincide with the observed one. Modified soil constants were evaluated by the curve fitting method for the observed settlement.

Under the modified condition, the ultimate settlement of

alluvial layer is calculated to be approximately 5.5 m, which is 1.0 m less than previously estimated.

This modification turned out to improve the forecast accuracy for the regular works.

On the other hand, the observed settlement of diluvial layers was approximately 1.5 m after 6 months from fill completion, and more in amount and/or faster in speed than designed. The comparison of many calculations under the various conditions regarding consolidation yield stress pc, coefficient of consolidation c_V and specification of drainage layer indicated that the observed settlement of diluvial layers coincided with the calculated under the condition where the entire diluvial sandy layers were specified as drainage layers. On the modified drainage condition, the settlement of diluvial layers was calculated to be 4 m at the airport opening and 5.5 m after 50 years from the opening.

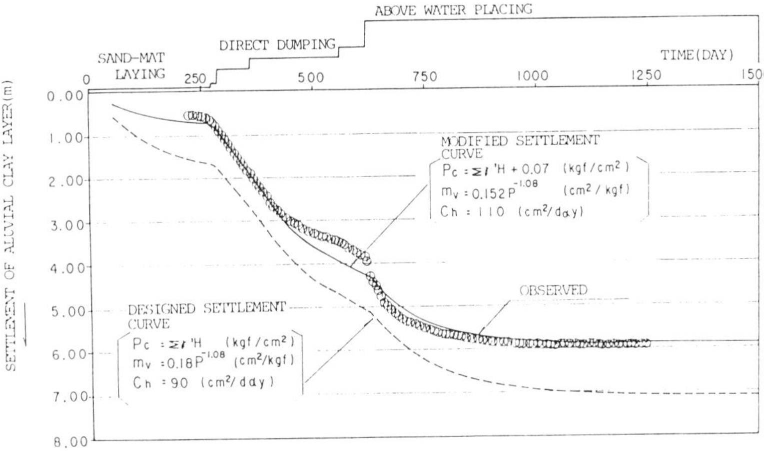

Fig.9 Comparison of calculated and observed settlement of alluvial clay layer

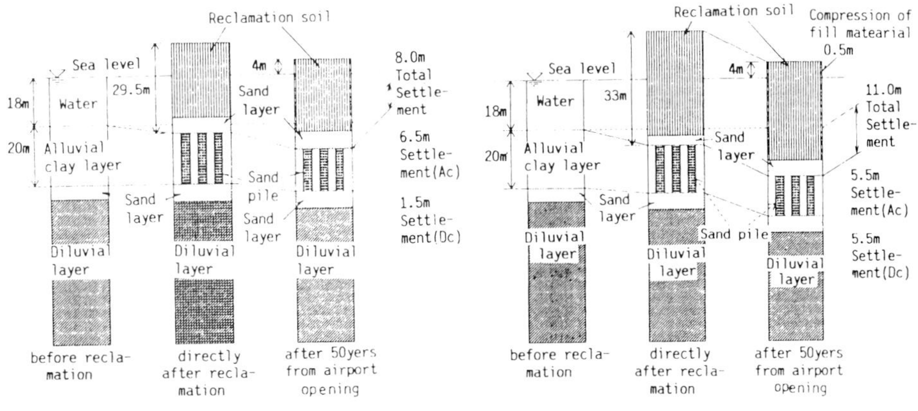

(in design stage) (after modification)

Fig.10 Settlement of seabed due to reclamation

The total settlement after 50 years from the airport opening under the modified consolidation condition of alluvial and diluvial layers is approximately 11.5 m, including a compression of fill material of 0.5 m, which is 3.5 m larger than originally calculated. Therefore, the specified reclamation surface elevation was modified to be 3.5 m higher than the designed elevation, not to be lower than C.D.L +4 m after 50 years from the opening, shown in Fig.10.

4.3 Mechanical positioning of barges

Since settlement occurs more quickly because of use of the sand drain method, the most important object is to prevent unequal settlement. To achieve this, reclamation soils need to be placed evenly to ensure uniform depth throughout the reclaimed area. Barges engaged in direct dumping should be directed to planned positions to ensure uniform ground levels. To confirm barge position, a radio-wave position indicator has been newly developed. This indicator is set on the push boat of the bottom hopper barge, and the operator can obtain information of the ship's actual position and specified position as shown in Fig.11, and can easily set the ship and dump the reclamation soils exactly onto the specified position.

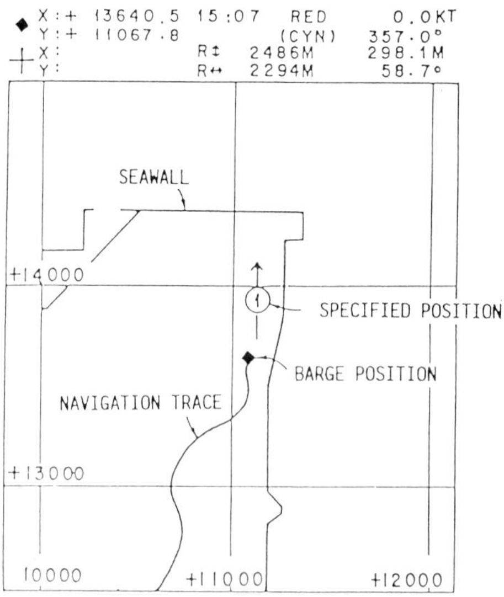

Fig.11 Radio-wave position indicator

4.4 Reclamation soil volume measuring system

A sufficient quantity of pit sand is now available for the project. As a large amount of reclamation pit sand must be transported from several pits, about 80 barges are used daily during the peak period. Measuring of a pit sand volume in each barge is usually done by leveling by use of surveying staff. However it is difficult to adopt this method in transportation of the reclamation pit sand, since the measuring of the sand volume is very frequent and proceeded also at night. This was the primary reason for our developing a mechanical measuring system.

Accuracy in measuring, safety and economy are required for the new measuring system. At first, two systems were proposed for these requirements. One was a system which used an optical beam on the base of a remote sensing technique and another was a system using photographs. Comparing the optical beam system with the photograph system, the former was cheaper than the latter in operational cost, and the latter had problems in use of the strobe light at night, which affected fishing around the site. Therefore, the optical beam system as shown in Fig.11, was selected and developed. In this system, sand surface monitoring by the optical beam is done under loading of the reclamation pit sand and it takes only 2 minutes to calculate the soil volume after the monitoring. A trial run showed that the measuring error of this system was under 1%.

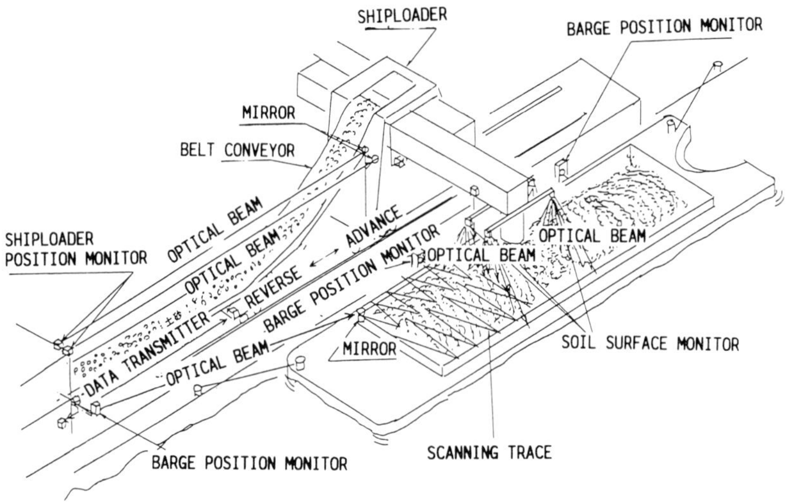

Fig.12 Conceptual view of soil volume measuring system

7. CONCLUSION

The access bridge is being built in parallel with the construction of the airport island, and airport facilities are to be constructed on the island in parallel with the reclamation works. These parallel and subsequent construction operations must be kept in line so that Kansai International Airport will be completed as scheduled. As of the end of June 1991, reclaimed land above sea is 392 ha(77% of total area) and pit sand volume dumped is 165 million m3 (90% of total volume). If the pit sand for reclamation can be secured in the necessary quantities, the airport island will be completed by December 1991.

Construction works on the airport island are noted for a

number of technical problems induced by various constraints: the particular soil conditions of the site and the compressed construction schedule. But all those concerned with the project are doing their best to overcome any obstacles to open the airport.

REFERENCES

1. Kansai International Airport Co.,Ltd.: Kansai International Airport Construction of Airport Island", 1986

2. Nakase,A: Kansai International Airport-Construction of Man-Made Island, Special Lecture in the 8th Asian Regional Conference on Soil Mechanics and Foundation Engineering(Kyoto), 1987

3. Maeda,S.,Higuchi,Y. and Furuichi,M: Large-scale Sand Drain Works for the Kansai International Airport, Proc. Airports into the 21st Century (Hong Kong), pp.485-496, 1990

4. Maeda,S.: Future Prospects of Soil Improvement Methods to be applied for the Construction of Large-scale Offshore Artificial Islands, Proc. JSCE, No.406, pp.1-15, 1989(in Japanese)

5. Oikawa,K.,Suzuki,S. and Yamagata,N.: Total Field Observation System for the Settlement and Stability Control of Large-scale Reclamation Work, Proc. Symposium on Investigation and Observation Technique, 1990(in Japanese)

6. Oikawa, K and Endo, H: Construction of a Large-Scale Man-Made Island for the Kansai International Airport, Kansai Int. Geotech Forum'90, 1990

Photo 2 The state of progress as of July 1991

causing a wave field with an average H_s of approximately 2.0 m. During the other months of the year the sea is very calm and the breakwater is not damaged during this period.

7. The three-hourly observations of the monsoon wave climate are modelled by the following Gumbel distribution (Fig.4):

$$F_{H_s}(x) = e^{-e^{-\frac{x-A}{B}}} \tag{2}$$

where: $A = 1.941$ *and* $B = 0.284$

The deep water wave steepness H_s/L_p defined on the spectral peak period is independent of the wave height and also modelled by a Gumbel distribution (Fig.5) with parameter values of A = 0.01 and B = 0.00167.

On the basis of a visual comparison the significant wave height can be equally well described by a Weibull distribution (Fig.6):

$$F_{H_s}(x) = 1 - e^{-\left(\frac{x-A}{B}\right)^C} \tag{3}$$

where: $A = 1.50$, $B = 0.675$ *and* $C = 1.866$

8. Every year during the monsoon the coast is assumed to be hit by one hurricane. The significant wave height during the hurricane that lasts 24 hours is modelled by a Frechet distribution (Fig.7):

$$F_{H_s}(x) = e^{-\left(\frac{x-A}{B}\right)^{-C}} \tag{4}$$

where: $A = 0.00$, $B = 2.498$ *and* $C = 3.388$

The wave steepness H_s/L_p follows a Normal distribution with μ = 0.028 and σ = 0.00724 (Fig.8) and is independent of the wave height.

9. The significant wave height at the breakwater is limited to 5.0 m by water depth.

10. The number of individual waves during the monsoon season amounts to 960,000. During a hurricane it is assumed that 10,800 individual waves hit the structure.

THE BREAKWATER DESIGN

11. The basic breakwater design criteria are as follows. The breakwater was designed for a lifetime of 50 years. The rock size for the berm was designed on the basis of a 1/100 year design significant wave height of H_s = 5.0 m.

A value of H_0 = 2.75, which is for berm breakwaters rather conservative , led, in combination with a Δ value of 1.57, to a choice of D_{n50} = 1.15 m (M_{50} = 4.0 tonnes).

12. The total longshore transport during the planned lifetime of the structure can be calculated by choosing an appropriate design wave and the number of waves in the lifetime N_{life}.

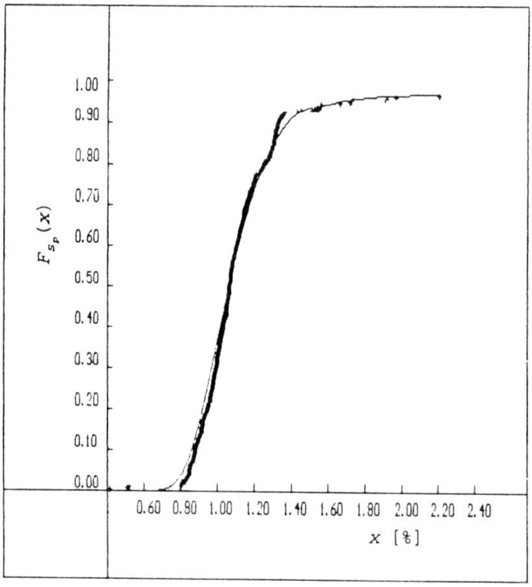

Fig. 5. The distribution of the deep water wave steepness during the monsoon

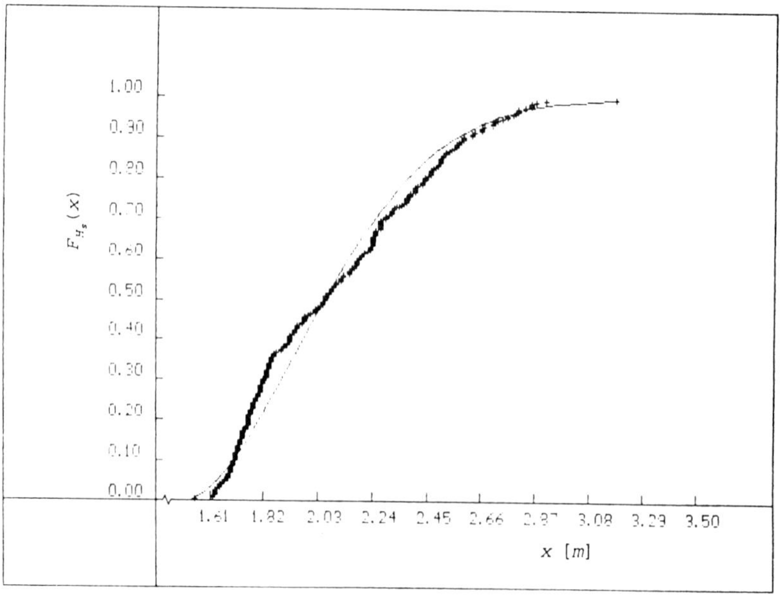

Fig. 6. The alternative distribution of the significant wave height during the monsoon

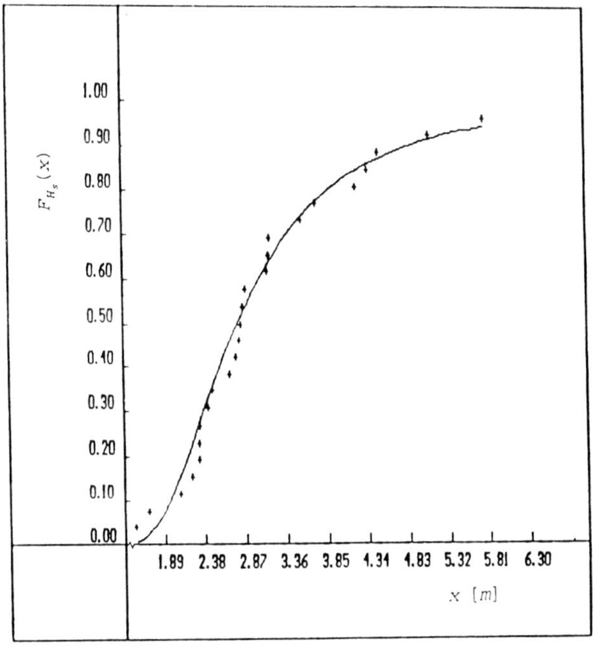

Fig. 7. The distribution of the significant wave height during the hurricane

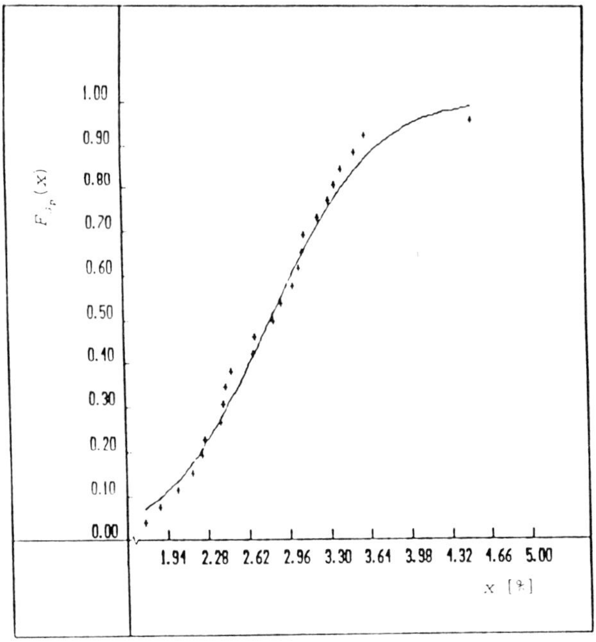

Fig. 8. The distribution of the deep water wave steepness during the hurricane

$$S_{life} = N_{life} * S \quad [stones] \tag{5}$$

13. The choice of the appropriate design wave for this case is far from easy. For the monsoon H_s and T_p values of 2.80 m and 13.3 s (1% wave steepness), a condition that is exceeded during 5% of the time, seem a reasonable choice. However, as for this combination H_0T_0 equals 38.85, the resulting longshore transport according to Equation (1) is zero.
The hurricane design condition of H_s = 5.0 m and T_p = 10.7 s is characterised by a value H_0T_0 of 86. As this value is smaller than 100 no transport is predicted for hurricane conditions either.
So, no provision for the nourishment of the berm at the weather side of the breakwater has to be made.
However, it is imaginable that the design values are exceeded or that the coefficients of the transport equation (1) differ from the most likely values. Then the berm at the weather side might be quickly eroded away.

PROBABILISTIC APPROACH; A FIRST STEP

14. When the probability distribution of the wave height is known, the expected value of the longshore transport, expressed as the number of stones passing through a certain cross-section during the lifetime, should be calculated according to:

$$E(S_{life}) = \int_0^{H_{s,max}} \int_0^{s_{p,max}} N_{life}\, A(H_0T_0-B)^C f_{H_s}(H)\, f_{s_p}(s)\, dH\, ds \tag{6}$$

where: $f_{Hs}(H)$ = the probability density function (pdf) of H_s
$f_{sp}(s)$ = the pdf of s_p, the wave steepness defined on T_p at deep water

15. The expected value of the number of stones transported during the monsoon period over the entire lifetime of the structure can be calculated by substituting the pdf's of significant wave height and wave steepness, related to the Gumbel distribution (2), in the integral (6). When the integration is performed with the transport relation given by (1) and with the upper limits $H_{s,max}$ = 5.0 m and $s_{p,max}$ = 0.05, the expected value of the transport is 435 stones. This amount increases to 551 stones, when the integration is extended to $H_{s,max}$ = 7.0 m.
In the longshore transport calculation of Equation (6) this H_s of 7.0 m is reduced to 5.0 m due to water depth limitations (see Paragraph 9). However, the influence on H_0T_0 of a deep water H_s > 5.0 m emerges if it is assumed that the deep water wave maintains its period while travelling to the breakwater location.

16. The difference between the result of the deterministic calculation of Paragraph 13 and the result reached here is caused by the fact that now the full range of wave height and steepness values are taken into account by the integration.

17. If the Gumbel pdf of the significant wave height is replaced by the pdf related to the Weibull distribution (3), the integration yields an expected value for the transport of only 4.6 stones during the lifetime. An increase in the upper limit of integration to 7.0 m now has a negligible effect.
So, the expected value $E(S_{life})$ of the transport appears to be very sensitive to the form of the pdf of the significant wave height (Gumbel or Weibull) and the upper limit of integration $H_{s,max}$, which may point at an important question point in the design.

18. The expected value $E(S_{life})$ of the transport is also dependent on the uncertain values of A and B in Equation (1). If for the coefficients in the transport relation the following safe values are chosen: $A = 7.2*10^{-5}$; $B = 80$, then the expected value of the transport for the Gumbel distributed wave height becomes 4,747 stones or $4{,}747 * M_{50} = 19{,}000$ tonnes.

19. A similar approach is followed for the hurricane wave climate by substituting in the integral (6) the Frechet pdf related to (4) for the significant wave height and the Normal pdf for the steepness. Also in this case the transport is sensitive to changes in the upper limit of integration and to the substitution of safe values in the transport relation:

Hurricane transport	*upper limit $H_{s,max}$*	
	15 m	*17 m*
average transport relation	*668*	*841*
"safe" transport relation	*3,625*	*3,837*

20. As the stone transport by the monsoon and by the hurricane have to be added to find the amount of nourishment at the weather side, the following matrix should form the basis of the design.

Total transport	*upper limits $H_{s,max}$*	
	low	*high*
average transport relation	*1,103*	*1,392*
"safe" transport relation	*8,372*	*8,978*

The designer is confronted with a difficult decision.

PROBABILISTIC APPROACH; A SECOND STEP

21. The correct solution for the problem posed above is found by introducing the uncertainties described into the design equations in the form of additional pdf's.
Moreover, the uncertainties in the Δ and D_{n50} values (hidden in the dimensionless quantity H_oT_o) that will be realised during construction, and the exact amount of nourishment that will be supplied, can be included.

22. By means of a Level II probabilistic calculation (ref.3) the influence of the uncertainties in Δ, D_{n50}, the coefficients A and B in the transport relation and the integration limits $H_{s,max}$ on the expected transport, can be calculated. If the amount of material at the weather side to nourish the longshore transport is known, the probability of failure can be established.

23. To this end the following reliability function is defined:

$$Z = Nourish - E(S_{life,mons.}) - E(S_{life,hurr.}) \quad [stones] \qquad (7)$$

where: $Nourish$ = nourishment at $t=0$ [stones]

$E(S_{life,mons.})$ = expected value of the transport over the lifetime due to monsoon waves [stones]

$E(S_{life,hurr.})$ = expected value of the transport over the lifetime due to hurricane waves [stones]

24. For the constants A and B of the transport relation normal distributions are adopted with the following parameters based on the scarce experimental data:

	μ	σ
A	$4.8*10^{-5}$	$2.4*10^{-5}$
B	100.0	20.0

25. The uncertainties of the stone size D_{n50} and the relative density Δ are also modelled by normal distributions. The values for the standard deviations are based on the experience of the authors:

	μ	σ
D_{n50}	1.15	0.058
Δ	1.57	0.060

26. The upper limits of the integration are set by the maximum significant wave heights that will occur during the monsoon and during the hurricane in the lifetime of the breakwater of 50 years. These maximum significant wave heights during monsoon and hurricane over the entire lifetime follow extreme value distributions.

The lifetime contains 50 years * 3 months * 30 days * 24 hours / 3 hours = 36,000 independent realisations of the monsoon wave climate described by the Gumbel distribution (2), so the highest significant wave is distributed according to:

$$F_{H_{s,extr.}}(x) = F_{H_s}(x)^N \qquad (8)$$

where: $N = 36,000$

Algebraically, the extreme value distribution is again a Gumbel distribution with the coefficients A'= A + Bln(N) and B' = B. The upper limit of the hurricane climate is modelled following the same reasoning for N = 50 independent storms during the lifetime. The exact extreme value distribution is, in this case, easily derived to be a Frechet distribution with parameters A' = A, B' = $B*N^{1/C}$ and C' = C.

27. The number of stones that is provided at the weather side of the breakwater at t=0 to provide nourishment for the transport occurring over the lifetime (the resistance side of the design problem), is modelled by a Normal distribution to account for construction tolerances:

Table 1. Result of Level II probabilistic calculation; the nourishment amounts to 2,500 stones

β = 0.550

Probability of failure = $2.9*10^{-1}$

Name	Type	A	B	C	μ	σ	X	α^2
Δ	N	0.000	0.000	0.000	1.570	0.060	1.564	0.037
D_{n50}	N	0.000	0.000	0.000	1.150	0.058	1.138	0.143
A	N	0.000	0.000	0.000	$4.8*10^{-5}$	$2.4*10^{-5}$	0.000	0.062
B	N	0.000	0.000	0.000	100.000	20.000	90.428	0.757
$H_{s,max,mons.}$	G	4.920	0.284	0.000	5.024	0.327	5.024	0.000
$H_{s,max,hurr.}$	F	0.000	7.926	3.388	8.832	3.001	8.832	0.000
Nourish	N	0.000	0.000	0.000	2500.000	200.000	2495.381	0.002

Z(X) = -0.2689 (X is the design point)

Number of iterations = 8

Table 2. Result of Level II probabilistic calculation; the nourishment amounts to 5,000 stones

β = 0.923

Probability of failure = $1.8*10^{-1}$

Name	Type	A	B	C	μ	σ	X	α^2
Δ	N	0.000	0.000	0.000	1.570	0.060	1.560	0.034
D_{n50}	N	0.000	0.000	0.000	1.150	0.058	1.131	0.132
A	N	0.000	0.000	0.000	$4.8*10^{-5}$	$2.4*10^{-5}$	0.000	0.059
B	N	0.000	0.000	0.000	100.000	20.000	83.761	0.774
$H_{s,max,mons.}$	G	4.920	0.284	0.000	5.024	0.327	5.024	0.000
$H_{s,max,hurr.}$	F	0.000	7.926	3.388	8.832	3.001	8.832	0.000
Nourish	N	0.000	0.000	0.000	5000.000	200.000	4996.381	0.000

Z(X) = 0.0018 (X is the design point)

Number of iterations = 12

	μ	σ	
Nourish	*2,500*	*100*	*[stones]*

28. The Level II calculation shows (see Table 1) that the probability of failure is 28% in the lifetime of the breakwater. In this case, failure implies that the extra amount of nourishment provided at the time of construction is completely taken away during the design lifetime. The Level II calculation also shows the contributions of the variables to the standard deviation of Z in the column α^2. The values in the column indicate that the greatest contribution comes from the B coefficient in the transport relation, with the D_{n50} in the second place.
The integration limits, formed by the maximum significant wave heights due to the monsoon and the hurricane during the lifetime, are relatively unimportant.

29. From the table the conclusion may be drawn that further research should be done to establish the transport relation, especially the start of the movement modelled by the coefficient B. During construction the D_{n50} of the rock going into the berm should be strictly controlled by quality assurance measures to prevent errors in this variable to the lower side.

CONCLUSIONS

30. The paper shows that although the design of the cross-section of a berm breakwater is well defined by the work of Baird and Hall (ref.4), Van der Meer (ref.1) and others, the longshore transport caused by angular wave attack may pose a threat that must be addressed by the designer and the researcher of this type of structure.

31. On the basis of scarce data from Burcharth and Frigaard (ref.2) and from tests performed for a project in which the authors were involved, a simple longshore transport relation for the armour rock of the berm has been derived. The influence of the angle of wave attack cannot be deduced from the data.

32. Statistical descriptions of the monsoon and the hurricane wave climates at the project location are given. A simple deterministic approach of the longshore transport of the berm material shows no transport.

33. A first step probabilistic approach, that takes into account the variability of the waves, proves, however, that the expected value of the longshore transport has a positive value and that nourishment has to be provided. It is shown that the decision on the amount of nourishment is far from simple in the light of the uncertainties in the transport relation and the maximum significant wave heights during monsoon and hurricane. The amount may vary from 1,103 stones to 8,978 stones depending on the optimism of the designer.

34. In a second step all uncertainties are included in a Level II probabilistic calculation. It is shown that a nourishment of 2,500 stones has a probability of failure of 29% in the lifetime. The Level II calculation shows further that the uncertainty of the transport relation contributes most to the failure probability. So, more research into this problem is needed.

Secondly, the stone size has to be carefully controlled during construction by quality assurance procedures.

35. If the nourishment is increased to 5,000 stones the probability of failure is reduced to 18% (see Table 2). The designer has now to decide on the basis of probabilities and risks.

REFERENCES

1. VAN DER MEER J.W. Rock slopes and gravel beaches under wave attack. Doctoral thesis, Delft University of Technology, Delft, the Netherlands, 1988. Also: Delft Hydraulics Communication No.396, the Netherlands, 1988.

2. BURCHARTH H.F. and FRIGAARD P. Reshaping breakwaters; on the stability of berm breakwater roundheads and trunk erosion in oblique waves. From: Berm breakwaters: unconventional rubble mound breakwaters; edited by D.H. Willis, W.F. Baird and O.T. Magoon, Canada, pp. 55-72, 1987.

3. CIRIA Rationalisation of safety and serviceability factors in structural codes. CIRIA report No.63, London, 1974.

4. BAIRD W.F. and HALL K.R. The design of armour systems for the protection of rubble mound breakwaters, Proc. Conf. on Breakwater design and construction, Inst. of Civ. Eng. London, 1983.

Discussion

F. VASCO COSTA. Consulmar

My comments concern difficulties to be faced when using 'partial coefficients' to design the armour of slender blocks. Such blocks are held in place not only by their own weight but also by the weight of the neighbouring blocks with which they are interlocked.

To remove a slender block from its position the waves will have to apply on the block not just a pulling force but also a torque, both acting in adequate directions. This can happen, but only now and then. How often will it occur cannot be forecast from average values or standard deviation values of the waves. The same can be said about the resistance of slender blocks against being removed.

H. LIGTERINGEN, Frederic R. Harris BV

Could Arai et al. give some more information on the causes of the settlement being higher than originally estimated?

Concerning Paper 30, if I understand correctly the location of the project which served as a case in the paper, the waves related to monsoon and hurricane come from different quadrants. As this would give conflicting transports, how could the two transports be added up in the analysis?

Given the problems which the client may have in providing a regular supply of rock for nourishment, would it not have been prudent to reduce further the risk of lateral transport by a slight increase of the stone diameter?

H. OUMERACI, University of Hannover

Recently I had the opportunity to be involved in three projects related to the design of shore protection of 'offshore airports'. One of the most important problems we encountered every time was the design of the filter between the rubble mound sea wall and the reclamation fill. This is because there are no filter rules yet available for the case of the cyclic flow which is induced at this boundary by the wave run-up and run down.[1]

In order to avoid the washing out of the fine material of the fill and the following subsidence which has been experienced by a number of reclamation fills, we generally adopted in addition to a conventional mineral filter a geotextile. How did Arai et al. solve this problem for the sea walls of the Kansai airport?

Reference:
1. OUMERACI H. and KOHLHASE S. Hydrogeotechnical aspects of filter design for coastal structures. Proc. Seminar on the use of filters in coastal structures. Colombo, Sri Lanka, 1989.

C. P. ORBELL-DURRANT, May Gurney Contractors

Concerning Paper 29, how were the 'no noise pollution' criteria defined by day and by night?

Osaka Bay is very nearly circular and surrounded by high ground. Under these circumstances one would expect still air conditions, particularly at night, to lead to the formation of a series of temperature inversion layers which are very efficient in propagating sound. How were these effects taken into account when selecting the site for the airport?

M. A. LOSADA, University of Cantabria

How do Vrijling et al. explain a longshore transport equation which does not depend on the wave incidence angle? Further, have they an explanation for the type of dependence on wave height and wave period they propose?

Have they compared their results with those obtained by the application of Van Hijum et al. for the longshore transport of gravel on beaches?

H. OUMERACI, University of Hannover

I would like to express my agreement with Professor Losada's remark. I fully share his worries about the physical background we are still lacking until we can adopt the approach proposed by Professor Vrijling. Perhaps we need a special conference to further discuss this topic.

I. MIRCHANDANI, W. S. Atkins

My comments are prompted by Paper 30 on berm breakwaters, but relate generally to this type of breakwater. Four or five papers at this conference discussed berm breakwaters as an important breakwater type, whereas at the 1983 Breakwater Conference only one paper (by Barra and Hall) touched upon this subject. In less than a decade a subject which was only of historical interest is now much discussed and researched.

Berm breakwaters were described earlier as 'primitive' breakwaters, but for some applications they could well be the appropriate solution, provided we have rational design rules which can be used with confidence. They seem particularly suited to remote locations or some developing nations, where importing large armour rock from Norway (as happens in the UK) is a luxury which cannot be considered. Regular monitoring is desirable for all breakwaters, but for dynamically stable breakwaters - which by definition have profiles which change with time - it seems essential.

J. N. WOLFE-BARRY, Sir William Halcrow & Partners
Paper 8 and Paper 16 suggest that it may be possible to relax the specification for the ratio D_{85}/D_{15} for armour rock grading from the low values traditionally met to values up to 5 or more in the case of berm breakwaters. Would Vrijling et al. comment on this in light of their work on longshore transport? May not broadening the grading of the armour lead to an unacceptable loss of the small sizes under oblique wave attack?

J. JUHL, Danish Hydraulic Institute
In other papers at this conference the title included 'dynamically stable breakwaters', which I find contradictory.

I prefer berm breakwater, and the design has to be based on stable reshaped profiles, as moving stones can be associated with an abrasion process or result in 'longshore transport' in the case of oblique waves. Additional research is needed in order to study under which conditions a berm breakwater trunk section exposed to oblique waves or a berm breakwater roundheads after some initial reshaping may develop into a stable slope that is not subject to continued erosion, or at least such slow erosion that it may be acceptable for a permanent structure.

A reshaped berm breakwater is a traditional breakwater with a flatter slope. A special feature of berm is that during construction advantage is taken of wave action in order to reach a stable reshaped profile.

In Mast II the three-dimensional effects on berm breakwaters are proposed for study. Two recently constructed breakwaters in Norway are being monitored after each major storm causing significant reshaping.

Y. ARAI, Paper 29
No noise pollution criteria represent the noise level of WECPNL 70 or less for residential areas as stipulated in Japan's environmental standards. WECPNL

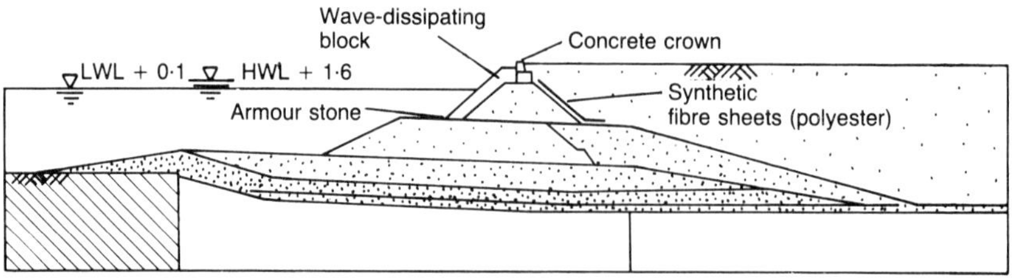

Fig. 1 . Rubble mound type seawall (wave-dissipating type)

weights the sounds at different times during the day. In Japan's case the number of aircraft is weighted as three times in the evening and as ten times in the night as in the daytime.

Osaka Bay almost always has temperature inversion layers, particularly at night. However, this fact was not taken into account when selecting the site for the airport, because the airport had to be constructed off the coast of Osaka Bay in order to avoid a noise problem.

To avoid washing out the reclamation fills, synthetic fibre sheets (polyester) were placed between the reclamation fills and armour stones (Fig. 1), and were placed around the water level where suctions were the most hard area.

In answer to Mr Ligteringen's request: in the design stage, we selected one case as an initial prediction among various consolidation conditions which was thought the most reliable. However, there still remained many unobvious factors in the consolidation conditions which were evaluated by soil investigations. So the settlement was observed in order to make sure the actual settlement was the same as the predicted one.

In May 1989 the observation of settlement began in the pilot work area. Three months later (August 1989) it showed that the settlement of an alluvial clay layer was smaller and the settlement of diluvial clay layer was higher than the design conditions. The prediction of settlement was modified and the final settlement was estimated to be 10.0 m - 2.0 m higher than at the design stage.

J. K. VRIJLING, Paper 30

In answer to Mr Ligteringen's question, the deep-water waves related to monsoon and hurricane coming from different directions; however, in front of the breakwater the direction of the waves is mainly governed by the depth contours of the sea bottom (refraction). The small influence of the incoming wave direction is neglected in view of the insensitivity of the transport phenomenon around an angle of 25°.

As the client might have problems in supplying rock for replenishment, the problem was to provide sufficient rock material at the weatherside of the breakwater to limit the probability of failure to an acceptable level. In a deterministic framework no replenishment was needed, but a probabilistic analysis showed a rather high probability of failure in that case.

In reply to Professor Losada, it should be added that the tests at Delft Hydraulics Laboratory were performed with 25° wave incidence. One test was performed at 50° and showed a much higher H_oT_o value for the start of transport. The tests of Burcharth and Frigaard[2] studied transport for 15 and 30°. Between these three directions no significant difference in transport could be detected. This means that the longshore transport is roughly similar in the domain between 10 and 35° wave incidence. Outside this domain the transport seems to be considerably less. Therefore the wave angle was omitted from the formula in Paper 30.

The dependence of the transport on wave period and wave height is directly derived from the work of Van der Meer.[1] The results of our tests ware compared with the results of van Hijum et al. They studied the 10 < H < 100 range. Our formula, that is only valid for very low transports, agrees with the start of the formula of van Hijum et al.

It should be stressed that the formula presented in Paper 30 is based on tests limited to the project under study and the tests published by Burcharth and Frigaard.[3] The formula mainly serves to illustrate the design approach.

I agree with Oumeraci and Losada that more research on longshore transport of large rock is needed. It was one of the aims of our paper to show that.

To Mr Mirchandani, regular monitoring of dynamically stable berm breakwaters is essential, especially if high $H = H_s/\Delta D_{n50}$ values are applied, say higher than 3.0.

To Mr Wolfe-Barry, according to laboratory tests[1] the ratio D_{85}/D_{15} has no significant influence on the profile up to values of 2.25. Normal gradings for berm breakwaters have D_{85}/D_{15} ratios < 2.0 It is not recommended that one apply higher values.

In reply to Mr Juhl: a berm breakwater is initially dynamically stable. After the first storms and reshaping it becomes more or less statically stable. Another advantage of a berm breakwater discerning it from a traditional one is the large porous mound that effectively dissipates wave energy. The berm breakwater studied in our paper is relatively stable as the $H_s/\Delta D_{n50}$ value is equal to 2.75. Still the probability of considerable longshore transport is non-negligible.

References:

1. VAN DER MEER J. W. Rock slopes and gravel beaches under wave attack. PhD thesis, delft, 1988.
2. BURCHARTH H. F. and FRIGAARD P. Reshaping breakwaters, on the stability of roundheads and trunk erosion in oblique waves. Berm breakwaters. ASCE, 1987.
3. VAN DER MEER J. W. Personal communication, 1992.

Analysis of rubble mound breakwaters

J. D. METTAM, Scott Bertlin

SYNOPSIS. Working Group 12 was set up by Permanent Technical Committee II of PIANC to consider the analysis of rubble mound breakwaters with a view to achieving a better understanding of safety aspects. This paper provides a very brief introduction to the work of the group. A practical way of evaluating safety has been developed (for use particularly with Hudson's formula or Van der Meer's formula for design of main armour stability) which enables a theoretical design to be evaluated more realistically in respect of varying probabilities of failure within specified design life. Its value has been demonstrated by examples based upon data collected. Methods are outlined for using partial coefficients to prepare preliminary designs, and then use hydraulic model testing to refine the preliminary design and provide the basis for final designs.

INTRODUCTION

1. After a considerable number of large breakwaters had failed or suffered severe damage, a working group established by PIANC's Permanent Technical Committee II produced a report, published as a supplement to Bulletin No 48 (1985), on the stability of rubble mound breakwaters. This summarised important details of design and construction of a large number of rubble mound breakwaters and concluded, inter-alia, that "it is not possible at the present time to determine risk with a satisfactory degree of accuracy".

2. In late 1986, PTC II set up a further group, Working Group 12, to consider the Analysis of Rubble Mound Breakwaters. The terms of reference were as follows:

Terms of Reference

3. The immediate need for designers of breakwaters is to achieve a better understanding of the overall safety aspects in the design of rubble mound breakwaters. This is important because the existing design tools are still very incomplete.

4. Practical approach of the group

(i) Identify and list the parameters to be con-

sidered in the design of rubble mound breakwaters.
Evaluate the relative importance and the quality of existing available knowledge related to these parameters.

(ii) Evaluate the safety aspects (risk of failure) and propose ways of dealing with the safety problems in practical ways. This should include a check of the proposed safety guidelines against the behaviour of selected existing breakwater structures.

MEMBERS OF THE WORKING GROUP

5. The working group comprised the following members:-

MR J D METTAM - Chairman Scott Bertlin	UNITED KINGDOM
PROF. H F BURCHARTH - Secretary Aalborg University	DENMARK
MR N W H ALLSOP Hydraulics Research Limited	UNITED KINGDOM
MR P BONAFOUS (retired 1990) Port Autonome du Havre	FRANCE
DIPL. ING M BORCHERT Prof. Dr. Lackner & Partners	FEDERAL REPUBLIC OF GERMANY
MR J E CLIFFORD Consultant	UNITED KINGDOM
DR T ENDO Applied Hydraulic Laboratory Nippon Tetrapod Co Ltd	JAPAN
PROF D FOSTER Unisearch Ltd Tasmania	AUSTRALIA
PROF L FRANCO Politecnico di Milano, Dipartimento di Ingegneria Idraulica	ITALY
PROF ENG P A HEDAR PA Hedar AB	SWEDEN
M. X LE BARS (from 1990) Port Autonome du Havre	FRANCE
MR H LIGTERINGEN Frederic R Harris Holland BV	THE NETHERLANDS
DR E P D MANSARD National Research Council of Canada Hydraulics Laboratory	CANADA

MR C A R M PITA PORTUGAL
WW Consultores de Hidraulica E Obras Maritimes LDA

MR L STADLER ISRAEL
Consulting Engineer

IR D VANDENBOSSCHE BELGIUM
Haecon NV

MR L VAN DAMME BELGIUM
Ministry of Public Works, Coastal Department

DR JENTSJE VAN DER MEER THE NETHERLANDS
Delft Hydraulics

DR R W WHALIN U.S.A.
US Army Corps of Engineers, Waterways Experiment Station

MR JOOP A ZWAMBORN SOUTH AFRICA
Division of Earth, Marine and Atmospheric Science and Technology, CSIR

METHOD OF WORKING

6. The working group decided to develop the practical application of risk analysis in the design of rubble mound breakwaters by using partial coefficients. Tasks were delegated to six subgroups.

7. Subgroup chairmen and terms of reference were as follows:-

Subgroup A Theories, Formulae
Chairman - Mr N W H Allsop
Identification of theories, formulae, empirical data from model tests etc. which give the functional relationship between the structural response and the environmental and structural parameters. Evaluation of the related statistical uncertainty.

Subgroup B Environmental data and extreme events
Chairman - Prof H F Burcharth
Evaluation of uncertainty related to environmental data including estimated extreme events.

Subgroup C Risk Analysis in Breakwater Design
Chairman - Dr J W van der Meer
Description and evaluation of practical ways of dealing with safety aspects (risk of failure) based on the uncertainties related to the parameters involved in the design.
Level I methods (safety factors, partial coefficients)
Level II methods (safety index, normal distributions)
Level III methods (full consideration of probability distributions)

Subgroup D Deviations from Design and Reliability of Construction

Chairman - Mr M Borchert
Identification (quantitative) of typical tolerances related to construction procedures.

Subgroup E Investigation of Selected Cases
Chairman - Mr H Ligteringen
Selection of existing breakwater structures (of types as given in Fig. 3.1) for which reliable data on environmental conditions, structural parameters and structural response (damage/no damage) are already available or can be made available. (Sub group E also analysed all selected cases using partial coefficients as far as possible).

Subgroup F Determination of Partial Coefficients
Chairman - Prof H F Burcharth
Develop and propose safety guidelines and check (calibrate) these against the documented behaviour of existing breakwaters.

8. The Main Report of the Working Group will be published by PIANC next year bound together in one volume with the full texts of the six subgroup reports.

WORK OF THE SUBGROUPS

9. The amount of work carried out by the subgroups was most impressive. Each subgroup chairman had help from a small number of members of the Working Group, and in addition was able to call on top experts from all over the world who gave freely of their knowledge. This high level of international co-operation is a feature of the work of PIANC which is difficult to appreciate without first hand experience.

10. The result will be seen in the reports. While the work of each of the Subgroups A-D was designed to provide inputs to the analysis carried out by Subgroup F, and Subgroup E has served chiefly to provide a practical check on the method developed by Subgroup F, each of the subgroup reports has become a valuable document in itself.

11. The relationship between the work of the different subgroups is very briefly described below and illustrated in Fig 1.

12. Subgroup A provided a review of the state of art of relevant analytical formulae. This has identified those failure mechanisms which are sufficiently well described by formulae to be used as a basis for application of the partial factor approach. The uncertainty attached to each of these formulae was also investigated and expressed where possible in terms of standard deviations.

13. Subgroup B studied the uncertainty related to environmental data as a function of methods of data collection, including estimates of standard deviations for different methods of determining significant wave heights and sea level variations. The estimation of uncertainty on extrapolation of design wave heights corresponding to extreme return periods was also included.

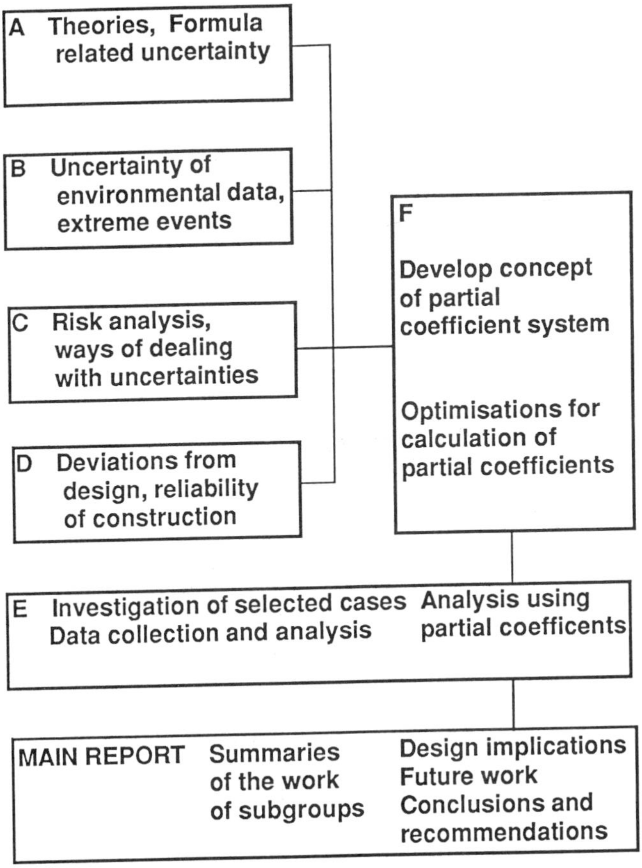

Fig 1 Method of working

14. Subgroup C reviewed different methods of determining the risk of failure. This led to the selection of the partial coefficient approach as a practical and relatively simple method of breakwater design, and the decision to derive values for the partial coefficients from higher level calculations incorporating the uncertainties in environmental data and construction obtained by Subgroups B and D, as well as uncertainties in the design formulae, derived by Subgroup A.

15. Subgroup D examined the importance of deviations from the design which will be found in the breakwater as built so that allowance for these uncertainties could also be made in the higher level analyses carried out to determine partial coefficients.

16. Subgroup E investigated the safety level of a variety of existing breakwaters with 10-20 year life for selected

breakwater types. Although data collection started, with the work of Subgroups A-D, at an early stage in the work of the group the results from Subgroup E were not, like those of Subgroups A-D, inputs to the work of subgroup F, but were designed to provide a means of "calibrating" the theoretical development of partial coefficients. Results are limited to the stability of main armour.

17. Subgroup F carried out an impressive amount probabilistic computation to develop and optimise the system of partial coefficients. This is described fully in the accompanying paper by Professor H F Burcharth.
It is therefore necessary only to state here that this work constitutes a fundamental advance in the design of rubble mound breakwaters, and to acknowledge the considerable input of time of staff and computers from the Universities of Aalborg and Delft without which it would not have been possible to conclude the work; financial aid from the Dutch Rijkswaterstaat and the Danish Technical Scientific Council is also gratefully acknowledged.

BRIEF DESCRIPTION OF PROPOSED DESIGN PROCEDURE

18. Within the scope of this paper it is possible only to give the very briefest account of the nature of the proposed design procedure. This will be described in relation to Hudson's formula for main armour.

19. Written in its original format, but using the notation of the Working Group 12 report, Hudson's formula is:

$$M = \frac{\rho_a \ H^3}{K_D \ \Delta^3 \ \mathrm{Cot}\ \alpha} \qquad (1)$$

(Notations are listed at the end of this paper)
This can be reorganised into a response function:

$$g = Z \ \Delta \left(\frac{M \ K_D \ \cot\alpha}{\rho_a} \right)^{\frac{1}{3}} - H \qquad (2)$$

Failure occurs for $g \leq 0$. The right hand term H is a load parameter, while the other term comprises resistance parameters.

20. In our analysis the parameter Z is a stochastic variable representing the reliability of the formula. Z has a mean value of 1.0. and normalised standard deviation $\sigma'_Z = 0.18$. K_D is treated as a deterministic coefficient of damage while all the other parameters involved are taken as stochastic variables.

21. Having decided for simplicity to introduce only two partial coefficients γ_z and γ_H (both greater than unity) the response function can be developed into a design equation

$$G = \frac{Z}{\gamma_z} \cdot \Delta \left(\frac{M \ K_D \ \cot\alpha}{\rho_a} \right)^{\frac{1}{3}} - \gamma_H \cdot H \qquad (3)$$

$$\text{or } G = \frac{Z}{\gamma_z} \cdot \Delta \ D_n \ (K_D \ \cot\alpha)^{\frac{1}{3}} - \gamma_H \cdot H \qquad (4)$$

22. The design condition is satisfied by G ≥0. It will be seen that the product of the two partial factors γ_Z γ_H represents a simple factor of safety in the design equation.

23. As described in Professor Burcharth's paper, formulae have been selected to calculate γ_Z and γ_H . These include coefficients whose optimum values for various design formulae have been determined by higher level analysis using standard deviations for all the stochastic variables.

24. Adopting those values the partial factors for any design can be calculated in terms of probabilities of failure within a given design lifetime. Preliminary designs may then be prepared using Hs values for storms of the appropriate return period, together with published K_D values such as those in the Shore Protection Manual. Examples of this system, applied to the case of Torshavn Breakwater are shown in Table 1.

Table 1: Design Examples, based on Torshavn

Case	1	2(a)	2(b)	3	4
Life in years	25	50	50	50	50
Damage level	Moderate	Moderate	Moderate	Moderate	None
S-Van der Meer	6	6	-	6	2
K_D-Hudson	-	-	3.4	-	-
Failure probability P_f	0.4	0.7	0.7	0.1	0.1
Partial factors					
γ_{Hs}	1.10	1.00	1.00	1.31	1.37
γ_z	1.02	1.01	1.01	1.06	1.06
Armour size					
D_{n50}-m	1.20	1.35	1.35	1.64	2.14
Av, Mass-t	4.8	7.0	7.0	12.3	27.4

25. The preliminary design must then normally be subjected to thorough hydraulic model testing, both to refine the main armour design and to check aspects not covered by the design formula. At this stage it will be necessary to consider the effect on the test procedure of having introduced partial factors in the design equation.

26. Current practice is to build a scale model representing the intended design as accurately as possible. With no partial factors this means that the response function is being tested directly and that the degree of damage represented by the chosen K_D value should be expected to be reproduced when testing under the design wave.

27. With a design incorporating partial factors the armour size Dn will have been increased by the factor of safety γ_Z . γ_H . The stability is thereby increased and the design wave will not produce the damage corresponding to the chosen K_D value. To confirm that the design has the

intended margin of safety it will be necessary either to increase the wave height (load parameter) or to reduce the armour size Dn, or some other resistance parameter, in proportion to $\gamma_Z \cdot \gamma_H$. Various options will be discussed in the report of the Working Group. The simplest solution for any particular case will vary depending, for example, whether waves are depth limited or not.

28. Having completed model testing the response function for the specific conditions of the breakwater may be modified and used to revise the design equation for the final design.

CONCLUSIONS

29. The more rational assessment of probabilities made possible by the partial coefficient system will be a valuable tool, but it can of course only be applied to failure mechanisms for which design formulae are available. It will not take into account other forms of failure such as geotechnical failure or seabed scour leading to increased wave attack.

30. Improvements are proposed in the statistical treatment of uncertainty in predicting extreme wave heights; this will help the design but will not obviate the need for further improvements in wave recording and use of wind data to supplement inadequate observations.

31. A design prepared using the proposed method of partial coefficients will result in a more objective assessment of probabilities of failure of the structure during its design life than has previously been possible. This can be used to assess economic damage levels and will not necessarily result in more expensive design although this will depend upon the nature and consequences of failure.

32. As will be emphasised in the Report to be published by PIANC next year. the results which have been achieved should not be regarded as a code of practice for design of breakwaters. Much more work is required to reach that stage; nevertheless the new system is seen as a major step forward, and engineers are urged to use the method to examine existing breakwaters and new designs. This will give useful insights and, with further research, will enable more logical design codes to be developed in future.

33. One problem, which has perplexed engineers using the Shore Protection Manual to prepare designs using the Hudson formula, has been the recommendation in the 1984 edition to use $H_{1/10}$ instead of Hs. In conjunction with K_D values previously adopted with Hs this resulted in the effective introduction of a factor of safety of 1.27, doubling the required weight of armour. Considered in relation to the partial coefficient system proposed by Working Group 12 this can be seen as a prudent response to the recent breakwater failures but it is now considered more rational to revert to the use of Hs as the design wave height (H in

equation 14) and use the system of partial coefficients to calculate the acceptable probabilities of failure.

34. With armour units which can break leading to rapid and progressive failure it is likely that it will be economic to design for significantly lower probabilities of failure than have been adopted in recent years and the new method may lead to results at least as conservative as the use of $H_{1/10}$. However in many simpler cases the partial factors will be found to be little more than unity. For example, with simple rock armouring, with no great difficulties in repair, the economic design may not differ significantly from the earlier designs using Hudson's formula with Hs.

35. The British Code (BS 6349 Part 7), which has been circulated for comment and has recently been published, has adopted a different approach by recommending the use of Hs and adopting K_D values from the Shore Protection Manual for rock armour but introducing more conservative K_D values for certain types of concrete armour as a method of introducing a margin of safety. This is a more rational approach than using $H_{1/10}$ for all types of armour but it should be seen as a temporary expedient to be replaced by the partial coefficient system when this has been further developed.

NOTATION

D_n	Nominal armour size, defined $(M/\rho_a)^{\frac{1}{3}}$
G	Design equation
g	Failure or response function
H	Wave height, trough to crest
H_s	Significant wave height, average of highest 1/3
$H_{1/10}$	Mean height of highest 1/10 waves
K_D	Empirical damage coefficient in Hudson's formula
M	Mass of armour unit
Z	Stochastic variable signifying the relaibility of a response function
α (alpha)	Breakwater front slope, angle to horizontal
γ (gamma)	Partial coefficient
γ_{Hs} γ_Z	Partal coefficients on H_s and Z
ρ_a (rho)	Mass density of armour units
ρ_w	Mass density of seawater
Δ (delta)	Relative density $(\rho_a/\rho_w - 1)$

Introduction of partial coefficients in the design of rubble mound breakwaters

H. F. BURCHARTH, University of Aalborg

INTRODUCTION

1. After a number of large breakwaters had failed or suffered severe damage, a working group established by PIANC's Permanent Technical Committee II produced a report on the stability of rubble mound structurs, published as a supplement to Bulletin No. 48, 1985. This report pointed out a large number of specific limitations in the general knowledge of the behaviour of many structures. In this situation there was an urgent need for guidelines on the implementation of a reasonable degree of safety in the designs.

2. In 1986 PIANC set up a further working group, WG12, to consider the Analysis of Rubble Mound Breakwaters. Mr. J.D. Mettam was nominated chairman of the group. The terms of reference were formulated as follows:

Scope of the work	The immediate need for designers of breakwaters is to achieve a better understanding of the overall safety aspects in the design of rubble mound breakwaters. This is important because the existing design tools are still very incomplete.
Practical approach of the group	1. Identify and list the parameters to be considered in the design of rubble mound breakwaters. Evaluate the relative importance and the quality of existing available knowledge related to these parameters.
	2. Evaluate the safety aspects (risk of failure) and propose ways of dealing with the safety problems in practical ways. This should include a check of the proposed safety guidelines against the behaviour of selected existing breakwater structures.

3. The group decided that the safety guidelines should be based on a "partial coefficient" system, because it is simple to apply and, in principle, well-known from other fields of civil engineering.

4. The tasks were delegated to six expert Sub-groups, A-E. The terms of reference and the chairmen for the sub-groups were as follows:

Sub-group A, chairman Mr. N.W.H. Allsop.
Identification of theories, formulae, empirical data from model tests etc., which give the functional relationship between the structural response and the environmental and structural parameters. Evaluation of the related statistical uncertainty.

Sub-group B, chairman Prof. H.F. Burcharth.
Evaluation of uncertainty related to environmental data including estimated extreme events.

Sub-group C, chairman Dr. J.W. van der Meer.
Description of and evaluation of practical ways of dealing with safety aspects (risk of failure) based on the uncertainties related to the parameters involved in the design.

Level 1 methods (safety factors). Level 2 methods (safety index). Level 3 methods (full consideration of probability distributions for involved parameters).

Sub-group D, chairman Mr. M. Borcherth.
Identification (quantitative) of typical tolerances related to construction procedures.

Sub-group E, chairman Mr. H. Ligteringen.
Selection of existing typical breakwater structures for which reliable data on environmental conditions, structural parameters and structural response (damage/no damage) are already available or can be made available.

Sub-group F, chairman Prof. H.F. Burcharth.
Proposal of safety guidelines and checking (calibration) of these against the documented behaviour of existing breakwaters. The work includes the synthesis of the work of the other sub-groups.

5. The full report of the maingroup and the sub-groups will be published by PIANC.
This paper describes mainly the work of Sub-group F, which resulted in a proposal for a partial coefficient system. The main contributors to this work are: From University of Aalborg, Prof. H.F. Burcharth, Dr. John

Dalsgaard Sørensen, MSc Jørgen Andersen, PhD-student Jan Pedersen. From Delft Univ. of Technology, Prof. J.K. Vrijling, MSc-student K. van Ort. From Delft Hydraulics, Dr. J.W. van der Meer. From Frederic R. Harris B.V., Mr. H. Ligteringen. The work was sponsered by University of Aalborg, The Danish Technical Scientific Council and Rijkswaterstaat, Holland.

GENERAL INTRODUCTION TO PARTIAL COEFFICIENTS

6. Evaluation of structural safety is always related to the structural response as defined by the failure modes. Neglect of an important failure mode will bias the estimation of the safety of the structure. Fig. 1 illustrates the failure modes for a conventional rubble mound breakwater with a capping wall.

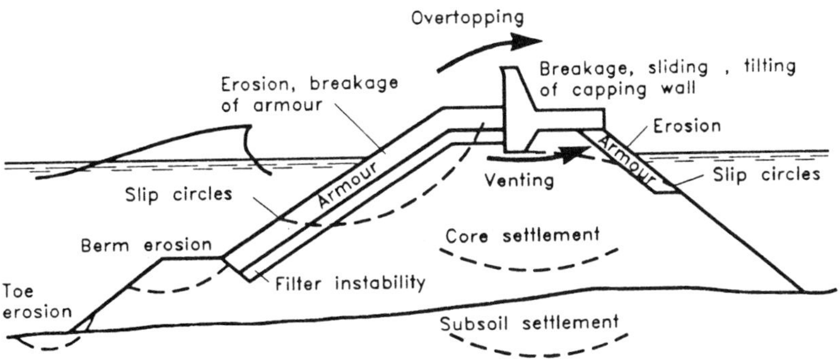

Fig. 1. Failure modes for a rubble mound breakwater.

Each failure mode must be described by a formula and the interaction (correlation) between the failure modes must be known. As an illustrative example let us consider only one failure mode, "hydraulic stability of the main armour layer", described by the Hudson formula

$$D_n^3 = \frac{H_s^3}{K_D \, \Delta^3 \, cot\alpha} \tag{1}$$

where D_n is the nominal block diameter, $\Delta = S - 1$, where S is the ratio of the block and water densities, α is the slope angle, H_S is the significant wave height and K_D is the coefficient signifying the degree of damage (movements of the blocks).

7. The formula can be split into load variables X_i^{load} and resistance variables, X_i^{res}. Whether a parameter is a load or a resistance parameter can be seen from the failure function. If a larger value results in a safer structure it is a resistance parameter and if a larger value results in a less safe structure it is a load parameter.

According to this definition one specific parameter can in one formula act as a load parameter while in another it can act as a resistance parameter. An example is the wave steepness in the Van der Meer formulae for rock, which is a load parameter in the case of surging waves but a resistance parameter in the case of plunging waves. The only load variable in eq. (1) is H_s while the others are resistance variables.

Eq. (1) is formulated as a **failure function**

$$g = Z \cdot \Delta \cdot D_n \left(K_D \, cot\alpha\right)^{1/3} - H_s = \begin{cases} \leq 0 & \text{failure} \\ > 0 & \text{no failure} \end{cases} \tag{2}$$

8. All the involved parameters are regarded as stochastic variables, X_i, except K_D, which signifies the "failure", i.e. a specific damage level chosen by the designer. The factor Z in eq. (2) is also a stochastic variable signifying the uncertainty of the formula. In this case the mean value of Z is 1.0.

9. The partial coefficients, γ_i, are related to characteristic values of the stochastic variables, $X_{i,ch}$. In conventional civil engineering codes the characteristic values of loads and other action parameters are often chosen to be an upper fractile (e.g. 5%), while the characteristic values of material strength parameters are chosen to be the mean values. The values of the partial coefficients are uniquely related to the applied definition of the characteristic values.

10. The partial coefficients, γ_i are usually larger than or equal to one. Consequently, if we define the variables as either load variables X_i^{load} (as for example H_s) or resistance variables X_i^{res} (as for example the block volume) then the related partial coefficients should be applied as follows to obtain the design values

$$\begin{aligned} X_i^{design} &= \gamma_i^{load} \cdot X_{i,ch}^{load} \\ X_i^{design} &= \frac{X_{i,ch}^{res}}{\gamma_i^{res}} \end{aligned} \tag{3}$$

11. The magnitude of γ_i reflects both the uncertainty on the related parameter X_i and the relative importance of X_i in the failure function. A large value, e.g. $\gamma_{H_s} = 1.4$, indicates a relatively large sensitivity of the failure probability to the significant wave height, H_S. On the other hand, $\gamma_i \simeq 1$ indicates no or negligible sensitivity in which case the partial coefficient should be omitted. It is to be stressed that the magnitude of γ_i is not – in a mathematical sense – a stringent measure of the sensitivity of the failure probability to the parameter, X_i.

12. When the partial coefficients are applied to the characteristic values of the parameters in eq. (2) we obtain the **design equation**, i.e.

the definition of how to apply the coefficients. The partial coefficients can be related either to each parameter or to combinations of the parameters (overall coefficients). In the first case we obtain the design equation

$$G = \frac{Z_{ch}}{\gamma_z}\frac{\Delta_{ch}}{\gamma_\Delta}\frac{D_{n,ch}}{\gamma_{Dn}}\left(K_D\frac{cot\alpha_{ch}}{\gamma_{cot\alpha}}\right)^{1/3} - \gamma_{H_s}H_{s,ch} \geq 0$$

or (4)

$$D_{n,ch} \geq \gamma_z\gamma_\Delta\gamma_{Dn}\gamma_{cot\alpha}^{1/3}\gamma_{H_s}\frac{H_s}{Z_{ch}\,\Delta_{ch}K_D\,cot\alpha_{ch}}$$

In the second case we could for example have only γ_{H_s} and an overall coefficient γ_z related to the first term on the right hand side of eq. (2). The design equation would then be

$$G = \frac{Z_{ch}}{\gamma_z}\Delta_{ch}\,D_{n,ch}\,(K_D cot\alpha)^{1/3} - \gamma_{H_s}H_s \geq 0$$

or (5)

$$D_{n,ch} \geq \gamma_z\,\gamma_{H_s}\frac{H_s}{Z_{ch}\,\Delta_{ch}\,K_D\,cot\alpha_{ch}}$$

Eqs. (4) and (5) expresses two different "code formats". By comparing the two equations it is seen that the product of the partial coefficients is independent of the chosen format, other things equal. It is desirable to have a system which is as simple as possible, i.e. as few partial coefficients as possible, but without invalidating the accuracy of the design equation beyond acceptable limits.

13. Fortunately, it is very often possible to use overall coefficients, like γ_z in eq. (5), without losing significant accuracy within the realistic range of combinations of parameter values. This is the case for the system proposed in this paper where only two partial coefficients, γ_{H_s} and γ_z, are used in each design formula.

14. Usually several failure modes are relevant to a design. The relationship between the failure modes are characterized either as series systems or parallel systems, cf. the Sub-group C report. A fault tree can be used to illustrate the complete system. The partial coefficients for failure modes being in a system with failure probability, P_f are different from the partial coefficients for the single failure modes with the same failure probability, P_f. Therefore, partial coefficients for single failure modes and multi failure mode systems are treated separately.

OVERALL CONCEPT OF THE PROPOSED PARTIAL COEFFICIENT SYSTEM

15. In existing civil engineering codes of practise, e.g. for steel and concrete structures, it is a characteristic of them that

- partial coefficients are related to combinations of basic variables rather than to each of them in order to reduce the number of coefficients.
- the partial coefficients reflect the safety level inherent in a large number of well proven designs. Two sets of coefficients covering permanent and preliminary structures are usually given, but the related average probabilities of failure are not specified. In other words, it is not possible by means of the normal structural codes to design a structure to a predetermined failure probability.

16. However, it is not advisable to copy this concept in safety recommendations for rubble mound breakwaters for the following reasons:

- For coastal structures and breakwaters there is no generally accepted tradition which reflects one or more levels of failure probability. On the contrary it is certain that the safety level of existing structures varies considerably and is often very low. Besides, it is very difficult to evaluate the safety level of existing coastal structures and breakwaters because of lack of information, especially on the environmental conditions, e.g. the water level variations and the wave climate. Consequently, it is not possible to produce sets of partial coefficients which, in a meaningful way, are calibrated against existing designs.
- Due to the very nature of coastal engineering where design optimization dictates considerable variations in the safety level of the various structures it is necessary (advisable) to have sets of partial coefficients which correspond to various failure probabilities. In other words the designer and the client decide on the basis of optimization and cost benefit analyses that the structure should be designed for a specific safety level (for example 20% probability of failure (P_f = 20%) within a structural lifetime of L = 80 years, where failure is defined as a certain degree damage). The code should then contain a set of partial coefficients corresponding to this failure probability.
- Because the quality of information about the long term wave climate (the dominating load) varies from very unreliable (uncertain)

wave statistics based on few uncertain data sets to very reliable statistics based on many years of high quality wave recordings and hindcast values it is necessary that the partial coefficients must be a function of the quality of the available information on the wave climate. This means that the statistical uncertainty due to limited number of wave data and errors in the wave data should be implemented.

17. Extensive calculations, performed at University of Aalborg, of partial coefficients for armour layer stability formulae demonstrated that it was possible to develop a concept which satisfies these demands. A first version of this concept was reported in Burcharth, 1990a. The present paper presents and recommends an optimized version.

METHOD OF DETERMINING THE PARTIAL COEFFICIENTS

18. The overall procedure for the development of a partial coefficient system is as follows

- Define the failure modes and the failure element structure (single element analysis and/or system analysis)
- Select the code format (design equations)
- Define intervals of the parameters, their statistical properties and combinations
- Select target probabilities of failure,(e.g. in terms of a target reliability index β_T)
- Calculate the partial coefficients
- Optimize and calibrate the system (optimization with respect to β_T incl. reduction of number of partial coefficients)
- Verify the partial coefficient system against the observed behaviour of existing structures.

19. The partial coefficients γ_i are determined from a so-called level II reliability analysis, cf. the Sub-group C report. The applied computer programmes BWREL (Break Water RELiability programme) and BWCODE (Break Water CODE) were developed at the University of Aalborg by Dr. John Dalsgaard Sørensen especially for the reliability analysis of breakwaters.
BWREL estimates the reliability (or P_f) of single failure modes (failure

elements) and of systems modelled as series systems. The basic variables defining the failure modes are modelled as stochastic variables. As a measure of the structural reliability the so-called reliability indices are determined using First Order Reliability Methods (FORM), see Madsen et al. (1986) and Thoft-Christensen et al. (1982). For series systems the system reliability index is determined and there is an option to determine an equivalent element with a linear safety margin, which can be used to determine partial coefficients. Parametric sensitivity factors, which are very important measures of the importance of the parameters used to define the failure modes or the distributions of the stochastic variables, can also be determined. BWCODE is used to calibrate partial coefficients.

20. The determination of the partial coefficients γ_i assumes that a characteristic value (e.g. the mean value μ_{X_i}) and the distribution type and statistical parameters (e.g. normal distribution, expected value μ_{X_i} and standard deviation σ_{X_i}) of the variable X_i are known. In some codes of practice the characteristic values of loads and other action parameters are chosen to be an upper fractile (e.g. 5%), while the characteristic values of material strength parameters are chosen to be the mean values. In this proposal the mean value μ_{X_i} is chosen as the characteristic value for all the involved basic random variables, except for the significant wave height where the T-year return period value H_s^T is used. The basic theoretical concepts of BWREL are described in the Sub-group F report.

21. The calibration of the partial coefficients can be described briefly as follows. For each failure mode the *failure function* (limit state function) is written

$$g(\overline{x},\overline{p},\overline{z}) = 0 \tag{6}$$

where $\overline{x} = (x_1,\ldots,x_n)$ is a realisation of $\overline{X} = (X_1,\ldots,X_n)$ modelling n stochastic variables describing the uncertain quantities. $\overline{p} = (p_1,\ldots,p_N)$ are N parameters (deterministic) and $\overline{z} = (z_1,\ldots,z_m)$ are m design variables (which are used to design the actual breakwater, for example the nominal diameter of the armour blocks.

The *code format* is described by the design equation (e.g. eq. (4))

$$G(\overline{x}^c,\overline{p},\overline{z},\overline{\gamma}) \geq 0 \tag{7}$$

where $\overline{x}^c = (x_1^c,\ldots,x_n^c)$ are characteristic values corresponding to the stochastic variables $\overline{X}$. $\overline{\gamma} = (\gamma_1,\ldots,\gamma_M)$ are M partial coefficients ($\gamma_i \geq 1, i = 1,\ldots,M$).

The application area for the code is described by the set I of L different vectors $\overline{p}_i, i = 1,\ldots,L$. The set I can for example contain different geometrical forms of the breakwater, different parameters for the stochastic variables and different statistical models for the significant wave height H_S.

The partial coefficients $\overline{\gamma}$ are calibrated such that the reliability indices corresponding to the L vectors $\overline{p}$ are as close as possible to a target reliability index β_T. This is formulated as the following optimization problem

$$\min_{\overline{\gamma}} \quad W(\overline{\gamma}) = \sum_{j=1}^{L} \left[w_j(\beta_j(\overline{\gamma}) - \beta_T)^2 + \delta_j \sum_{i=1}^{n} (\gamma_i - \gamma_{ji}^*)^2 \right] \tag{8}$$

$$\beta_i(\overline{\gamma}) \geq \beta_T^{\min} \quad , i = 1, \ldots, L \tag{9}$$

$$\gamma_i^l \leq \gamma_i \leq \gamma_i^u \quad , i = 1, \ldots, M \tag{10}$$

where $w_j, j = 1, \ldots, L$ and $\delta_j, j = 1, \ldots, L$ are weighting factors $\left(\sum_{j=1}^{L} w_j = 1\right)$. $\beta_j(\overline{\gamma})$ is the reliability index for combination j obtained as described below. γ_{ji}^* is an estimate of the partial coefficient obtained by considering combination j in isolation. (If only one combination is considered then $\gamma_{ji}^* = \frac{x_{ji}^*}{x_{ji}^c}$ where x_{ji}^* is the design point value of the stochastic variable x_i in combination j). The constraint (9) has the effect that no combination has a reliability index smaller than $\beta_T^{\min}$ and the constraints in (10) are simple bounds on the partial coefficients.

The reliability index $\beta_j(\overline{\gamma})$ for combination j is obtained as follows. First for given $\overline{\gamma}$ the design equation (7) is used. If the number of design variables is one the equation is simply solved giving z_1. If $m > 1$ then a design optimization problem is formulated (often related to the cost of the structure) and solved, giving $\overline{z}$. In this paper $m = 1$.

Next the reliability index $\beta(\overline{\gamma})$ and the corresponding design point $\overline{x}^*$ are estimated.

It should be noted that following the above described procedure, the partial coefficients are not uniquely determined. They will be functionally dependent, in the simplest case as a product which has to be equal to a constant. The second term in the objective function (8) is added due to this "problem" and has the effect of forcing the partial coefficients in the direction of the "simple" definition of partial coefficients: $\gamma = \frac{x^*}{x^c}$.

CONSIDERED FAILURE MODES

22. It was decided to study the types of rubble mound breakwaters shown in Fig. 2 and the related failure modes

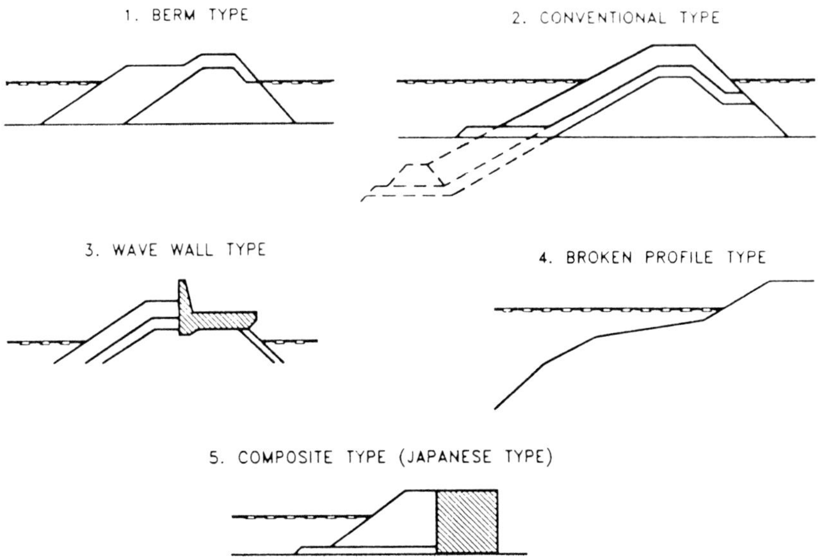

Fig. 2. Considered types of rubble mound breakwaters.

However, at the stage of the work, it was possible only for some failure modes related to the type 2 and the type 3 breakwaters to find sufficiently documented general formulae to be used for the calculation of partial coefficients, cf. the Sub-group A report. These failure modes are hydraulic stability of front face armour, hydraulic stability of low crested rock breakwaters, rock toe berm stability and run-up on rock armoured slopes.

FAILURE MODE FORMULAE AND RANGE OF PARAMETER VARIATIONS

23. Two examples of formulae and related range of parameter variations, used in the calculation of the partial coefficients, are shown below.

The following notations are used

$D(p)$	indicates a deterministic value, p.
$N(x_1, x_2)$	indicates a normally distributed parameter with mean value x_1 and standard deviation, x_2.
z	is the design parameter
$\bar{p}$	defines the ranges of application of the code for this failure mode.
$\hat{X}$	expected value (mean value) of X.
FH_s	error function on H_s, cf. paragraph 28.
$\hat{H}_s^T$	central estimate of the significant wave height which on average is exceeded once every T years.

24. *Stability of conventional armour (examples).*

Hudson formula $\dfrac{H_s}{\Delta D_n} = (K_D cot\alpha)^{1/3}$

Failure function: $g = Z\Delta D_n(K_D cot\alpha)^{1/3} - (H_S FH_s)$ (0.11)

Design equation : $G = \dfrac{1}{\gamma_Z}\hat{\Delta}\hat{D}_n(K_D cot\hat{\alpha})^{1/3} - \gamma_{H_S}\hat{H}_s^T \geq 0$ (0.12)

Parameters for the stochastic variables:

parameter	distribution	variation of $\bar{p}$
Δ	N(p_1, p_2)	$(p_1, p_2) = (1.4, 0.03), (1.6, 0.06)$
D_n	N(z, pz)	$p = 0.01,\ 0.05$
$cot\alpha$	N(p, 0.1)	$p = 1.5,\ 2,\ 3$
K_D	D(6)	
FH_s	N(1,p)	$p = \sigma_{FH_s} = (0, 0.10, 0.20)$
Z	N(1.0, 0.14)	
H_s	Extreme distribution fitted to local wave data	

25. Van der Meer formula for rock armour.

Plunging waves, $Ir \leq Ir_c$: $\dfrac{H_s}{\Delta D_n} = 6.2 P^{0.18} Ir^{-0.5} \left(S/\sqrt{N}\right)^{0.2}$

where $Ir = tan\alpha\, S_m^{-0.5} \leq Ir_c = \left[6.2\, P^{0.31} (tan\alpha)^{0.5}\right]^{(P+0.5)^{-1}}$

Failure function: $g = Z S^{0.2} P^{0.18} \Delta D_n cot\alpha^{0.5} S_m^{0.25} N^{-0.1} - (H_S F H_s)$ (0.13)

Design equation:

$$G = \frac{1}{\gamma_Z} \hat{S}^{0.2} \hat{P}^{0.18} \hat{\Delta} \hat{D}_n cot\hat{\alpha}^{0.5} \hat{S}_m^{0.25} \hat{N}^{-0.1} - \gamma_{H_S} \hat{H}_S^T \geq 0 \qquad (0.14)$$

Surging waves, $Ir > Ir_c$: $\dfrac{H_s}{\Delta D_N} = 1.0 P^{-0.13} Ir^P \left(S/\sqrt{N}\right)^{0.2} cot\alpha^{0.5}$

Failure function:

$$g = Z\, S^{0.2} P^{-0.13} \Delta D_n cot\alpha^{(0.5-P)} S_m^{-0.5P} N^{-0.1} - (H_S F H_s) \qquad (0.15)$$

Design equation:

$$G = \frac{1}{\gamma_Z} \hat{S}^{0.2} \hat{P}^{-0.13} \hat{\Delta} \hat{D}_n cot\hat{\alpha}^{(0.5-P)} \hat{S}_m^{-0.5P} \hat{N}^{-0.1} - \gamma_{H_S} \hat{H}_S^T \geq 0 \qquad (0.16)$$

Parameters for the stochastic variables:

Parameter	distribution	variation of $\bar{p}$
Z plunging waves	N(6.2, 0.4)	
Z surging waves	N(1.0, 0.08)	
S	D(5)	
P	N(p, 0.05)	$p = 0.5, 0.3$
Δ	N(1.6, 0.06)	
D_n	N(z, 0.05z)	
$cot\alpha$	N(1.5, 0.1)	
S_m	N(0.025, 0.005)	
N	N(5000, 1000)	
FH_s	N(1, p)	$p = \sigma_{FH_s} = (0, 0.10, 0.20)$
H_s	Extreme distributions fitted to local wave data	

STATISTICAL MODELS FOR THE INVESTIGATED EXTREME DISTRIBUTIONS OF THE SIGNIFICANT WAVE HEIGHT

26. For the development of the partial coefficient system the statistical model for H_s is described by three of the commonly used theoretical extreme distributions: Weibull, Gumbel and Exponential. The distributions are given below expressing the non-exceedence probability within T years. λ is the average number of H_s-data per year and N is the total number of data available for fitting the distribution.

The statistical uncertainty of the distributions is included through the parameters α and β which are modelled as stochastic normally distributed variables with variances based on the maximum likelihood estimates of α and β. It should be noted that no quality measure (correlation coefficient or X^2-test) of the fit of a distribution to a data sample, is included in the analyses that were carried out.

27. The considered distribution functions are listed below. $N(x_1, x_2)$ indicates a normally distributed parameter with mean value x_1 and standard deviation, x_2.

Gumbel $$F_{H_s^T}(h) = [exp(-exp(-\alpha(h-\beta)))]^{\lambda T}$$

$$\alpha : N\left(\alpha, \alpha\sqrt{\frac{0.608}{N}}\right) \qquad \beta : N\left(\beta, \frac{1}{\alpha}\sqrt{\frac{1.109}{N}}\right)$$

Weibull $$F_{H_s^T}(h) = \left[1 - exp\left(-\left(\frac{h-H_s'}{\beta}\right)^{\alpha}\right)\right]^{\lambda T}$$

$$\alpha : N\left(\alpha, \sqrt{\frac{1}{N}}\right) \qquad \beta : N\left(\beta_1\left(\frac{\beta}{N^2}\left[\frac{\Gamma(1+2/\alpha)}{\Gamma^2(1+1/\alpha)} - 1\right]\right)^{\frac{1}{2}}\right)$$

$var[\alpha] = \frac{1}{N}$ is an assumption since it has not been possible to find an analytical expression. Γ is the gamma function.

Exponential $$F_{H_s^T}(h) = \left[1 - exp\left(-\frac{h-H_s'}{\alpha}\right)\right]^{\lambda T} \qquad \alpha : N\left(\alpha, \alpha\sqrt{\frac{1}{N}}\right)$$

28. The statistical uncertainty described through the variance of α and β does not include uncertainties due to

- lack of knowledge about the true extremal distribution
- climatological changes

- measurement errors
- variability due to imperfect calculations of H_s and short term randomness

The last two points are incorporated in the analysis by the multiplication term FH_s on H_s. FH_s is modelled as a normally distributed variable with a mean value of unity and a specified coefficient of variation, σ'_{FH_s}, the size of which depends on the quality of the available information, cf. Table 2 reproduced from the Sub-group B report. The table provides typical overall values of the coefficient of variation and possible bias. It might be used in cases where no more detailed information is available. In the table bias is normalized with respect to the true value. The uncertainty related to the direction of wave propagation is given as the standard deviation in degrees.

The first two points cannot be treated through FH_s, but in a design situation the designer must try the different models for the extreme wave height and thereby select the most appropriate. A partial coefficient system cannot take these problems into account.

Moreover, it is assumed inherent in the analysis that the N values of H_s represent the statistical population to which H_s belongs. This sets limits to minimum length of the period of observation N/λ and N in order to prevent seasonal changes from biasing the results.

29. The H_s data samples used in the analysis are real deep water and shallow water data set from the North Sea, the Atlantic Ocean, the Bay of Biscaya and the Mediterranean Sea. Table 1 shows the distribution parameters for the data sets.

Table 1. Distribution parameters for H_s-data samples.

			Gumbel		Weibull		Exp.	
	N	λ	α	β	α	β	H'_s	α
Bilbao	50	4.17	1.95	5.55	1.39	1.06	4.9	0.97
Sines	15	1.25	0.88	8.75	1.78	2.53	7.1	2.27
Tripoli	15	0.75	0.74	5.06	1.83	3.24	2.9	2.91
North Sea	30	1.88	1.30	6.65	1.28	1.48	5.7	1.39
Follonica	46	5.94	3.14	5.73	1.14	0.58	2.69	0.55
Pozzallo	22	6.94	3.62	4.68	1.05	0.48	2.20	0.47

Fig. 3 shows the distribution fitted to the data samples. The related correlation coefficients are also shown.

Table 2. Typical overall uncertainties related to sea state parameters.

Parameter	Methods of determination	estimated typical values σ'	Bias	Comments
Significant wave height, offshore	Accelerometer buoy, pressure cell, vertical radar	0.05-0.1	~ 0	
	Horizontal radar	0.15	~ 0	
	Hindcast, num. models	0.1-0.2	0-0.1	Very dependent on quality of weather maps.
	Hindcast, SMB method	0.15-0.2	?	Valid only for storm conditions in restricted sea basins.
	Visual observations from ships	0.2	0.05	
Mean wave period offshore on condition of fixed significant wave height	Accelerometer buoy records	0.02-0.05	~ 0	
	Estimates from amplitude spectra	0.15	~ 0	
	Hindcast, num. models	0.1-0.2	~ 0	
Duration of sea state with significant wave height exceeding a specific level	Direct measurements	0.02	~ 0	
	Hindcast, num. models	0.05-0.1	~ 0	
Spectral peak frequency offshore	Measurements	0.05-0.15	~ 0	
	Hindcast, num. models	0.1-0.2	~ 0	
Spectral peakedness offshore	Measurements and hindcast, num. models	0.4	~ 0	
		Degrees		
Mean direction of wave propagation offshore	Pitch - roll buoy	5°		
	Measurements μ, u, v and p, u, v	10°		
	Hindcast, num. models	15 – 30°		
		σ'		
Significant wave height nearshore determined from offshore significant waveheight taking into account typical shallow water effects (refraction, diffraction, shoaling, ...)	Numerical models	0.1-0.20	0.1	σ' can be much larger in some cases
	Manual calculations	0.15-0.35		
Astro tides	Prediction from constants	0.001-0.07	~ 0	
Storm surge	Numerical models	0.1-0.25	±0.1	

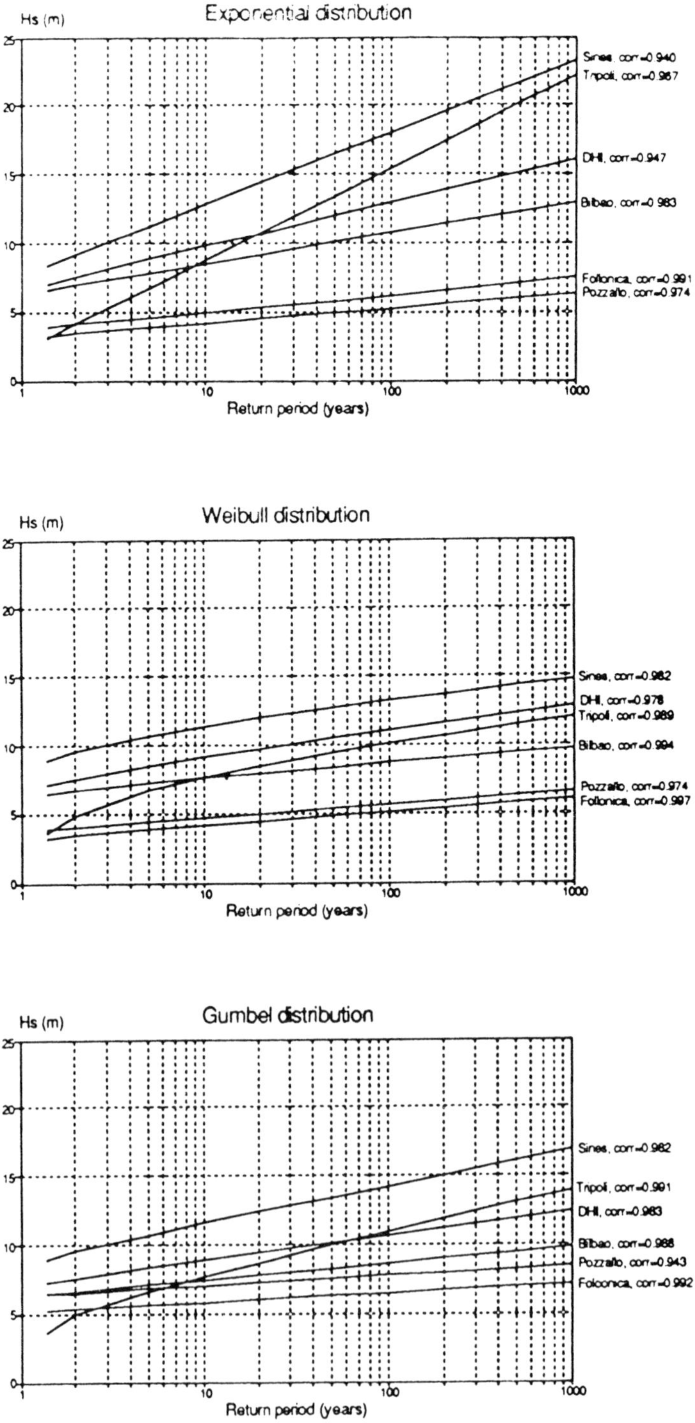

Fig. 3. Distribution fitted to the H_s-data samples.

FORMAT FOR SINGLE FAILURE MODES

30. For each failure mode only two partial coefficients γ_{H_s} and γ_z are used, cf. the example given by eq. 5 and the design equations in paragraphs 24 and 25. Three different models for the formulae have been evaluated and the following were chosen as being acceptable with respect to deviations from the target probability of failure.

$$\gamma_{H_s} = \frac{\hat{H}_s^{T_{P_f}}}{\hat{H}_s^T} + \sigma'_{FH_s} \left(1 + \left(\frac{\hat{H}_s^{3T}}{\hat{H}_s^T} - 1\right) k_\beta P_f\right) + \frac{k_s}{\sqrt{P_f N}} \tag{17}$$

$$\gamma_z = 1 - k_\alpha \ln P_f \tag{18}$$

where

$\hat{H}_s^T$ is the central estimate of the T-year return period value of H_s, where T is the structural lifetime (T = 20, 50 and 100 years were used for the code calibration). γ_{H_s} is applied to $\hat{H}_s^T$ (the characteristic value of H_s, cf. the design equations).

$\hat{H}_s^{3T}$ is the central estimate of the $3T$-year return period value of H_s.

$\hat{H}_s^{T_{P_f}}$ is the central estimate of H_s corresponding to an equivalent return period T_{P_f} defined as the return period corresponding to a probability P_f that $\hat{H}_s^{T_{P_f}}$ will be exceeded during the structural lifetime T. T_{P_f} is calculated from the encounter probability formula $T_{P_f} = \left(1 - (1 - P_f)^{\frac{1}{T}}\right)^{-1}$, cf. Fig. 4.
For the code calibration the following values were used: P_f = 0.01, 0.05, 0.10, 0.20, 0.40 corresponding to target reliability indices β_T = 2.33, 1.65, 1.28, 0.84 and 0.25.

σ'_{FH_s} is the variational coefficient of a function FH_s modelled as a factor on H_s. FH_s signifies the measurement errors and short term variability of H_s and has the mean value 1.0. σ'_{FH_s} is equal to σ' for H_s in Table 2. The values σ'_{FH_s} = 0.00, 0.10 and 0.20 were used in the code calibration. The statistical uncertainty on H_s is not included in FH_s.

N is the number of H_s data, used for fitting the extreme distributions. The statistical uncertainty depends on this parameter.

k_α, k_β and k_s are coefficients which are determined by the optimization procedure, eqs. (8) - (10). $k_s \simeq 0.05$ for all failure modes. Examples of the k_α and the k_β values are given in Table 3.

31. The first term in eq. (17) gives the correct γ_{H_s} provided no statistical uncertainty and measurement errors related to H_s are present. The middle term in eq. (17) signifies the measurement errors and the short term variability related to the wave data. The last term in eq. (17) signifies the statistical uncertainty of the estimated extreme distribution of H_s. The statistical uncertainty depends on the total number of wave data, N, but not on the length of the period of observation, as might be expected. The 10 largest values of H_s over a 15 years period provides a much more reliable estimate of the extreme distribution than the 10 largest values of H_s over 1 year. However, in the statistical analysis it is assumed that the data samples are equally representative of the true distribution. In other words it is assumed that the data, besides being non-correlated, are sampled with a frequency and over a length of time which ensures that periodic variations (e.g. seasonal) are not biasing the sample. The designer must be aware of these restrictions.

If the extreme wave statistics is not based on N wave data, but for example on estimates of H_s from information about water level variations in shallow water, then the last term in eq. (17) disappears and instead the value of σ'_{FH_s} must account for the inherent uncertainty.

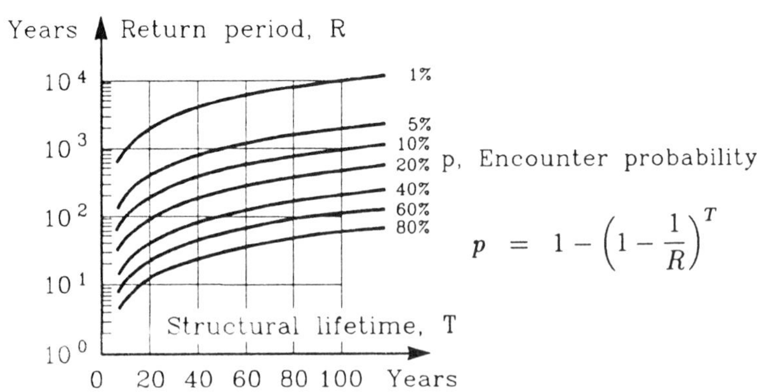

Fig. 4. Encounter probability, i.e. the probability p that the R-year return period event will be exceeded during a T-year structural life.

FORMAT FOR MULTI FAILURE MODES

32. A simple series system is considered. The reliability of the system depends on the correlation between the failure modes. Two factors contribute to the correlation, namely the physical correlation, e.g. the erosion of a toe berm triggering a slide in the main armour layer, and the correlation through common parameters like H_s. The physical cor-

relations are not yet generally known. Consequently, only the common parameter correlations have been implemented in the present work.

33. A simple system is to treat each failure mode, $i = 1, 2, 3, \ldots, n$ separately using the single failure mode models. The upper and lower bounds of probability of the system, P_f^s could then be estiamted as

Upper bound $\quad P_f^s = 1 - (1 - P_f^1)(1 - P_f^2)\ldots(1 - P_f^n)$ (19)

Lower bound $\quad P_f^s = \max P_f^i$ (20)

Max P_f^i is the largest of the failure probabilities of the failure modes.

Eqs. (19) and (20) correspond to no correlation and full correlation, respectively. Due to the common parameters there will always be a correlation of some size. However, closer bounds must await further work on correlation between failure modes.

THE OPTIMIZED PARTIAL COEFFICIENTS AND THEIR ACCURACY

34. The outcome of the calculations are optimized values of the coefficients k_α , k_β and k_s in eqs. (17) and (18). It turned out that a constant value of $k_s = 0.05$ can be used with good accuarcy for all the consideres failure modes. As to k_α and k_β one set of values has to be applied to each failure mode formula.

As examples the optimized coefficient values for for the formula (12), (14) and (16) are given in Table 3. Values for more formulae are given in the Sub-group F report.

Table 3. Examples of coefficients to be applied in the partial coefficient formulae (eqs. 11 and 12).

Failure mode	Design formula	k_α	k_β
Rock armour stability	Hudson	0.036	151
Rock armour stability	Van der Meer (plunging waves) Van der Meer (surging waves)	0.027 0.031	38 38

35. The accuracy of the system is illustrated for one failure mode formula (eq. 14) in Table 4 which provides comparisons between target values of the failure probability, P_f^T, and the values P_f, calculated by the proposed system. P_f is given by the maximum value, P_f^{max}, the mean value, μ_{P_f} and the standard deviation, σ_{P_f}.

In order to quantify the deviations between P_f^T and P_f they are also given in terms of deviations in the related partial coefficients. γ_T is the target coefficient corresponding to P_f^T and γ, the coefficient corresponding to P_f. γ_T and γ are estimates found from eqs. (17) and (18). γ^{min}, which is the absolute smallest value found in the calculations, corresponds to P_f^{max}. $\gamma^{\mu-\sigma}$ and $\gamma^{\mu+\sigma}$ correspond to the confidence limits: mean plus/minus one standard deviation. The relative deviation, denoted % dev., is defined as $\frac{\gamma-\gamma_T}{\gamma}\cdot 100$. Negative values of % dev. correspond to unsafe design values.

Table 4. Typical deviations between target values and calculated values of failure probability and related total partial coefficients ($=\gamma_z\cdot\gamma_{H_s}$).

P_f^T	σ_{FH_s}	P_f^{max}	μ_{P_f}	σ_{P_f}	γ^T	γ^{min}	% dev.	$\gamma^{\mu-\sigma}$	% dev.	$\gamma^{\mu+\sigma}$	% dev.
0.01	0	0.015	0.008	0.003	1.45	1.40	−3	1.52	5	1.43	−1
	0.10	0.012	0.007	0.002	1.55	1.53	−1	1.63	5	1.56	1
	0.20	0.022	0.013	0.002	1.66	1.56	−6	1.65	−1	1.61	−3
0.10	0	0.138	0.101	0.017	1.21	1.18	−2	1.23	2	1.19	−2
	0.10	0.107	0.096	0.006	1.27	1.26	−1	1.28	1	1.27	0
	0.20	0.116	0.109	0.004	1.35	1.32	−2	1.34	−1	1.33	−1
0.40	0	0.38	0.37	0.004	1.07	1.07	0	1.08	1	1.07	0
	0.10	0.38	0.38	0.007	1.08	1.08	0	1.09	1	1.08	0
	0.20	0.40	0.38	0.020	1.11	1.11	0	1.12	1	1.11	0

36. In the case of armour stability failure mode the deviations in γ can be interpreted as deviations in the nominal diameter of the armour blocks. It is seen from Table 4 that deviations are small, especially if they are evaluated on the background of the large sensitivity of the nominal diameter to variations in H_s. The variations in the H_s-design values, which appear by fitting two different theoretical extreme distributions, are generally larger, cf. Fig. 3.

EXAMPLE OF THE USE OF PARTIAL COEFFICIENT SYSTEM

37. The following example illustrate how the partial coefficient system is applied for design purpose.

Objective:

Determination of the average mass, or the nominal diameter D_{n50}, of quarry rock armour corresponding to the following design conditions:

Case 1. Moderate damage with a probability $P_f = 0.2$ within a structural life of $T = 50$ years.

Case 2. Severe damage with a probability $P_f = 0.2$ within a structural life of $T = 100$ years.

Case 3. Moderate damage with a probability $P_f = 0.1$ within a structural life of $T = 100$ years.

The Van der Meer formulae given by design equations (14) and (16) are assumed valid.

Design parameters:

Densities: Rock 2.8 t/m^3, water 1.03 t/m^3, $\Delta = 1.72$

Slope: $cot\alpha = 1.5$, porosity $P = 0.4$

Wave climate: Weibull distribution of H_s with the site specific coefficients $(\alpha, \beta, H_s') = (1.39, 1.06, 0.44)$ determined by fitting to a hindcasted H_s-data set consisting of the $N = 50$ largest values within a 12 years period, i.e. $\lambda = 50/12 = 4.17$, cf. paragraph 27. σ'_{FH_s} is estimated to 0.2 for the hindcasted H_s values. Wave steepness $S_m = 0.4$, number of waves $N = 2500$.

Damage: Moderate to severe damage $S = 6$,
very severe damage $S = 14$.

Procedure:

The procedure and the partial coefficent formulae described in paragraph 30 are used.

Calculations:

In case of a Weibull distribution the central estimate of the significant wave height with an average return period of T years is given by

$$\begin{aligned} \hat{H}_s^T &= H_s' + \beta\,(exp\,[ln(ln(\lambda T))/\alpha]) \\ &= 0.44 + 1.06\,(exp\,[ln(ln(4.17T))/1.39]) \end{aligned}$$

The equivalent return period is given by

$$T_{P_f} = \left(1 - (1 - P_f)^{\frac{1}{T}}\right)^{-1}$$

From this is obtained

Case	T (year)	P_f	T_{P_f} (year)	$\hat{H}_s^T$ (m)	$\hat{H}_s^{3T}$ (m)	$\hat{H}_s^{T_{P_f}}$ (m)
1	50	0.2	225	3.98	4.49	4.67
2	100	0.2	449	4.30	4.80	4.83
3	100	0.1	950	4.30	4.80	5.29

From Table 3 (for plunging waves)

$k_\alpha = 0.027$, $k_\beta = 38$

From the formulae

$$\gamma_{H_s} = \frac{\hat{H}_s^{T_{P_f}}}{\hat{H}_s^T} + \sigma'_{FH_s}{}^{\left(1+\left(\frac{\hat{H}_s^{3T}}{\hat{H}_s^T}-1\right)k_\beta P_f\right)} + \frac{0.05}{\sqrt{P_f N}}$$

$$\gamma_Z = 1 - k_\alpha ln\, P_f$$

and the Van der Meer design equation (14) is obtained

Case	γ_{H_s}	γ_Z	D_{n50} (m)	Average mass (t)
1	1.23	1.04	1.58	11.0
2	1.19	1.04	1.38	7.4
3	1.35	1.06	1.91	19.5

38. The example illustrates how easy it is to calculate the size of the armour for various design conditions. The system facilitates economical optimization of a design.

The system can be used also for the evaluation of the failure probability of existing structures. Such an evaluation will be implemented in the final report of the working group.

FURTHER DEVELOPMENT OF THE PARTIAL COEFFICIENT SYSTEM

39. The proposed partial coefficient system will be updated currently with the appearance of applicable formulae for other failure modes.

For multi-failure modes more narrow bounds than those given in paragraph 33 will be developed when better knowledge about the failure mode correlations is available.

The significance of the proposed system is that the partial coefficients can be determined for given failure probability within given structural life. This system can be applied also to other structures where the wave height is the dominating load parameter.

ACKNOWLEDGEMENT

40. The inspiring cooperation with all members of the PIANC PTCII working group and the related sub-groups is greatly acknowledged.

REFERENCES

BURCHARTH, H.F. 1990a. Safety of Marine Structures. From failure experience to the development of design standards for rubble mound breakwaters. First Int. Conf. on Ports and Marine Structures. Teheran, Iran, 1990.

MADSEN, H.O., KRENK, S., LIND, N.C. 1986. Methods of Structural Safety. Prentice-Hall, 1986.

REPORT OF SUB-GROUPS A - F under the PIANC PTC II Working Group 12, Analysis of rubble mound structures. 1990.

Sub-group A: Failure mode formulae. Allsop, N.W.H., 1989.

Sub-group B: Uncertainties related to environmental data and estimated extreme events. Burcharth, H.F., 1990b.

Sub-group C: Risk analysis in breakwater design. Van der Meer, J. 1990.

Sub-group D: Construction deviation and reliability of construction. Borchert, M., 1990.

Sub-group E: Description of selected breakwaters. Ligteringen, H., 1989.

Sub-group F: Introduction of a partial coefficients system for the design of rubble mound breakwaters. Burcharth, H.F., 1990b.

THOFT-CHRISTENSEN, P., BAKER, M.J. 1982. Structural Reliability Theory and Its Applications. Springer-Verlag, 1982.

Discussion

J. A. ZWAMBORN, CSIR

Monitoring of coastal structures is general, and breakwaters in particular, is of the utmost importance, and, if properly managed, can be made cost-effective. Three different types of measurement should be considered.[1]

As-built surveys. Contracts should include proper as-built measurements such as: profile measurements by levelling (above water) and/or measurements by the 'crane and ball' technique with a ball or cage diameter of $0.8D_{n50}$; aerial (helicopter), boat or crane photographic surveys; diver survey and, possibly, side-scan survey.

An example where strict prescriptions for these measurements were included in the contract agreements was for the Ashdod and Rutenberg breakwater repairs (Israel).

Performance monitoring. This is relatively low-level monitoring to check on a regular basis on the integrity and the serviceability of the main protection works, such as main breakwaters. The result are used to identify and plan necessary repair works. Such surveys could include: photographic survey from a boat and/or a crane or helicopter (photographs from fixed positions); levelling of profiles, as above; underwater inspections by diver.

Such monitoring has been carried out since 1988 for all seven of South Africa's main harbours (Richards Bay, Durban, East London, Port Elizabeth, Mossel Bay, Cape Town and Saldanha Bay) at a fairly nominal cost of less than 100 000 per year.

Monitoring for correlation studies. Such monitoring is especially designed to improve our knowledge and check our design assumptions. It could, inter alia, include all of the above, plus: recording of armour unit movements (stereo photo pairs); armour unit stress measurements; water pressure measurements, etc.; wave-recording at the specific site - this is an

Grassy Bay (Australia), constructed 1971/73

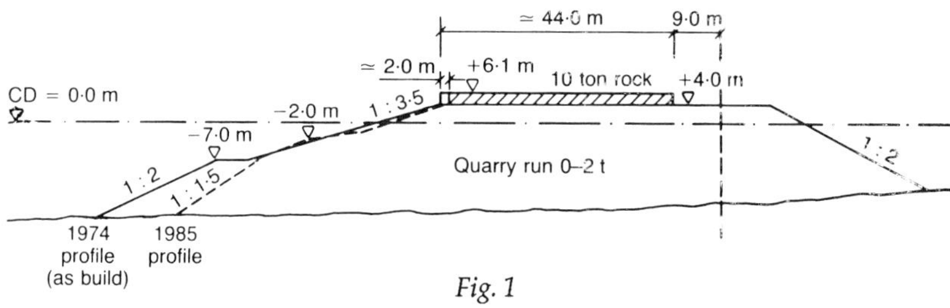

Fig. 1

Codroy (Canada), constructed 1983/84

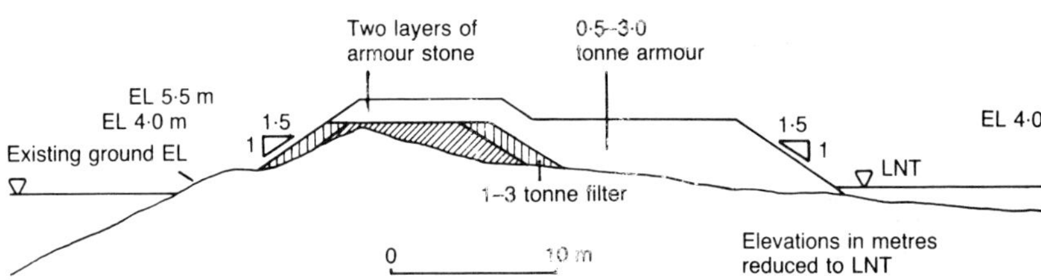

Fig. 2

Crescent City (USA), constructed 1974

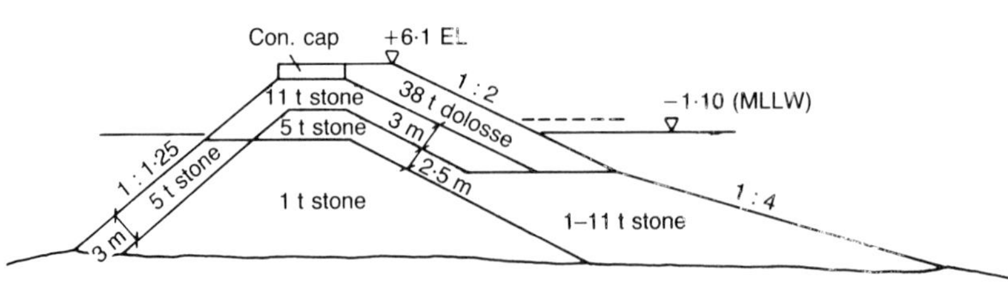

Fig. 3

Visby (Sweden), constructed 1980

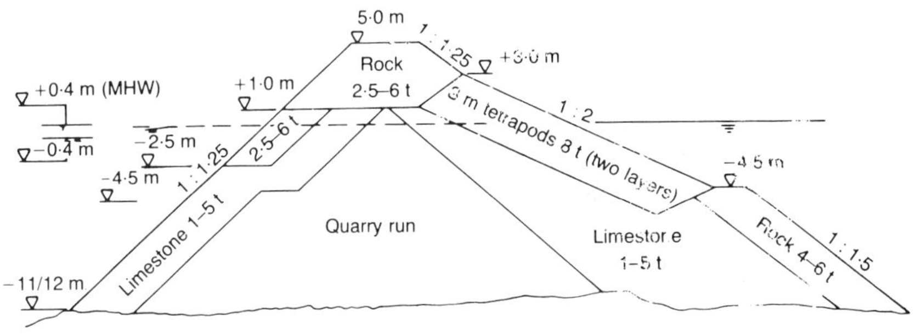

Fig. 4

Le Havre West Breakwater (France), constructed 1972/73

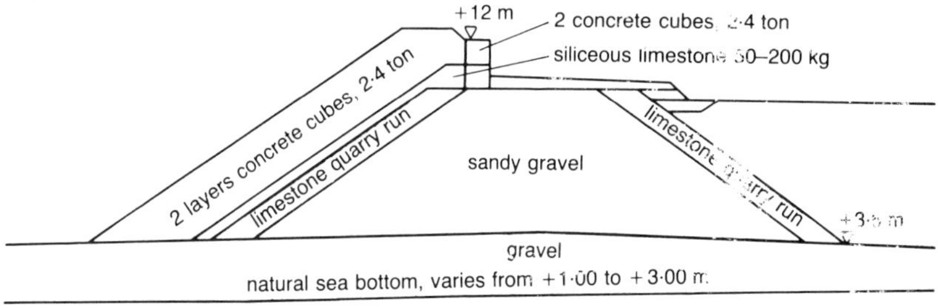

Fig. 5

essential requirement. Examples of such special measurement programmes are the Crescent City dolos project, the Zeebrugge project and the Table Bay breakwater project.

Reference:
1. ZWAMBORN J. A. et al. Redesign, repair and monitoring of the Table Bay breakwater. Proc. PIANC Congress, Osaka, 1990.

H. OUMERACI, University of Hannover
In the case of a rubble mound breakwater, we unfortunately do not have the probability density function (pdf) of the load or that of the resistance as shown by Professor Burcharth, but actually the joint probability density function (jpdf) of the environmental factor (H, T, d) and the pdf of the response of the structure, which is directly obtained by hydraulic model testing (or through empirical formulae); i.e. we still have a kind of black box

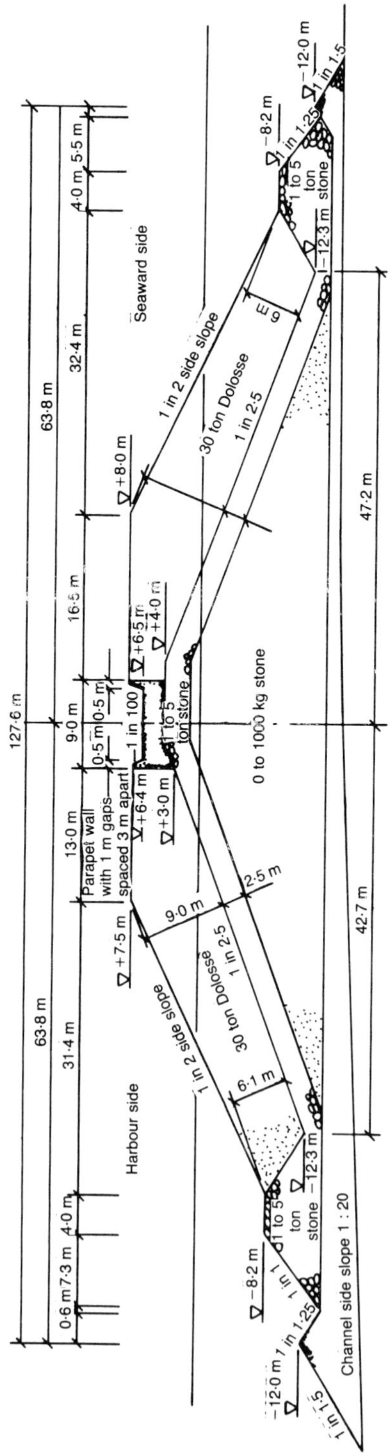
127·6 m
63·8 m
63·8 m
32·4 m
31·4 m
16·5 m
9·0 m
13·0 m
4·0 m
5·5 m
0·6 m
7·3 m
0·5 m
Seaward side
Harbour side
Parapet wall
with 1 m gaps
spaced 3 m apart
1 in 100
+8·0 m
+7·5 m
+6·5 m
+6·4 m
+4·0 m
+3·0 m
1 in 2 side slope
30 ton Dolosse
1 in 2·5
9·0 m
2·5 m
6·1 m
6 m
1 to 5
ton stone
0 to 1000 kg stone
−8·2 m
−12·0 m
−12·3 m
1 in 1·25
1 in 1·5
47·2 m
42·7 m
Channel side slope 1 : 20

Fig. 6

Torshavn (Faroe Islands), constructed 1968/73

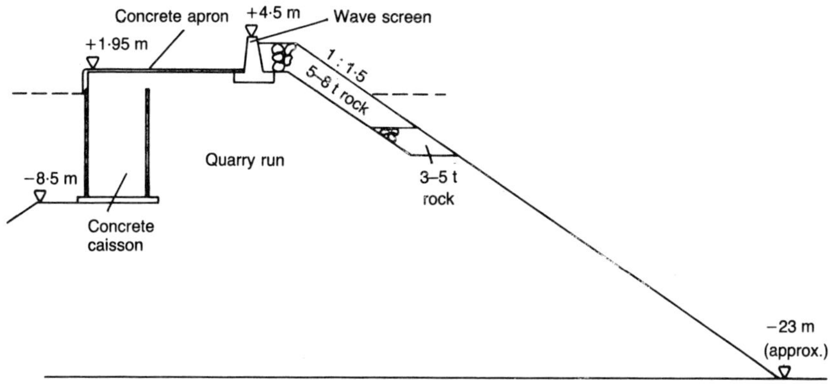

Fig. 7

Sakata North Breakwater (Japan), constructed 1975

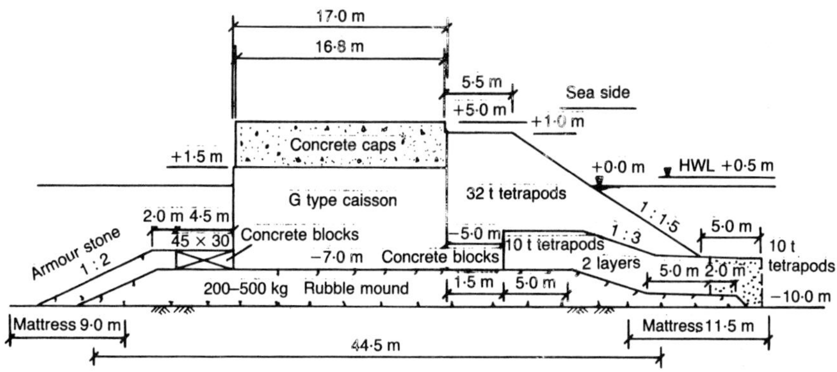

Fig. 8

Naha Northern Breakwater (Japan), constructed 1980

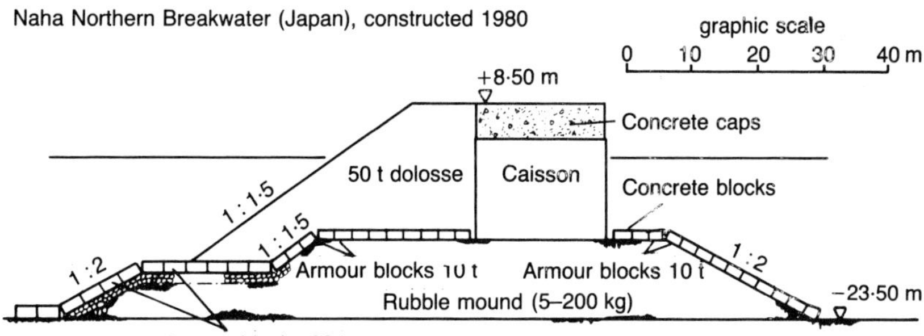

Fig. 9

system. We could better understand the physics and the failure mechanisms if we could also determine the pdf of the load.

In Professor Burcharth's example, a return period of 475 years was suggested. I do not think there is at present any theory in the statistics of extremes which can allow us to extrapolate from, say, 10-50 years data to a 500 year design wave.

F. B. BARENDS, Delft Geotechics
There are effects of time which should be looked at in a probability analysis. One is related to the service time of a structure. Another is related to the duration of the critical loading during the service time. Another is related to the change in the resistance during the service time. The latter effect has not been addressed at all.

The design procedures suggest a probability of failure during the service time. In fact a failure probability up to the last day of that service time. In reality, however, the reliability of a structure always decreases with time because of deterioration due to natural causes (erosion, abrasion, suffocation, settlement), and due to human causes (damage, negligence).

One would imagine that with growing age, monitoring and maintenance becomes more important. Such is the case with the Dutch dikes. In principle a decrease of resistance with time can be handled in a probability analysis (like in nuclear power station design). Something of this kind should be mentioned in the breakwater probabilistic design approach.

H. LIGTERINGEN, Frederic R. Harris BV
As Chairman of Subgroup E of Working Party 12, I am

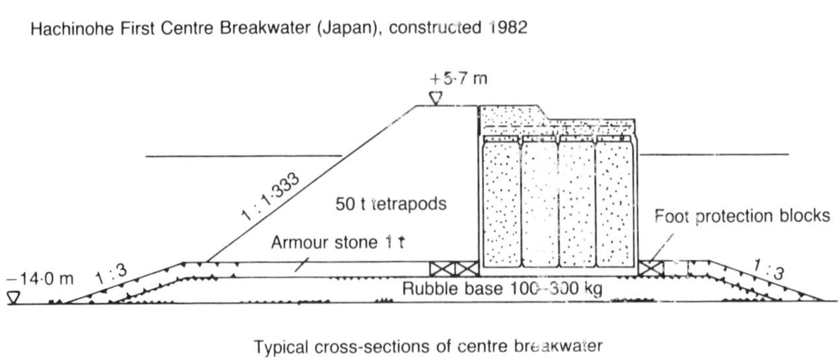

Fig. 10

pleased to give you a brief overview of the approach we followed in selecting and analysing the breakwater cases, and a summary of the results.

The objective of this work was to provide a basis for testing the design approach, - explained to you by John Mettam and Hans Burcharth. As a Working Group we found it of essence to try the partial coefficient system out on a number of breakwaters, which had to meet the following requirements

(a) 10-20 years in existence (to judge the performance)

(b) sufficient documented data available on design, construction and service conditions.

The analysis would give the Working Group members a check and validation of the proposed method, necessary when presenting such a fundamental change to the profession. It is also aimed at giving the design engineers, who will start applying the partial coefficient system in their future projects, a reference base.

In order to identify cases which complied with the above requirements, an inventory was organized. For about 50 breakwater cases a brief description was obtained on location, water depth, performance, data availability, etc. Although many of these cases would qualify for further analysis, a final selection was made of 10 breakwaters which covered best the range of types, of water depths, and which comprised damaged and undamaged cases. Appendix 1 gives these 10 cases and some characteristics.

For these 10 breakwaters, detailed data were collected by subgroup members (often themselves previously involved in the design or construction of the case) by means of a questionnaire. Information requested included geometrical data, hydraulic and geotechnical design and information on the construction process and behaviour after construction. Figs 1-10 give typical sections of these selected cases.

The data obtained were sieved and checked, condensed and presented as a first draft of the subgroup report in 1987. It was clear that information was even more limited than originally envisaged, which influenced the scope of the probabilistic analyses in Subgroup F.

In Subgroup E this work was continued with more extensive analysis of the cases, using both deterministic design and the newly developed partial coefficient system. The analysis was necessarily confined to stability of primary armour. The design formulae applied were the Hudson and Van der Meer formulae. These computations provided a deterministic check of the design and on possible causes of damage or failure. They gave the level of probability of failure as realized in the design or related to the period of actual service, taking into account the case-specific uncertainties. The correlation between the two approaches gave the desired 'calibration' of the partial coefficient system.

Conclusions drawn from the work of Subgroup E are as follows.

(a) Data availability was generally below expectation. There is a need for more systematic documentation of relevant data, which should be kept and updated by the Owner of the Work.

(b) Very little systematic monitoring is carried out of completed breakwaters. The Working Group strongly recommends the carrying out of monitoring as part of the breakwater maintenance and repair. Minimum requirements for monitoring should be established.

(c) Comparisons of deterministic design results with partial coefficients design show only small differences for failure probabilities of P_f = 0.5-0.6, and a greater effect on armour mass for decreasing P_f.

(d) In cases where failure occurred, this was not due to a wrong application of design formulae or model test results, but either to changes of design conditions or to the fact that the failure mechanism was overlooked in the design.

(e) The use of H1/10 in the Hudson formula (Shore Protection Manual 1984) leads to an unrealistic high mass of primary armour. The partial coefficient system allows a more objective definition of the desired safety in the design.

The report of Subgroup E, which will be issued together with the Working Group Main Report, presents the data on each of the breakwater cases and the results of the deterministic and partial coefficient analyses.

Appendix 1

Berm type

Grassy Bay (Australia), 1973 m, 20 m depth, Hs=9.1 m, 2-2000 kg stone. Model tests during design, profiles taken after readjustment, no damage.

Codroy (Newfoundland, Canada), 1984, 8 m depth, Hs=6.5 m, rocks larger than 0.5 t largely ranging between 0.5-3.0 t. Maximum stone sizes range up to 8-9 t. Model tests during design, profiles taken after construction, no damage.

Trapezoidal

Crescent City (USE), 1974, 15 m depth. Depth limited waves, 38 t unreinforced dolosse. No model tests of this particular section. Considerable damage including breakage of dolosse, information on damage available.

Visby (Sweden), 1980, 12 m depth. Hs=4.5 m, tetrapods 8 t. Design based on Hedar formulae and checked in model tests. No significant damage, some tetrapods lost and broken.

Trapezoid with crownwall

Le Havre West (France), 1972, 7 m depth. Depth limited waves, Antifer Cubes 2.4 t. Original design service lifetime extended, considerable damage, profiles taken after damage.

Richards Bay (South Africa), 1976, 18 m depth. Design wave height 7.2 m, 20 t dolosse on trunk, 30 t on the head. Many data on wave conditions and performance of the structure, minor damage.

Torshavn (Faroe Islands), 1973, 15 m depth. Design wave height H=4 m, 5-8 t rock. Model tests of original design, prototype measurements on waves and damage profiles, comparison with model test.

Composite type

Sakata North (Japan), 1975, 10 m depth, Hs=6.8 m, 32 t tetrapods. Wave data and profiles available. Settlement of main structure and rubble mound.

Naha North (Japan), 1980, 15 m depth, Hs=10.7 m, 50 t (reinforced) dolosse. Settlement of the slope after storms, wave data available.

Hachinohe First Centre (Japan), 1982, 12 m depth, Hs=6.7 m, 50 t tetrapods. Breakage of tetrapods and readjustment of the slope after storms, wave data available.

J. METTAM, Scott Bertlin

Mr Zwamborn has rightly drawn attention to the importance of monitoring the performance of breakwaters. This cannot be emphasized too strongly. Wave recordings are, as he says, essential for correlation studies, but they should also be a feature for performance monitoring so that a realistic estimate can be made of the likely variations in the amount of repair work which will be required.

Mr Ligteringen has contributed a valuable summary of the work of Subgroup E which formed a very important part of the study. As he makes clear in the conclusions taken from the Subgroup E report, more systematic documentation and monitoring of breakwaters is strongly recommended. Fuller records would have enabled more meaningful analyses of breakwater performance to be carried out. The Working Group report also recommends that engineers responsible for other existing breakwaters should use the newly developed partial coefficient system to analyse their performance, and to publish the results to help build up a body of information on the results which can be achieved. In many cases it will be found that more monitoring, including wave recording, would have assisted the analysis, but a study of the full report, when this becomes available, will show how to make the best use of existing information. If this exercise also helps engineers to formulate plans for better monitoring and to put them into practice this will be of great value.

H. F. BURCHARTH, University of Aarlborg
In reply to Mr Zwamborn, I certainly agree on the importance of monitoring of coastal structures (cf. my reply to Mr Lacey, Session 2).

In reply to Dr Oumeraci, his first statement is illustrated in Fig. 3 of my paper. It is true that direct determination of the actual loads will lead to more rational and generally applicable design calculations. However, determination of the loads related to wave-structure interactions on rubble mound slopes is very complicated, but research is proceeding in this field.

As to Dr Oumeraci's comment on extrapolation from 10-50 years' data to events of say 500 year return period, it is clear that such extrapolation is based on the assumption that the 10-50 years' data belong to the same statistical population as the 500 year event. For such conditions the theoretical concept for extrapolation to extreme events is clear. Also available are theories for the calculation of confidence bands, taking into account the varying sources of uncertainties - except those related to climatological changes. The theories of extreme statistics are pure mathematics and do not automatically take into account physical limitations. For example, depth limitations will occur at some stage when extrapolating wave heights in shallow water. This kind of limitation must be implemented in the specific cases to account for local conditions.

The subject is further discussed in the reports of subgroups B and F of the PIANC Working Group 12 on Rubble Mound Breakwaters.

In reply to Dr Barends, I agree that both the duration of the critical loading during the service time of a structure and the change in resistance of the structure should be taken into account in a probability analysis. I refer to the paper 'Stochastic design of rubble mound breakwaters' by Soren Nielsen and H. F. Burcharth (Proc. 11th IFIP Conference on system modelling and optimization, Copenhagen, July 1983), where for the first time these aspects were implemented in a probabilistic reliability analysis of a rubble mound breakwater. With respect to the work presented by the PIANC Working Group 12, the same two aspects are included in the following somewhat pragmatic ways.

Duration of the critical load is implemented through

the number of waves, which is a parameter in some of the failure mode formulae.

The effect of the deterioration of armour can be implemented by choosing a lower stability coefficient in the formulae (e.g. corresponding to rounded stones instead of angular quarry stones). In the case of concrete armour unit breakage, which is not dealt with in the PIANC reports, fatigue is taken into account by reducing the strength of the concrete, which is a parameter in the design formulae.

The probabilistic analysis and the partial coefficient system presented by the PIANC Working Group are limited to the failure mode cases where formulae exist including information about related uncertainty.

As was also mentioned by Dr Barends, it is in principle not difficult to implement service life load history and deterioration in probabilistic reliability evaluations of structures. It is, however, sometimes very difficult to provide reliable data input for the calculations. More research in this field is needed.

Closing address

Y. GODA, Yokohama National University

The conference began with the call by Professor Sorensen for preparation for the expected climate change. It then proceeded to the review and data compilation of wave overtopping and transmission of rubble mound structures by Mr Ahrens and Dr van der Meer. Professor Burcharth made a very detailed review of the innovation of rubble mound breakwater designs.

These papers were followed by another detailed review by Dr Hettiarachchi on the mathematical analyses of the flow within a rubble mound. The first day of the conference was thus dedicated to the discussion of rubble mound breakwaters, except for Mr Bouchet's paper on very innovative breakwater concepts for the principality of Mónaco.

On the second day of the conference, topics on vertical-face breakwaters were actively taken up. This is an apt response to the increase of engineers' interest in this type of structure which has been rather neglected in Western countries. By taking the privilege of being a delegate from Japan, I am pleased to say that we Japanese engineers do welcome the keen interest expressed in the conference by the UK and European colleagues, and will do our best to exchange our experiences with them. An important contribution of the second day was the proposal for quarry rock specifications by Mr Simon and his colleagues.

The third day of the conference was mainly dedicated to discussion of sea defence works including maintenance management, stability of blocks, use of bituminous materials, and so on. We also had the opportunity to hear the results of the PIANC Working Group on the analysis of rubble mound breakwaters.

During the conference many site-specific problems and their solutions were presented. They caught the attention of the delegates and provoked lively discussions.

I must say that the paper on Alderney breakwater gave me a sort of pleasant surprise, because it demonstrated the importance of the pore pressure increase caused by cyclic wave loading. The problem of pore pressure in the sand layer under wave loading is being recognized in various gravity-type structures. It was first raised in connection with the development of concrete platforms for oil exploitation in the North Sea. Then some of the vertical-face breakwaters were found to be susceptible to the liquefaction phenomenon in the foundations. The dislocation of Niigata breakwater is one example, because it was rested upon a foundation composed of multi-layers of loose sand and silty clay. The scour at the toe of rubble mound foundation in front of vertical breakwaters can also be accelerated by the liquefaction phenomenon. It is my belief that the wave-structure and soil interaction problem requires much attention and should be pursued further.

At the opening of this conference, Professor Severn, the past president of the Institute of Civil Engineers, cited the needs of research and development efforts. He specifically mentioned the topics of research funding, organization, personnel, and dissemination. I would like to add one more topic; what problems should be subject to research works? Or in other words, who will define and propose the subjects of research?

I found this conference quite unique in the sense that the majority of delegates are practising engineers, working at consulting firms, contractors, port authorities, municipalities and others. By examining the list of delegates, I noticed that the university and hydraulic laboratory people make up less than 20% of the total. On the other hand, two-thirds of the first authors of papers belong to research institutions and only one-third are practising engineers. Indeed, the selection of a good research subject is a quite delicate and difficult procedure if it is left to research people alone. Engineers in charge of planning, design and construction are in a better position to define their problems to be solved by researchers.

I would give one example from my experience in service to the Ministries of Transport, Japan. During the past 20 years or so, breakwaters in Japan have moved into deeper water with more severe wave conditions. It then became apparent that the critical factor in caisson breakwater design was not sliding but the bearing capacity of the rubble mound foundation. In the customary design procedure, the toe pressure had been kept below 40 to 50 tonnes/m^2. The threshold value was gradually raised to 60 tonnes/m^2, then to 80 tonnes/m^2 in some designs, but there was no theoretical basis for using the toe pressure as a measure of the bearing capacity. Such a design procedure also resulted in very conservative design from the viewpoint of sliding safety. Thus, the group of design engineers in the Bureau for Ports and Harbours strongly asked the researchers in soil mechanics and foundation engineering of the same ministry for clarification of the phenomenon, with the hope of economizing the breakwater design.

In the beginning, the soil mechanics people were rather reluctant to tackle the problem, saying that it was not so interesting from the academic point of view. However, under repeated requests and pressure by fellow engineers, they eventually took up the study, including large-scale triaxial tests of rubble stones. The study yielded a new method for analysing the bearing capacity of rubble mound and ground foundations, which uses the Bishop circular slip calculation, and has succeeded in economizing breakwater design in deep water. Details can be found in the paper by Dr Masaki Kobayashi, which is referred to in Paper 4 of this conference.

I have a feeling that there exists a gap between researchers and practitioners with regard to the understanding of engineering problems to be solved, and that the gap seems to be widening. Whenever I attend conferences and symposiums related to coastal engineering, the feeling is reconfirmed.

Now let me say this. Is it not the time for practising engineers to define their unsolved problems concretely and specifically, so that researchers can understand what are the problems in need of solution? Then practising engineers could have closer communications with research people and obtain the results they really want. I believe the Institution of Civil Engineers is the most appropriate agency to carry out this proposal.